T. N. Farris

Contact problems in the classical theory of elasticity

Monographs and textbooks on mechanics of solids and fluids
editor-in-chief: G. Æ. Oravas

Mechanics of elastic and viscoelastic solids

Contact problems in the classical theory of elasticity

by

G.M.L. Gladwell
Professor of Civil Engineering and
of Applied Mathematics
University of Waterloo, Ontario, Canada

SIJTHOFF & NOORDHOFF 1980
Alphen aan den Rijn, The Netherlands
Germantown, Maryland, USA

ISBN 90 286 0440 5
ISBN 90 286 0760 9 (pbk)

Printed in The Netherlands

To: N.I. Muskhelishvili who inspired me.
A.C. Stevenson who taught me.
I.N. Sneddon who showed me how it could be done.

PREFACE

Problems concerning the contact between elastic bodies have provided a challenge to applied mathematicians ever since the work of Heinrich Hertz in the 1860's. This book is concerned with idealized versions of these problems as they appear in the classical theory of elasticity. This theory is widely regarded as one of the most beautiful and challenging fields of classical linear mechanics. During the last hundred years it has been the cradle in which a number of powerful mathematical methods have grown. Amongst these are the complex variable method conceived by Kolosoff in the early years of the twentieth century and brought to maturity by Muskhelishvili and his coworkers in Tbilisi. Muskhelishvili's great contribution was the introduction of a systematic, direct method to replace the guesswork by which the complex potentials had previously been chosen. Muskhelishvili's work is described briefly in Chapters III and IV of this present book. It is hoped that these chapters will inspire the reader to study Muskhelishvili's treatises (1953a), (1953b) on elasticity and singular integral equations.

The second powerful mathematical tool which has been sharpened by its use in elasticity theory is the integral transform. Integral transforms were developed in piecemeal fashion during the nineteenth century but, in the author's opinion, it was the seminal work of I.N. Sneddon in his Fourier Transforms (1951) that showed how they could be used for the actual solution of the difficult boundary value problems of elasticity theory. In particular it was he who rescued Ida Busbridge's work on dual integral equations from oblivion and who reworded it to make it accessible to applied mathematicians. Through his writings, his students and his coworkers, his influence can be traced in much of the modern research on classical contact problems.

In writing this book I have been continually aware of the gaps in its coverage. As if the limitation to classical problems were not enough, I have excluded almost entirely a discussion of uniqueness and existence of solutions in elasticity theory, preferring to concentrate on mathematical methods which can be used to find a solution (hopefully the unique one) if one exists. This omission may be viewed as just one aspect of a larger omission of mathematical rigour. Applied mathematicians with pure mathematical inclinations may be inclined to dismiss much of the analysis as *formal*, but I believe that they will find that the analysis may be hedged about with provisos that will make it acceptable in a pure mathematical sense and moreover leave it essentially the same.

My aim in writing the book has been to present a fairly complete account of the subject which may be read without extensive reference to other books. Thus in the book there will be found accounts of the theory of elasticity (Chapter I), complex variable theory (Chapter III), integral transforms (Chapter V) and elliptic functions (Chapter XII). All these accounts start from fundamentals and aim to provide the reader with all the results that he will need to understand the accompanying analysis of contact problems. On the other hand, it is expected that the reader will consult the original papers which provide the details and extensions of the theory described in the text. These papers are listed in the bibliography. This list attempts to give a fairly complete coverage of the Western literature, and of the Soviet literature which has appeared in the more easily accessible Soviet journals. Many of the papers which are cited in the books on Contact Problems in Elasticity Theory by Shtaerman (1949) and Galin (1961) do not fall into this category. I did not think it was worthwhile to include them just to make the list complete.

The reader may supplement his reading with solution of the Examples which are to be found at the end of most sections. Many of these examples rely on particular integral transforms and the reader would do well to acquire handbooks such as Abramowitz and Stegun (1972) and Gradshteyn and Ryzhik (1965).

Preface

The whole of the book was read in manuscript by A.H. England of Nottingham University and L.M. Keer of Northwestern University. They eliminated many of the rough spots; I am responsible for those that remain. Early versions of the manuscript were typed by Jan Daum, Debbie Mustin and Jane Skinner. The final copy was produced in the office of the Solid Mechanics Division of the University of Waterloo. My heartfelt thanks are due to Solid Mechanics Division's Publication Officer D.E. Grierson, Assistant Pam McCuaig, typists Cynthia Jones and Linda Strouth. Louise Adamson cheerfully undertook the Herculean tasks of proofreading successive versions of the manuscript, checking the literature entries and preparing the index.

The book was made possible through a research grant from the National Science and Engineering Council of Canada.

I would be most grateful to be notified of any errors in the text and Examples, and any omissions in the bibliography.

G.M.L. Gladwell
Waterloo, Ontario, Canada
March, 1980

CONTENTS

Contents

page

Contents

page

page

In the list of references the first number refers to the journal so numbered in the journal list. The last sequence of numbers, separated by commas, indicates the page(s) of the book where this reference is cited. Thus:

Abramian, B.L. and Arutiunian, N.Kh., On some contact problems for composite half-spaces.
80,Vol.31(1967)1001-1008[1007-1013]. 570

appears in Russian in Journal no. 80, namely *Prikladnaia matematika i mekhanika*, pages 1001-1008. The English translation appears in *Journal of Applied Mathematics and Mechanics*, pages 1007-1013. The article is cited on page 570 of this book.

CHAPTER I - THE CLASSICAL THEORY OF ELASTICITY

"Where shall I begin, please your Majesty"? he asked. "Begin at the beginning" the King said, very gravely, "and go on till you come to the end: then stop."

- *Alice in Wonderland*

1.1 *Introduction*

This book is concerned with a certain class of problems, so-called contact problems, which appear within the theory of elasticity. In this chapter we present an account of those parts of the infinitesimal theory of elasticity which are necessary for the understanding of the problems which will appear later. For a more comprehensive account of this theory the reader should refer to the classical treatises. Love (1927) gives an exhaustive account with numerous historical references which has formed the basis for much of the literature on elasticity in the West. Muskhelishvili's treatise (1953a) is the basis of much of the Russian work, particularly that using complex variable methods. Lur'e (1964) gives a detailed presentation with copious bibliographical notes covering the Russian literature up to the 1950's, and is one of the few treatises dealing explicitly with the effective solutions of 3-dimensional problems. The Handbuch der Physik articles by Sneddon and Berry (1958) and the more recent one by Gurtin (1972) are authoritative. Timoshenko and Goodier (1951) give a presentation of the theory more suited to the engineer, which may prove a useful starting point for those new to the field. Sokolnikoff (1946) emphasizes the tensorial character of the equations of elasticity and is particularly valuable for the chapters on general variational theorems. Green and Zerna (1954) also employ a tensorial approach to 3-dimensional problems, but develop the complex variable methods for plane problems.

The literature on contact problems has been reviewed by a number of authors. Shtaerman (1949), Galin (1961), Ufliand (1965), Rvachev (1967) and Abramian (1971) are concerned mostly with the Soviet literature. Kalker (1975) emphasizes non-classical and rolling contact phenomena which are not discussed in this book. Kalker (1977a) reviewed both classical and non-classical contact phenomena. Hetenyi (1946), (1964), Vlasov and Leont'ev (1966), Popov (1972b), Gorbunov-Posadov and Malikova (1973), and Iyer (1973) are concerned with beams and plates on elastic foundations. References to Hertzian contact may be found in Morton and Close (1922), Way (1940), Timoshenko and Goodier (1951, p. 372f), Dinnik (1952) and Krolevets (1966).

No attempt has been made to include in the bibliography all the references listed in the works cited above. This book is concerned primarily with the study of contact problems as a topic of applied mathematics. We have therefore ignored many researches treating important practical contact problems when we have felt that they are concerned primarily with computing approximate results, rather than illustrating a new mathematical method. (Unfortunately, it cannot be claimed that the cited works all illustrate new mathematical principles. Too many papers have been written in which known, even if only recently discovered, mathematical methods are applied to a problem which differs from previously solved problems only in some small detail.) On the other hand we have paid little attention to pure mathematical questions such as the existence and uniqueness of solutions to contact problems. In general these questions have not been fully answered, but see e.g. Turteltaub and Sternberg (1967), Kalker (1968).

1.2 *The analysis of strain*

In accordance with our limited objective stated in the previous section we shall endeavour to present an analysis of strain which introduces as few complications as possible.

Consider a *continuum* occupying a certain region in 3-dimensional space, and suppose that, by some means, the continuum is *deformed* so that its points undergo *displacement*. Let the displacement of the point P which had coordinates (x,y,z) referred to axes Ox,Oy,Oz fixed in space be $\vec{d} = (u,v,w)$. This is often denoted $\vec{u} = (u,v,w)$. After deformation P therefore occupies the position $P' \equiv (x+u,\ y+v,\ z+w)$, as shown in Figure 1.2.1. If we assume that u,v,w are differentiable functions of the coordinates

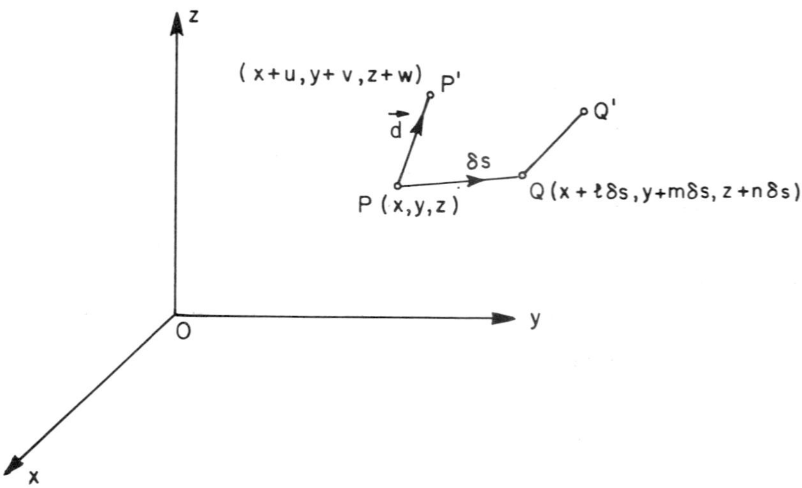

Figure 1.2.1 - The points P,Q *are displaced to* P',Q'.

x,y,z, then the displacement of a neighbouring point $Q(x+\ell\delta s,\ y+m\delta s,\ z+n\delta s)$ distant δs from P along a line with direction cosines (ℓ,m,n) will be

$$u + \delta u = u(x+\ell\delta s,\ y+m\delta s,\ z+n\delta s) = u(x,y,z) + \delta s\left\{\ell\frac{\partial u}{\partial x} + m\frac{\partial u}{\partial y} + n\frac{\partial u}{\partial z} + \ldots\right\}, \tag{1.2.1}$$

$$v + \delta v = v(x+\ell\delta s,\ y+m\delta s,\ z+n\delta s) = v(x,y,z) + \delta s\left\{\ell\frac{\partial v}{\partial x} + m\frac{\partial v}{\partial y} + n\frac{\partial v}{\partial z} + \ldots\right\}, \tag{1.2.2}$$

$$w + \delta w = w(x+\ell\delta s,\ y+m\delta s,\ z+n\delta s) = w(x,y,z) + $$
$$+ \delta s\left\{\ell \frac{\partial w}{\partial x} + m \frac{\partial w}{\partial y} + n \frac{\partial w}{\partial z} + \ldots\right\}, \tag{1.2.3}$$

where all the derivatives are evaluated at (x,y,z). Provided that δs is sufficiently small, the higher terms in (1.2.1)-(1.2.3) which involve powers of δs may be neglected.

We now restrict our analysis to *infinitesimal* deformations by assuming that the products of derivatives of u,v,w may be neglected in comparison with the linear terms. The distance between the displaced points P',Q' will be $\delta s + \Delta s$, where

$$(\delta s + \Delta s)^2 = (\delta s)^2\left[\left\{\ell\left(1 + \frac{\partial u}{\partial x}\right) + m \frac{\partial u}{\partial y} + n \frac{\partial u}{\partial z}\right\}^2 + \right.$$
$$+ \left\{\ell \frac{\partial v}{\partial x} + m\left(1 + \frac{\partial v}{\partial y}\right) + n \frac{\partial v}{\partial z}\right\}^2 +$$
$$\left. + \left\{\ell \frac{\partial w}{\partial x} + m \frac{\partial w}{\partial y} + n\left(1 + \frac{\partial w}{\partial z}\right)\right\}^2\right].$$

Under the restriction to infinitesimal strain the proportional increase in the length of the line PQ due to the deformation is

$$\frac{\Delta s}{\delta s} = \ell^2 \frac{\partial u}{\partial x} + m^2 \frac{\partial u}{\partial y} + n^2 \frac{\partial w}{\partial z} + mn\left(\frac{\partial v}{\partial z} + \frac{\partial w}{\partial y}\right) +$$
$$+ n\ell\left(\frac{\partial w}{\partial x} + \frac{\partial u}{\partial z}\right) + \ell m\left(\frac{\partial u}{\partial y} + \frac{\partial v}{\partial x}\right).$$

The *coefficients* in this expression are called the *components of strain* and are denoted by

$$e_{xx} = \frac{\partial u}{\partial x}, \qquad e_{yy} = \frac{\partial v}{\partial y}, \qquad e_{zz} = \frac{\partial w}{\partial z}, \tag{1.2.4}$$
$$e_{yz} = \frac{1}{2}\left(\frac{\partial v}{\partial z} + \frac{\partial w}{\partial y}\right),\ e_{zx} = \frac{1}{2}\left(\frac{\partial w}{\partial x} + \frac{\partial u}{\partial z}\right),\ e_{xy} = \frac{1}{2}\left(\frac{\partial u}{\partial y} + \frac{\partial v}{\partial x}\right). \tag{1.2.5}$$

Thus

$$\frac{\Delta s}{\delta s} = \ell^2 e_{xx} + m^2 e_{yy} + n^2 e_{zz} + 2mn e_{yz} +$$
$$+ 2n\ell e_{zx} + 2\ell m e_{xy}. \tag{1.2.6}$$

Note that Love (1927) omits the factor 1/2 in the *shear strains* e_{yz}, e_{zx}, e_{xy}; many authors omit the second suffix in the *normal strains* e_{xx}, e_{yy}, e_{zz}: Timoshenko and Goodier (1951) denote them by $\varepsilon_x, \varepsilon_y, \varepsilon_z$. Note that the shear strains can be treated as *symmetric* quantities: i.e. we introduce e_{zy}, e_{xz}, e_{yx}, where

$$e_{zy} = e_{yz}, \quad e_{xz} = e_{zx}, \quad e_{yx} = e_{xy}. \tag{1.2.7}$$

If the displacement of the body is due to a combination of *rigid body displacement*, in which each point receives the same displacement, and an infinitesimal *rigid body rotation* $(\omega_1, \omega_2, \omega_3)$, in which $\omega_1, \omega_2, \omega_3$ represent the angles of rotation about the three axes Ox,Oy,Oz respectively, then equations (1.2.1)-(1.2.3) become

$$u + \delta u = u - \omega_3 m\delta s + \omega_2 n\delta s,$$

$$v + \delta v = v - \omega_1 n\delta s + \omega_3 \ell\delta s,$$

$$w + \delta w = w - \omega_2 \ell\delta s + \omega_1 m\delta s.$$

Under such a deformation all the components of strain are zero since, for example,

$$\frac{\partial u}{\partial y} = -\omega_3 = -\frac{\partial v}{\partial x}.$$

This means that if we introduce three *components of rotation*, defined by

$$\left.\begin{aligned} \omega_{yz} &= \frac{1}{2}\left(\frac{\partial w}{\partial y} - \frac{\partial v}{\partial z}\right), \\ \omega_{zx} &= \frac{1}{2}\left(\frac{\partial u}{\partial z} - \frac{\partial w}{\partial x}\right), \\ \omega_{xy} &= \frac{1}{2}\left(\frac{\partial v}{\partial x} - \frac{\partial u}{\partial y}\right), \end{aligned}\right\} \tag{1.2.8}$$

then equations (1.2.1)-(1.2.3) may be written

$$u + \delta u = u + \delta s\{\ell e_{xx} + m e_{xy} + n e_{xz} - m\omega_{xy} + n\omega_{zx}\},$$

$$v + \delta v = v + \delta s\{\ell e_{yx} + m e_{yy} + n e_{yz} - n\omega_{yz} + \ell\omega_{xy}\},$$

$$w + \delta w = w + \delta s\{\ell e_{zx} + m e_{zy} + n e_{zz} - \ell\omega_{zx} + m\omega_{yz}\}.$$

The strains may be given physical interpretations. By putting $m = 0 = n$ we see that e_{xx} represents the proportional change of length of the line joining $P(x,y,z)$ to $Q(x+\ell\delta s, y,z)$. Again, consider two points $Q(x+\ell\delta s_1, y,z)$ and $R(x, y+m\delta s_2, z)$ neighbouring P as in Figure 1.2.2; initially PQ and PR are orthogonal.

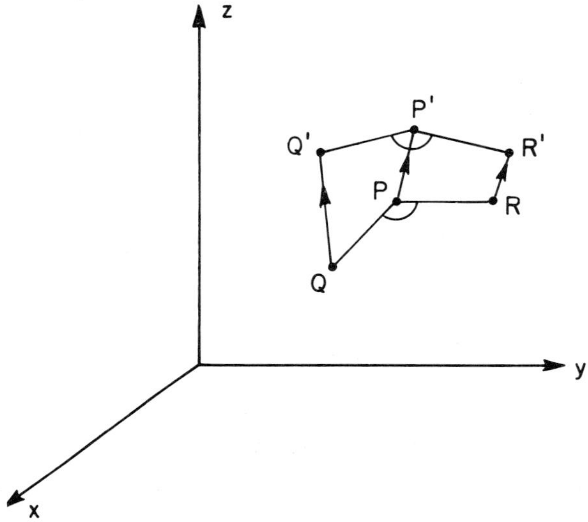

Figure 1.2.2 - The right-angle QPR *is changed to* Q'P'R'.

Under deformation P goes to $P'(x+u, y+v, z+w)$ while Q,R go respectively to Q',R' with coordinates

$$(x+u+\ell\delta s_1(1+e_{xx}),\ y+v+\ell\delta s_1(e_{yx}+\omega_{xy}),\ z+w+\ell\delta s_1(e_{zx}-\omega_{zx}))$$

$$(x+u+m\delta s_2(e_{xy}-\omega_{xy}),\ y+v+m\delta s_2(1+e_{yy}),\ z+w+n\delta s_2(e_{zy}-\omega_{yz})).$$

The cosine of the angle $(\pi/2-\alpha)$ between P'Q' and P'R' is given to the first order by

$$\cos(\pi/2-\alpha) = \sin\alpha = \alpha = 1(e_{xy}-\omega_{xy}) + 1(e_{yx}+\omega_{xy}) = 2e_{xy}.$$

Thus $2e_{xy}$ represents the decrease in the angle between the two lines PQ,PR initially at right angles.

The strain components $e_{xx}, e_{xy}, e_{xz}, e_{yx}, e_{yy}, e_{yz}, e_{zx}, e_{zy}, e_{zz}$ form the 9 components of a tensor of rank 2, (see e.g. Sokolnikoff

(1946) or Sneddon and Berry (1958)). This means that if Ox', Oy', Oz' is a second Cartesian frame the axes of which have direction cosines (ℓ_1, m_1, n_1), (ℓ_2, m_2, n_2) and (ℓ_3, m_3, n_3) with respect to Ox, Oy, Oz, then by using an argument similar to that leading to equation (1.2.6) we find that the normal and shear strains in the new coordinate frame are $e_{x'x'}, e_{y'y'}, e_{z'z'}, e_{y'z'}, e_{z'x'}, e_{x'y'}$, where

$$e_{x'x'} = \ell_1^2 e_{xx} + m_1^2 e_{yy} + n_1^2 e_{zz} + 2m_1 n_1 e_{yz} + 2n_1 \ell_1 e_{zx} + 2\ell_1 m_1 e_{xy}, \tag{1.2.9}$$

$$e_{y'z'} = \ell_2 \ell_3 e_{xx} + m_2 m_3 e_{yy} + n_2 n_3 e_{zz} + (m_2 n_3 + m_3 n_2) e_{yz} + (n_2 \ell_3 + n_3 \ell_2) e_{zx} + (\ell_2 m_3 + \ell_3 m_2) e_{xy}, \text{ etc.} \tag{1.2.10}$$

The quantity

$$\varepsilon = e_{xx} + e_{yy} + e_{zz} \equiv \operatorname{div} \vec{d} \tag{1.2.11}$$

is called the *cubical dilatation*, and is sometimes denoted by Δ; it is equal to the proportional increase in the size of a small volume situated at P. The dilatation is *invariant* with respect to coordinate changes. Indeed equation (1.2.9) and the corresponding equations for $e_{y'y'}$ and $e_{z'z'}$ show that

$$e_{x'x'} + e_{y'y'} + e_{z'z'} = e_{xx} + e_{yy} + e_{zz}. \tag{1.2.12}$$

1.3 *Strains in cylindrical and spherical polar coordinates*

It is possible to derive expressions for the strain components in general orthogonal curvilinear coordinates (Love (1927)). We content ourselves with the appropriate expressions in the commonly used coordinates.

Let P have coordinates (r, θ, z) in cylindrical polar coordinates, as shown in Figure 1.3.1. Then

$$\vec{OP} = r\hat{r} + z\hat{k}.$$

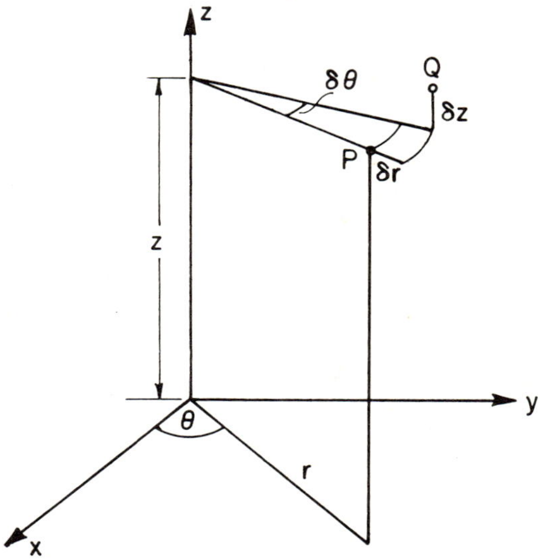

Figure 1.3.1 - Neighbouring points P,Q *in cylindrical polar coordinates.*

The neighbouring point Q(r+δr, θ+δθ, z+δz) has position vector

$$\vec{OQ} = (r+\delta r)\hat{r} + r\delta\theta\hat{\theta} + (z+\delta z)\hat{k}.$$

Thus if the length PQ is δs then the direction cosines of $\vec{PQ}$ referred to the directions $\hat{r},\hat{\theta},\hat{k}$ at P are

$$\ell = \delta r/\delta s, \quad m = r\delta\theta/\delta s, \quad n = \delta z/\delta s. \tag{1.3.1}$$

The points P,Q will be displaced to P',Q' where

$$\vec{OP'} = r\hat{r} + z\hat{k} + u_r\hat{r} + u_\theta\hat{\theta} + u_z\hat{k},$$

$$\vec{OQ'} = (r+\delta r)\hat{r} + r\delta\theta\hat{\theta} + (z+\delta z)\hat{k} + (u_r+\delta u_r)\hat{r} + u_r\delta\theta\hat{\theta} +$$

$$+ (u_\theta+\delta u_\theta)\hat{\theta} - u_\theta\delta\theta\hat{r} + (u_z+\delta u_z)\hat{k}.$$

Thus the length P'Q' is δs + Δs where

$$(\delta s + \Delta s)^2 = \left\{\delta r\left(1 + \frac{\partial u_r}{\partial r}\right) + r\delta\theta\left(\frac{\partial u_r}{r\partial\theta} - \frac{u_\theta}{r}\right) + \delta z\,\frac{\partial u_r}{\partial z}\right\}^2 +$$

$$+\left\{\delta r\,\frac{\partial u_\theta}{\partial r} + r\delta\theta\left(1 + \frac{u_r}{r} + \frac{\partial u_\theta}{r\partial\theta}\right) + \delta z\,\frac{\partial u_\theta}{\partial z}\right\}^2 +$$

$$+\left\{\delta r\,\frac{\partial u_z}{\partial r} + r\delta\theta\,\frac{\partial u_z}{r\partial\theta} + \delta z\left(1 + \frac{\partial u_z}{\partial z}\right)\right\}^2 .$$

Making use of equation (1.3.1) and neglecting products of the derivatives we find that

$$\frac{\Delta s}{\delta s} = \ell^2 e_{rr} + m^2 e_{\theta\theta} + n^2 e_{zz} + 2mne_{\theta z} + + 2n\ell e_{zr} + 2\ell m e_{r\theta}, \tag{1.3.2}$$

where

$$e_{rr} = \frac{\partial u_r}{\partial r}, \quad e_{\theta\theta} = \frac{1}{r}\frac{\partial u_\theta}{\partial\theta} + \frac{u_r}{r}, \quad e_{zz} = \frac{\partial u_z}{\partial z}, \tag{1.3.3}$$

$$e_{\theta z} = \frac{1}{2}\left(\frac{\partial u_z}{r\partial\theta} + \frac{\partial u_\theta}{\partial z}\right), \quad e_{zr} = \frac{1}{2}\left(\frac{\partial u_r}{\partial z} + \frac{\partial u_z}{\partial r}\right),$$

$$e_{r\theta} = \frac{1}{2}\left(\frac{\partial u_\theta}{\partial r} - \frac{u_\theta}{r} + \frac{\partial u_r}{r\partial\theta}\right). \tag{1.3.4}$$

When there is no possibility of confusion, the displacements u_r, u_θ, u_z will sometimes be denoted u,v,w respectively. The components of rotation are

$$\omega_{\theta z} = \frac{1}{2}\left(\frac{\partial u_z}{r\partial\theta} - \frac{\partial u_\theta}{\partial z}\right), \tag{1.3.5}$$

$$\omega_{zr} = \frac{1}{2}\left(\frac{\partial u_r}{\partial z} - \frac{\partial u_z}{\partial r}\right), \tag{1.3.6}$$

$$\omega_{r\theta} = \frac{1}{2}\left(\frac{\partial u_\theta}{\partial r} + \frac{u_\theta}{r} - \frac{\partial u_r}{r\partial\theta}\right). \tag{1.3.7}$$

In spherical polar coordinates, Figure 1.3.2, the direction cosines referred to coordinate directions $\hat{R}, \hat{\theta}, \hat{\phi}$ at P of the line joining two neighbouring points $P(R,\theta,\phi)$ and $Q(R+\delta R,\ \theta+\delta\theta,\ \phi+\delta\phi)$, are

$$\ell = \delta R/\delta s, \quad m = R\delta\theta/\delta s, \quad n = R\sin\theta\,\delta\phi/\delta s, \tag{1.3.8}$$

and the proportional deformation of P'Q' is

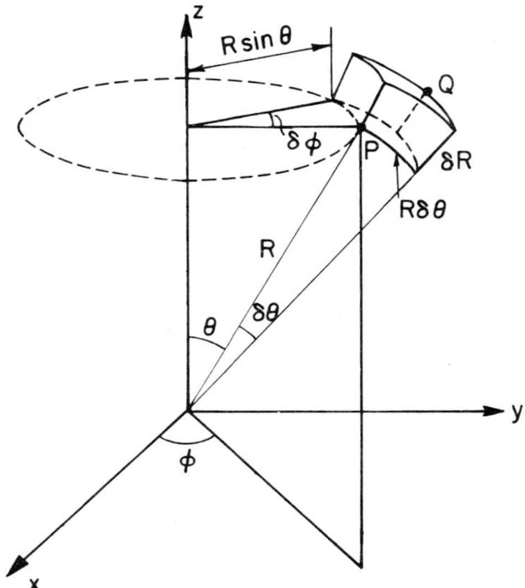

Figure 1.3.2 - Neighbouring points P,Q *in spherical polar coordinates.*

$$\frac{\Delta s}{\delta s} = \ell^2 e_{RR} + m^2 e_{\theta\theta} + n^2 e_{\phi\phi} + 2mne_{\theta\phi} + $$
$$+ 2n\ell e_{\phi R} + 2\ell m e_{R\theta}, \tag{1.3.9}$$

where

$$e_{RR} = \frac{\partial u_R}{\partial R}, \quad e_{\theta\theta} = \frac{1}{R}\frac{\partial u_\theta}{\partial \theta} + \frac{u_R}{R}, \tag{1.3.10}$$

$$e_{\phi\phi} = \frac{1}{R\sin\theta}\frac{\partial u_\phi}{\partial \phi} + \frac{u_\theta}{R}\cot\theta + \frac{u_R}{R}, \tag{1.3.11}$$

$$e_{\theta\phi} = \frac{1}{2R}\left(\frac{\partial u_\phi}{\partial \theta} - u_\phi \cot\theta\right) + \frac{1}{2R\sin\theta}\frac{\partial u_\theta}{\partial \phi}, \tag{1.3.12}$$

$$e_{\phi r} = \frac{1}{2}\left\{\frac{1}{R\sin\theta}\frac{\partial u_R}{\partial \phi} + \frac{\partial u_\phi}{\partial R} - \frac{u_\phi}{R}\right\}, \tag{1.3.13}$$

$$e_{R\theta} = \frac{1}{2}\left\{\frac{\partial u_\theta}{\partial R} - \frac{u_\theta}{R} + \frac{1}{R}\frac{\partial u_R}{\partial \theta}\right\}. \tag{1.3.14}$$

The components of rotation are

$$\omega_{\theta\phi} = \frac{1}{2R^2\sin\theta}\left\{\frac{\partial}{\partial\theta}(R_{u\phi}\sin\theta) - \frac{\partial}{\partial\phi}(R_{u\theta})\right\}, \tag{1.3.15}$$

$$\omega_{\phi R} = \frac{1}{2R \sin\theta}\left\{\frac{\partial u_R}{\partial \phi} - \frac{\partial}{\partial R}(Ru_\phi \sin\theta)\right\}, \tag{1.3.16}$$

$$\omega_{R\theta} = \frac{1}{2R}\left\{\frac{\partial}{\partial R}(Ru_\theta) - \frac{\partial u_R}{\partial \theta}\right\}. \tag{1.3.17}$$

Example 1.3

1. Show that the cubical dilatation ε may be written

$$\varepsilon = \frac{1}{r}\frac{\partial}{\partial r}(ru_r) + \frac{1}{r}\frac{\partial u_\theta}{\partial \theta} + \frac{\partial u_z}{\partial z}$$

$$\varepsilon = \frac{1}{R^2 \sin\theta}\left\{\frac{\partial}{\partial R}(R^2 u_R \sin\theta) + \frac{\partial}{\partial \theta}(Ru_\theta \sin\theta) + \frac{\partial}{\partial \phi}(Ru_\phi)\right\}$$

in cylindrical and spherical coordinates respectively.

1.4 *The analysis of stress*

To specify the state of stress at a point P in a continuum we consider an element Δs of a surface passing through P. Let $\hat{n}$ be the unit normal to Δs at P. The forces exerted *by the material on side 2 of* Δs *on the material on side 1* may be reduced to a force $\Delta\vec{T}$ at P and a moment $\Delta\vec{G}$ as shown in Figure 1.4.1. We shall *assume*, and this assumption may be considered as having an experimental basis, that

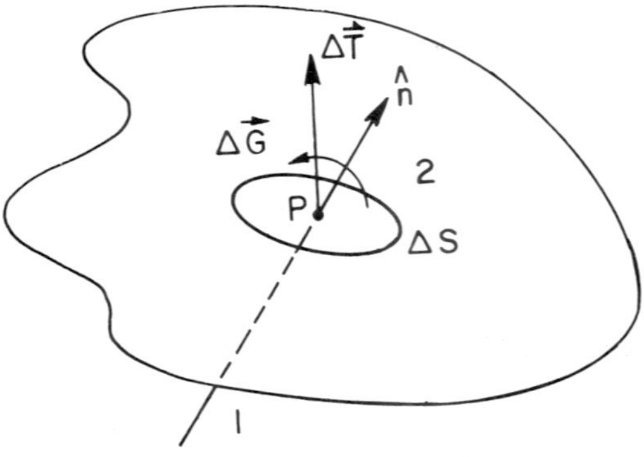

Figure 1.4.1 - The normal $\hat{n}$ *is drawn from region 1 to region 2.*

$$\lim_{\Delta s \to 0}(\Delta \vec{T}/\Delta s) = \vec{T}_n, \qquad \lim_{\Delta s \to 0}(\Delta \vec{G}/\Delta s) = 0, \tag{1.4.1}$$

where $\vec{T}_n$ is a vector, called the *traction* or the *stress vector* across the surface at P. In general $\vec{T}_n$ will not be in the normal direction $\hat{n}$, but will have components in each of the coordinate directions. We write

$$\vec{T}_n = \tau_{nx}\hat{i} + \tau_{ny}\hat{j} + \tau_{nz}\hat{k}. \tag{1.4.2}$$

It is found that the forces acting on a volume element ΔV of a continuum may be divided into *surface forces* caused by the tractions described above, and *body forces*. The most common example of the latter is provided by gravity, which exerts a downward force $\rho g \Delta V$ on an element of volume ΔV, where ρ is the mass density of the continuum. We shall assume that if the body forces acting on a small volume element surrounding a point P are equivalent to a force $\Delta\vec{R}$ at P and a moment $\Delta\vec{M}$ then

$$\lim_{\Delta V \to 0} \frac{\Delta \vec{R}}{\Delta V} = \vec{F}, \qquad \lim_{\Delta V \to 0} \frac{\Delta \vec{M}}{\Delta V} = 0. \tag{1.4.3}$$

The vector $\vec{F}$ is called the body force per unit volume. We shall assume that $\vec{F}$ is a continuous function of the coordinates (x,y,z) and write

$$\vec{F} = F_x\hat{i} + F_y\hat{j} + F_z\hat{k}. \tag{1.4.4}$$

When a medium is undergoing time-dependent deformation the *inertia forces* may be viewed as providing a body force per unit volume of $-\rho\partial^2\vec{d}/\partial t^2$.

Consider a small rectangular parallelopiped with sides 2a,2b, 2c parallel to the axes and centred on P. The moment in the z-direction of the forces acting on the element is provided by the moment of the body forces ΔM_z and the moment of those surface stresses shown in Figure 1.4.2. Thus equilibrium of the element demands that

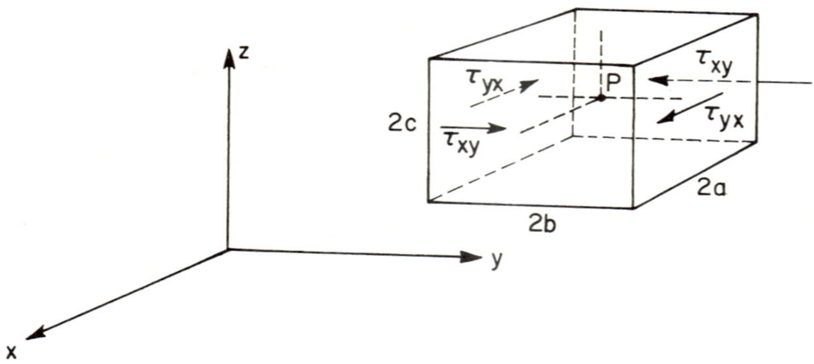

Figure 1.4.2 - The stresses τ_{xy} *and* τ_{yx} *contribute to the moment about the z-axis.*

$$\Delta M_z - b(4ac\tau_{yx}) - (-b)(-4ac\tau_{yx}) + (-a)(-4bc\tau_{xy}) + a(4bc\tau_{xy}) = 0.$$

But the volume of the element is $V = 8abc$ so that

$$(\Delta M_z/\Delta V) - \tau_{yx} + \tau_{xy} = 0,$$

and as ΔV tends to zero the first term tends to zero on account of the assumption (1.4.3), giving

$$\tau_{xy} = \tau_{yx}.$$

The symmetry of the other *shear stresses* may be proved in a similar way, so that

$$\tau_{yz} = \tau_{zy}, \quad \tau_{zx} = \tau_{xz}, \quad \tau_{xy} = \tau_{yx}. \tag{1.4.5}$$

Now consider a small rectangular tetrahedron element PABC with three sides parallel to the coordinate planes and the fourth with normal $\hat{n}$, as shown in Figure 1.4.3. Consider the forces acting on the element in the y-direction. If the area ABC is ΔS, and the height PN is h, then the areas PBC, PCA and PAB will be $\ell\Delta S$, $m\Delta S$ and $n\Delta S$ and the volume of the tetrahedron will be $\Delta V = (1/3)h\Delta S$. Equilibrium of the element in the y-direction demands that

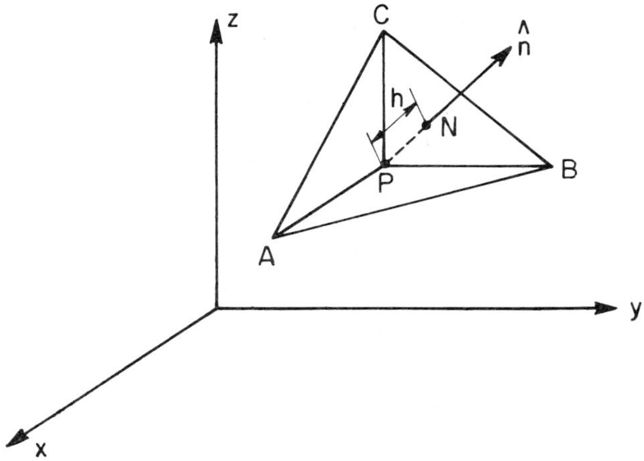

Figure 1.4.3 - The unit vector n̂ *is normal to the triangle* ABC.

$$\Delta R_y + \Delta S(\tau_{ny} - \ell\tau_{xy} - m\tau_{yy} - n\tau_{zy}) = 0. \tag{1.4.6}$$

The surface stresses may, when h and ΔS are small, be approximated by their values at P. Divide equation (1.4.6) by ΔS and consider the limit as ΔV tends to zero.

$$\lim_{\Delta V\to 0} \frac{\Delta R_y}{\Delta S} = \lim_{\Delta V\to 0} \frac{\Delta R_y}{\Delta V} \cdot \frac{\Delta V}{\Delta S} = 0\,.$$

Thus we may obtain the three equations

$$\left.\begin{aligned} \tau_{nx} &= \ell\tau_{xx} + m\tau_{yx} + n\tau_{zx}, \\ \tau_{ny} &= \ell\tau_{xy} + m\tau_{yy} + n\tau_{zy}, \\ \tau_{nz} &= \ell\tau_{xz} + m\tau_{yz} + n\tau_{zz}. \end{aligned}\right\} \tag{1.4.7}$$

The normal stresses $\tau_{xx},\tau_{yy},\tau_{zz}$ are sometimes (Timoshenko and Goodier (1951)) denoted by $\sigma_x,\sigma_y,\sigma_z$ respectively. Love (1927) and Muskhelishvili (1953a) use X_z etc. for τ_{zx} etc. Equation (1.4.7) may be used to prove that the 9 components $\tau_{xx},\tau_{xy},\tau_{xz},\tau_{yx},\tau_{yy}$, $\tau_{yz},\tau_{zx},\tau_{zy},\tau_{zz}$, like the 9 components of strain, form a tensor of rank 2. Indeed equation (1.4.2) shows that the component of

$\vec{T}_n$ in the direction $\hat{n}$, denoted by τ_{nn}, is

$$\begin{aligned}\tau_{nn} &= \hat{n}\cdot\vec{T}_n = \ell\tau_{nx} + m\tau_{ny} + n\tau_{nz}\\ &= \ell^2\tau_{xx} + m^2\tau_{yy} + n^2\tau_{zz} + 2mn\tau_{yz} +\\ &+ 2n\ell\tau_{zx} + 2\ell m\tau_{xy}.\end{aligned} \tag{1.4.8}$$

Now consider the equation governing the (time-dependent) displacement of an arbitrary volume V bounded by a surface S in a continuum. The continuum will be in dynamic equilibrium under the action of the body forces and the surface forces; thus

$$\int_S \vec{T}_n dS + \int_V \vec{F} dV = \int_V \rho \frac{\partial^2\vec{d}}{\partial t^2} dV. \tag{1.4.9}$$

Consider the x-component of this equation and use equation (1.4.2); the x-component of $\vec{T}_n$ is

$$\tau_{nx} = \ell\tau_{xx} + m\tau_{yx} + n\tau_{xz},$$

so that

$$\tau_{nx} dS = (\tau_{xx}\hat{i} + \tau_{xy}\hat{j} + \tau_{xz}\hat{k})\cdot dS\hat{n} = \vec{T}_x\cdot\vec{d}S.$$

Here we have used the symmetry of the shear stresses. Now apply Gauss' divergence theorem:

$$\int_S \tau_{nx} dS = \int_S \vec{T}_x\cdot\vec{d}S = \int_V \mathrm{div}\,\vec{T}_x dV.$$

Thus

$$\int_V \left\{\mathrm{div}\,\vec{T}_x + F_x - \frac{\rho\partial^2 u}{\partial t^2}\right\} dV = 0, \tag{1.4.10}$$

and since the volume V is arbitrary the integrand must be zero throughout the continuum. Since

$$\mathrm{div}\,\vec{T}_x = \frac{\partial\tau_{xx}}{\partial x} + \frac{\partial\tau_{xy}}{\partial y} + \frac{\partial\tau_{xz}}{\partial z},$$

we obtain the first of the three equations

$$\frac{\partial \tau_{xx}}{\partial x} + \frac{\partial \tau_{xy}}{\partial y} + \frac{\partial \tau_{xz}}{\partial z} + F_x = \rho \frac{\partial^2 u}{\partial t^2}, \qquad (1.4.11)$$

$$\frac{\partial \tau_{yx}}{\partial x} + \frac{\partial \tau_{yy}}{\partial y} + \frac{\partial \tau_{yz}}{\partial z} + F_y = \rho \frac{\partial^2 v}{\partial t^2}, \qquad (1.4.12)$$

$$\frac{\partial \tau_{zx}}{\partial x} + \frac{\partial \tau_{zy}}{\partial y} + \frac{\partial \tau_{zz}}{\partial z} + F_z = \rho \frac{\partial^2 w}{\partial t^2}. \qquad (1.4.13)$$

These equations are called the *stress equations of equilibrium.*

Examples 1.4

1. Ox', Oy', Oz' is a Cartesian frame, the axes of which have direction cosines $(\ell_1, m_1, n_1), (\ell_2, m_2, n_2), (\ell_3, m_3, n_3)$ with respect to $O(x,y,z)$. Show that

$$\tau_{y'z'} = \ell_2\ell_3\tau_{xx} + (\ell_2 m_3 + \ell_3 m_2)\tau_{xy} + (\ell_2 n_3 + \ell_3 n_2)\tau_{xz}$$
$$+ m_2 m_3 \tau_{yy} + (m_2 n_3 + m_3 n_2)\tau_{yz} + n_2 n_3 \tau_{zz},$$

   and find the corresponding expression for $\tau_{x'x'}$.

2. Show that $\tau_{x'x'} + \tau_{y'y'} + \tau_{z'z'} = \tau_{xx} + \tau_{yy} + \tau_{zz}$. This is described by saying that $\Theta \equiv \tau_{xx} + \tau_{yy} + \tau_{zz}$ is a *stress invariant.*

3. Tensor notation gives a concise formulation of elasticity theory. The interested reader may consult Prager (1961). What is the tensor form of equations (1.4.11)-(1.4.13)?

1.5 *Stress in cylindrical and spherical polar coordinates*

In order to define stress components in cylindrical coordinates we write the stress vector acting across a plane with normal vector $\hat{r}$, for instance, as

$$\vec{T}_r = \tau_{rr}\hat{r} + \tau_{r\theta}\hat{\theta} + \tau_{rz}\hat{k}. \qquad (1.5.1)$$

The symmetry of the shear stresses (i.e. $\tau_{r\theta} = \tau_{\theta r}$ etc.) may be proved as before, and the stress equations of equilibrium are

$$\frac{\partial \tau_{rr}}{\partial r} + \frac{1}{r}\frac{\partial \tau_{r\theta}}{\partial \theta} + \frac{\partial \tau_{rz}}{\partial z} + \frac{\tau_{rr} - \tau_{\theta\theta}}{r} + F_r = \rho \frac{\partial^2 u_r}{\partial t^2} \qquad (1.5.2)$$

$$\frac{\partial\tau_{\theta r}}{\partial r} + \frac{1}{r}\frac{\partial\tau_{\theta\theta}}{\partial\theta} + \frac{\partial\tau_{\theta z}}{\partial z} + \frac{2\tau_{\theta r}}{r} + F_\theta = \rho\frac{\partial^2 u_\theta}{\partial t^2}, \tag{1.5.3}$$

$$\frac{\partial\tau_{zr}}{\partial r} + \frac{1}{r}\frac{\partial\tau_{z\theta}}{\partial\theta} + \frac{\partial\tau_{zz}}{\partial z} + \frac{\tau_{zr}}{r} + F_z = \rho\frac{\partial^2 u_z}{\partial t^2}, \tag{1.5.4}$$

where the body force per unit volume is

$$\vec{F} = F_r\hat{r} + F_\theta\hat{\theta} + F_z\hat{k}. \tag{1.5.5}$$

The corresponding equations in spherical polar coordinates are

$$\frac{\partial\tau_{RR}}{\partial R} + \frac{1}{R}\frac{\partial\tau_{R\theta}}{\partial\theta} + \frac{1}{R\sin\theta}\frac{\partial\tau_{R\phi}}{\partial\phi} + \frac{1}{R}(2\tau_{RR}-\tau_{\theta\theta}-\tau_{\phi\phi}+\tau_{R\theta}\cot\theta) + + F_R = \rho\frac{\partial^2 u_R}{\partial t^2}, \tag{1.5.6}$$

$$\frac{\partial\tau_{\theta R}}{\partial R} + \frac{1}{R}\frac{\partial\tau_{\theta\theta}}{\partial\theta} + \frac{1}{R\sin\theta}\frac{\partial\tau_{\theta\phi}}{\partial\phi} + \frac{1}{R}\{(\tau_{\theta\theta}-\tau_{\phi\phi})\cot\theta + 3\tau_{\theta R}\} + + F_\theta = \rho\frac{\partial^2 u_\theta}{\partial t^2}, \tag{1.5.7}$$

$$\frac{\partial\tau_{\phi R}}{\partial R} + \frac{1}{R}\frac{\partial\tau_{\phi\theta}}{\partial\theta} + \frac{1}{R\sin\theta}\frac{\partial\tau_{\theta\phi}}{\partial\phi} + \frac{1}{R}(3\tau_{\phi R}+2\tau_{\phi\theta}\cot\theta) + + F_\phi = \rho\frac{\partial^2 u_\phi}{\partial t^2}. \tag{1.5.8}$$

Love (1927) uses the notation $\widehat{r\theta}$ etc. for $\tau_{r\theta}$ etc.

Examples 1.5

1. Use equations (1.4.7), (1.4.8) to express the cylindrical polar stress components in terms of the Cartesian components. Show that

$$\tau_{rr} = \cos^2\theta\tau_{xx} + \sin^2\phi\tau_{yy} + 2\cos\theta\sin\theta\tau_{xy}, \tag{1.5.9}$$

$$\tau_{\theta\theta} = \sin^2\theta\tau_{xx} + \cos^2\theta\tau_{yy} - 2\cos\theta\sin\theta\tau_{xy}, \tag{1.5.10}$$

$$\tau_{r\theta} = \cos\theta\sin\theta(\tau_{yy}-\tau_{xx}) + (\cos^2\theta-\sin^2\theta)\tau_{xy}, \tag{1.5.11}$$

$$\tau_{rz} = \cos\theta\tau_{xz} + \sin\theta\tau_{yz},\quad \tau_{\theta z} = -\sin\theta\tau_{xz} + \cos\theta\tau_{yz}. \quad (1.5.12)$$

1.6 *The relation between stress and strain*

This matter is considered at great length in all of the standard treatises on Elasticity mentioned in the Introduction and there is no need to repeat the discussion here. The outline is as follows.

Hooke, in 1678, stated that *strain is proportional to stress.* This statement may be taken to be the definition of an *ideal*, or *linearly elastic*, medium; it implies that if the stresses are removed, the strains will vanish also. It is found for instance that a stress τ_{xx} applied to the ends of a small cylindrical element (Figure 1.6.1) produces a strain e_{xx} given by

$$e_{xx} = \tau_{xx}/E.$$

Figure 1.6.1 - The stress τ_{xx} is applied at the ends of a cylindrical element.

E is called *Young's modulus* and is a characteristic of the material in the continuum; Young's modulus has the dimensions of *stress.* In addition, it is found that such a normal stress produces *lateral contractions*

$$e_{yy} = -\nu\tau_{xx}/E, \quad e_{zz} = -\nu\tau_{yy}/E,$$

where ν is a dimensionless constant called *Poisson's ratio.* Experimental evidence shows that the strains produced by the action of a combination of normal stresses can be obtained by superimposing the three equations of the type shown for the three co-ordinate directions. Thus for an *isotropic* elastic body, that is, one for which the elastic properties are the same for each direction emerging from a given point P of the continuum,

$$e_{xx} = [\tau_{xx} - \nu(\tau_{yy} + \tau_{zz})]/E, \tag{1.6.1}$$

$$e_{yy} = [\tau_{yy} - \nu(\tau_{zz} + \tau_{xx})]/E, \tag{1.6.2}$$

$$e_{zz} = [\tau_{zz} - \nu(\tau_{xx} + \tau_{yy})]/E. \tag{1.6.3}$$

It is found also that the shear strains are directly proportional to the shear stresses and that, for an isotropic continuum,

$$e_{yz} = \frac{(1+\nu)}{E}\,\tau_{yz}, \quad e_{zx} = \frac{(1+\nu)}{E}\,\tau_{zx}, \quad e_{xy} = \frac{(1+\nu)}{E}\,\tau_{xy}. \tag{1.6.4}$$

Unless the contrary is explicitly stated, it will be assumed throughout this book that the elastic continuum under consideration is isotropic, linearly elastic and *homogeneous*, that is, the elastic constants are independent of the position of the point in the continuum.

Equations (1.6.1)-(1.6.3) may be inverted to give

$$\tau_{xx} = \frac{E}{(1+\nu)(1-2\nu)}\{\nu\varepsilon + (1-2\nu)e_{xx}\} \tag{1.6.5}$$

where ε is the cubical dilatation

$$\varepsilon = e_{xx} + e_{yy} + e_{zz}. \tag{1.6.6}$$

The coefficients in this equation are called Lamé's constants and denoted by λ, μ; thus

$$\lambda = \frac{E\nu}{(1+\nu)(1-2\nu)}, \quad \mu = \frac{E}{2(1+\nu)} \tag{1.6.7}$$

and (1.6.5) may be written

$$\tau_{xx} = \lambda\varepsilon + 2\mu e_{xx}. \tag{1.6.8}$$

Throughout this book the two basic elastic constants will be taken to be μ, the *shear modulus* (sometimes denoted by G), and Poisson's ratio ν. The symbol λ will be used for the Lamé coefficient in *this* chapter, but in the remainder of the book it will be freed to denote some parameter of local significance. Since

$$E = 2(1 + \nu)\mu, \qquad \lambda = 2\mu\nu/(1 - 2\nu), \tag{1.6.9}$$

the standard form for the strain-stress equations is

$$e_{xx} = \frac{\tau_{xx} - \nu(\tau_{yy} + \tau_{zz})}{2(1+\nu)\mu}, \qquad e_{yz} = \frac{\tau_{yz}}{2\mu}, \text{ etc.} \tag{1.6.10}$$

while that for the stress-strain equations is

$$\tau_{xx} = 2\mu(e_{xx} + \nu\varepsilon/(1-2\nu)), \qquad \tau_{yz} = 2\mu e_{yz}, \text{ etc.} \tag{1.6.11}$$

It is shown in most of the treatises that the *stress-strain* equation for an isotropic medium *must* have the form given above if the relations are to be *invariant* with respect to the choice of Cartesian coordinate axes. Thus, if Ox',Oy',Oz' is another Cartesian frame, then the strain-stress relations in that frame must have the form

$$e_{x'x'} = [\tau_{x'x'} - \nu(\tau_{y'y'} + \tau_{z'z'})]/E, \text{ etc.}$$

$$e_{y'z'} = (1 + \nu)\tau_{y'z'}/E. \tag{1.6.12}$$

For an isotropic body the stress-strain equations in other orthogonal coordinate systems will be exactly like those given above. Thus for cylindrical coordinates

$$\tau_{rr} = 2\mu\{\nu\varepsilon + (1-2\nu)e_{rr}\}/(1-2\nu), \qquad \tau_{r\theta} = 2\mu e_{r\theta}, \text{ etc.} \tag{1.6.13}$$

where again

$$\varepsilon = e_{rr} + e_{\theta\theta} + e_{zz}. \tag{1.6.14}$$

Three other abbreviations will be used throughout the book for combinations of the elastic constants which appear often: these are

$$\vartheta = \frac{1 - \nu}{\mu} \equiv \frac{2(1-\nu^2)}{E}, \quad \varphi = \frac{1 - 2\nu}{2(1-\nu)}, \quad \kappa = 3 - 4\nu. \qquad (1.6.15)$$

1.7 *Strain energy and Betti's reciprocal theorem*

The concept of *strain energy* is important because of its use in proofs of the uniqueness of solution for certain types of boundary value problems, and also because of its place in numerical methods based on variational principles.

Consider an elastic continuum occupying a region V with surface S, as shown in Figure 1.7.1, and suppose that initially, at time t = 0, it is in a *natural state*, i.e. at rest, with zero displacement and uniform temperature. Now suppose that it undergoes infinitesimal displacement due to surface forces $\vec{T}_n$ and body forces $\vec{F}$. The rate at which these forces do work is

$$\frac{dE}{dt} = \int_V \vec{F}\cdot\dot{\vec{d}}dV + \int_S \vec{T}_n\cdot\dot{\vec{d}}dS, \qquad (1.7.1)$$

where $\dot{\vec{d}} \equiv \partial\vec{d}/\partial t$. The second integral may be transformed into a volume integral by using Gauss's theorem.

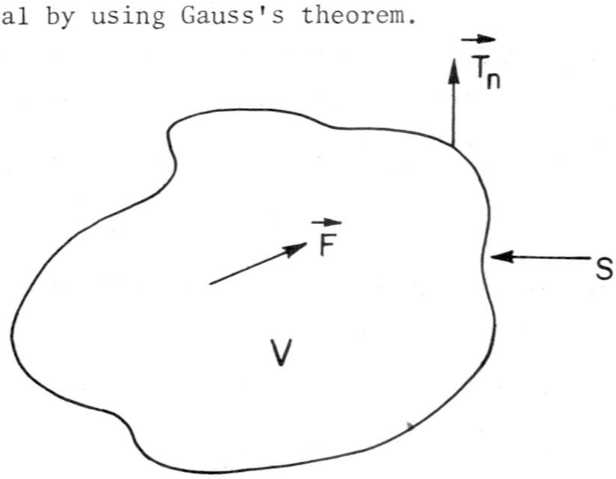

Figure 1.7.1 - The elastic medium occupies the region V with surface S.

$$\vec{T}_n \cdot \dot{\vec{d}} = \tau_{nx}\dot{u} + \tau_{ny}\dot{v} + \tau_{nz}\dot{w},$$

$$\tau_{nx}\dot{u} = (\ell\tau_{xx}+m\tau_{yz}+n\tau_{zx})\dot{u} = \vec{T}_x\dot{u}\cdot\hat{n}.$$

Thus

$$\int_S \vec{T}_n \cdot \dot{\vec{d}} dS = \int_S (\vec{T}_x\dot{u}+\vec{T}_y\dot{v}+\vec{T}_z\dot{w})\cdot \vec{dS} = \int_V \text{div}(\vec{T}_x\dot{u}+\vec{T}_y\dot{v}+\vec{T}_z\dot{w})\,dV,$$

and

$$\text{div}(\vec{T}_x\dot{u}+\vec{T}_y\dot{v}+\vec{T}_z\dot{w}) = \dot{u}\,\text{div}\,\vec{T}_x + \dot{v}\,\text{div}\,\vec{T}_y + \dot{w}\,\text{div}\,\vec{T}_z + $$
$$+ \vec{T}_x\cdot\vec{\nabla}\dot{u} + \vec{T}_y\cdot\vec{\nabla}\dot{v} + \vec{T}_z\cdot\vec{\nabla}\dot{w}, \qquad (1.7.2)$$

where $\vec{\nabla} = \hat{i}\frac{\partial}{\partial x} + \hat{j}\frac{\partial}{\partial y} + \hat{k}\frac{\partial}{\partial z}$.

Now the first of the three terms involving the $\vec{\nabla}$ operator may be written

$$\vec{T}_x\cdot\vec{\nabla}\dot{u} = \tau_{xx}\frac{\partial\dot{u}}{\partial x} + \tau_{xy}\frac{\partial\dot{u}}{\partial y} + \tau_{xz}\frac{\partial\dot{u}}{\partial z},$$

and using the expressions for the strains and rotations derived in §1.2 we have

$$\vec{T}_x\cdot\vec{\nabla}\dot{u} = \tau_{xx}\dot{e}_{xx} + \tau_{xy}\{\dot{e}_{xy}-\dot{\omega}_{xy}\} + \tau_{xz}\{\dot{e}_{xz}-\dot{\omega}_{xz}\}.$$

Because the components of rotation are antisymmetrical while the components of strain are symmetrical we find on addition that

$$A \equiv \vec{T}_x\cdot\vec{\nabla}\dot{u} + \vec{T}_y\cdot\vec{\nabla}\dot{v} + \vec{T}_z\cdot\vec{\nabla}\dot{w} = \tau_{xx}\dot{e}_{xx} + \tau_{yy}\dot{e}_{yy} + \tau_{zz}\dot{e}_{zz}$$
$$+ 2\tau_{yz}\dot{e}_{yz} + 2\tau_{zx}\dot{e}_{zx} + 2\tau_{xy}\dot{e}_{xy}.$$

For a linearly elastic isotropic body the stresses are given by equations (1.6.1)-(1.6.4) and

$$A \equiv \lambda\varepsilon(\dot{e}_{xx}+\dot{e}_{yy}+\dot{e}_{zz}) + 2\mu(e_{xx}\dot{e}_{xx}+e_{yy}\dot{e}_{yy}+e_{zz}\dot{e}_{zz}+$$
$$+2e_{yz}\dot{e}_{yz}+2e_{zx}\dot{e}_{zx}+2e_{xy}\dot{e}_{xy}).$$

This is the time derivative of the function W given by

$$2W = \lambda\varepsilon^2 + 2\mu(e_{xx}^2+e_{yy}^2+e_{zz}^2+2e_{yz}^2+2e_{zx}^2+2e_{xy}^2) \tag{1.7.3}$$

and which is called the *strain energy density* function. This function must always be positive, whatever the strains, and this implies that $\lambda > 0$, $\mu > 0$; these inequalities imply that the Poisson's ratio ν satisfies $0 < \nu < 1/2$. Return to equation (1.7.1) and use the results of equations (1.7.2) and (1.7.3).

$$\frac{dE}{dt} = \iint_V \left\{\dot{u}(\mathrm{div}\,\vec{T}_x+F_x) + \dot{v}(\mathrm{div}\,\vec{T}_y+F_y) + \dot{w}(\mathrm{div}\,\vec{T}_z+F_z) + \frac{\partial W}{\partial t}\right\}dV.$$

But the surface forces and body forces satisfy equations (1.4.11)-(1.4.13) so that

$$\mathrm{div}\,\vec{T}_x + F_x \equiv \frac{\partial\tau_{xx}}{\partial x} + \frac{\partial\tau_{xy}}{\partial y} + \frac{\partial\tau_{xz}}{\partial z} + F_x = \rho\ddot{u}$$

and therefore

$$\frac{dE}{dt} = \iint_V \left\{\rho(\dot{u}\ddot{u} + \dot{v}\ddot{v} + \dot{w}\ddot{w}) + \frac{\partial W}{\partial t}\right\}dV = \frac{dT}{dt} + \frac{dU}{dt} \tag{1.7.4}$$

where

$$T = \frac{1}{2}\int \rho(\dot{u}^2 + \dot{v}^2 + \dot{w}^2)dV \tag{1.7.5}$$

is the kinetic energy of the body, and

$$U = \int_V W dV \tag{1.7.6}$$

is the *strain energy* of the body. Integrating equation (1.7.4) from $t = 0$ until $t = t$, and noting that T and U both vanish in the natural state, we deduce that

$$E = T + U. \tag{1.7.7}$$

This equation states that the work done by the external and body forces is equal to the sum of the kinetic energy T of the body and the strain energy U.

Betti's reciprocal theorem (Betti (1872)) relates to two *states* of a given elastic body, having a volume V bounded by the surface S. The first state corresponds to elastic displacements u,v,w and is maintained by a system of body forces F_x,F_y,F_z and surface tractions $\vec{T}_n$. The second state is specified by the corresponding starred quantities u*, v*, w*, F_x^*, F_y^*, F_z^* and $\vec{T}_n^*$. The theorem states that the work done by the body forces, kinetic reactions and surface tractions of the first state acting over the displacements of the second is equal to the work done by the forces of the second state acting over the displacements of the first. Symbolically

$$\int_V \left(\vec{F} - \rho \frac{\partial^2 \vec{d}}{\partial t^2}\right)\cdot\vec{d}^* dV + \int_S \vec{T}_n\cdot\vec{d}^* dS = \int_V \left(\vec{F}^* - \rho \frac{\partial^2 \vec{d}^*}{\partial t^2}\right)\cdot\vec{d}\, dV + \int_S \vec{T}_n^*\cdot\vec{d}\, dS. \qquad (1.7.8)$$

By using the argument which led to equation (1.7.2) we may show that

$$\int_S \vec{T}_n\cdot\vec{d}^* dS = \int_V \mathrm{div}(\vec{T}_x u^* + \vec{T}_y v^* + \vec{T}_z w^*)\, dV.$$

Thus the terms in u* on the left hand side of (1.7.8) are

$$\int_V \left\{\left[- \frac{\partial\tau_{xx}}{\partial x} - \frac{\partial\tau_{xy}}{\partial y} - \frac{\partial\tau_{xz}}{\partial z}\right]u^* + \mathrm{div}(\vec{T}_x u^*)\right\} dV$$

$$= \int_V \left\{\tau_{xx}\frac{\partial u^*}{\partial x} + \tau_{xy}\frac{\partial u^*}{\partial y} + \tau_{xz}\frac{\partial u^*}{\partial z}\right\} dV.$$

Grouping the terms corresponding to u*,v*,w* we find that the left hand side of (1.7.8) is

$$\int_V \{\tau_{xx}e^*_{xx} + \tau_{yy}e^*_{yy} + \tau_{zz}e^*_{zz} + 2\tau_{yz}e^*_{yz} + 2\tau_{zx}e^*_{zx} + 2\tau_{xy}e^*_{xy}\} dV.$$

It may easily be verified that the expression in braces is symmetrical in starred and unstarred quantities and is therefore equal to the right hand side of (1.7.8).

Examples 1.7

1. Verify the analysis leading from (1.7.1)-(1.7.4).
2. For an isotropic elastic medium the strain energy density must be invariant under coordinate changes. Show that the bracketed quantity in (1.7.3) is such an invariant. It may be shown that this quantity and ε^2 are the only two independent quadratic invariants of the strains.
3. Establish equation (1.7.8) by verifying the details of the argument given in brief.

1.8 *Anisotropic media*

For anisotropic media it is appropriate to use tensor analysis. We recall (see §1.2 and 1.4) that the strain components $e_{xx}, e_{xy}, e_{xz}, e_{yx}, e_{yy}, e_{yz}, e_{zx}, e_{zy}, e_{zz}$ and the corresponding stress components $\tau_{xx}, \tau_{xy}, \ldots, \tau_{zz}$ form tensors of rank 2. It is therefore convenient to denote these components by e_{ij}, τ_{ij} where i,j = 1,2,3. The most general linear relationship between these quantities is

$$\tau_{ij} = c_{ijk\ell} e_{k\ell} \qquad (i,j = 1,2,3). \tag{1.8.1}$$

Here we have made use of Einstein's summation convention that a *repeated* suffix (here k,ℓ are both repeated) implies a summation over the range of the values; here the range is 1,2,3. Thus equation (1.8.1) is an abbreviation for

$$\tau_{ij} = e_{ij11} e_{11} + c_{ij12} e_{12} + \ldots + c_{ij33} e_{33}.$$

The 81 quantities $c_{ijk\ell}$ define the elastic properties of the medium. We shall assume that the medium is *homogeneous* and then these quantities will be constants. Since both stress components and the strain components are symmetric, i.e.

$$\tau_{ij} \equiv \tau_{ji}, \quad e_{ij} \equiv e_{ji}, \tag{1.8.2}$$

the $c_{ijk\ell}$ satisfy

$$c_{ijk\ell} = c_{jik\ell}, \quad c_{ijk\ell} = c_{ij\ell k}. \tag{1.8.3}$$

This reduces the 81 constants to 36, and suggests the following less cumbersome notation

$$\left.\begin{array}{lll} e_{11} = e_1, & e_{22} = e_2, & e_{33} = e_3 \\ e_{23} = e_{32} = \frac{1}{2} e_4, & e_{31} = e_{13} = \frac{1}{2} e_5, & e_{12} = e_{21} = \frac{1}{2} e_6 \\ \tau_{11} = \tau_1 & \tau_{22} = \tau_2, & \tau_{33} = \tau_3 \\ \tau_{23} = \tau_{32} = \tau_4, & \tau_{31} = \tau_{13} = \tau_5, & \tau_{12} = \tau_{21} = \tau_6. \end{array}\right\} \tag{1.8.4}$$

The factors of 1/2 are introduced so that the terms $c_{ij12}e_{12}$ + $c_{ij21}e_{21}$ etc. may be condensed into $c_{ij12}e_6$. Now we may introduce just a set of double suffix constants c_{ij} and write the stress-strain relations as

$$\tau_i = c_{ij}e_j \quad (i = 1,6) \tag{1.8.5}$$

and the repeated suffix implies that j is summed over 1,2,...,6.

It may be proved (see Love (1927), §65) that for *isothermal* processes the 36 constants c_{ij} are *symmetric*, i.e. $c_{ij} = c_{ji}$ so that there are only 21 different constants. In this case there is a *strain energy density function* W such that

$$\tau_i = \frac{\partial W}{\partial e_i} \tag{1.8.6}$$

and

$$W = \frac{1}{2} c_{ij}e_ie_j. \tag{1.8.7}$$

It may be proved that equation (1.7.7), namely

$$E = T + U, \tag{1.8.8}$$

holds for such an anisotropic elastic body. This is the subject of Ex. 1.8.1.

The general relation (1.8.5) may be specialized for media with symmetry properties. Thus, for example, if the properties of the medium are symmetric with respect to the plane z = 0, the transformation

$$x' = x, \quad y' = y, \quad z' = -z \tag{1.8.9}$$

should leave equations (1.8.5) invariant. But the transformation (1.8.9) leaves the stress and strain components unchanged, except that

$$e_4' = -e_4, \quad e_5' = -e_5, \quad \tau_4' = -\tau_4, \quad \tau_5' = -\tau_5.$$

The equations for $\tau_1 = \tau_1'$ in the old and new coordinates are

$$\tau_1 = c_{11}e_1 + c_{12}e_2 + c_{13}e_3 + c_{14}e_4 + c_{15}e_5 + c_{16}e_6,$$

$$\tau_1' = c_{11}e_1 + c_{12}e_2 + c_{13}e_3 - c_{14}e_4 - c_{15}e_5 + c_{16}e_6.$$

If these are to be the same then

$$c_{14} = 0 = c_{15}.$$

It may be shown similarly that

$$c_{24} = 0 = c_{25} = c_{34} = c_{35} = c_{46} = c_{56}$$

so that there are only 13 different constants.

An *orthotropic* medium is one which has elastic properties that are symmetric about three mutually orthogonal planes. Such media, of which wood is an example, have only 9 elastic constants. A classification of symmetry properties of various crystals, which are of current interest for the analysis of piezoelectric and magnetostrictive materials, may be found in Love (1927, §107-114). Useful discussions of anisotropy, with various examples, appear in Hearmon (1961), Lekhnitskii (1963a) and Milne-Thomson (1960).

Examples 1.8

1. Show that equation (1.7.7) holds for an elastic body which has a strain energy density given by equations (1.8.6), (1.8.7).
2. Let (x',y',z') be the coordinate axes obtained by rotating (x,y,z) about the z-axis through an angle θ and let e_i, τ_i $(i = 1,2,\dots,6)$ and e_i', τ_i' $(i = 1,2,\dots,6)$ be the strains and stresses in the unprimed and primed coordinates respectively.

Use equation (1.2.8),(1.2.9),(1.4.8) etc. to express τ_i' in terms of τ_j, e_i in terms of e_j' and hence show that the elastic constants c_{ij}' in the new coordinates are related to those in the old by the equations

$$c_{11}' = c_{11}\cos^4\theta + 2c_{12}\cos^2\theta\sin^2\theta + c_{22}\sin^4\theta +$$

$$+ 4c_{16}\cos^3\theta\sin\theta + 4c_{26}\cos\theta\sin^3\theta + 4c_{66}\cos^2\theta\sin^2\theta,$$

$$c_{12}' = c_{11}\cos^2\theta\sin^2\theta + c_{12}(\cos^4\theta + \sin^4\theta) -$$

$$- 2(c_{16} - c_{26})\cos\theta\sin\theta(\cos^2\theta - \sin^2\theta) -$$

$$- 4c_{66}\cos^2\theta\sin^2\theta,$$

$$c_{13}' = c_{13}\cos^2\theta + c_{23}\sin^2\theta + 2c_{63}\cos\theta\sin\theta,$$

$$c_{14}' = \cos\theta(c_{14}\cos^2\theta + c_{24}\sin^2\theta + 2c_{64}\cos\theta\sin\theta) -$$

$$- \sin\theta(c_{15}\cos^2\theta + c_{25}\sin^2\theta + 2c_{65}\cos\theta\sin\theta),$$

$$c_{15}' = \cos\theta(c_{15}\cos^2\theta + c_{25}\sin^2\theta + 2c_{65}\cos\theta\sin\theta) +$$

$$+ \sin\theta(c_{14}\cos^2\theta + c_{24}\sin^2\theta + 2c_{64}\cos\theta\sin\theta),$$

$$c_{16}' = -c_{11}\cos^3\theta\sin\theta + c_{22}\cos\theta\sin^3\theta +$$

$$+ c_{12}\cos\theta\sin\theta(\cos^2\theta - \sin^2\theta) +$$

$$+ c_{16}\cos^2\theta(\cos^2\theta - 3\sin^2\theta) + c_{26}\sin^2\theta(3\cos^2\theta - \sin^2\theta) +$$

$$+ 2c_{66}\cos\theta\sin\theta(\cos^2\theta - \sin^2\theta), \text{ etc.}$$

3. Use the results of Ex. 1.8.2 to show that if the medium is symmetric about the z-axis then the elastic constants c_{ij} satisfy

$$c_{14} = 0 = c_{15} = c_{16} = c_{24} = c_{25} = c_{26} = c_{34} = c_{35} = c_{36},$$

$$c_{56} = 0 = c_{64} = c_{45},$$

$$c_{11} = c_{22}, \quad c_{13} = c_{23}, \quad c_{55} = c_{44}, \quad c_{66} = \frac{1}{2}(c_{11} - c_{22}),$$

so that there are only 5 independent constants $c_{11}, c_{12}, c_{13}, c_{33}, c_{44}$. Such a medium is called *transversely isotropic*.

4. Show that if the elastic properties of a body are symmetrical with respect to each of the two planes $x = 0$, $y = 0$, then it has 9 elastic constants and is symmetrical with respect to the plane $z = 0$. Such a body is said to be orthotropic.
5. Show that if a medium possesses three *axes* of symmetry at right angles to each other it is isotropic.

1.9 *Uniqueness of solution for static problems*

The fundamental equations governing the static deformation of a linearly elastic isotropic body are equations (1.2.4),(1.2.5) linking the strains to the displacements, equations (1.4.11)-(1.4.13) expressing the equilibrium of the surface and body forces (and now $\ddot{\vec{d}} \equiv 0$), and equations (1.6.6),(1.6.7) linking the stresses and strains. When collected together therefore, the complete set of equations is

$$e_{xx} = \frac{\partial u}{\partial x}, \qquad e_{yy} = \frac{\partial v}{\partial y}, \qquad e_{zz} = \frac{\partial w}{\partial z}, \tag{1.9.1}$$

$$e_{yx} = \frac{1}{2}\left(\frac{\partial v}{\partial z} + \frac{\partial w}{\partial y}\right), \quad e_{zx} = \frac{1}{2}\left(\frac{\partial w}{\partial x} + \frac{\partial u}{\partial z}\right), \quad e_{xy} = \frac{1}{2}\left(\frac{\partial u}{\partial y} + \frac{\partial v}{\partial x}\right); \tag{1.9.2}$$

$$\tau_{xx} = \lambda\varepsilon + 2\mu e_{xx}, \quad \tau_{yy} = \lambda\varepsilon + 2\mu e_{yy}, \quad \tau_{zz} = \lambda\varepsilon + 2\mu e_{zz}, \tag{1.9.3}$$

$$\tau_{yz} = 2\mu e_{yz}, \qquad \tau_{zx} = 2\mu e_{zx}, \qquad \tau_{xy} = 2\mu e_{xy}; \tag{1.9.4}$$

$$\frac{\partial \tau_{xx}}{\partial x} + \frac{\partial \tau_{xy}}{\partial y} + \frac{\partial \tau_{xz}}{\partial z} + F_x = 0, \tag{1.9.5}$$

$$\frac{\partial \tau_{yx}}{\partial x} + \frac{\partial \tau_{yy}}{\partial y} + \frac{\partial \tau_{yz}}{\partial z} + F_y = 0, \tag{1.9.6}$$

$$\frac{\partial \tau_{zx}}{\partial x} + \frac{\partial \tau_{zy}}{\partial y} + \frac{\partial \tau_{zz}}{\partial z} + F_z = 0. \tag{1.9.7}$$

In order to prove that, under certain conditions, these equations admit only one solution, we need *Clapeyron's Theorem*, namely: If an elastic body is in equilibrium under the action of body forces $\vec{F}$ and surface forces $\vec{T}_n$, then the strain energy of deformation is equal to one-half the work that would be done by the external forces (of the equilibrium state) acting through the (final) displacements $\vec{d}$ of the equilibrium state.

This theorem asserts that

$$\int_V \rho\vec{F}\cdot\vec{d}dV + \int_S \vec{T}_n\cdot\vec{d}dS = 2\int_V W dV, \tag{1.9.8}$$

and may be proved by transforming the surface integral as was done in the previous section. Thus

$$\int_S \vec{T}_n\cdot\vec{d}dS = \int_S (\tau_{nx}u + \tau_{ny}v + \tau_{nz}w)dS$$

$$\int_S \tau_{nx}udS = \int_S (\ell\tau_{xx} + m\tau_{yx} + n\tau_{zx})udS = \int_S \vec{T}_x u\cdot d\vec{S},$$

$$= \int_V \mathrm{div}(\vec{T}_x u)dV = \int_V (u \;\mathrm{div}\; \vec{T}_x + \vec{T}_x\cdot\vec{\nabla}u)dV,$$

and when we use the equilibrium equations (1.8.5)-(1.8.7) we find that the left hand side of equation (1.9.8) may be written

$$\int_V (\vec{T}_x\cdot\vec{\nabla}u + \vec{T}_y\cdot\vec{\nabla}v + \vec{T}_z\cdot\vec{\nabla}w)dV,$$

and the integrand may be transformed as before into $2W$.

Following Muskhelishvili (1953a) we now define the three *fundamental boundary value problems* for the static deformation of an elastic body. Consider an elastic body occupying a region V bounded by a surface S and suppose that the body forces $\vec{F}$ are given functions of (x,y,z) throughout V.

The first fundamental, or stress boundary value problem. Find the elastic equilibrium state of the body subject to given surface stresses $\vec{T}_n$ on the surface S.

The second fundamental, or displacement boundary value problem. Find the elastic equilibrium state of the body subject to given *displacements* $\vec{d}$ on the surface S.

Uniqueness of solution for static problems

The fundamental mixed problem. Find the elastic equilibrium state of the body subject to given surface forces $\vec{T}_n$ on a part S_1 of the surface S and given displacements $\vec{d}$ on the remainder S_2.

The following proofs of uniqueness are due essentially to Kirchhoff and date from 1859 (see Love (1927)). Suppose that a certain problem, which may be of type I, II or mixed, has two solutions which will be distinguished by one and two primes respectively. Thus one solution, given by $u', v', w', e'_{xx}, \ldots, \tau'_{xx}, \ldots, \tau'_{xy}$ satisfies equations (1.9.1)-(1.9.7), and the other, $u'', v'', w'', e''_{xx}, \ldots, \tau''_{xx}, \ldots, \tau''_{xy}$, satisfies the same equations, for the same body forces F_x, F_y, F_z. Denote the difference of the two solutions by unprimed quantities; thus because of the linearity of the equations we have

$$u = u' - u'', \ \ldots, \ e_{xx} = e'_{xx} - e''_{xx} \text{ etc.}$$

so that $u, v, w, e_{xx}, \ldots, \tau_{xx}, \ldots, \tau_{xy}$ will satisfy the equations (1.9.1)-(1.9.7) with *zero values of the body forces*: in other words the unprimed solution will be an elastic equilibrium state of the body corresponding to zero body forces. Now apply Clapeyron's Theorem to the unprimed state; since $\vec{F} \equiv 0$ we obtain

$$\int_S \vec{T}_n \cdot \vec{d} dS = 2 \int_V W dV.$$

Consider Problem I. The surface stresses $\vec{T}_n$ are given over the whole of S so that they will have the same values for the two solutions ' and "; for the unprimed solution, therefore, $\vec{T}_n = 0$ on S so that

$$\int_V W dV = 0. \tag{1.9.9}$$

But W is a positive quantity so that equation (1.9.9) implies

$$W \equiv 0 \text{ throughout } V.$$

The strain energy density W is a positive definite quadratic form in the strains so that W can be zero only when each strain is zero. Therefore, for the unprimed solution

$$e_{xx} = 0 = e_{yy} = e_{zz} = e_{yz} = e_{zx} = e_{xy}. \tag{1.9.10}$$

This means that the strains, and therefore the stresses, corresponding to the states ' and '' are identical, i.e. the strains and stresses are uniquely determined. Equation (1.9.10) also implies that the displacements u',v',w' and u'',v'',w'' differ by at most a rigid-body translation and rotation.

For Problem II the integral on the left of equation (1.9.8) is again zero, this time because the displacement vector $\vec{d}$ for the unprimed state will be zero on S. Again W must be identically zero so that the strains and stresses are uniquely determined. Now, in addition, since $\vec{d}$ is specified on S, the displacement state will also be uniquely determined, i.e. there is no possibility of an undetermined rigid-body displacement.

In the Mixed Problem the integral over S may be divided into two,

$$\int_S \vec{T}_n \cdot \vec{d} dS = \int_{S_1} \vec{T}_n \cdot \vec{d} dS + \int_{S_2} \vec{T}_n \cdot \vec{d} dS,$$

and $\vec{T}_n$ will be zero in the first, while $\vec{d}$ will be zero in the second. Again the stresses, strains and displacements will be uniquely determined.

Some sophisticated extensions and generalizations of these uniqueness theorems to time-dependent states may be found in Gurtin (1972, p. 102f).

1.10 *Solutions of the displacement equations*

The equilibrium equations (1.4.11)-(1.4.13) may be expressed directly in terms of the displacements by using the stress-strain equations (1.6.6)-(1.6.7) and the expressions (1.2.4)-(1.2.5) for the strains. Thus

$$\vec{T}_x \equiv \tau_{xx}\hat{i} + \tau_{yx}\hat{j} + \tau_{zx}\hat{k} = \lambda\varepsilon\hat{i} + 2\mu(e_{xx}\hat{i} + e_{xy}\hat{j} + e_{xz}\hat{k}),$$

where $\varepsilon \equiv e_{xx} + e_{yy} + e_{zz} = \text{div}\,\vec{d}$, and

$$e_{xx}\hat{i} + e_{xy}\hat{j} + e_{xz}\hat{k} = \frac{1}{2}\left(\hat{i}\frac{\partial}{\partial x} + \hat{j}\frac{\partial}{\partial y} + \hat{k}\frac{\partial}{\partial z}\right)u + \frac{1}{2}\frac{\partial}{\partial x}(u\hat{i} + v\hat{j} + w\hat{k}) = \frac{1}{2}\vec{\nabla}u + \frac{1}{2}\frac{\partial\vec{d}}{\partial x},$$

so that the first equilibrium equation may be written

$$\text{div}\,\vec{T}_x + F_x \equiv \lambda\frac{\partial\varepsilon}{\partial x} + \mu\nabla^2 u + \mu\frac{\partial\varepsilon}{\partial x} + F_x = \rho\frac{\partial^2 u}{\partial t^2},$$

or

$$(\lambda + \mu)\frac{\partial\varepsilon}{\partial x} + \mu\nabla^2 u + F_x = \rho\frac{\partial^2 u}{\partial t^2};$$

and since equations (1.6.7) show that $(\lambda + \mu)/\mu = 1/(1 - 2\nu)$, we obtain the first component of the equation

$$\nabla^2\vec{d} + \frac{\vec{\nabla}\varepsilon}{(1-2\nu)} + \frac{\vec{F}}{\mu} = \frac{\rho}{\mu}\frac{\partial^2\vec{d}}{\partial t^2}. \qquad (1.10.1)$$

This is *Navier's* equation governing the elastic displacements.

There are a number of complete solutions of the static version of this equation, the most important of which is due to Papkovich (1932a),(1932b) and Neuber (1934); the idea may be found in Boussinesq (1885), and is described by Todhunter and Pearson (1893).

In the static case equation (1.10.1) may be written

$$2\mu\nabla^2\vec{d} = -2\vec{F} - \frac{2\mu\vec{\nabla}\varepsilon}{1-2\nu}, \qquad (1.10.2)$$

which exhibits $\nabla^2\vec{d}$ as the combination of two fields, $\vec{F}$ and $\vec{\nabla}\varepsilon$. It is therefore natural to seek $\vec{d}$ of the same form, i.e.

$$2\mu\vec{d} = 4(1-\nu)\vec{\psi} - \vec{\nabla}\chi, \qquad (1.10.3)$$

where $\vec{\psi}(x,y,z)$ and $\chi(x,y,z)$ are a vector and a scalar field respectively. Substitution into (1.10.2) yields

$$4(1-\nu)\nabla^2\vec{\psi} - \nabla^2\vec{\nabla}\chi = -2\vec{F} - 2\mu\vec{\nabla}\varepsilon/(1-2\nu). \qquad (1.10.4)$$

Choose $\vec{\psi}$ so that

$$2(1-\nu)\nabla^2\vec{\psi} = -\vec{F}; \tag{1.10.5}$$

then χ must satisfy

$$\nabla^2\chi = 2\mu\varepsilon/(1-2\nu). \tag{1.10.6}$$

Equation (1.10.3) gives

$$2\mu\varepsilon = 2\mu \operatorname{div} \vec{d} = 4(1-\nu)\operatorname{div} \vec{\psi} - \nabla^2\chi, \tag{1.10.7}$$

and the identity $\nabla^2(\vec{r}\cdot\vec{\psi}) = \vec{r}\cdot\nabla^2\vec{\psi} + 2 \operatorname{div} \vec{\psi}$, together with (1.10.5), yields

$$\begin{aligned} 2\mu\varepsilon &= 2(1-\nu)[\nabla^2(\vec{r}\cdot\vec{\psi}) - \vec{r}\cdot\nabla^2\vec{\psi}] - \nabla^2\chi, \\ &= 2(1-\nu)\nabla^2(\vec{r}\cdot\vec{\psi}) + \vec{r}\cdot\vec{F} - \nabla^2\chi, \end{aligned} \tag{1.10.8}$$

which when substituted into (1.10.6) gives

$$2(1-\nu)\nabla^2[\chi - (\vec{r}\cdot\vec{\chi})] = (\vec{r}\cdot\vec{F}). \tag{1.10.9}$$

If then $\phi(x,y,z)$ satisfies

$$2(1-\nu)\nabla^2\phi = (\vec{r}\cdot\vec{F}), \tag{1.10.10}$$

then $\chi = (\vec{r}\cdot\vec{\psi}) + \phi$ and

$$2\mu\vec{d} = 4(1-\nu)\vec{\psi} - \vec{\nabla}\{(\vec{r}\cdot\vec{\psi}) + \phi\}. \tag{1.10.11}$$

This is the Papkovich-Neuber solution (Papkovich (1932a),(1932b), Neuber (1934)). The potentials $\vec{\psi},\phi$ corresponding to a given $\vec{F}$ are not unique. In fact it may easily be verified (Ex. 1.10.4) that if a field $\vec{d}$ has one representation in terms of $\vec{\psi},\phi$, then $\vec{\psi}^*,\phi^*$ will give the same representation provided that

$$\vec{\psi}^* = \vec{\psi} + \vec{\nabla}\xi, \quad \phi^* = \phi + 4(1-\nu)\xi - \vec{r}\cdot\vec{\nabla}\xi, \quad \nabla^2\xi = 0. \tag{1.10.12}$$

This means, in particular, that it is always *possible* (though not always desirable) to choose one of the components of $\vec{\psi}$ to be zero; this result is due to Eubanks and Sternberg (1956). For a proof of the completeness of the Papkovich-Neuber solution, and details of other displacement solutions, see Gurtin (1972, p. 138ff); for further comments on and reference to other displacement solutions see Mindlin (1936), Lur'e (1964), particularly the appendix to Chapter 1, Stippes (1971), Golecki (1976), and Benthem (1979).

When there are no body forces, then $\vec{F} = \vec{0}$ and $\vec{\psi},\phi$ satisfy

$$\nabla^2\vec{\psi} = 0, \qquad \nabla^2\phi = 0. \tag{1.10.13}$$

If $\vec{F}$ is a conservative field, it may be written

$$\vec{F} = -\vec{\nabla}V, \tag{1.10.14}$$

where V is a potential. If V itself is written $2V = \text{div}\,\vec{W}$, then $2F = -\vec{\nabla}\text{div}\,\vec{W}$. The general solution of equation (1.10.2) may be expressed as the sum of two parts: the general solution of (1.10.2) with $\vec{F} \equiv 0$, and that is given by (1.10.11) with harmonic $\vec{\psi},\phi$, and one solution of (1.10.2). To seek one such solution, when $\vec{F}$ is conservative, choose $\vec{d}$ so that curl $\vec{d} = 0$; then $\nabla^2\vec{d} \equiv \vec{\nabla}\,\text{div}\,\vec{d} - \text{curl curl}\,\vec{d} = \vec{\nabla}\,\text{div}\,\vec{d}$, and equation (1.10.2) may be written

$$\vec{\nabla}\,\text{div}\left\{2\mu\vec{d} + \frac{2\mu\vec{d}}{1-2\nu} - \vec{W}\right\} = 0, \tag{1.10.15}$$

which has the solution

$$2\mu\vec{d}_0 = \varphi\vec{W}. \tag{1.10.16}$$

This will therefore be a solution of (1.10.2) provided that

$$\text{curl}\,\vec{W} = 0. \tag{1.10.17}$$

We conclude that the general solution of (1.10.2) is

$$2\mu\vec{d} = 2(1-\nu)\vec{\psi} - \vec{\nabla}\{(\vec{r}\cdot\vec{\psi}) + \phi\} + \varphi\vec{W} \tag{1.10.18}$$

where

$$\nabla^2\vec{\psi} = 0 = \nabla^2\phi, \qquad \text{curl}\,\vec{W} = 0, \qquad 2\vec{F} = -\vec{\nabla}\text{div}\,\vec{W}. \tag{1.10.19}$$

Examples 1.10

1. Show that if $\vec{\omega} = (\omega_{yz},\omega_{zx},\omega_{xy})$ (see (1.2.8)) then equation (1.10.1) may be written

$$\frac{2(1-\nu)}{1-2\nu}\vec{\nabla}\varepsilon - 2\,\text{curl}\,\vec{\omega} + \frac{\vec{F}}{\mu} = \frac{\rho}{\mu}\frac{\partial^2\vec{d}}{\partial t^2}. \tag{1.10.20}$$

2. Write down the special form of (1.10.4) in cylindrical polar coordinates when all displacements and body forces are independent of the coordinate θ.
3. Show that if $\vec{\psi}$ is an arbitrary harmonic function then $\vec{d} = \partial\vec{\psi}/\partial z - z/(3-4\nu)\vec{\nabla}$ div $\vec{\psi}$ is a static solution of equation (1.10.1) in the absence of body forces.
4. Show that if $\vec{\psi},\phi,\vec{\psi}^*,\phi^*$ are related by (1.10.12) then the fields $\vec{d},\vec{d}^*$ given by (1.10.11) are identical.

1.11 *Plane strain and plane stress*

The solution of the equations of the three-dimensional elasticity is very difficult. In many problems the variations of some of the elastic quantities are negligible in one of the coordinate directions and the problems can be considered to be approximately two-dimensional ones. There are three important classes of such problems, denoted by plane stain, plane stress and generalized plane stress.

In *plane strain* it is assumed that the displacement w is zero and the displacements u,v are independent of z, so that

$$e_{xz} = 0 = e_{yz} = e_{zz}.$$

In plane strain, plane sections of the body perpendicular to the z-axis remain plane and perpendicular to this axis. A cylinder with generators parallel to the z-axis and acted on by body forces and surface tractions which are independent of z and act in the x-y plane will be in a state of plane strain. Note that, although

$$\tau_{xz} = 0 = \tau_{yz},$$

the normal stress τ_{zz} is not zero, but is

$$\tau_{zz} = \lambda(e_{xx} + e_{yy}) = \lambda(e_{xx} + e_{yy}) = \lambda\varepsilon. \qquad (1.11.1)$$

To be kept in plane strain the cylinder mentioned above would have to be constrained by the surface tractions (1.11.1) over its end faces. The remaining stresses are given by

$$\left.\begin{aligned} (1-2\nu)\tau_{xx} &= 2\mu[(1-\nu)e_{xx} + \nu e_{yy}], \\ (1-2\nu)\tau_{yy} &= 2\mu[(1-\nu)e_{yy} + \nu e_{xx}], \\ \tau_{xy} &= 2\mu e_{xy}. \end{aligned}\right\} \tag{1.11.2}$$

Now there are just two equilibrium equations, namely

$$\frac{\partial\tau_{xx}}{\partial x} + \frac{\partial\tau_{xy}}{\partial y} + F_x = \rho\,\frac{\partial^2 u}{\partial t^2}, \tag{1.11.3}$$

$$\frac{\partial\tau_{xy}}{\partial x} + \frac{\partial\tau_{yy}}{\partial y} + F_y = \rho\,\frac{\partial^2 v}{\partial t^2}, \tag{1.11.4}$$

since the third is satisfied automatically.

The equations of static plane strain have the two-dimensional Papkovich-Neuber solution (1.10.18), i.e.

$$2\mu u = 4(1-\nu)\psi_1 - \frac{\partial}{\partial x}\{x\psi_1 + y\psi_2 + \phi\} + \varphi W_1, \tag{1.11.5}$$

$$2\mu v = 4(1-\nu)\psi_2 - \frac{\partial}{\partial y}\{x\psi_1 + y\psi_2 + \phi\} + \varphi W_2, \tag{1.11.6}$$

where ψ_1,ψ_2,ϕ are all plane *harmonic* functions, i.e.

$$\nabla_1^2(\psi_1,\psi_2,\phi) = 0; \qquad \nabla_1^2 \equiv \frac{\partial^2}{\partial x^2} + \frac{\partial^2}{\partial y^2}. \tag{1.11.7}$$

It is possible to take one of ψ_1,ψ_2 zero, but a more elegant theory may be derived by choosing ψ_1,ψ_2 so that

$$\frac{\partial\psi_1}{\partial x} = \frac{\partial\psi_2}{\partial y}. \tag{1.11.8}$$

This is always possible because of the flexibility there is in the choice of $\vec{\psi},\phi$ (see Ex. 1.10.4). Since ψ_1,ψ_2 are harmonic, equation (1.11.8) implies

$$\frac{\partial}{\partial x}\left(\frac{\partial\psi_1}{\partial y} + \frac{\partial\psi_2}{\partial x}\right) = \frac{\partial^2\psi_1}{\partial x\partial y} - \frac{\partial^2\psi_2}{\partial y^2} = \frac{\partial}{\partial y}\left(\frac{\partial\psi_1}{\partial x} - \frac{\partial\psi_2}{\partial y}\right) = 0,$$

$$\frac{\partial}{\partial y}\left(\frac{\partial \psi_1}{\partial y} + \frac{\partial \psi_2}{\partial x}\right) = -\frac{\partial^2 \psi_1}{\partial x^2} + \frac{\partial^2 \psi_2}{\partial y \partial x} = -\frac{\partial}{\partial x}\left(\frac{\partial \psi_1}{\partial x} - \frac{\partial \psi_2}{\partial y}\right) = 0,$$

so that

$$\frac{\partial \psi_1}{\partial y} = -\frac{\partial \psi_2}{\partial x}. \tag{1.11.9}$$

Equations (1.11.8),(1.11.9) are the Cauchy-Riemann equations (3.1.5) and imply that $\psi_1(x,y)$, $\psi_2(x,y)$ are the real and imaginary parts of a complex function of $z = x + iy$; this function is usually denoted by $\phi(z)$ or $\Omega(z)$, so that

$$\psi_1(x,y) + i\psi_2(x,y) = \phi(z). \tag{1.11.10}$$

The terms $\partial\phi/\partial x$ and $-\partial\phi/\partial y$ (in (1.11.5), not to be confused with the $\phi(z)$ of (1.11.10)) also satisfy the Cauchy-Riemann equations because

$$\frac{\partial}{\partial x}\left(\frac{\partial \phi}{\partial x}\right) = \frac{\partial}{\partial y}\left(-\frac{\partial \phi}{\partial y}\right), \quad \frac{\partial}{\partial y}\left(\frac{\partial \phi}{\partial x}\right) = -\frac{\partial}{\partial x}\left(-\frac{\partial \phi}{\partial y}\right), \tag{1.11.11}$$

so that $\partial\phi/\partial x$ and $-\partial\phi/\partial y$ are the real and imaginary parts of another function, denoted by $\psi(z)$ or $\omega(z)$; thus

$$\frac{\partial \phi(x,y)}{\partial x} - i\,\frac{\partial \phi(x,y)}{\partial y} = \psi(z). \tag{1.11.12}$$

(It is unfortunate that the symbol ϕ is used in the Papkovich-Neuber and the complex variable solutions, but since these two solutions are seldom used together, it is unlikely that there will be confusion).

Combine equations (1.11.5),(1.11.6) into one complex equation

$$2\mu D = 2\mu(u+iv) = (3-4\nu)(\psi_1+i\psi_2) - x\,\frac{\partial \psi_1}{\partial x} - y\,\frac{\partial \psi_2}{\partial y} -$$

$$- ix\,\frac{\partial \psi_1}{\partial y} - iy\,\frac{\partial \psi_2}{\partial y} - \frac{\partial \phi}{\partial x} - i\,\frac{\partial \phi}{\partial y} + \varphi(W_1+iW_2). \tag{1.11.13}$$

In order to simplify this equation we consider the operators $\partial/\partial z$ and $\partial/\partial\bar{z}$. Let $f(x,y)$ be a function of x,y; it can also be written as a function of $z = x + iy$ and $\bar{z} = x - iy$, and

$$\frac{\partial f}{\partial z} = \frac{\partial f}{\partial x}\frac{\partial x}{\partial z} + \frac{\partial f}{\partial y}\frac{\partial y}{\partial z}, \qquad \frac{\partial f}{\partial \bar{z}} = \frac{\partial f}{\partial x}\frac{\partial x}{\partial \bar{z}} + \frac{\partial f}{\partial y}\frac{\partial y}{\partial \bar{z}}. \tag{1.11.14}$$

But $x = (z + \bar{z})/2$, $y = i(\bar{z} - z)/2$, so that

$$\frac{\partial x}{\partial z} = \frac{1}{2} = \frac{\partial x}{\partial \bar{z}}, \qquad \frac{\partial y}{\partial z} = -\frac{i}{2} = -\frac{\partial y}{\partial \bar{z}},$$

$$\frac{\partial f}{\partial z} = \frac{1}{2}\frac{\partial f}{\partial x} - i\frac{\partial f}{\partial y}, \qquad \frac{\partial f}{\partial \bar{z}} = \frac{1}{2}\frac{\partial f}{\partial x} + i\frac{\partial f}{\partial y}. \tag{1.11.15}$$

The function $\phi(z)$ of (1.11.10) is a function of z alone, so that its complex conjugate $\overline{\phi(z)} \equiv \bar{\phi}(\bar{z})$ will be a function of $\bar{z}$ alone and

$$\frac{\partial\bar{\phi}(\bar{z})}{\partial\bar{z}} = \frac{d\bar{\phi}(\bar{z})}{d\bar{z}} = \bar{\phi}'(\bar{z}) = \frac{1}{2}\left(\frac{\partial}{\partial x} + i\frac{\partial}{\partial y}\right)(\psi_1 - i\psi_2)$$
$$= \frac{1}{2}\left(\frac{\partial\psi_1}{\partial x} + \frac{\partial\psi_2}{\partial y} + i\frac{\partial\psi_1}{\partial y} - i\frac{\partial\psi_2}{\partial x}\right) = \frac{\partial\psi_1}{\partial x} - i\frac{\partial\psi_2}{\partial x}. \tag{1.11.16}$$

The middle terms in (1.11.13) may be written

$$x\frac{\partial\psi_1}{\partial x} + y\frac{\partial\psi_2}{\partial x} + ix\frac{\partial\psi_1}{\partial y} + iy\frac{\partial\psi_2}{\partial y} = (x+iy)\frac{\partial\psi_1}{\partial x} - i(x+iy)\frac{\partial\psi_2}{\partial x}$$
$$= z\bar{\phi}'(\bar{z}). \tag{1.11.17}$$

Equation (1.11.12) shows that the last two terms in (1.11.13) may be written as $-\overline{\psi(z)} = -\bar{\psi}(\bar{z})$, so that altogether

$$2\mu D = \kappa\phi(z) - z\bar{\phi}'(\bar{z}) - \bar{\psi}(\bar{z}) + \varphi W, \tag{1.11.18}$$

where $W(z,\bar{z}) = W_1(x,y) + iW_2(x,y)$. This is the Kolosov-Muskhelishvili solution (Muskhelishvili (1953a)) which will be used in Chapter IV.

In *plane stress* it is assumed that there is no stress on planes parallel to the z-axis, i.e.

$$\tau_{xz} = 0 = \tau_{yz} = \tau_{zz}. \tag{1.11.19}$$

In plane stress the components u,v,w are not in general independent of z and the strain e_{zz} is not zero; in fact

$$e_{zz} = -\frac{\lambda}{\lambda+2\mu}(e_{xx} + e_{yy}). \qquad (1.11.20)$$

Plane stress is a first approximation to the state of stress in a thin plate acted on by forces in its own plane. The relation (1.11.19) may be used to eliminate e_{zz} from the stress-strain relations; thus

$$\tau_{xx} = \lambda(e_{xx} + e_{yy}) + \lambda e_{zz} + 2\mu e_{xx}$$

may be rewritten

$$\tau_{xx} = \left\{\lambda - \frac{\lambda^2}{\lambda+2\mu}\right\}(e_{xx} + e_{yy}) + 2\mu e_{xx}$$

$$= \lambda^*(e_{xx} + e_{yy}) + 2\mu e_{xx}, \qquad (1.11.21)$$

where

$$\lambda^* = 2\mu\lambda/(\lambda + 2\mu). \qquad (1.11.22)$$

The relation for τ_{yy} is, similarly,

$$\tau_{yy} = \lambda^*(e_{xx} + e_{yy}) + 2\mu e_{yy}. \qquad (1.11.23)$$

Comparing these relations with those for plane strain, namely (1.11.2), in which $\varepsilon = e_{xx} + e_{yy}$, we see that they have exactly the same form, but a modified Lamé constant λ^*. Alternatively, the stresses may be written in the form

$$\tau_{xx} = \frac{2\mu}{1-2\nu^*}[(1-\nu^*)e_{xx} + \nu^* e_{yy}],$$

$$\tau_{yy} = \frac{2\mu}{1-2\nu^*}[(1-\nu^*)e_{yy} + \nu^* e_{xx}], \qquad (1.11.24)$$

where $\nu^* = \nu/(1 + \nu)$.

This means that the stress-strain relation is the same as for plane strain, except that ν is replaced by ν^*. Note however that the displacements, strains and stresses are functions of z.

For plane stress the first two equilibrium equations are (1.11.3), (1.11.4), while the third is

$$F_z = \rho \frac{\partial^2 w}{\partial t^2} . \tag{1.11.25}$$

This shows that F_z must be zero for static problems.

An important generalization of plane stress was introduced by Filon (1903). In *generalized plane stress* it is assumed that the body is bounded by plane faces $z = \pm h$ and that

$$\tau_{xz} = 0 = \tau_{yz} = \tau_{zz} \text{ on } z = \pm h, \tag{1.11.26}$$

and the mean values of these stresses are also zero, thus

$$\tau^*_{xz} = 0 = \tau^*_{yz} = \tau^*_{zz}, \tag{1.11.27}$$

where, for instance,

$$\tau^*_{xz}(x,y) = \frac{1}{2h}\int_{-h}^{h} \tau_{xz}(x,y,z)dz. \tag{1.11.28}$$

Now we concentrate attention on the mean values of the displacements, strains and stresses. Because of equation (1.11.26) the relations between mean displacements, strains and stresses will be exactly the same as those occurring in plane stress. In addition the equilibrium equations will take the same form since, when ρ = constant,

$$\frac{1}{2h}\int_{-h}^{h} \left\{ \frac{\partial \tau_{xx}}{\partial x} + \frac{\partial \tau_{xy}}{\partial y} + \frac{\partial \tau_{xz}}{\partial z} + F_x - \rho \frac{\partial^2 u}{\partial t^2} \right\} dz = 0$$

may be written

$$\frac{\partial \tau^*_{xx}}{\partial x} + \frac{\partial \tau^*_{xy}}{\partial y} + F^*_x = \rho \frac{\partial^2 u^*}{\partial t^2} , \tag{1.11.29}$$

because

$$\frac{1}{2h}\int_{-h}^{h} \frac{\partial \tau_{xz}}{\partial z} dz = \frac{1}{2h}[\tau_{xz}]_{-h}^{h} = 0.$$

The last equilibrium equation, when integrated, becomes

$$F_z^* = \rho \frac{\partial^2 w^*}{\partial t^2}, \tag{1.11.30}$$

so that F_z^* must be zero for static problems.

We conclude that the equations governing *mean* displacements, strains and stresses in the x-y plane in generalized plane stress are identical to those governing the *actual* displacements, strains and stresses in the x-y plane in plane strain, provided only that the ν in plane strain is replaced by $\nu^* = \nu/(1 + \nu)$. In particular, generalized plane stress problems may be solved by using the Papkovich-Neuber solution (1.11.5)-(1.11.6), or the Kolosov-Muskhelishvili solution (1.11.18), in each case replacing ν by ν^*. In this book 'plane' problems will usually be described as 'plane strain' problems.

The conditions (1.11.26),(1.11.27) may be relaxed. It is sufficient to assume that

$$\tau_{xz} = 0 = \tau_{yz} \text{ on } z = \pm h, \tag{1.11.31}$$

and

$$\tau_{zz}^* = 0. \tag{1.11.32}$$

In this case the preceding analysis will all hold except for equation (1.11.30). If the third equilibrium equation is integrated it becomes

$$\frac{\partial \tau_{xz}^*}{\partial x} + \frac{\partial \tau_{yz}^*}{\partial y} + \frac{1}{2h}[\tau_{zz}]_{-h}^{h} + F_z^* = \rho \frac{\partial^2 w^*}{\partial t^2}. \tag{1.11.33}$$

This equation may be considered quite separately from the stress-strain relations for the mean stresses and mean strains and the two other integrated equilibrium equations of the form (1.11.29). The latter equations involve only $u^*, v^*, e_{xx}^*, e_{yy}^*, e_{xy}^*, \tau_{xx}^*, \tau_{yy}^*, \tau_{xy}^*$, while (1.11.32) involves only $\tau_{xz}^*, \tau_{yz}^*, \tau_{zz}^*(\pm h) F_z^*$ and w^*. A recent discussion of generalized plane stress will be found in England (1971a, p.25f).

Examples 1.11

1. Show that the stresses corresponding to the displacements (1.11.5),(1.11.6) are

$$\tau_{xx} = 2(1-2\nu)\frac{\partial\psi_1}{\partial x} + 2\nu\left(\frac{\partial\psi_1}{\partial x} + \frac{\partial\psi_2}{\partial y}\right) - x\frac{\partial^2\psi_1}{\partial x^2} - y\frac{\partial^2\psi_2}{\partial x^2} - \frac{\partial^2\phi}{\partial x^2},$$

$$\tau_{yy} = 2(1-2\nu)\frac{\partial\psi_2}{\partial y} + 2\nu\left(\frac{\partial\psi_1}{\partial x} + \frac{\partial\psi_2}{\partial y}\right) - x\frac{\partial^2\psi_1}{\partial y^2} - y\frac{\partial^2\psi_2}{\partial y^2} - \frac{\partial^2\phi}{\partial y^2},$$

$$\tau_{xy} = (1-2\nu)\left(\frac{\partial\psi_1}{\partial y} + \frac{\partial\psi_2}{\partial x}\right) - x\frac{\partial^2\psi_1}{\partial x\partial y} - y\frac{\partial^2\psi_2}{\partial x\partial y} - \frac{\partial^2\phi}{\partial x\partial y}.$$

2. Show that if e_{xx}, e_{yy}, e_{xy} are to be the strains in a plane strain field, so that there exists a displacement field u,v such that $e_{xx} = \partial u/\partial x$, $e_{yy} = \partial v/\partial y$, $e_{xy} = 1/2(\partial u/\partial y + \partial v/\partial x)$, then e_{xx}, e_{yy}, e_{xy} must satisfy

$$\frac{\partial^2 e_{xx}}{\partial y^2} + \frac{\partial^2 e_{yy}}{\partial x^2} = 2\frac{\partial^2 e_{xy}}{\partial x\partial y}.$$

This is an example of a compatibility equation. For a 3-D elastic field this is one of 6 compatibility equations. (See Muskhelishvili (1953a, p. 49).

3. Show that if ψ_1, ψ_2 are *conjugate*, in the sense of obeying (1.11.8),(1.11.9), and there are no body forces, then τ_{xx} τ_{xy}, τ_{yy} may be written

$$\tau_{xx} = \frac{\partial^2 U}{\partial y^2}, \quad \tau_{xy} = -\frac{\partial^2 U}{\partial x\partial y}, \quad \tau_{yy} = \frac{\partial^2 U}{\partial x^2},$$

where $U = x\psi_1 + y\psi_2 + \phi$. The junction U is called the Airy stress function. Use Ex. 1.11.2 to show that U is *biharmonic*, i.e. $\nabla_1^4 U = 0$. The Airy stress function has important historical significance (Love (1927, p. 88)), but has severe analytical disadvantages: if U is known then the displacements must be found by integration; the Papkovich-Neuber solution gives the displacements explicitly.

4. Show that in the complex notation of (1.11.10)-(1.11.18), $\mathcal{U}$ may be written $2\mathcal{U} = \bar{z}\phi(z) + z\overline{\phi(\bar{z})} + \chi(z) + \overline{\chi(\bar{z})}$, where $\psi(z) = \chi'(z)$.

5. Show that the polar strains $e_{rr}, e_{\theta\theta}, e_{r\theta}$ of equations (1.3.10), (1.3.14) must satisfy the compatibility equation

$$\frac{\partial}{\partial r}\left(r^2 \frac{\partial e_{\theta\theta}}{\partial r}\right) + \left(\frac{\partial^2}{\partial \theta^2} - r\frac{\partial}{\partial r}\right)e_{rr} = 2\frac{\partial^2}{\partial r \partial \theta}(e_{r\theta}).$$

6. Show that in plane strain or plane stress, when the stresses are expressed in terms of the Airy stress function $\mathcal{U}$ as in Ex. (1.11.3), then

$$\tau_{rr} = \frac{1}{r}\frac{\partial \mathcal{U}}{\partial r} + \frac{1}{r^2}\frac{\partial^2 \mathcal{U}}{\partial \theta^2}, \quad \tau_{\theta\theta} = \frac{\partial^2 \mathcal{U}}{\partial r^2}, \quad \tau_{r\theta} = -\frac{\partial}{\partial r}\left(\frac{1}{r}\frac{\partial \mathcal{U}}{\partial \theta}\right).$$

Deduce that $\tau_{rr} + \tau_{\theta\theta} = \nabla_1^2 \mathcal{U}$ and that $\nabla_1^2(\tau_{rr} + \tau_{\theta\theta}) = 0$.

CHAPTER II - ELEMENTARY PROBLEMS

> "I only took the regular course" the Mock Turtle replied, "the different branches of Arithmetic - Ambition, Distraction, Uglification and Derision."
>
> - *Alice in Wonderland*

The term *elementary problems* is an abbreviation for *problems which may be solved by elementary or direct methods.* The terms *elementary* or *direct* are used to exclude methods which are based on sophisticated manipulations of the equations of motion and, in particular, to exclude methods based on the theory of Cauchy integrals and transform methods. The solutions for plane problems start from Flamant's solution for a point force on the boundary; those for three-dimensional problems start from the results of Boussinesq and Cerruti.

2.1 *Basic solutions for plane problems*

In this section we shall develop some Papkovich-Neuber solutions appropriate to the half-plane $y > 0$ when the boundary $y = 0$ is subject to normal and shearing stresses, and there are no body forces. In the solution (1.11.5)-(1.11.6), put $\psi_1 = 0$, $\psi_2 = \psi$, $W = 0$; then

$$2\mu u = -\frac{\partial}{\partial x}(y\psi+\phi), \qquad 2\mu v = \kappa\psi - y\frac{\partial\psi}{\partial y} - \frac{\partial\phi}{\partial y}, \tag{2.1.1}$$

and the stresses $\tau_{xx}, \tau_{xy}, \tau_{yy}$ are

$$\left.\begin{aligned} \tau_{xx} &= 2\nu\frac{\partial\psi}{\partial y} - y\frac{\partial^2\psi}{\partial x^2} - \frac{\partial^2\phi}{\partial x^2}, \\ \tau_{xy} &= (1-2\nu)\frac{\partial\psi}{\partial x} - y\frac{\partial^2\psi}{\partial x\partial y} - \frac{\partial^2\phi}{\partial x\partial y}, \end{aligned}\right\} \tag{2.1.2}$$

$$\tau_{yy} = 2(1-\nu)\frac{\partial\psi}{\partial y} - y\,\frac{\partial^2\psi}{\partial y^2} - \frac{\partial^2\phi}{\partial y^2}, \tag{2.1.3}$$

so that the stresses on the boundary $y = 0$ are

$$\tau_{xy} = (1-2\nu)\frac{\partial\psi}{\partial x} - \frac{\partial^2\phi}{\partial x\partial y}, \quad \tau_{yy} = 2(1-\nu)\frac{\partial\psi}{\partial y} - \frac{\partial^2\phi}{\partial y^2}, \tag{2.1.4}$$

provided that

$$\lim_{y\to 0}\left(y\,\frac{\partial^2\psi}{\partial x\partial y},\ y\,\frac{\partial^2\psi}{\partial y^2}\right) = 0. \tag{2.1.5}$$

If now the function ϕ is chosen so that

$$\frac{\partial\phi}{\partial y} = (1-2\nu)\psi, \tag{2.1.6}$$

then $\tau_{xy}(x,0)$ will be identically zero and

$$v(x,0) = \vartheta\psi(x,0), \qquad \tau_{yy}(x,0) = \frac{\partial\psi}{\partial y}(x,0), \tag{2.1.7}$$

while, in general,

$$\tau_{xx} = \frac{\partial\psi}{\partial y} - y\,\frac{\partial^2\psi}{\partial x^2}, \quad \tau_{xy} = -y\,\frac{\partial^2\psi}{\partial x\partial y}, \quad \tau_{yy} = \frac{\partial\psi}{\partial y} - y\,\frac{\partial^2\psi}{\partial y^2}. \tag{2.1.8}$$

The function

$$\psi = -\frac{F}{2\pi}\,\ell n\left[\frac{(x^2+y^2)}{a^2}\right] \tag{2.1.9}$$

is harmonic in any region excluding the origin; a is an arbitrary length included to make the argument of ℓn dimensionless. For this ψ the conditions (2.1.5) are fulfilled and the stresses given by (2.1.8) are

$$\tau_{xx} = -\frac{2Fx^2y}{\pi r^4}, \qquad \tau_{xy} = -\frac{2Fxy^2}{\pi r^4}, \qquad \tau_{yy} = -\frac{2Fy^3}{\pi r^4}, \tag{2.1.10}$$

where $r^2 = x^2 + y^2$. This stress distribution was derived by Flamant. The stresses in polar coordinates (Ex. 1.5.1) are

$$\tau_{rr} = -\frac{2F \sin\theta}{\pi r}, \qquad \tau_{r\theta} = 0 = \tau_{\theta\theta}, \tag{2.1.11}$$

as shown in Figure 2.1.1. The sense in which a stress acts is stated at the beginning of §1.4. The total force produced by the stresses (2.1.11) on the semi-circular region around the origin has components in the x- and y- directions given by

$$\int_0^\pi \tau_{rr} r \cos\theta d\theta = 0, \qquad \int_0^\pi \tau_{rr} r \sin\theta d\theta = -F. \tag{2.1.12}$$

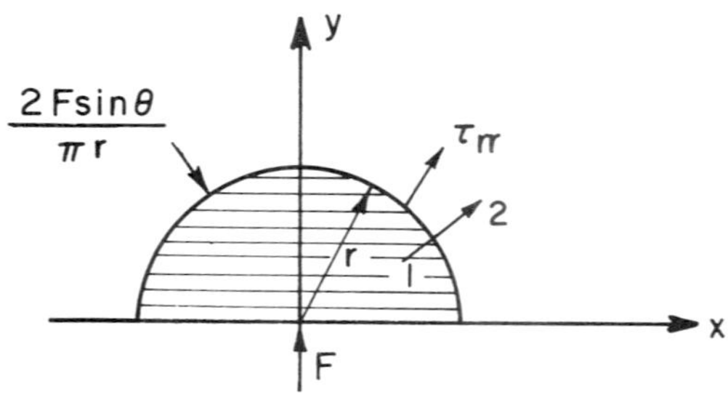

Figure 2.1.1 - Stresses acting on the boundary of a semi-circular region near the origin.

Therefore, in order for the region to be in equilibrium there must be a vertical force F at the origin as shown. Equation (2.1.10) therefore yields the stress field due to this vertical force.

In order to find the displacements u,v we must find the potential ϕ. We have

$$\int \ell n\left(\frac{x^2+y^2}{a^2}\right)dy = y\ \ell n\left(\frac{x^2+y^2}{a^2}\right) - \int \frac{2y^2}{x^2+y^2}\,dy,$$

$$= y\ \ell n\left(\frac{x^2+y^2}{a^2}\right) - 2y + 2x^2\int \frac{dy}{x^2+y^2}. \tag{2.1.13}$$

But $\frac{\partial}{\partial y}[\tan^{-1}(y/x)] = \frac{x}{x^2+y^2}$, so that

$$\int \ell n\left(\frac{x^2+y^2}{a^2}\right)dy = y\ \ell n\left(\frac{x^2+y^2}{a^2}\right) - 2y + 2x\tan^{-1}(y/x) + C. \quad (2.1.14)$$

However, the function $\tan^{-1}(y/x)$ is not continuous in the upper half-plane. If $\tan^{-1}(t)$ is defined, as usual, as that angle α in $-\pi/2 < \alpha \leq \pi/2$ such that $\tan\alpha = t$, then the boundary values of $\tan^{-1}(y/x)$ as the point (x,y) approaches the x- and y- axes are shown in Figure 2.1.2(a): there is a jump of π as (x,y) crosses the y-axis. We therefore define a new function $\mathrm{Tan}^{-1}(y/x)$, continuous in the half-plane $y > 0$ such that

$$\mathrm{Tan}^{-1}(y/x) = \begin{cases} \tan^{-1}(y/x), & x \geq 0, \\ \pi + \tan^{-1}(y/x), & x < 0. \end{cases} \quad (2.1.15)$$

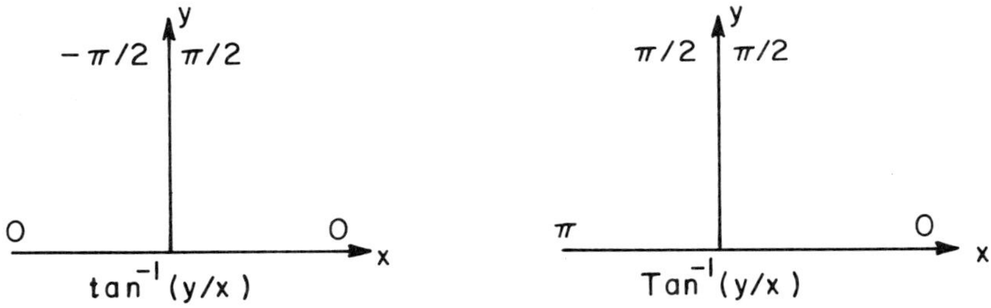

Figure 2.1.2 - Boundary values of $\tan^{-1}(y/x)$ *and* $\mathrm{Tan}^{-1}(y/x)$.

Then we may write

$$\phi = -\frac{F(1-2\nu)}{2\pi}\left\{y\ \ell n\left(\frac{x^2+y^2}{a^2}\right) - 2y + 2x\,\mathrm{Tan}^{-1}(y/x)\right\}, \quad (2.1.16)$$

and the displacements u,v have the form

$$u(x,y) = G_{12}(x,y)F, \quad v(x,y) = G_{22}(x,y)F, \quad (2.1.17)$$

where

$$2\pi\mu G_{12}(x,y) = xyr^{-2} + (1-2\nu)\mathrm{Tan}^{-1}(y/x), \quad (2.1.18)$$

$$2\pi\mu G_{22}(x,y) = -x^2r^{-2} - 2(1-\nu)\ell n(r/a). \quad (2.1.19)$$

These expressions have no meaning at the origin. The expression for $G_{12}(x,y)$ tends to the limiting value

$$\lim G_{12}(x,y) = (2\pi\mu)^{-1}[\sin\alpha\cos\alpha + (1-2\nu)\alpha] \tag{2.1.20}$$

as x,y tend to infinity along the line $y = x\tan\alpha$. The expression for G_{22} tends to infinity at infinity. This behaviour of the displacements is one of the inherent inadequacies of plane elasticity theory.

Equations (2.1.17)-(2.1.19) may be used to find the displacements due to a distributed normal loading on the x-axis, i.e. the displacements corresponding to the boundary conditions

$$\tau_{xy} = 0, \qquad \tau_{yy} = -p(x) \text{ on } y = 0. \tag{2.1.21}$$

They are

$$u(x,y) = \int_{-\infty}^{\infty} G_{12}(x-x',y)p(x')dx' \tag{2.1.22}$$

$$v(x,y) = \int_{-\infty}^{\infty} G_{22}(x-x',y)p(x')dx'. \tag{2.1.23}$$

These displacements, and those given in Ex. 2.1.1 may be supplemented by an arbitrary rigid-body translation and rotation. Kalker (1971),(1972) has developed a theory of 'elastic line contact' in which the loading is made over a slender region of an elastic half-space; the two-dimensional results given above then appear as a limiting case.

Examples 2.1

1. Show that if ψ,ϕ are chosen so that $2(1-\nu)\psi = \partial\phi/\partial y$ in the solution (2.1.1) the stresses $\tau_{xx},\tau_{xy},\tau_{yy}$ will be

$$\tau_{xx} = 2\frac{\partial\psi}{\partial y} - y\frac{\partial^2\psi}{\partial x^2}, \quad \tau_{xy} = -\frac{\partial\psi}{\partial x} - y\frac{\partial^2\psi}{\partial x\partial y}, \quad \tau_{yy} = -y\frac{\partial^2\psi}{\partial y^2}.$$

Hence show that if $\psi = -(S/\pi)\mathrm{Tan}^{-1}(y/x)$, then

$$\tau_{xx} = -\frac{2Sx^3}{\pi r^4}, \qquad \tau_{xy} = -\frac{2Sx^2y}{\pi r^4}, \qquad \tau_{yy} = -\frac{2Sxy^2}{\pi r^4},$$

and that these are the stresses in the upper half-plane due to a concentrated shearing force S acting at the origin.

2. Show that the displacements due to a distributed shearing force s(x) in the x-direction, i.e. those associated with the boundary condition $\tau_{xy} = -s(x)$, $\tau_{yy} = 0$ on $y = 0$, are

$$u(x,y) = \int_{-\infty}^{\infty} G_{11}(x-x',y)s(x')dx',$$

$$v(x,y) = \int_{-\infty}^{\infty} G_{21}(x-x',y)s(x')dx',$$

where

$$G_{11}(x,y) = (2\pi\mu)^{-1}[x^2r^{-2} - 2(1-\nu)\ln(r/a)],$$

$$G_{21}(x,y) = (2\pi\mu)^{-1}[xyr^{-2} - (1-2\nu)\mathrm{Tan}^{-1}(y/x)].$$

3. Show that

$$\lim_{y\to 0+}\int_a^b \frac{ys(x')dx'}{(x-x')^2+y^2} = \pi s(x)$$

and hence show that the values on $x = 0$ of the displacement derivatives due to a distribution of normal and shearing stress on an interval $[a,b]$ are given by

$$2\mu\frac{\partial u}{\partial x} = -(1-2\nu)p(x) - \frac{2(1-\nu)}{\pi}\int_a^b \frac{s(x')dx'}{x'-x},$$

$$2\mu\frac{\partial v}{\partial x} = (1-2\nu)s(x) - \frac{2(1-\nu)}{\pi}\int_a^b \frac{p(x')dx'}{x'-x}.$$

4. On the line $y = 0$, $\mathrm{Tan}^{-1}(y/x) = \pi/2 - \pi/2\ \mathrm{sign}(x)$. Show that apart from a constant the displacements on the boundary are

$$u(x) = -\int_{-\infty}^{\infty} \left\{\frac{\vartheta}{\pi} \ln\left|\frac{x-x'}{a}\right| s(x') + \frac{1-2\nu}{4\mu} \operatorname{sign}(x-x')p(x')\right\} dx',$$

$$v(x) = \int_{-\infty}^{\infty} \left\{\frac{1-2\nu}{4\mu} \operatorname{sign}(x-x')s(x') - \frac{\vartheta}{\pi} \ln\left|\frac{x-x'}{a}\right| p(x')\right\} dx'.$$

5. Show that the two equations in Ex. 2.1.4 may be combined to give

$$u(x) + iv(x) = \vartheta\int_{-\infty}^{\infty} \left\{\frac{1}{\pi} \ln\left|\frac{x-x'}{a}\right| - \right.$$

$$\left. - \frac{i\varphi}{2} \operatorname{sign}(x-x')\right\} (s(x') + ip(x'))dx'.$$

6. Show that the normal displacement on the boundary due to a uniform load p over [-a,a] satisfies

$$v(x) - v(0) = -\frac{\vartheta pa}{\pi}\left[\left(1 - \frac{x}{a}\right)\ln\left|1 - \frac{x}{a}\right| + \left(1 + \frac{x}{a}\right)\ln\left|1 + \frac{x}{a}\right|\right].$$

2.2 *Simple plane punch problems*

The elastic displacements in the upper half-plane due to a distribution of normal and shearing stress on the boundary were found in §2.1 - in equations (2.1.22), (2.1.23) and Ex. 2.1.2. A general expression for the displacements will first be found; afterwards this will be applied to a number of contact problems.

Since there are two displacements u,v and two stresses s(x'), p(x'), it is convenient to combine the four equations giving u,v in terms of s,p into one complex equation expressing $D = u + iv$ in terms of $q(x') = s(x') + ip(x')$.

$$D(x,y) = \int_{-\infty}^{\infty} [G_{11}(x-x',y) + iG_{21}(x-x',y)]s(x') +$$

$$+ [G_{12}(x-x',y) + iG_{22}(x-x',y)]p(x')dx'. \qquad (2.2.1)$$

Now if $r^2 = (x-x')^2 + y^2$, then

$$\left.\begin{aligned} 2\,\ell n\,\frac{r}{a} &= \ell n\left(\frac{z-x'}{a}\right) + \ell n\left(\frac{\bar{z}-x'}{a}\right), \\ 2i\,\mathrm{Tan}^{-1}\,\frac{y}{x-x'} &= \ell n\left(\frac{z-x'}{a}\right) - \ell n\left(\frac{\bar{z}-x'}{a}\right), \end{aligned}\right\} \tag{2.2.2}$$

$$2(x-x')[(x-x') + iy]r^{-2} = 2 + (z-\bar{z})(\bar{z}-x')^{-1}, \tag{2.2.3}$$

so that

$$4\pi\mu D(x,y) = -\kappa\int_{-\infty}^{\infty} \ell n\left(\frac{z-x'}{a}\right)q(x')dx' - \int_{-\infty}^{\infty} \ell n\left(\frac{\bar{z}-x'}{a}\right)q(x')dx' +$$

$$+ (z-\bar{z})\int_{-\infty}^{\infty} (\bar{z}-x')^{-1}\bar{q}(x')dx' + \left[2\int_{-\infty}^{\infty} \bar{q}(x')dx'\right]. \tag{2.2.4}$$

At this point we should check that the displacement has the form given in (1.11.18). Comparing (2.2.4) with (1.11.18) we see that

$$\phi(z) = -(2\pi)^{-1}\int_{-\infty}^{\infty} \ell n\left(\frac{z-x'}{a}\right)q(x')dx', \tag{2.2.5}$$

$$\psi(z) = (2\pi)^{-1}\int_{-\infty}^{\infty} \ell n\left(\frac{z-x'}{a}\right)\bar{q}(x')dx' +$$

$$+ (2\pi)^{-1}z\int_{-\infty}^{\infty} (z-x')^{-1}q(x')dx'. \tag{2.2.6}$$

The last term in (2.2.4) has been neglected since it is constant and thus represents a rigid body translation.

The explicit evaluation of the integrals in (2.2.5),(2.2.6) may be difficult for some functions $q(x')$, but can be carried out simply for some special functions $q(x')$. We shall now state some results which are proved in §2.3. Let $q(x')$ be zero outside the interval $[-a,a]$, and within the interval be given by $q(x') = a_0(a^2 - x'^2)^{-1/2}$ where a_0 is a constant, which may be complex, corresponding to a distribution of both normal and shearing stresses; then $\phi(z)$ may be written

$$\phi(z) = -\frac{1}{2} a_0 A_0(z), \tag{2.2.7}$$

where

$$A_0(z) \equiv \pi^{-1}\int_{-a}^{a} (a^2-x'^2)^{-1/2} \ell n\left(\frac{z-x'}{a}\right) dx' \tag{2.2.8}$$

may be shown to have the value

$$A_0(z) = -\ell n[2\{z - (z^2-a^2)^{1/2}\}/a]. \tag{2.2.9}$$

The derivative of $A_0(z)$, which appears in (2.2.6), is

$$A_0'(z) \equiv \pi^{-1}\int_{-a}^{a} (a^2-x'^2)^{-1/2}(z-x')^{-1}dx' = (z^2-a^2)^{-1/2}. \tag{2.2.10}$$

Before going further we must consider how the functions such as $A_0(z)$ and $(z^2 - a^2)^{-1/2}$ are defined. Consider the function $(z^2 - a^2)^{1/2}$. As shown in textbooks on functions of a complex variable, this function can be defined unambiguously only if a cut is made between the points -a and +a so that the points $z = x + i\varepsilon$, $z = x - i\varepsilon$, where $-a \le x \le a$ and ε is vanishingly small, are considered to be different points, labelled $z = x+$ and $z = x-$ respectively. The reason for this is as follows: Let ρ_1, θ_1 and ρ_2, θ_2 be the polar coordinates of z referred to a and -a respectively with the positive x-axis as datum line, as shown in Figure 2.2.1. Then

$$z^2 - a^2 = (z-a)(z+a) = \rho_1 \exp(i\theta_1)\cdot\rho_2 \exp(i\theta_2) \tag{2.2.11}$$

and

$$(z^2 - a^2)^{1/2} = (\rho_1\rho_2)^{1/2}\exp[i(\theta_1 + \theta_2)/2]. \tag{2.2.12}$$

Consider the limiting value of this as $z \to x+$ where $x \ge a$. Here $\theta_1 \to 0$, $\theta_2 \to 0$ and

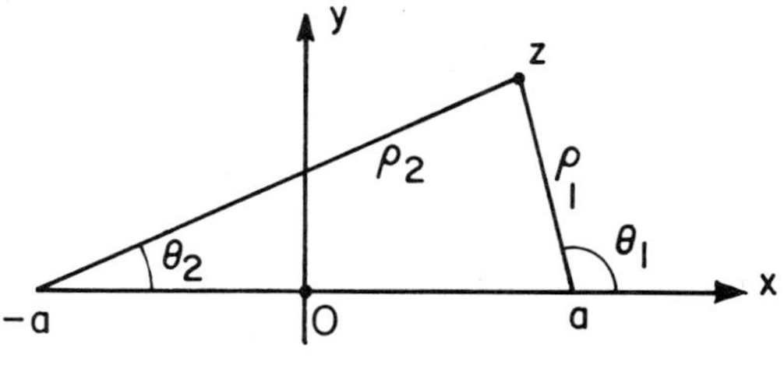

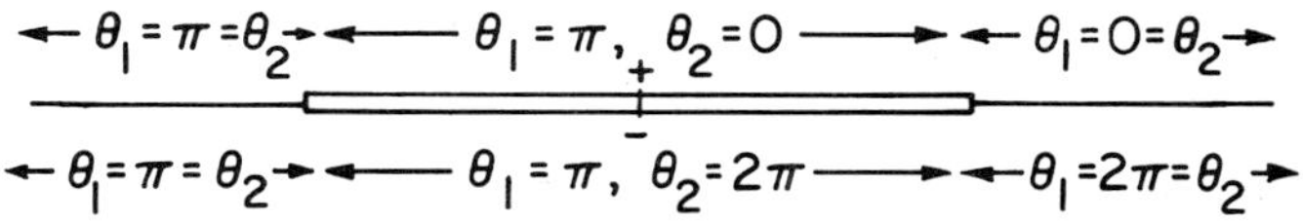

Figure 2.2.1 - Boundary values of the angles θ_1, θ_2.

$$(z^2-a^2)^{1/2} \rightarrow (\rho_1\rho_2)^{1/2} = (x^2-a^2)^{1/2}.$$

As $z \rightarrow x-$ when $x \geq a$ then $\theta_1 \rightarrow 2\pi$, $\theta_2 \rightarrow 2\pi$ and again $(z^2 - a^2)^{1/2}$ tends to the same value. Thus $(z^2 - a^2)^{1/2}$ is continuous as z crosses the x-axis. On the part of the x-axis between -a and a the result is different. As $z \rightarrow x+$, $\theta_1 \rightarrow \pi$, $\theta_2 \rightarrow 0$ and $(z^2 - a^2)^{1/2} \rightarrow (\rho_1\rho_2)^{1/2}\exp(i\pi/2) = i(a^2 - x^2)^{1/2}$. On the other hand, as $z \rightarrow x-$, $\theta_1 \rightarrow \pi$, $\theta_2 \rightarrow 2\pi$ and $(z^2 - a^2)^{1/2} \rightarrow (\rho_1\rho_2)^{1/2}\exp(3i\pi/2) = -i(a^2 - x^2)^{1/2}$. These boundary values, which are shown in Figure 2.2.2, refer to one *branch* of $(z^2 - a^2)^{1/2}$. This particular way of choosing a branch of such a function will be followed throughout the book.

Now return to $A_0(z)$. On the x-axis, when $x \geq a$, $A_0(x)$ is real and

$$A_0(x) = -\ln[2\{x - (x^2-a^2)^{1/2}\}/a], \qquad x \geq a. \tag{2.2.13}$$

On the side of [-a,a] in the upper half-plane

$$A_0(x,0+) \equiv \overset{+}{A}_0(x) = -\ln[2\{x - i(a^2-x^2)^{1/2}\}/a]. \tag{2.2.14}$$

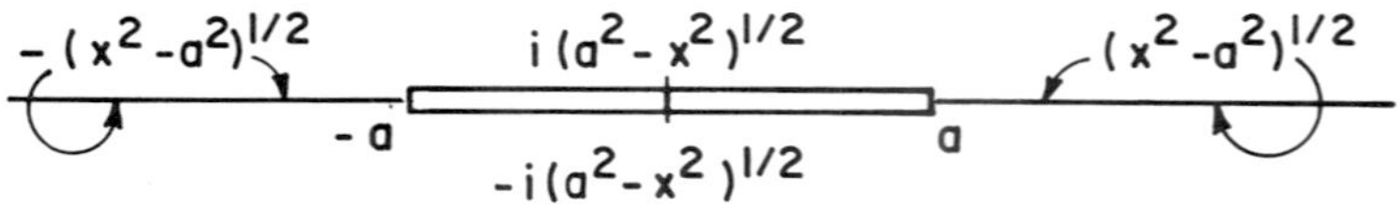

Figure 2.2.2 - Boundary values of the function $(z^2 - a^2)^{1/2}$.

To find the value of this complex logarithm, put $x = a \cos \theta$, $(a^2 - x^2)^{1/2} = a \sin \theta$; then $x - i(a^2 - x^2)^{1/2} = a[\cos \theta - i \sin \theta] = a \exp(-i\theta)$, and

$$A_0^+(x) = -\ln[2 \exp(-i\theta)] = -\ln 2 + i\theta, \quad |x| < a. \tag{2.2.15}$$

On the remainder of the x-axis, where $x \leq -a$,

$$A_0(x) = -\ln[2\{x + (x^2-a^2)^{1/2}\}/a], \quad x \leq -a. \tag{2.2.16}$$

The displacements in the upper half-plane are now given by (2.2.4). On the x-axis, where $z - \bar{z} \equiv 2iy = 0$, the displacement will be given by the first two terms of (2.2.4) if the last, rigid-body displacement, term is neglected. The value of a_0 is related to the total force applied over $[-a,a]$. Thus

$$X + iY = \int_{-a}^{a} (s + ip)dx' = \int_{-a}^{a} q(x')dx' = \pi a_0. \tag{2.2.17}$$

If a normal force P_0 is applied, then $a_0 = iP_0/\pi$, and the displacement for $-a \leq x \leq a$ is given by

$$-4\pi\mu D(x,0+) = iP_0\{\kappa(-\ln 2 + i\theta) + (-\ln 2 - i\theta)\}, \tag{2.2.18}$$

so that

$$D(x,0+) = u(x,0+) + iv(x,0+)$$

$$= \frac{P_0}{\pi}\left\{\left(\frac{1-2\nu}{2\mu}\right)\theta + i\left(\frac{1-\nu}{\mu}\right)\ln 2\right\}. \tag{2.2.19}$$

This equation should be studied carefully. It shows, first, that within [-a,a], where the pressure is applied, the normal displacement v is constant. This means that the displacement corresponds to indentation of the upper half-plane by a flat rigid punch, as shown in Figure 2.2.3. The equation *appears* to show that the penetration, i.e. $v(x,0+)$, is proportional to the applied force, but this is not the case, since the D in (2.2.19) may be supplemented by an arbitrary rigid-body displacement and an arbitrary rigid-body rotation. In fact the penetration *cannot be found*; this is true for every problem in which a punch, rigid or elastic, indents an elastic half-plane. Physically one would expect that the *lateral* displacement, u, would be antisymmetric about the y-axis, in the sense that $u(-x,0) = -u(x,0)$; this can be achieved by superimposing a lateral translation on (2.2.19) to give

$$D(x,0+) = \pi^{-1}\vartheta P_0\left\{-\varphi\left(\frac{\pi}{2} - \theta\right) + i\ \ell n\ 2\right\}, \qquad \varphi = \frac{1-2\nu}{2(1-\nu)}. \tag{2.2.20}$$

The term $\pi/2 - \theta$ is antisymmetric in x. The boundary displacements for $|x| \geq a$ are given in Ex. 2.2.2.

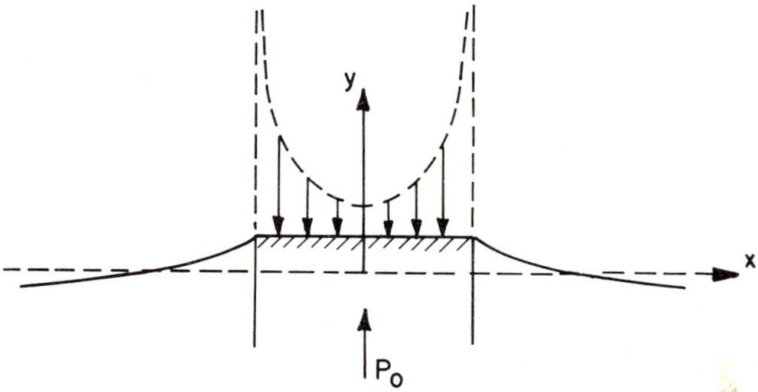

Figure 2.2.3 - Indentation of the upper half-plane by a rigid plane punch.

For large $|z|$, the dominant terms in (2.2.4) are the first two, so that

$$-4\mu D(x,y) = \kappa a_0 A_0(z) + a_0\overline{A_0(z)} + O(1). \tag{2.2.21}$$

But

$$A_0(z) = \ln(z/a) + O(|z|^{-2}) \text{ for large } |z|, \qquad (2.2.22)$$

so that

$$-4\mu D(x,y) = \kappa a_0 \ln(z) + a_0 \overline{\ln(z)} + O(1), \qquad (2.2.23)$$

and we note that $D(x,y)$ has no rotatory component for large $|z| \equiv r$, in the sense that

$$\lim_{r\to\infty} r^{-1} D(x,y) = 0. \qquad (2.2.24)$$

In fact, in the class of displacements satisfying this condition, the displacement is the unique solution for the flat punch, apart from a rigid-body translation.

For a second example, take

$$q(x') = 2iP_0(a^2-x'^2)^{1/2}/\pi a^2. \qquad (2.2.25)$$

This corresponds to a distribution of normal stress over $[-a,a]$ which has resultant P_0 and which vanishes at the ends of the contact interval.

The integrals in (2.2.4) may now be evaluated by using the identity

$$(a^2-x'^2)^{1/2} = \frac{a^2-x'^2}{(a^2-x'^2)^{1/2}} = \frac{(a^2-2x'^2)}{2(a^2-x'^2)^{1/2}} + \frac{a^2}{2(a^2-x'^2)^{1/2}}. \qquad (2.2.26)$$

The integrals involving the second term may be expressed in terms of $A_0(z)$ (see (2.2.8)) while the integrals involving the first term may be found by using the result

$$A_2(z) = \pi^{-1}\int_{-a}^{a} (a^2-x'^2)^{-1/2}\, \frac{2x'^2-a^2}{a^2}\, \ln\left(\frac{z-x'}{a}\right) dx ,$$

$$= -\frac{1}{2}[\{z - (z^2-a^2)^{1/2}\}/a]^2. \qquad (2.2.27)$$

The displacement is now somewhat more complicated, but its value in the contact region $[-a,a]$ may again be found easily from the first two terms of (2.2.4). Thus, in that region,

$$4\pi\mu D(x,0+) = 2i\kappa P_0 \left\{ \frac{1}{2} A_2^+(x) - \frac{1}{2} A_0^+(x) \right\} +$$

$$+ 2iP_0 \left\{ \frac{1}{2} \overline{A_2^+(x)} - \frac{1}{2} \overline{A_0^+(x)} \right\} . \qquad (2.2.28)$$

Now $A_2^+(x) = -\frac{1}{2}[\{x - i(a^2-x^2)^{1/2}\}/a]^2 =$

$$= -\frac{2x^2-a^2}{2a^2} + \frac{2ix(a^2-x^2)^{1/2}}{2a^2} , \qquad (2.2.29)$$

so that, apart from the constant term due to $A_0^+(x)$, the normal displacement is given by

$$4\pi\mu v(x,0+) = -(\kappa+1)P_0(2x^2-a^2)/2a^2 , \qquad |x| \leq a . \qquad (2.2.30)$$

This displacement, of which only the first term is significant, corresponds to a punch with a circular (actually parabolic) base of radius R as shown in Figure 2.2.4; thus

$$v(x,0+) = -\frac{\vartheta P_0 x^2}{\pi a^2} = -\frac{x^2}{2R} , \qquad (2.2.31)$$

and the total force P_0 is linked to the contact length 2a and the radius of curvature, R, of the punch, by the equation

$$P_0 = \pi a^2/(2\vartheta R) . \qquad (2.2.32)$$

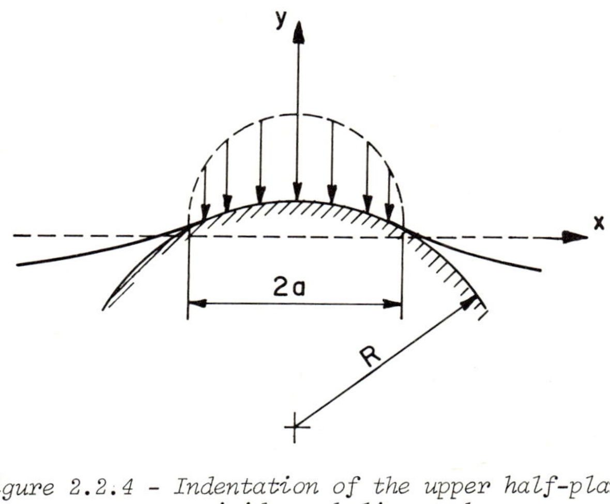

Figure 2.2.4 - Indentation of the upper half-plane by a rigid parabolic punch.

This equation may also be viewed as one giving the value of a for a given force P_0. As P_0 increases, so does a, and a has the property that the normal stress, which has resultant P_0, just vanishes at the ends of the contact interval. Such contact is called incomplete (or unbonded); it is characterized by the square root $(a^2 - x^2)^{1/2}$ form for the contact pressure. It may be contrasted with the complete penetration of the flat punch in Figure 2.2.3, which has a square root singularity $(a^2 - x^2)^{-1/2}$ in the contact pressure. Note that in both the problems that have been considered there is lateral displacement of points of the half-space boundary immediately under the punch. Such displacement can be restrained only if some adhesive shear stresses act on the boundary. (See §4.6.)

An important application of the foregoing solution is to the two-dimensional contact of two bodies with circular profiles of radii R_1, R_2 so large that they can be treated as half-planes; this is termed the two-dimensional *Hertzian* contact problem and is shown in Figure 2.2.5. Let the bodies 1,2 occupy the upper and lower half-plane respectively. If the profiles of the undeformed bodies are

$$y_1 = \frac{x^2}{2R_1}, \qquad y_2 = \frac{x^2}{2R_2}, \tag{2.2.33}$$

referred to axes Ox, Oy_1, Ox, Oy_2 respectively, then the distance between two opposing points, one on the surface of each body, is $y_1 + y_2$. When the two bodies are pressed together the distance between two such points is $y_1 + v_1 + y_2 + v_2$, and this must be constant in the contact region so that

$$v_1 + v_2 = C - \frac{x^2}{2R_1} - \frac{x^2}{2R_2}, \qquad -a \le x \le a. \tag{2.2.34}$$

Equilibrium requires that

$$p_1(t) = p_2(t), \qquad -a \le t \le a. \tag{2.2.35}$$

Thus again

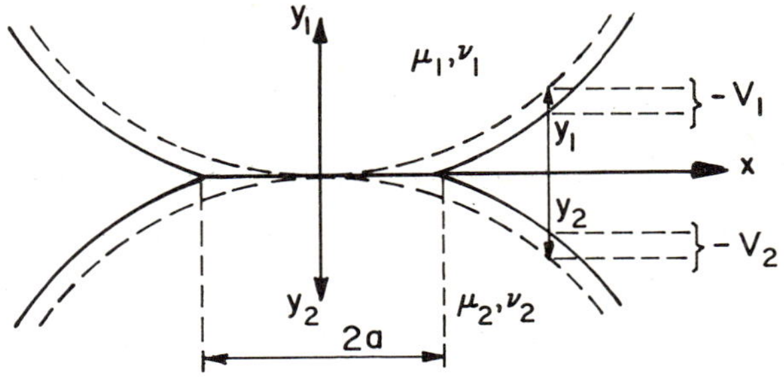

Figure 2.2.5 - Two slightly curved elastic bodies, initially just touching ---, are pressed together and come into contact over $-a \le x \le a$.

$$p_1(t) = p_2(t) = \frac{2P_0}{\pi a^2}(a^2-t^2)^{1/2}, \qquad (2.2.36)$$

giving

$$v_1 = C_1 - \frac{\vartheta_1 P_0 x^2}{\pi a^2}, \quad v_2 = C_2 - \frac{\vartheta_2 P_0 x^2}{\pi a^2}, \qquad (2.2.37)$$

so that

$$(\vartheta_1+\vartheta_2)\frac{P_0}{\pi a^2} = \frac{1}{2R_1} + \frac{1}{2R_2}, \qquad (2.2.38)$$

which may be viewed as an equation for the contact length 2a as a function of $\vartheta_1, \vartheta_2, R_1, R_2$ and the total force P_0. Note that it is not possible to obtain the *approach* of the two bodies; also, as is expected in half-plane problems, the displacements are infinite at infinity. This solution was obtained early in the century in the Soviet Union. Poritsky (1950) obtained explicit expressions and tabulated results for displacements and stresses inside the bodies; an extraneous factor of 2 appears in his analysis which can be resolved by taking his F as $P_0/2$.

Examples 2.2

1. Show that if $A_0(z)$ is defined in the plane cut along [-a,a] then its boundary values in the upper half-plane are as follows:

$A_0(t) = -\ell n\ 2 + \phi$, for $t = a \cosh \phi$, i.e. for $t > a$,

$A_0^+(t) = -\ell n\ 2 + i\theta$, for $t = a \cos \theta$, i.e. for $-a < t < a$,

$A_0(t) = -\ell n\ 2 + \phi + i\pi$, for $t = -a \cosh \phi$, i.e. for $t < -a$.

2. Show that the normal displacement $v(x,0)$ outside the contact region corresponding to (2.2.18) is given by

$$v(x,0) - v(0,0) = -\pi^{-1}\vartheta P_0 \cosh^{-1}(x/a), \quad x > a.$$

2.3 *Chebyshev polynomials*

During recent years the subject of polynomial approximation of functions has been revolutionized by the use of Chebyshev polynomials. These polynomials, defined below, lead to a very accurate and concise means of approximating a large class of functions; see for example Fox and Parker (1968). Their introduction into plane elasticity has the added advantage that many of the integrals, such as (2.2.4), may be evaluated explicitly when the load $q(x')$ is approximated in terms of these polynomials. They will be used to solve some plane elasticity problems in §2.4.

The Chebyshev polynomials $T_n(x), U_n(x)$, of the first and second kinds respectively, are defined initially for x in $[-1,1]$ by the equations

$$T_n(x) = \cos n\theta, \quad U_n(x) = \frac{\sin(n+1)\theta}{\sin\theta}, \quad x = \cos\theta. \tag{2.3.1}$$

The first few are

$$T_0(x) = 1, \quad T_1(x) = x, \quad T_2(x) = 2x^2 - 1, \tag{2.3.2}$$

$$U_0(x) = 1, \quad U_1(x) = 2x, \quad U_2(x) = 4x^2 - 1. \tag{2.3.3}$$

The functions T_1, T_2, T_3 are shown in Figure 2.3.1. They satisfy the recurrence relations

$$T_{n+1}(x) - 2xT_n(x) + T_{n-1}(x) = 0, \tag{2.3.4}$$

$$U_{n+1}(x) - 2xU_n(x) + U_{n-1}(x) = 0, \tag{2.3.5}$$

and are related to each other by the equations

$$T_n(x) - T_{n+2}(x) = 2(1-x^2)U_n(x), \quad T_n'(x) = nU_{n-1}(x). \tag{2.3.6}$$

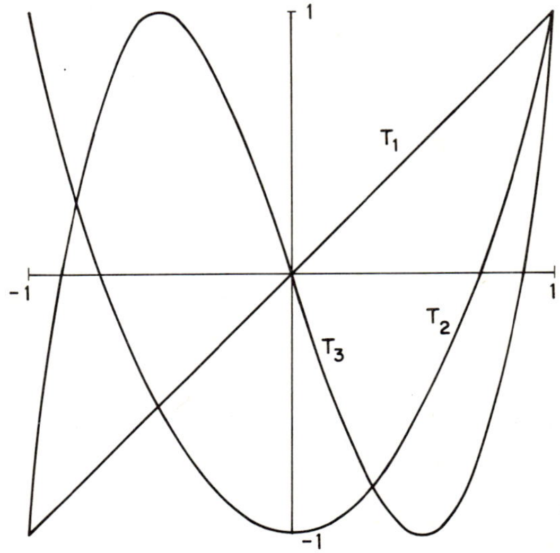

Figure 2.3.1 - The Chebyshev polynomials T_1, T_2, T_3.

The recurrence relation (2.3.4) and the values of the first two polynomials provide a convenient algorithm for computing higher polynomials.

Chebyshev polynomials are orthogonal polynomials. In fact the well known orthogonality of the cosines over $[0,\pi]$ leads to an orthogonality condition for the $T_n(t)$. Thus the relation

$$\int_0^{\pi} \cos m\theta \cos n\theta d\theta = \begin{cases} 0, & n \neq m \\ \pi/2, & n = m \neq 0 \\ \pi, & n = m = 0 \end{cases} \tag{2.3.7}$$

leads immediately to

$$\int_{-1}^{1} (1-t^2)^{-1/2} T_m(t) T_n(t)\,dt = \begin{cases} 0, & n \neq m \\ \pi/2, & n = m \neq 0 \\ \pi, & n = m = 0. \end{cases} \tag{2.3.8}$$

This *continuous* orthogonality condition is important, but not so important in computational work as the *discrete* orthogonality condition satisfied by the polynomials. This may be established by starting from the identity (Ex. 2.3.3)

$$1 + \cos t + \cos 2t + \ldots + \cos Nt = \frac{1}{2} + \frac{1}{2}\cos Nt + \\ + \frac{1}{2}\sin Nt \cot\frac{t}{2}\,, \tag{2.3.9}$$

which may be written

$$\sum_{k=0}^{N}{}'' \cos kt = \frac{1}{2}\sin Nt \cot\frac{t}{2}\,, \tag{2.3.10}$$

where " means that the first and last terms of the sum are halved. Now combine (2.3.10) with the identity

$$2\cos mt \cos nt = \cos(m+n)t + \cos(m-n)t$$

and note that the right hand side of (2.3.10) vanishes when t is an integral multiple of π/N. Thus if m,n ($m \neq n$) are positive integers and $\theta_k = k\pi/N$ then

$$\sum_{k=0}^{N}{}'' \cos m\theta_k \cos n\theta_k = \frac{1}{2}\sum_{k=0}^{N}{}'' \{\cos(m+n)\theta_k + \cos(m-n)\theta_k\}$$

$$= \frac{1}{2}\sum_{k=0}^{N}{}'' \left\{\cos k\left[\frac{(m+n)\pi}{N}\right] + \cos k\left[\frac{(m-n)\pi}{N}\right]\right\} = 0.$$

The final result is that if $m \leq N$, $n \leq N$ then

$$\sum_{k=0}^{N}{}'' \cos m\theta_k \cos n\theta_k = \begin{cases} 0, & m \neq n \\ N/2, & m = n \neq 0, N \\ N, & m = n = 0, N. \end{cases} \tag{2.3.11}$$

The corresponding result for Chebyshev polynomials may be obtained by writing

$$\cos m\theta_k = T_m(x_k), \qquad x_k = \cos(k\pi/N), \tag{2.3.12}$$

$$\sum_{k=0}^{N}{}'' T_m(x_k)T_n(x_k) = \begin{cases} 0, & m \neq n \\ N/2, & m = n \neq 0,N \\ N, & m = n = 0,N. \end{cases} \tag{2.3.13}$$

This result may be used to find the polynomial $f_N(x)$ of degree N which is equal to a given function f(x) at each of the (N + 1) points x_k. Let the polynomial be

$$f_N(x) = \sum_{n=0}^{N}{}'' a_n T_n(x); \tag{2.3.14}$$

then

$$f(x_k) = \sum_{n=0}^{N}{}'' a_n T_n(x_k), \qquad k = 0,1,\ldots,N,$$

so that

$$\sum_{k=0}^{N}{}'' f(x_k)T_m(x_k) = \sum_{k=0}^{N}{}'' T_m(x_k) \sum_{n=0}^{N}{}'' a_n T_n(x_k) = \frac{N}{2}\, a_m \tag{2.3.15}$$

and it may be noted that this equation holds for all m = 0,1,...,N. Equation (2.3.15) shows that the coefficients a_m are obtained explicitly in terms of the values of f(x) at the *Chebyshev points* x_k.

For the application of Chebyshev polynomials to plane problems it is convenient to introduce the integrals

$$R_n(z) = -\frac{1}{\pi}\int_{-1}^{1} \frac{(1-\tau^2)^{1/2} U_{n-1}(\tau)\,d\tau}{\tau - z}, \qquad n = 1,2,\ldots \tag{2.3.16}$$

$$G_{n-1}(z) = \frac{1}{\pi}\int_{-1}^{1} \frac{(1-\tau^2)^{-1/2} T_n(\tau)\,d\tau}{\tau - z}, \qquad n = 0,1,2\ldots \tag{2.3.17}$$

where z = x + iy. The functions $R_n(z), G_{n-1}(z)$ are functions of the complex variable z. The detailed study of such functions will be deferred to Chapter III, but a number of important results may be obtained without much difficulty by considering the boundary values of functions of a complex variable.

The definition of a branch of the function $(z^2 - 1)^{1/2}$ was discussed in §2.2. It may easily be shown (see the note at the end of this section, and the Examples) that these functions also must be defined in the plane cut along [-1,1] and that

$$R_n(z) = [z - (z^2-1)^{1/2}]^n, \qquad n = 1,2,\ldots, \tag{2.3.18}$$

$$G_{n-1}(z) = -(z^2-1)^{-1/2}R_n(z), \qquad n = 1,2,\ldots. \tag{2.3.19}$$

It is convenient to define $R_0(z) \equiv 1$; then equations (2.3.18), (2.3.19) hold for $n = 0$ also. The values of the functions on the x-axis outside $|x| = 1$ are

$$R_n(x) = [x - \text{sign}(x)(x^2-1)^{1/2}]^n, \qquad n = 0,1,2,\ldots, \tag{2.3.20}$$

$$G_{n-1}(x) = -\text{sign}(x)(x^2-1)^{-1/2}R_n(x), \qquad n = 0,1,2,\ldots, \tag{2.3.21}$$

while the values on the + and - sides of the axis for $|x| < 1$ are

$$R_n^{\pm}(x) = T_n(x) \mp i(1-x^2)^{1/2}U_{n-1}(x), \tag{2.3.22}$$

$$G_{n-1}^{\pm}(x) = U_{n-1}(x) \pm i(1-x^2)^{-1/2}T_n(x), \tag{2.3.23}$$

where $U_{n-1}(x)$ for $n = 0$ must be interpreted as zero.

Equations (2.3.22),(2.3.23) give the boundary values of the integrals $R_n(z), G_{n-1}(z)$ on $|x| < 1$; it is possible also to consider the values of the integrals (2.3.16),(2.3.17) when $z = x$ and $|x| < 1$, i.e. the values of $R_n(x)$ and $G_{n-1}(z)$ *on* the axis, between -1 and 1. In this case the integrands are *singular* at $\tau = x$ and care has to be taken in order to define the integrals unambiguously. The usual definition for such integrals is the Cauchy Principal Value, namely

$$(P)\int_{-1}^{1} \frac{f(\tau)d\tau}{\tau - x} = \lim_{\varepsilon \to 0}\left\{\int_{-1}^{x-\varepsilon} \frac{f(\tau)d\tau}{\tau - x} + \int_{x+\varepsilon}^{1} \frac{f(\tau)d\tau}{\tau - x}\right\}. \tag{2.3.24}$$

It may be shown that $R_n(x), G_{n-1}(x)$, defined as the principal values of the integrals (2.3.16),(2.3.17) for $|x| \leq 1$, are given by

$$R_n(x) = -\frac{(P)}{\pi}\int_{-1}^{1} \frac{(1-\tau^2)^{1/2}U_{n-1}(\tau)d\tau}{\tau - x} = T_n(x), \quad n = 1,2,..,(2.3.25)$$

$$G_{n-1}(x) = \frac{(P)}{\pi}\int_{-1}^{1} \frac{(1-\tau^2)^{-1/2}T_n(\tau)d\tau}{\tau - x} = U_{n-1}(x), \quad n = 0,1,..,(2.3.26)$$

where, as before, $U_{-1}(x) \equiv 0$, and (2.3.25) does not hold for $n = 0$. These results may be described by saying that the polynomials $T_n(x), U_{n-1}(x)$ are the principal values for $|x| \leq 1$ of the complex functions $R_n(z), G_{n-1}(z)$ respectively. This result will be used in §2.4.

It is difficult to unravel the history of the results obtained in this section. Equation (2.3.26), written in the form

$$\frac{1}{\pi}\int_0^{\pi} \frac{\cos n\theta d\theta}{\cos\theta - \cos\phi} = \frac{\sin n\phi}{\sin\phi}, \quad n = 0,1,..., \qquad (2.3.27)$$

is Glauert's integral (Glauert (1948, p. 92)). The expression (2.3.18) appears in Klubin (1952) (see also Pykhteev (1959)), while (2.3.25),(2.3.26) appeared in the 1964 edition of Abramowitz and Stegun, although it was not until the second (1970) edition that (2.3.25) was given correctly. The integrated version of (2.3.26), given in Ex. 2.3.8, has been used extensively by Popov (1961a),(1963a),(1969).

The results stated in this section form the subject matter of the Examples; as a starting point the integral $G_{n-1}(z)$ is evaluated. Write $\tau = \cos\theta$; then

$$G_{n-1}(z) = \frac{1}{\pi}\int_0^{\pi} \frac{\cos n\theta d\theta}{\cos\theta - z}. \qquad (2.3.28)$$

Form the sum

$$G_{n+1}(z) + G_{n-1}(z) - 2zG_n(z)$$

$$= \frac{1}{\pi}\int_0^\pi \frac{\cos(n+2)\theta + \cos n\theta - 2z\cos(n+1)\theta}{\cos\theta - z}\,d\theta$$

$$= \frac{1}{\pi}\int_0^\pi \frac{2\cos(n+1)\theta[\cos\theta - z]}{\cos\theta - z}\,d\theta = 0, \quad n = 0,1,2,\ldots.$$

Thus $G_n(z)$ satisfies the 'Chebyshev' recurrence relation (see (2.3.4) and (2.3.5))

$$G_{n+1}(z) + G_{n-1}(z) - 2zG_n(z) = 0, \qquad n = 0,1,\ldots. \tag{2.3.29}$$

The first integrals are

$$G_{-1}(z) = \frac{1}{\pi}\int_0^\pi \frac{d\theta}{\cos\theta - z}, \qquad G_0(z) = 1 + zG_{-1}(z). \tag{2.3.30}$$

The integral $G_{-1}(z)$ is the basic one on which all the $G_n(z)$ depend; it may be evaluated by contour integration, but it must be defined in the plane cut along [-1,1]. It may be written

$$G_{-1}(z) = \frac{1}{2\pi}\int_0^{2\pi} \frac{2\exp(i\theta)d\theta}{\exp(2i\theta) + 1 - 2\exp(i\theta)z}$$

$$= \frac{1}{\pi i}\int_\Gamma \frac{d\zeta}{\zeta^2 + 1 - 2\zeta z}, \tag{2.3.31}$$

where Γ is the unit circle in the ζ-plane. The integrand has two poles at $\zeta = \zeta_1, \zeta_2$ where

$$\zeta_1 = z - (z^2-1)^{1/2}, \quad \zeta_2 = z + (z^2-1)^{1/2}.$$

If that branch of $(z^2 - 1)^{1/2}$ is chosen which satisfies

$$(z^2-1)^{1/2} = z + O(z^{-2}) \quad \text{for large } |z|$$

(this is the branch shown in Figure 2.2.1) then ζ_1 lies inside Γ, ζ_2 outside and

$$G_{-1}(z) = \frac{1}{\pi i(\zeta_1-\zeta_2)}\int_\Gamma\left\{\frac{1}{\zeta-\zeta_1} - \frac{1}{\zeta-\zeta_2}\right\}d\zeta$$

$$= \frac{2}{\zeta_1-\zeta_2} = -(z^2-1)^{-1/2}. \qquad (2.3.32)$$

From this value of $G_{-1}(z)$, the corresponding value of $G_0(z)$ given by (2.3.30), and the recurrence (2.3.29) we deduce (2.3.19). Equation (2.3.18) may be established similarly.

Examples 2.3

1. Use (2.3.8) to show that if

$$f(t) = {\sum_{n=0}^{N}}' a_n T_n(t), \qquad |t| \leq 1,$$

and ' indicates that the first term is halved, then

$$a_n = \pi^{-1}\int_{-1}^{1} (1-t^2)^{-1/2} f(t)T_n(t)dt.$$

2. Derive the orthogonality condition

$$\frac{2}{\pi}\int_{-1}^{1} (1-t^2)^{1/2} U_n(t)U_m(t)dt = \delta_{n,m}$$

where $\delta_{n,m}$ is the Kronecker delta

$$\delta_{n,m} = \begin{cases} 0 \ , & n \neq m \\ 1 \ , & n=m. \end{cases}$$

3. Starting from the geometric progression

$$1 + \exp(it) + \exp(2it) + \ldots + \exp(Nit) = (\exp[(N+1)it] - 1)/(\exp(it) - 1),$$

prove the identity (2.3.9).

4. Another branch of $(z^2 - 1)^{1/2}$ may be found by taking θ_1, θ_2 in Figure 2.2.1 to lie in the ranges $-\pi \le \theta_i \le \pi$. Find the boundary values of this branch.

5. Prove that the $R_n(z)$ defined by (2.3.16) satisfy

$$R_{n+1}(z) + R_{n-1}(z) - 2zR_n(z) = 0 \qquad \text{(A)}$$

for n = 2,3,.... By writing $\tau = \cos\theta$ in (2.3.16) show that

$$2R_n(z) = G_n(z) - G_{n-2}(z), \qquad n = 1,2,\ldots,$$

and hence prove (2.3.18) with the aid of (2.3.19). Show also that if $R_0(z) = 1$ then the recurrence (A) holds for n = 1 also.

6. Establish the results (2.3.20)-(2.3.23) and (2.3.25).

7. Show that $R_n(z) = T_n(z) - (z^2 - 1)^{1/2}U_{n-1}(z)$ and $G_{n-1}(z) = U_{n-1}(z) - (z^2 - 1)^{-1/2}T_n(z)$.

8. Show by integrating (2.3.26) with respect to x that

$$-\pi^{-1}\int_{-1}^{1} (1-\tau^2)^{-1/2} \ln|t-\tau| T_n(\tau)d\tau = \begin{cases} n^{-1}T_n(t) & \text{if } n \ne 0 \\ \ln 2 & \text{if } n = 0. \end{cases}$$

This result provides and example of what Popov (1969) terms a *polynomial kernel*; see §6.11.

9. Show that $A_n(z)$, defined by

$$A_n(z) = \pi^{-1}\int_{-1}^{1} (1-t^2)^{-1/2}T_n(t)\ln(z-t)dt,$$

satisfies $A_n'(z) = -G_{n-1}(z) = -R_n'(z)/n$, $n \ne 0$, and hence deduce $A_n(z) = -R_n(z)/n$, $n \ne 0$. Show that $A_0'(z) = -G_{-1}(z) = (z^2 - 1)^{-1/2}$ and hence deduce

$$A_0(z) = -\ln[2\{z - (z^2-1)^{1/2}\}].$$

2.4 *Indentation by a frictionless punch of general profile*

There are two distinct classes of problems relating to indentation by a frictionless punch. In the first kind of indentation, called *complete penetration*, there is complete contact between the punch and the half-plane over a specified contact region, in the sense that the normal displacement of the half-plane at the boundary matches the profile of the punch. Such problems are characterized by a contact pressure which has a (square root) singularity at the ends of the contact region. In the second kind, called *incomplete penetration*, the extent of the contact region, i.e. the extent of the region over which the normal displacement of the half-plane matches the profile of the punch, is initially unknown. Cases of incomplete penetration are characterized by a contact pressure which is zero at the ends of the contact region.

First we obtain some general results which can be applied to a number of different situations. Suppose that contact occurs over the segment $-b \leq x \leq a$ of the x-axis. Introduce new co-ordinates (ξ,η) such that

$$x = \ell\xi + c, \quad y = \ell\eta, \quad \ell = (a+b)/2, \quad c = (a-b)/2; \qquad (2.4.1)$$

then the contact region is given by $-1 \leq \xi \leq 1$. Suppose that, relative to the new coordinates (ξ,η), the profile of the punch may be approximated by

$$\eta = y/\ell = \sum_{n=0}^{N} b_n T_n(t), \qquad |t| \leq 1. \qquad (2.4.2)$$

If y is given initially as a function of x then the dimensionless parameters b_n will be real functions of a,b, i.e. of ℓ,c. The boundary conditions for the half-plane are then

$$v(x,0+)/\ell = -d + \sum_{n=0}^{N} b_n T_n(\xi), \qquad |\xi| \leq 1; \qquad (2.4.3)$$

$$\tau_{yy}(t,0) = 0, \quad |t| > 1; \qquad \tau_{xy}(t,0) \text{ for all } t. \qquad (2.4.4)$$

Indentation by a frictionless punch of general profile

Now assume that the normal loading over [-1,1] has the form

$$q(t) = ip(t) = i(1-t^2)^{-1/2} \sum_{n=0}^{N} a_n T_n(t), \tag{2.4.5}$$

where $x = \ell t + c$. The displacement will be given by equation (2.2.4); the constant a in that equation will be taken to be ℓ. The first integral in (2.2.4) corresponding to the term $i(1 - t^2)^{-1/2} a_n T_n(t)$ will be

$$\int_{-b}^{a} \ln\left(\frac{z-x'}{\ell}\right) q(x')dx' = i\ell a_n \int_{-1}^{1} (1-t^2)^{-1/2} T_n(t) \ln(\zeta - t)dt$$

$$= i\ell a_n \pi A_n(\zeta), \tag{2.4.6}$$

while the complete displacement may be written

$$4\mu i D(x,y)/\ell = \sum_{n=0}^{N} a_n \{\kappa A_n(\zeta) + \overline{A_n(\zeta)} -$$

$$- (\zeta - \bar{\zeta}) G_{n-1}(\zeta)\} + C^*, \tag{2.4.7}$$

where $\zeta = \xi + i\eta$. In order to match the normal displacement with that given by (2.4.3) we need the boundary values of $A_n(z)$. Equation (2.3.22) and Ex. 2.3.9 show that

$$A_n^+(\xi) = -[T_n(\xi) - i(1-\xi^2)^{1/2} U_{n-1}(\xi)]/n, \quad n = 1,2,\ldots, \tag{2.4.8}$$

while Ex. 2.2.1 shows that for $\xi = \cos\theta$, $A_0^+(\xi) = -\ln 2 + i\theta$. Thus the normal displacement on $|\xi| \leq 1$ is given by

$$v(x,0+)/\ell = C + \vartheta \sum_{n=1}^{N} n^{-1} a_n T_n(\xi), \tag{2.4.9}$$

and the boundary condition (2.4.3) will be satisfied provided that

$$b_n = \vartheta n^{-1} a_n, \quad n = 1,2,\ldots,N. \tag{2.4.10}$$

The coefficient a_0 does not relate to the shape of the punch, but to the total load applied, according to

$$P_0 = -\int_{-b}^{a} \tau_{yy} dx = \pi \ell a_0. \tag{2.4.11}$$

The moment applied by the punch is

$$M = -\int_{-b}^{a} x\tau_{yy} dx = \ell\int_{-1}^{1} (\ell t+c)(1-t^2)^{-1/2}(a_0+a_1T_1(t))dt$$

$$= \pi\ell(2ca_0+\ell a_1)/2. \tag{2.4.12}$$

We conclude therefore that when the punch is in complete contact over [-b,a], so that equation (2.4.3) holds, then the a_n, b_n are related by (2.4.10), and a_0 is related to the total load P_0 by (2.4.11); the contact stress has a square root singularity at the ends $|\xi| = 1$.

Now consider incomplete penetration of a symmetrical punch. Suppose that contact occurs over [-a,a], where a is initially unknown, and that the punch is pressed into the half-plane by a known force P_0. It is preferable to reverse the roles of known and unknown, to assume that a is known and to find the magnitude of the force P_0 required to cause incomplete penetration over [-a,a].

Assume therefore that the profile of the punch may be approximated by

$$y/a = \sum_{n=0}^{N} b_n T_n(t), \tag{2.4.13}$$

where, if the punch is symmetrical, the odd b_n are all zero, and again b_0 may be taken to be zero. The contact pressure is given by (2.4.5) so that, explicitly,

$$p(t) = -\tau_{yy} = (1-t^2)^{-1/2}[P_0/(\pi a) + \vartheta^{-1} \sum_{n=1}^{N} nb_nT_n(t)]. \tag{2.4.14}$$

The condition that the contact pressure be zero at the ends $t = \pm 1$ is therefore

$$\vartheta P_0 = -\pi a \sum_{n=1}^{N} nb_n, \quad n \text{ even}. \tag{2.4.15}$$

This is the relation linking P_0 to the contact length 2a.

As an example take the punch with a parabolic base of radius R, considered in §2.2. Then the profile is given by

$$-\frac{y}{a} = \frac{x^2}{2aR} = \frac{at^2}{2R} = \frac{a(2t^2-1)}{4R} + \frac{a}{4R} = \frac{a}{4R}[T_0(t) + T_2(t)]. \tag{2.4.16}$$

Thus $b_2 = -a/(4R)$, the pressure is

$$p(t) = -\tau_{yy} = 2P_0(1-t^2)^{1/2}/(\pi a), \tag{2.4.17}$$

and the force required is

$$P_0 = \pi a^2/(2\vartheta R). \tag{2.4.18}$$

As a further illustration consider the flat punch simultaneously pressed and twisted as shown in Figure 2.4.1. If contact occurs over [-b,a] then the profile will be

$$y/\ell = -d + \alpha x, \tag{2.4.19}$$

so that $b_1 = \alpha$. The stress distribution will be given by (2.4.5); that is

$$\vartheta\tau_{yy} = -(1-t^2)^{-1/2}[\vartheta\pi^{-1}P_0/\ell + \alpha T_1(t)], \tag{2.4.20}$$

and the moment M about the origin x = 0 = y will be

$$M = cP_0 + \pi\ell^2\alpha/(2\vartheta). \tag{2.4.21}$$

This equation can be inverted to give α in terms of M and P; then τ_{yy} may be written

$$\tau_{yy} = -(\pi\ell)^{-1}(1-t^2)^{-1/2}[P + 2(M-cP)t/\ell]. \tag{2.4.22}$$

If M = 0 the contact will occur over [-a,a], so that b = a and c = 0. Now consider the effect of steadily increasing M. At first, contact will still occur over [-a,a]; thus b will still be equal to a and c = 0; there will still be a square root singularity in τ_{yy} at each end of the punch. However, when M reaches

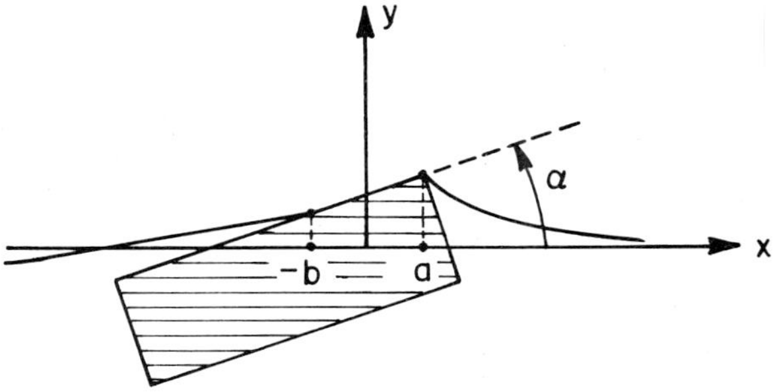

Figure 2.4.1 - Flat punch pressed unsymmetrically into the half-plane.

the value $\ell P/2 = aP/2$ (remember $b = a = \ell$) the numerator of τ_{yy} will be zero at $t = -1$ so that τ_{yy} will be zero there. For this value of M the punch begins to lose contact with the half-space at $t = -1$ ($x = -a$). As M is increased further, then b will decrease, the contact region will recede and the extent of the contact will be given by

$$P = 2(M-cP)/\ell, \quad \text{i.e. } b = 3a - 4M/P. \tag{2.4.23}$$

When M/P reaches the value a, then $b = -a$ and the punch upsets. Clearly, considerable care must be exercised when applying these results to an actual physical punch. The results are based on infinitesimal elasticity theory and the boundary conditions are applied on the plane $y = 0$, not at the interface between punch and half-space.

Examples 2.4

1. Use the boundary values of $A_2(z)$ and $A_0(z)$ to deduce that for $t > 1$ the normal displacement of the boundary corresponding to the contact pressure distribution (2.4.17) is

$$v = C + \frac{\vartheta P_0}{\pi}\left\{\ell n\ w + \frac{1}{2w^2}\right\}, \quad w = [t + (t^2-1)]^{1/2}.$$

2.5 *The equations of Boussinesq and Cerruti*

Two of the basic problems considered by Boussinesq and Cerruti, as well as by other researchers in the nineteenth century, were the determination of the elastic displacement fields in the half-space $z \geq 0$ when

either a) $\tau_{xz}, \tau_{yz}, \tau_{zz}$ *or* b) u, v, w

were specified on $z = 0$. The solutions to these problems were obtained independently by Boussinesq and Cerruti in 1882. Historical accounts of the subject may be found in Todhunter and Pearson (1893, p. 1489 et seq) and Love (1927, p. 16 and Chapter X).

Boussinesq's solution for problem a) in 3-dimensions may be considered to be a special case of the solution (1.10.11) in the absence of body forces. First choose $\phi, \vec{\psi}$ so that $\psi_1 = 0 = \psi_2$, $\psi_3 = \psi$; then

$$2\mu u = -\frac{z\partial\psi}{\partial x} - \frac{\partial\phi}{\partial x}, \qquad 2\mu v = -\frac{z\partial\psi}{\partial y} - \frac{\partial\phi}{\partial y}, \tag{2.5.1}$$

$$2\mu w = 4(1-\nu)\psi - \left(\frac{z\partial\psi}{\partial z} + \psi + \frac{\partial\phi}{\partial z}\right). \tag{2.5.2}$$

Suppose that a portion S of the plane $z = 0$, bounded by a closed curve, is subjected to a normal pressure $p(x,y)$ so that

$$\tau_{zz}(x,y,0) = -p(x,y) \text{ for } (x,y) \text{ in } S, \tag{2.5.3}$$

and suppose that there are no other tractions applied to the boundary, so that

$$\tau_{xz}(x,y,0) = 0 = \tau_{yz}(x,y,0), \tag{2.5.4}$$

and

$$\tau_{zz}(x,y,0) = 0 \text{ when } (x,y) \text{ lies outside } S. \tag{2.5.5}$$

The stresses corresponding to the displacements (1.10.11) are

$$\tau_{xz} = \frac{(1-2\nu)\partial\psi}{\partial x} - \frac{z\partial^2\psi}{\partial x\partial z} - \frac{\partial^2\phi}{\partial x\partial z}, \tag{2.5.6}$$

$$\tau_{yz} = \frac{(1-2\nu)\partial\psi}{\partial y} - \frac{z\partial^2\psi}{\partial y\partial z} - \frac{\partial^2\phi}{\partial y\partial z}, \tag{2.5.7}$$

$$\tau_{zz} = \frac{2(1-\nu)\partial\psi}{\partial z} - \frac{z\partial^2\psi}{\partial z^2} - \frac{\partial^2\phi}{\partial z^2}. \tag{2.5.8}$$

Now the analysis proceeds as in the plane case discussed in §2.1. Thus, provided that

$$\lim_{z\to 0}\left\{\frac{z\partial^2\psi}{\partial x\partial z}, \frac{z\partial^2\psi}{\partial y\partial z}, \frac{z\partial^2\psi}{\partial z^2}\right\} = 0, \tag{2.5.9}$$

the tractions τ_{xz}, τ_{yz} will be zero on $z = 0$ only if

$$(1-2\nu)\psi = \frac{\partial\phi}{\partial z}, \tag{2.5.10}$$

and the normal stress will then be

$$\tau_{zz} = \frac{\partial\psi}{\partial z} - \frac{z\partial^2\psi}{\partial z^2}, \tag{2.5.11}$$

The problem has therefore been reduced to that of finding a function ψ such that its derivative has a specified value on $z = 0$. Boussinesq's solution of this problem is phrased in terms of integrals of the form

$$\psi(x,y,z) = \frac{1}{2\pi}\iint_S \frac{f(x',y')dx'dy'}{R}, \tag{2.5.12}$$

where $R = \{(x-x')^2 + (y-y')^2 + z^2\}^{1/2}$ is the distance from point (x,y,z) of the medium to the point $(x',y',0)$ of the surface. It is necessary to show that the integral (2.5.12) yields a harmonic function satisfying conditions (2.5.3)-(2.5.5). Formally

$$\left.\begin{aligned} \frac{2\pi\partial\psi}{\partial z} &= -z\iint_S \frac{f(x',y')dx'dy'}{R^3} \\ \frac{2\pi\partial^2\psi}{\partial x^2} &= -1\iint_S \frac{f(x',y')dx'dy'}{R^3} + 3\iint_S \frac{(x-x')\ f(x',y')dx'dy'}{R^5} \\ \frac{2\pi\partial^2\psi}{\partial x\partial z} &= 3z\iint_S \frac{(x-x')f(x',y')dx'dy'}{R^5}. \end{aligned}\right\} \tag{2.5.13}$$

The three equations of the form (2.5.13), when added together, show that ψ is harmonic. When $(x,y,0)$ is outside S then R is never zero, the integrals in $\partial\psi/\partial z$, $z\partial^2\psi/\partial x\partial z$, $z\partial^2\psi/\partial y\partial z$, $z\partial^2\psi/\partial z^2$ are all well defined, and all these four quantities are zero on $z = 0$, as is required by equations (2.5.8) and (2.5.9). To verify the conditions within S we proceed heuristically, and leave the reader to make the argument rigorous. Let $P^*(x,y,z)$ with $z > 0$ be a point with the same x,y coordinates as $P(x,y,0)$. We consider the stresses at P^* and find the limiting values of these stresses as z tends to zero. (Note that one should take P^* to be (x^*,y^*,z) and find the limit as $(x^*,y^*,z) \to (x,y,0)$; this is left to the reader.) Let S_ε be a circle of radius ε, centre P lying wholly within S as in Figure 2.5.1. Since P lies outside S-S_ε, the contributions to the stresses $\tau_{xz},\tau_{yz},\tau_{zz}$ at P arising from the integrals over S-S_ε will be zero, by the argument used above. It is therefore sufficient to consider the contributions from S_ε. Introduce polar coordinates r,θ,z with origin at P; the contribution to $\partial\psi/\partial z$ from S_ε will be given by

$$\frac{2\pi\partial\psi}{\partial z} = -zf(x,y)\int_0^{2\pi} d\theta \int_0^{\varepsilon} \frac{rdr}{(r^2+z^2)^{3/2}}$$

$$= -2\pi f(x,y)z\{z^{-1} - (\varepsilon^2+z^2)^{-1/2}\}, \qquad (2.5.14)$$

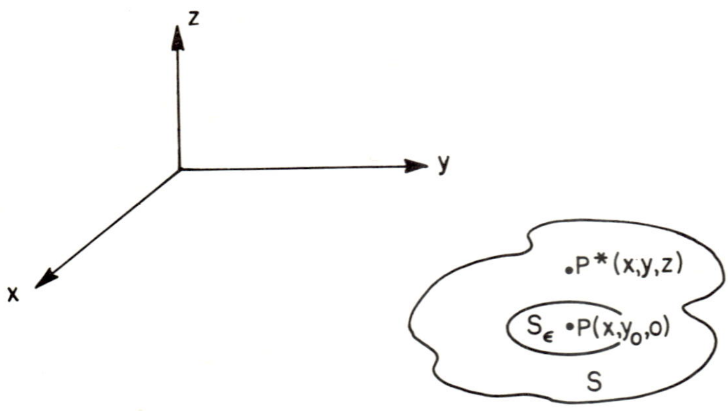

Figure 2.5.1 - The circle S_ε, centre P, radius ε, lies within S on the plane $z = 0$.

so that, when $\varepsilon > 0$,

$$\lim_{z\to 0}\left(\frac{\partial\psi}{\partial z}\right) = -f(x,y). \tag{2.5.15}$$

This means that, provided that $\lim_{z\to 0}(z\partial^2\psi/\partial z^2) = 0$ for (x,y) in S, the condition (2.5.3) may be satisfied by taking

$$f(x,y) = p(x,y). \tag{2.5.16}$$

Finally, it may be shown that the limits (2.5.9) are indeed all zero, even when (x,y) is inside S, and this completes the verification. Thus

$$\psi(x,y,z) = \frac{1}{2\pi}\iint_S \frac{p(x',y')dx'dy'}{R} \tag{2.5.17}$$

and integration of (2.5.17) with respect to z gives

$$\phi(x,y,z) = \frac{(1-2\nu)}{2\pi}\iint_S \ell n(z+R)p(x',y')dx'dy'. \tag{2.5.18}$$

Love (1929) states some further limiting properties of ψ,ϕ, the displacements and stresses, and these are important in applications.

The potentials given by (Todhunter and Pearson (1893, p. 244))

$$\left.\begin{aligned} &2(1-\nu)\psi_1 = -\frac{\partial\chi}{\partial y}, \quad 2(1-\nu)\psi_2 = \frac{\partial\chi}{\partial x}, \quad 2(1-\nu)\psi_3 = \frac{\partial\theta}{\partial z}, \\ &\phi = \theta - x\psi_1 - y\psi_2 , \end{aligned}\right\} \tag{2.5.19}$$

$$\left.\begin{aligned} \theta &= -\frac{(1-\nu)}{\pi}\iint \frac{(x-x')f_1(x',y') + (y-y')f_2(x',y')}{z+R}\,dx'dy' \\ \chi &= \frac{1}{2\pi}\iint \frac{(x-x')f_2(x',y') - (y-y')f_1(x',y')}{z+R}\,dx'dy' \end{aligned}\right\} \tag{2.5.20}$$

correspond to a loading of the region S by shearing forces

$$\tau_{xz}(x,y,0) = -f_1(x,y), \quad \tau_{yz}(x,y,0) = -f_2(x,y). \tag{2.5.21}$$

The equations of Boussinesq and Cerruti

We conclude this section by giving explicit expressions for the displacements due to a distribution of tractions over part of the plane $z = 0$. First it is convenient to obtain the displacements due to a point load applied at the origin. Let the force have components (X,Y,Z) so that

$$\tau_{xz}(x,y,0) = -X\delta(x)\delta(y), \quad \tau_{yz}(x,y,0) = -Y\delta(x)\delta(y), \tag{2.5.22}$$

$$\tau_{zz}(x,y,0) = -Z\delta(x)\delta(y), \tag{2.5.23}$$

where $\delta(x)$ is the Dirac delta function with the properties

$$\delta(x) = 0 \quad x \neq 0, \quad \int_{-\infty}^{\infty} \delta(x)dx = 1. \tag{2.5.24}$$

Then if $T = (xX + yY)/R$ where $R = (x^2 + y^2 + z^2)^{1/2}$, the displacements are given by

$$4\pi\mu u = \frac{2(1-\nu)X}{z+R} + \frac{2\nu Tx}{(z+R)^2} - \frac{(1-2\nu)Zx}{R(z+R)} + \\ + \frac{z}{R}\left\{\frac{X}{z+R} + \frac{Tx(z+2R)}{R(z+R)^2} + \frac{Zx}{R^2}\right\}, \tag{2.5.25}$$

$$4\pi\mu v = \frac{2(1-\nu)Y}{z+R} + \frac{2\nu Ty}{(z+R)^2} - \frac{(1-2\nu)Zy}{R(z+R)} + \\ + \frac{z}{R}\left\{\frac{Y}{z+R} + \frac{Ty(z+2R)}{R(z+R)^2} + \frac{Zy}{R^2}\right\}, \tag{2.5.26}$$

$$4\pi\mu w = \frac{2(1-\nu)Z}{R} + \frac{(1-2\nu)T}{z+R} + \frac{z}{R}\left\{\frac{Zz}{R^2} + \frac{T}{R}\right\}. \tag{2.5.27}$$

An important application of these results is to the displacements on the surface $z = 0$ due to a distribution of tractions on $z = 0$. Thus the displacements due to a normal loading $\tau_{zz}(x,y,0) = -p(x,y)$ are given by

$$2\pi u(x,y,0) = -\varphi\vartheta\iint_S \frac{(x-x')p(x',y')dx'dy'}{\{(x-x')^2+(y-y')^2\}}, \tag{2.5.28}$$

$$2\pi v(x,y,0) = -\varphi\vartheta\iint_S \frac{(y-y')p(x',y')dx'dy'}{\{(x-x')^2+(y-y')^2\}}, \tag{2.5.29}$$

$$2\pi w(x,y,0) = \vartheta\iint_S \frac{p(x',y')dx'dy'}{\{(x-x')^2+(y-y')^2\}^{1/2}} \tag{2.5.30}$$

where ϑ,φ are given in (1.6.15).

Examples 2.5

1. Show that the limits (2.5.9) are all zero, even when P lies within S.
2. Establish the results (2.5.25)-(2.5.27).
3. Establish (2.5.17) by showing that if $\Phi(x,y,z)$ is harmonic in $z \geq 0$, is bounded appropriately at infinity, and zero on $z = 0$, then Φ is identically zero.
4. Show that the solution to problem b) mentioned at the beginning of the section may be obtained from the general solution

   $$\vec{d} = \frac{\partial\vec{\psi}}{\partial z} - \frac{z}{3-4\nu}\vec{\nabla}\,\mathrm{div}\,\vec{\psi}$$

   obtained in Ex. 1.10.3. Proceed as in the treatment of problem a) and deduce that if $\vec{d} = \vec{d}_0(x,y)$ on $z = 0$ and if $|\vec{d}| = O(1/R)$ at infinity then

   $$\vec{\psi}(x,y,z) = -\frac{1}{2\pi}\iint_S \frac{\vec{d}_0(x',y')dx'dy'}{\{(x-x')^2+(y-y')^2+z^2\}^{1/2}}$$

   where the range of integration is the whole of the plane $z = 0$.
5. If $H_{ij}(x-x',y-y')$ denotes the displacement $u_i(x,y)$ ($u_1 = u$, $u_2 = v$, $u_3 = w$) due to a unit point load at (x',y') in direction $j(1,2,3)$, use Betti's Reciprocal Theorem to prove that

   $$H_{ij}(x-x',y-y') = H_{ji}(x'-x,y'-y).$$

   The quantities H_{ij} are called *influence coefficients* and may be derived from (2.5.25)-(2.5.27).

2.6 *Axisymmetric loading over a circular region*

In §2.5 we derived expressions for the displacements in an elastic half-space due to a point force applied on the surface at the origin. The displacements due to an axisymmetric distribution of loads will themselves be axisymmetric. Thus the displacements u_r, u_z due to an axisymmetrical distribution of loads applied over the circle of radius a may be determined as the u,w displacements respectively at the point (r,0) on the x-axis, as shown in Figure 2.6.1. Thus the normal displacement due to the normal loads p(s) may be obtained from (2.5.30) as

$$u_z(r) \equiv w(r) = (\vartheta/2\pi)\int_0^a sp(s)ds\int_0^{2\pi}(r^2+s^2-2rs\cos\theta)^{-1/2}d\theta. \quad (2.6.1)$$

The integral is of elliptic type; we write

$$\begin{aligned} r^2 + s^2 - 2rs\cos\theta &= r^2 + s^2 - 2rs[2\cos^2(\theta/2) - 1] \\ &= (r + s)^2 - 4rs\cos^2(\theta/2) \end{aligned}$$

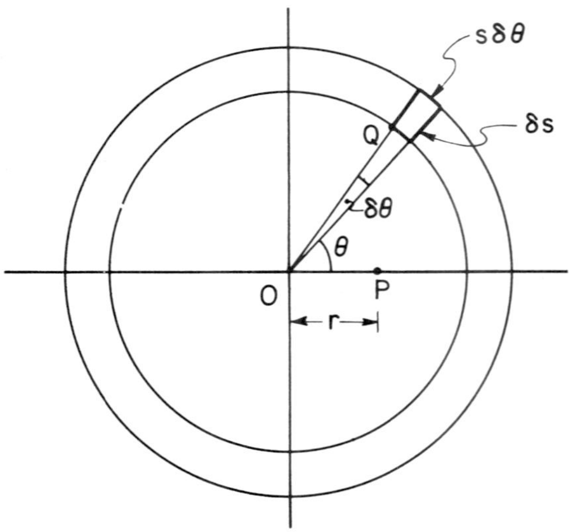

Figure 2.6.1 - The distance PQ *is* $(r^2+s^2-2rs\cos\theta)^{1/2}$.

so that

$$\int_0^{2\pi} (r^2+s^2-2rs\cos\theta)^{-1/2} d\theta = \frac{4}{(r+s)}\int_0^{\pi/2} \left[1 - \frac{4rs\cos^2\phi}{(r+s)^2}\right]^{-1/2} d\phi$$

and

$$w(r) = \frac{\vartheta}{2\pi}\int_0^a \frac{4s}{(r+s)} K\left(\frac{2(rs)^{1/2}}{r+s}\right) p(s)\,ds \tag{2.6.2}$$

where $K(\cdot)$ denotes the complete elliptic integral of the first kind.

If, in addition, there is a distribution of shearing stress τ_{rz} on $z = 0$, so that

$$\tau_{rz}(r,0) = -t(r), \quad \tau_{zz}(r,0) = -p(s), \quad r < a, \tag{2.6.3}$$

then

$$u_r(r) \equiv u(r) = \int_0^a s k_{11}(r,s) t(s)\,ds + \int_0^a s k_{13}(r,s) p(s)\,ds, \tag{2.6.4}$$

$$u_z(r) \equiv w(r) = \int_0^a s k_{31}(r,s) t(s)\,ds + \int_0^a s k_{33}(r,s) p(s)\,ds, \tag{2.6.5}$$

where the values of the k_{ij} for $r < s$ are given by

$$k_{11}(r,s) = \frac{2\vartheta}{\pi r}\{K(r/s) - E(r/s)\}, \tag{2.6.6}$$

$$k_{13}(r,s) = 0, \quad k_{31}(r,s) = -\vartheta r/s, \tag{2.6.7}$$

$$k_{33}(r,s) = \frac{2\vartheta}{\pi s} K(r/s), \tag{2.6.8}$$

while these for $r > s$ may be found by using the relation

$$k_{ij}(r,s) = k_{ji}(s,r). \tag{2.6.9}$$

$E(\cdot)$ denotes the complete elliptic integral of the second kind. The equivalence of the two expressions for k_{33} given in (2.6.2) and (2.6.8) is the subject of Ex. 2.6.5. Alternative derivations of these results may be found in Noble and Spence (1971) and Spence (1975a).

Two important consequences of equations (2.6.4),(2.6.5) are that

a) an axisymmetric shear stress distribution over $s \le a$ produces no surface normal displacement outside $r = a$

b) an axisymmetric normal stress distribution over $s \le a$ produces a surface radial displacement outside $r = a$ which is proportional to $1/r$. (See Ex. 2.6.7).

Klubin (1952) and Popov (1962) have shown that if the normal stress distribution $p(s)$ is expressed in terms of modified Legendre polynomials then the integral in (2.6.1) may be evaluated in closed form. This is a particular case of a very general result which is proved in §6.11. The Legendre polynomials $P_n(x)$ are defined by the initial values

$$P_0(x) = 1, \qquad P_1(x) = x, \tag{2.6.10}$$

and the recurrence relation

$$(n+1)P_{n+1}(x) = (2n+1)xP_n(x) - nP_{n-1}(x), \tag{2.6.11}$$

and Popov's result is simply that if the pressure distribution

$$p(r) = (1-r^2/a^2)^{-1/2}P_{2n}[(1-r^2/a^2)^{1/2}] \tag{2.6.12}$$

is applied over $r \le a$ then the displacement $w(r)$, written $r \le a$, is given by

$$w(r) = \frac{\pi}{2}\vartheta a[P_{2n}(0)]^2 P_{2n}[(1-r^2/a^2)^{1/2}], \tag{2.6.13}$$

where

$$P_{2n}(0) = (-)^n(2n)!/[2^{2n}(n!)^2]. \tag{2.6.14}$$

This may be used to give the distribution of normal loading over the circle $r \le a$ which will yield a specified displacement

$$w(r) = \frac{\pi}{2}\vartheta a \sum_{n=0}^{N} a_n [P_{2n}(0)]^2 P_{2n}[(1-r^2/a^2)^{1/2}] \tag{2.6.15}$$

over $r \leq a$; the pressure is

$$p(r) = (1-r^2/a^2)^{-1/2} \sum_{n=0}^{N} a_n P_{2n}[(1-r^2/a^2)^{1/2}], \tag{2.6.16}$$

while the total load is

$$P = 2\pi \int_0^a sp(s)ds = 2\pi a^2 a_0, \tag{2.6.17}$$

where the orthogonality condition in Ex. 2.6.6 has been used.

For an example consider the incomplete contact of a rigid spherical cap pressed against the half-space by a normal force P If the sphere has radius R and is in contact with the half-space over a circle of radius a (unknown at first), as shown in Figure 2.6.2, then

$$w(r) = \delta - r^2/2R, \qquad r \leq a. \tag{2.6.18}$$

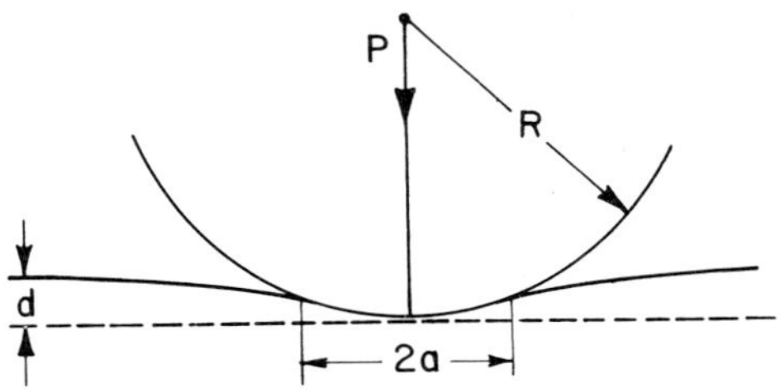

Figure 2.6.2 - Spherical punch pressed into a half-space.

Writing

$$w(r) = \frac{\pi}{2}\vartheta a\{a_0[P_0(0)]^2 + a_1[P_2(0)]^2 P_2[(1-r^2/a^2)^{1/2}]\}$$

$$= \frac{\pi}{2}\vartheta a\{a_0 + \frac{a_1}{4}[3(1-r^2/a^2) - 1]/2\},$$

we find

$$\frac{3\pi\vartheta a_1}{16} = \frac{a}{2R}, \qquad \frac{\pi\vartheta a_0}{2} = \frac{\delta}{a} - \frac{a}{3R}.$$

The radius of contact is determined by the condition $p(a) = 0$, i.e.

$$\sum_{n=0}^{N} a_n P_{2n}(0) = 0. \tag{2.6.19}$$

In the example then $2a_0 - a_1 = 0$ or $\delta = a^2/R$, giving $\pi\vartheta a_0/2 = 2a/(3R)$ and the total load

$$P = 8a^3/(3\vartheta R) \tag{2.6.20}$$

which is the known result for Hertzian contact. The pressure distribution is

$$p(r) = 3P(a^2-r^2)^{1/2}/(2\pi a^3). \tag{2.6.21}$$

England and Shail (1977) showed that the displacement on the surface $z = 0$ outside $r = a$ can be expressed in terms of the associated Legendre functions Q_n of the second kind, and that the displacement corresponding to (2.6.15) and (2.6.16) is (see §10.5)

$$w(r) = \vartheta a i \sum_{n=0}^{N} a_n [P_{2n}(0)]^2 Q_{2n}[i(r^2/a^2-1)^{1/2}], \quad r > a \tag{2.6.22}$$

where iQ_{2n} may be computed from

$$iQ_{2n}(i\xi) = (-)^n \xi \int_0^1 \frac{x^{2n}(1-x^2)^n}{(\xi^2+x^2)^{n+1}}\, dx. \tag{2.6.23}$$

In particular we note that

$$iQ_0[i(r^2/a^2-1)^{1/2}] = \sin^{-1}(a/r) \tag{2.6.24}$$

so that the first terms of (2.6.15),(2.6.22) yield the familiar displacement

$$w(r) = \begin{cases} 1, & r \le a \\ \frac{2}{\pi}\sin^{-1}(a/r), & r > a \end{cases} \tag{2.6.25}$$

for a flat-ended circular punch.

Examples 2.6

1. By using (2.6.4) and (2.6.8) show that the displacement due to a constant load $p = P/(\pi a^2)$ applied over a circle of radius a is given by (Terazawa (1916))

$$\frac{\pi^2 a w(r)}{2\vartheta P} = \begin{cases} E(r/a), & r < a \\ (r/a)[E(a/r) - (1-a^2/r^2)K(a/r)], & r > a. \end{cases}$$

Note that the displacement is continuous at $r = a$, but that the derivative $\partial w/\partial r$ is infinite there.

2. Define the quantity $q = aw/(\vartheta P)$ corresponding to the displacement due to a total load P. For the flat-ended punch show that $q = 1/4$, and for the distribution of Ex. 2.6.1,

$$q(0) = 1/\pi, \qquad q(a) = 2/\pi^2, \qquad q_{mean} = 2/(3\pi).$$

3. Use Betti's reciprocal relation (§1.7) to prove the reciprocity relation (2.6.9).

4. Use the identity

$$\frac{-s(r-s\cos\theta)}{r^2+s^2-2rs\cos\theta} = -\frac{s}{2r} + \frac{s\exp(i\theta)}{2r}\left\{\frac{1}{\exp(i\theta)-r/s} - \frac{1}{\exp(i\theta)-s/r}\right\}$$

to derive the expressions for k_{13}, k_{31} given in (2.6.7).

5. Show the equivalence of the expressions for k_{33} given in (2.6.2) and (2.6.8).

6. Starting from the orthogonality condition

$$\int_{-1}^{1} P_n(x)P_m(x)dx = 2\delta_{nm}/(2n+1)$$

for the Legendre polynomials, deduce that

$$a^{-2}\int_{-1}^{1} r(1-r^2/a^2)^{-1/2}P_{2n}[(1-r^2/a^2)^{1/2}]P_{2m}[(1-r^2/a^2)^{1/2}]dr$$
$$= \delta_{nm}/(4n+1),$$

and hence deduce (2.6.17).

7. Show that the radial displacement due to a force P uniformly distributed over a circle of radius a is

$$u_r = \frac{-\vartheta\varphi Pr}{2\pi a^2}, \quad r < a; \qquad u_r = \frac{-\vartheta\varphi P}{2\pi r}, \quad r > a.$$

2.7 *Loading applied over an elliptical region*

The results of §2.5 allow one to write down, in integral form, the displacement field due to an arbitrary distribution of normal and shearing stress on part of the boundary of the half-space. In this section we shall *state* the more important simple results relating to some particular distributions applied over an elliptical region. The derivation of the results will be deferred until §12.2 when the Fourier transform technique will enable us to derive some much more general results.

Equation (2.5.30) shows that the normal displacement $w(x,y,0) \equiv w(x,y)$ on $z = 0$ due to a distribution of pressure $p(x',y')$ applied over a region S of the plane $z = 0$ is

$$w(x,y) = \frac{\vartheta}{2\pi}\int_S \frac{p(x',y')dS'}{R} \tag{2.7.1}$$

where $R^2 = (x-x')^2 + (y-y')^2$ and

$$\vartheta = (1 - \nu)/\mu. \tag{2.7.2}$$

The principal results are as follows. Let S be the interior of the ellipse $L \equiv (1 - x^2/a^2 - y^2/b^2)^{1/2} = 0$, and write $L' \equiv (1 - x'^2/a^2 - y'^2/b^2)^{1/2}$. The normal displacement w *at a point inside* S due to a total load P_0 distributed over S according to

$$p(x',y') = \frac{P_0}{2\pi abL'} \tag{2.7.3}$$

is constant, and has the value

$$w(x,y) = \vartheta P_0 K(e)/(2\pi a), \tag{2.7.4}$$

where $e = (1 - b^2/a^2)^{1/2}$ and $K(e)$ is the complete elliptic integral of the first kind. This result shows that the distribution (2.7.3) is that corresponding to a flat punch with elliptical cross section pressed, without friction, against the half-space.

The generalization of equation (2.6.18) is as follows: if

$$p(x',y') = p_0 L' \tag{2.7.5}$$

then inside S

$$w(x,y) = \frac{1}{2}\vartheta b p_0\{K(e) - D(e)x^2/a^2 - B(e)y^2/b^2\}, \tag{2.7.6}$$

where

$$D(e) = \frac{K(e)-E(e)}{e^2}, \qquad B(e) = \frac{E(e)-(1-e^2)K(e)}{e^2}, \tag{2.7.7}$$

and $E(e)$ is the complete elliptic integral of the second kind. The limiting values of $D(e),B(e)$ as $e \to 0$ are $D(0) = B(0) = \pi/4$.

One further result is that the displacement corresponding to a tilted punch with plane face, namely

$$w(x,y) = \delta + \alpha x + \beta y \quad \text{inside } S, \tag{2.7.8}$$

is associated with the stress distribution

$$p(x',y') = \frac{1}{\vartheta b L'}\left\{\frac{\delta}{K(e)} + \frac{\alpha x'}{D(e)} + \frac{\beta y'}{B(e)}\right\}. \tag{2.7.9}$$

For a circular punch this reduces to

$$p(x',y') = \frac{2(a^2-r^2)^{-1/2}}{\pi\vartheta}\{\delta + 2\alpha x' + 2\beta y'\}. \tag{2.7.10}$$

Note that if the punch is to remain in contact with the half-space over the whole punch face, i.e. if $p(x',y')$ is to be positive throughout S, then δ must be sufficiently large compared to α,β to ensure this (Ex. 2.7.1).

Examples 2.7

1. Show that if the circular punch under the loading (2.7.10) is to remain in contact with the half-space everywhere over

the circle of radius a then $\delta > a(\alpha^2 + \beta^2)^{1/2}$.

2.8 *The use of Betti's reciprocal theorem*

For static problems in the absence of body forces, Betti's theorem (1.7.8) states that

$$\int_S \vec{T}_n \cdot \vec{d}^* dS = \int_S \vec{T}^*_n \cdot \vec{d} dS. \tag{2.8.1}$$

We recall that the theorem relates to two elastic states of the body; the displacement $\vec{d}$ and surface fraction $\vec{T}_n$ correspond to the first state, $\vec{d}^*$, $\vec{T}^*_n$ to the second. The theorem may be applied to the elastic half-space if both fields satisfy $|\vec{d}| = 0(1/r)$ and $|\vec{T}_n| = 0(1/r^2)$ at infinity, and in this case S may be interpreted as the boundary plane $z = 0$. It may be proved also that the theorem holds even if the stresses have the type of square root singularity at the edge of the loaded region exemplified by equation (2.7.3).

Following Mossakovskii (1953), Galin (1961, p. 183 et seq) and Shield (1967) we show that Betti's theorem may be used to give important information concerning the elastic field produced by a rigid punch of elliptical cross-section, but arbitrary profile, when it is pushed into an elastic half-space in the absence of friction. The results were obtained also by Galin (1948) and Mossakovskii (1951) by using ellipsoidal coordinates; see §12.3 and Galin (1961, p. 183 et seq.).

Suppose that the profile of the punch is given by $z = f(x,y)$; then when it is pushed into the elastic half-space by a normal force P the equilibrium field on the surface will satisfy

$$w(x,y) = \delta + f(x,y) \quad \text{within } S, \quad \tau_{zz} = 0 \quad \text{outside } S \tag{2.8.2}$$

$$\tau_{xz} = 0 = \tau_{yz} \quad \text{for } z = 0 \tag{2.8.3}$$

where S is the area of contact, and δ is the depth of penetration.

For the starred field we take the dimensionless form of that for a flat punch of elliptical cross section, for which

$$w^*(x,y) = 1 \quad \text{within } S, \quad \tau^*_{zz} = 0 \quad \text{outside } S, \tag{2.8.4}$$

$$\tau^*_{zz}(x',y') = -1/[\vartheta b K(e) L'] \quad \text{within } S, \tag{2.8.5}$$

$$\tau^*_{xz} = 0 = \tau^*_{yz} \quad \text{for } z = 0. \tag{2.8.6}$$

Then, applying the reciprocal theorem, we find that the total force applied by the (curved) punch is

$$P = -\int_S \tau_{zz} \cdot 1 dS = -\int_S \tau^*_{zz} \cdot w dS$$

$$= \frac{1}{\vartheta b K(e)} \int_S \left\{ \frac{\delta + f(x,y)}{L} \right\} dS,$$

or $$P = \frac{2\pi a \delta}{\vartheta K(e)} + \frac{1}{\vartheta b K(e)} \int_S \frac{f(x,y) dS}{L} . \tag{2.8.7}$$

This equation determines P (or δ) when δ (or P) is known.

If the punch is tilted, then the displacement over the contact ellipse will have the form

$$w(x,y) = \delta + \alpha x + \beta y + f(x,y), \tag{2.8.8}$$

where α,β are the inclinations of the punch with respect to the x- and y-axes. The relations between α,β and the applied moments may be found by using the fields corresponding to the tilted flat punch. Thus we introduce

$$w^o(x,y) = x \quad \text{within } S, \quad \tau^o_{zz} = 0 \quad \text{outside } S, \tag{2.8.9}$$

$$\tau^o_{zz}(x',y') = \frac{-x'}{\vartheta b D(e) L'} \quad \text{within } S, \tag{2.8.10}$$

$$\tau^o_{xz} = 0 = \tau^o_{yz} \quad \text{on } z = 0, \tag{2.8.11}$$

and find

$$M_2 = -\int_S \tau_{zz}\cdot x dS = -\int_S \tau^o_{zz}(\delta + \alpha x + \beta y + f(x,y))dS$$

$$= \frac{1}{\vartheta bD(e)}\int_S \frac{x}{L}(\delta + \alpha x + \beta y + f(x,y))dS. \qquad (2.8.12)$$

Because of the symmetry of the ellipse, two of the integrals are zero, leaving

$$M_2 = \frac{1}{\vartheta bD(e)}\int_S \frac{\{\alpha x^2 + xf(x,y)\}dS}{L},$$

$$M_2 = \frac{2\pi a^2}{3\vartheta D(e)}\alpha + \frac{1}{\vartheta bD(e)}\int_S \frac{xf(x,y)dS}{L}. \qquad (2.8.13)$$

Similarly if

$$M_1 = -\int_S \tau_{zz}\cdot y dS, \qquad (2.8.14)$$

then

$$M_1 = \frac{2\pi b^2}{3\vartheta B(e)}\beta + \frac{1}{\vartheta bB(e)}\int_S \frac{yf(x,y)dS}{L}. \qquad (2.8.15)$$

These equations allow M_1, M_2 (or α,β) to be determined from α,β (or M_1, M_2).

An example of equation (2.8.7) is provided by a smooth rigid punch in the form of a sphere of radius R. Then, approximately,

$$f(x,y) = f(r) = -r^2/(2R), \qquad (2.8.16)$$

and equation (2.8.7) gives

$$P = 4\vartheta^{-1}(a\delta - a^3/(3R)). \qquad (2.8.17)$$

Equation (2.8.7) may be used to find the radius of the contact area between an axially symmetric punch which, like a sphere or a paraboloid, has no abrupt edge. The radius, a, of the contact is determined by the condition that

$$\tau_{zz}(a,0) = -p(a) = 0. \qquad (2.8.18)$$

For an axisymmetric punch with contact radius t the total load P(t) is

$$P(t) = 2\pi \int_0^t rp(r)dr, \qquad (2.8.19)$$

so that $P'(t) = 2\pi tp(t)$ and $p(a) = 0$ implies

$$P'(a) = 0. \qquad (2.8.20)$$

For the spherical indenter, equation (2.8.17) gives

$$P'(a) = 4\vartheta^{-1}(\delta - a^2/R),$$

giving $a = (\delta R)^{1/2}$, in agreement with (2.6.19), and

$$P = 8a^3/(3\vartheta R). \qquad (2.8.21)$$

The depth of penetration in the general axisymmetric case may be deduced from (2.8.7) which is now

$$P(t) = \frac{4}{\vartheta}\left[t\delta + \int_0^t \frac{rf(r)dr}{(t^2-r^2)^{1/2}}\right]. \qquad (2.8.22)$$

Since $f(0) = 0$ the second term may be integrated by parts to give

$$P(t) = \frac{4}{\vartheta}\left[t\delta + \int_0^t (t^2-r^2)^{1/2} f'(r)dr\right] \qquad (2.8.23)$$

so that

$$P'(t) = \frac{4}{\vartheta}\left[\delta + t\int_0^t (t^2-r^2)^{-1/2} f'(r)dr\right]$$

and

$$\delta = -a\int_0^a (a^2-r^2)^{-1/2} f'(r)dr. \qquad (2.8.24)$$

Finally, we apply the reciprocal theorem to estimate the displacement produced by a normal force P in a punch with arbitrary cross-section and plane face (Galin (1961, p. 192)). Suppose the punch occupies the region E of the plane $z = 0$. Describe an ellipse about E, denote the interior of the ellipse by S, and let the portion of S not occupied by E be F. Then the displacement on $z = 0$ due to the punch will satisfy

$$w(x,y) = \delta \quad \text{in } E, \qquad w(x,y) = g(x,y) < \delta \quad \text{in } F$$

$$\tau_{zz} = 0 \quad \text{outside } E.$$

Apply the reciprocal theorem to this field and the field (2.8.4). Then

$$P = -\int_E \tau_{zz}\cdot 1dS = -\int_S \tau_{zz}\cdot 1dS = -\int_S \tau^*_{zz}\cdot wdS$$

$$= \frac{1}{\vartheta bK(e)}\int_S \frac{wdS}{L} < \frac{\delta}{\vartheta bK(e)}\int_S \frac{dS}{L} = \frac{2\pi a\delta}{\vartheta K(e)},$$

so that

$$\frac{P}{\delta} < \frac{2\pi a}{\vartheta K(e)}, \tag{2.8.25}$$

where a,e refer to any ellipse described about the punch cross-section.

By using an argument due to Pólya and Szegö (1945), it is possible to prove that a punch with plane base and arbitrary cross-section has, for a given force, a smaller displacement than a punch with a circular cross-section of the same area. Thus, using equation (2.7.31), we may deduce that for a punch of area A

$$\frac{P}{\delta} > \frac{4}{\vartheta}\left(\frac{A}{\pi}\right)^{1/2}. \tag{2.8.26}$$

Equations (2.8.25) and (2.8.26) provide upper and lower bounds for P/δ.

2.9 *Hertz's theory of contact*

Hertz (1882a), (1882b) was the first to formulate and solve problems of contact between two elastic bodies. The theory was expanded by Boussinesq (see Todhunter and Pearson (1893, p. 257 et seq.)) and has been specialized to important applications by numerous writers, particularly Russian ones. An extended account with bibliographical references is given by Lur'e (1964, Ch. V).

We suppose that, in the unstressed state, the bodies are in contact at 0, and that, near 0, the bodies are goemetrically *smooth*, as in Figure 2.9.1. Then taking the common tangent plane at 0 as x-y plane, we may express the shapes of the two bodies near 0 in the form

$$\left.\begin{aligned} z_1 &= A_1x^2 + B_1y^2 + 2H_1xy, \\ z_2 &= A_2x^2 + B_2y^2 + 2H_2xy, \end{aligned}\right\} \qquad (2.9.1)$$

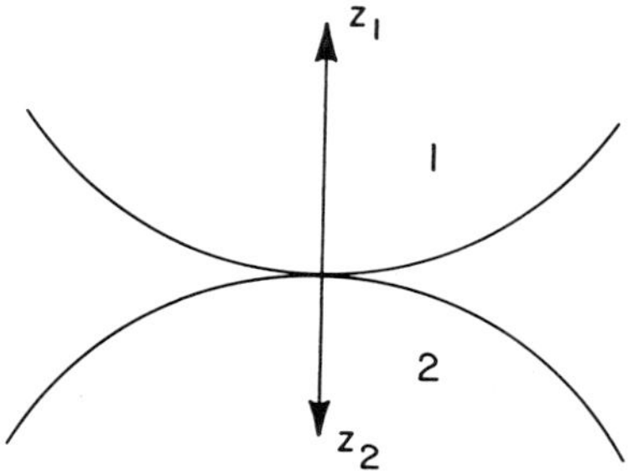

Figure 2.9.1 - Axes used in Hertzian contact.

where z_1, z_2 are directed into the bodies. The distance between two points in the unstressed state, measured along the common normal, is

$$z_1 + z_2 = (A_1+A_2)x^2 + 2(H_1+H_2)xy + (B_1+B_2)y^2.$$

This must be positive whichever way the axes of x and y are chosen, and we may choose them so that $H_1 + H_2 = 0$. Then, in the unstressed state,

$$z_1 + z_2 = Ax^2 + By^2,$$

where $A = A_1 + A_2$, $B = B_1 + B_2$. When the bodies are pressed together there will be displacement of both. Take the displacements of body 1 to be (u_1,v_1,w_1) relative to (x,y,z_1) and that of body 2 to be (u_2,v_2,w_2) relative to (x,y,z_2). Within the area of contact, therefore,

$$z_1 + w_1 = -(z_2 + w_2) + \delta,$$

where δ is the *approach* of the two bodies. Therefore, within the contact area S,

$$w_1 + w_2 = \delta - z_1 - z_2 = \delta - Ax^2 - By^2. \tag{2.9.2}$$

Outisde the contact area

$$w_1 + w_2 > \delta - Ax^2 - By^2, \tag{2.9.3}$$

in order that the surfaces should be separated from each other.

Now assume that the two bodies can be replaced by two elastic half-spaces which are in frictionless contact over the area S in the common tangent plane. Torvik (1967) notes that the errors incurred by assuming that the spheres are half-spaces are smaller than those due to assuming infinitesimal strains. It was shown in §2.7 that a displacement of the form (2.9.1) may be formed in an elliptical area S on the surface of a half-space by a pressure distribution $p(x',y') = p_0(1 - x'^2/a^2 - y'^2/b^2)^{1/2}$ acting over S. We therefore assume that S is the ellipse $x^2/a^2 + y^2/b^2 = 1$ and that the pressure has this form. Note that Lur'e wrongly, and most misleadingly, states [equation (5.11.25)] that the displacement $w_1 + w_2$ is constant on the boundary of the contact region. This is so only when the region is circular. The displacements w_1,w_2 in S will be given by

$$2\pi w_1 = \vartheta_1\int_S \frac{p\,dS'}{R}, \qquad 2\pi w_2 = \vartheta_2\int_S \frac{p\,dS'}{R}, \tag{2.9.4}$$

where $\vartheta_i = (1 - \nu_i)/\mu_i$ and ν_i,μ_i are the elastic constants of body i, and $R^2 = (x - x')^2 + (y - y')^2$. Combining equations (2.9.2), (2.9.4) we find that

$$w_1 + w_2 = \frac{\vartheta_1+\vartheta_2}{2\pi}\int_S \frac{pdS'}{R} = \delta - Ax^2 - By^2 \quad \text{in } S. \tag{2.9.5}$$

Using the value of the integral given in equation (2.7.6) we find

$$\frac{1}{2}(\vartheta_1+\vartheta_2)p_0 b\left\{K(e) - D(e)\frac{x^2}{a^2} - B(e)\frac{y^2}{b^2}\right\}$$

$$\equiv \delta - Ax^2 - By^2 \quad x,y \text{ in } S. \tag{2.9.6}$$

Comparing coefficients of x^2 and y^2 we find

$$\frac{(1-e^2)D(e)}{B(e)} = \frac{A}{B}, \tag{2.9.7}$$

from which e may be computed. Once e is known, a and δ may be expressed in terms of the parameters

$$\alpha_a = [3D(e)]^{1/3}, \qquad \alpha_\delta = \left[\frac{9}{8D(e)}\right]^{1/3} K(e), \tag{2.9.8}$$

and the total contact force

$$P = 2\pi ab p_0/3. \tag{2.9.9}$$

They are

$$a = [(\vartheta_1+\vartheta_2)P/(4\pi A)]^{1/3}\alpha_a, \tag{2.9.10}$$

$$\delta = [(\vartheta_1+\vartheta_2)P(2A)^{1/2}/(2\pi)]^{2/3}\alpha_\delta. \tag{2.9.11}$$

Table 2.9.1 presents the values of $e, \alpha_a, \alpha_\delta$ for values of A:B ranging from 0.1 to 1.

In order to verify that the separation condition (2.9.2) is satisfied we need to be able to evaluate the integral (2.9.4) for (x,y) outside S; this will be deferred to §12.3.

Hertz's analysis is based upon the assumption that $z_1 + z_2$ has the form $Ax^2 + By^2$. There are a number of important cases in which this assumption is not valid. One of these, in which the surfaces have curvature in one plane only, is considered in Ex. 2.9.2. Here the expression for δ, the approach of the bodies, breaks down. The plane strain problem of two parallel cylinders pressed together is considered in §8.4; analytically, it is more difficult than the Hertz problem. Shtaerman (1939),(1940),(1941), (1943) considers a number of other special cases, e.g. when $z_1 + z_2$ has the form $Ax^4 + By^4$. Cooper (1969) gives design data on Hertzian contact coefficients.

Table 2.9.1

A/B	e	α_a	α_δ
0.1	0.9749	1.8017	2.4252
0.2	0.9382	1.6726	2.2362
0.3	0.8928	1.5911	2.1218
0.4	0.8390	1.5309	2.0392
0.5	0.7761	1.4831	1.9744
0.6	0.7025	1.4434	1.9211
0.7	0.6150	1.4096	1.8758
0.8	0.5072	1.3801	1.8365
0.9	0.3620	1.3540	1.8018

Examples 2.9

1. Show that if R_i, R_i' are the principal radii of curvature of body i at 0 (i = 1,2), and ψ is the angle between the normal planes containing the curvatures $1/R_1$ and $1/R_2$ then the constants A,B of (2.9.1) are given by

$$A + B = \frac{1}{2}(S_1 + S_2), \quad B - A = \frac{1}{2}(D_1^2 + D_2^2 + 2D_1D_2 \cos 2\psi)^{1/2}$$

where $S_i = 1/R_i + 1/R_i'$, $D_i = 1/R_i - 1/R_i'$.

2. Consider the limiting case of Hertz's solution when the bodies are curved in one direction only, so that $1/R_1' \to 0$, $1/R_2' \to 0$. Show that $A = 0$, $a/b \to \infty$ and that if P_0 is defined to be the load in the plane $x = 0$ so that $P_0 = \pi b p_0/2$ then b is equal to the value of a given by (2.2.38). See

Timoshenko and Goodier (1951, p. 381). Note that δ is unbounded as $a/b \to \infty$.

This limiting case was the subject of a long controversy in the Soviet Union. In 1909 Dinnik (1952) considered the contact of two parallel cylinders in contact along their common generators by using this limit. He correctly obtained the contact width (2b) and the maximum contact stress. The approach δ he found to be unbounded. Belyaev (1957) questioned his procedure, but it was vindicated by Krolevets (1966). The 'approach' of two cylinders must be found by using a procedure such as that discussed in §8.4. (See (8.4.35)-(8.4.37)). Further references may be found in Kil'chevskii and Kostyuk (1966); the issue of the journal containing this paper is largely denoted to Dinnik.

3. Show that the contact length 2a and the maximum pressure p_0 when an elastic cylinder of radius R is pressed against a rigid flat plate are given by

$$\frac{2a}{R} = 2\left(\frac{2\vartheta P_0}{\pi R}\right)^{1/2}, \qquad \vartheta p_0 = \left(\frac{2\vartheta P_0}{\pi R}\right)^{1/2}.$$

4. Show that the radius a of the circle of contact and the approach δ for two spheres of radii R_1, R_2 are given by

$$a = \left[\frac{3}{8}\frac{(\vartheta_1+\vartheta_2)P}{(1/R_1+1/R_2)}\right]^{1/3}, \qquad \delta = \frac{3}{8}\frac{P(\vartheta_1+\vartheta_2)}{a}.$$

2.10 *Beams on simple foundations*

The classical theory of beams on elastic foundations, as described, for instance, in all but the last chapter of the treatise of Hetenyi (1946), is based on the two fundamental assumptions:

i) The reactive pressure exerted by the foundation is proportional at every point to the deflection v of the beam at that point, i.e.

$p(x) = kv(x)$.

Such a foundation, introduced by Winkler (1867) and Zimmermann (1888), is equivalent to a set of closely spaced independent springs; a liquid foundation; a so-called 'Gibson' soil. The latter is an incompressible medium ($\nu = 1/2$) with a shear modulus μ varying linearly with depth, i.e. $\mu = mz$. This, and other non-homogeneous media, are considered in §6.10.

ii) The beam everywhere remains in contact with the foundation.

In this section we outline just enough of the classical theory so that we may describe some recent attempts at relaxing these assumptions. For results regarding the behaviour of beams under particular loading distributions the reader is referred to Hetenyi (1946). Korenev (1960b) reviews the Soviet literature for the period 1917-1957. To enter into the literature on the modelling of civil engineering foundations the reader may consult Hetenyi (1964), Gibson (1974), Poulos and Davis (1974) and Selvadurai (1979).

Consider a straight uniform beam of flexural rigidity EI resting on a Winkler foundation, as shown in Figure 2.10.1. The differential equation governing small bending deflections of the beam under a distributed applied load $q(x)$ and a concentrated force P applied at the origin is

$$EI\,\frac{d^4v}{dx^4} = -p(x) + q(x) + P\delta(x), \tag{2.10.1}$$

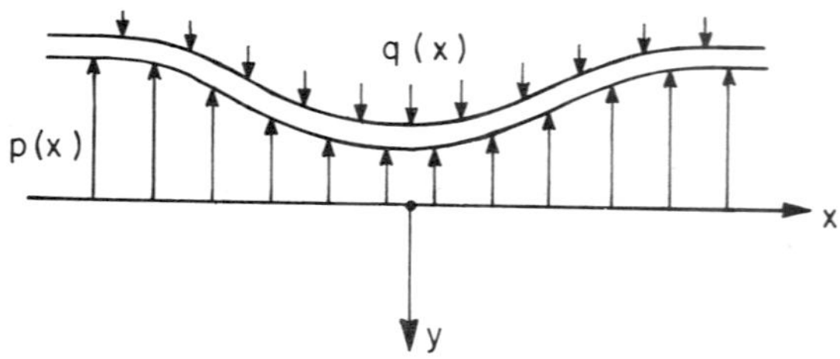

Figure 2.10.1 - Beam resting on a simple foundation.

where δ is the Dirac delta function. Denote

$$\gamma = [k/(4EI)]^{1/4}, \tag{2.10.2}$$

and introduce dimensionless quantities

$$\xi = \gamma x, \qquad u = \gamma v, \qquad \sigma(\xi) = q(x)/(4\gamma^3 EI),$$
$$F = P/(4\gamma^2 EI). \tag{2.10.3}$$

Then equation (2.10.1) becomes

$$\frac{1}{4} v^{(4)}(\xi) = -u(\xi) + \sigma(\xi) + F\delta(\xi), \tag{2.10.4}$$

where $v^{(n)}(\xi)$ denotes $d^n v/d\xi^n$. The general solution of the homogeneous equation, with $\sigma(\xi) \equiv 0 \equiv F$ is

$$u(\xi) = \exp(-\xi)(A \cos \xi + B \sin \xi) +$$
$$+ \exp(\xi)(C \cos \xi + D \sin \xi), \tag{2.10.5}$$

where A,B,C,D are arbitrary constants. For a beam of infinite extent, loaded by a concentrated force at the origin, the deflection $u(\xi)$ will be even in ξ and, for $\xi > 0$, will be given by

$$u(\xi) = \exp(-\xi)(A \cos \xi + B \sin \xi), \quad \xi \geq 0, \tag{2.10.6}$$

where A,B are determined from the conditions

$$u^{(1)}(0+) = 0, \qquad u^{(3)}(0+) = 2F, \tag{2.10.7}$$

and therefore

$$u(\xi) = (F/2)\exp(-\xi)(\cos \xi + \sin \xi), \ \xi > 0. \tag{2.10.8}$$

For a beam under the combined effect of a concentrated load F and a uniformly distributed load σ, which may be identified with the weight of the beam, the deflection is

$$u(\xi) = \sigma + (F/2)\exp(-\xi)(\cos \xi + \sin \xi), \quad \xi \geq 0. \tag{2.10.9}$$

For tabulated results relating to such a beam, refer to Hetenyi (1946).

If σ, F are chosen arbitrarily, this displacement will assume both positive and negative values, i.e. the reaction exerted by the foundations will be somewhere compressive and elsewhere tensile. Some foundations can react in compression only and, for such, if $u(\xi)$ becomes negative at some point, the beam will simply lose contact with the foundation and there will be no foundation pressure at that point. Such foundations have been discussed by Schindler (1956), Tsai and Westmann (1967) and Weitsman (1972). Provided $|F|$ is small compared to σ (assumed positive), $u(\xi)$ will be everywhere positive; as $|F|$ is increased, contact will be lost with the foundation first when the minimum value of $u(\xi)$ is zero. If this minimum occurs at $\hat{\xi}$ then

$$\exp(-\hat{\xi})(\cos \hat{\xi} + \sin \hat{\xi}) = -2\sigma/F, \tag{2.10.10}$$

$$\sin \hat{\xi} = 0. \tag{2.10.11}$$

Equation (2.10.11) yields $\hat{\xi} = n\pi$ while (2.10.10) gives $F = 2\sigma(-)^{n-1}\exp(n\pi)$. If $n = 0$, so that contact is about to be lost at the origin, $F = -2\sigma$ and the load points upward as indicated by the negative sign. If $n = 1$, $\hat{\xi} = \pi$ and $F = 2\sigma \exp(\pi)$. We conclude the the condition for the beam to remain everywhere in contact with the foundation is that $-2\sigma \leq F \leq 2\sigma \exp(\pi)$ or, in terms of dimensional quantities,

$$-2 \leq P\gamma/q \leq 2 \exp(\pi). \tag{2.10.12}$$

When the concentrated load lies outside this range the contact between beam and foundation will be lost, to be regained further from the origin by the downward effect of the weight. Schindler (1956) and Weitsman (1972) analyzed such cases by dividing the beam into regions of contact and separation and obtaining a transcendental equation to determine the points of separation (Ex. 2.10.2). Tsai and Westmann (1967) set up an integral equation for the contact pressure which they solve numerically. They conclude that if $P_{crit} = 2q \exp(\pi)/\gamma$ and $0 \leq P/P_{crit} \leq 4$ the deflected shape in the neighbourhood of the origin does not differ greatly from the classical result (2.10.9). Note that for a given ratio $P\gamma/q$ the contact of the beam and the foundation is *receding* in the sense of §4.10, so that 1) the contact region is *independent of the level of the loading*, 2) the passage from zero loading to non-zero loading is *discontinuous*, 3) the contact depends on the ratio $P\gamma/q$ so that it is not possible to apply a superposition law. All these results may easily be derived directly by analyzing (2.10.9).

Kerr (1964),(1965),(1976) reviewed several attempts to introduce some continuity in the foundation without resorting to the elastic half-space model which, in any case, fails to describe adequately many actual foundations. Filonenko-Borodich (1945) assumed that a membrane, or a taut string with tension T in two dimensions, is connected to the ends of the springs so that

$$p(x) = kv - T\frac{d^2v}{dx^2}. \qquad (2.10.13)$$

Hetenyi (1946),(1950) assumed that a beam is imbedded between two layers of Winkler material. Pasternak (1954) assumed that the ends of the springs are connected to a beam consisting of incompressible vertical elements which can deform only by transverse shear. Weitsman (1970) has considered a plane version of Reissner's (1958) foundation in which there are only τ_{xx} and τ_{xy} stresses and has given a detailed solution for various cases of continuous and discontinuous contact under a combination of point and distributed loads. Sneddon, Gladwell and Coen (1975) developed a general analysis, with several particular examples, for a beam in contact with a foundation for which the pressure p(x) may be expressed in the form

$$p(x) = L\{v(t);x\}, \qquad (2.10.14)$$

where L is a linear operator.

Vlasov and Leont'ev (1966) have developed a comprehensive method, based on previous work by themselves and others, by means of which a large number of foundation models can be generated and reduced to the solution of a set of ordinary differential equations. The essentials of their method for plane problems are as follows:

1) The foundation, of finite depth H, is assumed to consist of one or more layers in which the stress-strain relations of plane strain or plane stress apply.
2) The displacements $u(x,y), v(x,y)$ in each layer are assumed to have the form

$$\left.\begin{aligned} u(x,y) &= \sum_{i=1}^{m} U_i(x)\phi_i(y), \\ v(x,y) &= \sum_{k=1}^{m} V_k(x)\psi_k(y), \end{aligned}\right\} \qquad (2.10.15)$$

where the functions $\phi_i(y), \psi_k(y)$ are known.
3) The equations governing the unknown functions $U_i(x)$, $V_k(x)$ are obtained by applying the Principle of Virtual Work to the stresses obtained from the displacements u,v and the stress-strain relations, with the virtual displacements $\phi_i(y), \psi_k(y)$. This process leads to $(m + n)$ second-order ordinary differential equations for $U_i(x)$, $V_k(x)$.

Perhaps the simplest case is that in which there is just one layer,

$$u(x,y) = 0, \qquad v(x,y) = V_1(x)\psi_1(y); \qquad (2.10.16)$$

the foundation lies on a rigid base, so that $\psi_1(H) = 0$, there is no distributed load $F_y(x,y)$, but there is an applied load $\tau_{yy}(x,0) = -p(x)$. Then there is just one equation, namely

$$\mu r_{11}V_1^{(2)}(x) - 2\mu(1-\nu)/(1-2\nu)s_{11}V_1(x) + p(x)\psi_1(0) = 0, \qquad (2.10.17)$$

where $v^{(n)}$ denotes $d^n v/dx^n$ and

$$r_{11} = \int_0^H \psi_1^2(y)dy, \qquad s_{11} = \int_0^H [\psi_1^{(1)}(y)]^2 dy. \qquad (2.10.18)$$

This foundation is called a two-parameter model (see Ting (1973a), Robinson (1975), Selvadurai (1979)). Note that foundations governed by equation (2.10.17), unlike Winkler foundations, have the property that they are able to distribute surface loads, i.e. there will be a deflection of the foundation at points outside the region in which a load is applied. The model proposed by Filonenko-Borodich (1945) may be considered to be a particular case of (2.10.17). The family of models proposed by Vlasov and Leont'ev has great versatility: different foundations may be obtained by choosing different functions $\phi_i(y), \psi_k(y)$. The models proposed by Pasternak (1954), Hetenyi (1946) and Weitsman (1970) mentioned earlier may be considered as special cases of the general theory. They suggest $\psi_1(y) = (H - y)/H$ for a shallow foundation and $\psi_1^*(y) = \sinh\gamma(H - y)/\sinh\gamma H$ for a deep foundation. Here γ is a coefficient depending on the elastic properties of the foundation and determining the rate of decrease of the displacements over the depth of the foundation. The model with $\psi_1(y)$ will have a normal stress τ_{yy} which does not vary with depth while the stress derived from $\psi_1^*(y)$ behaves as $\cosh\gamma(H - y)$; the latter model can be used in the limit $H \to \infty$ to give a simple approximate model for an elastic half-space (Ex. 2.10.3).

The equation governing the deflection of a beam on a two-parameter model may be obtained by combining equations (2.10.1) and (2.10.17). If $\psi_1(y)$ is normalized so that $\psi_1(0) = 1$, then $v_1(x)$ will be the surface deflection of the half-space and will be equal to the beam deflection in the contact region. The equation governing the deflection is therefore

$$EIv^{(4)}(x) - \mu r_{11}v^{(2)}(x) + 2\mu(1-\nu)/(1-2\nu)S_{11}v(x)$$
$$= q(x) + P\delta(x), \qquad (2.10.19)$$

which differs from the equation corresponding to a Winkler foundation by the presence of the second derivative term. Vlasov and Leont'ev use the notation

$$\frac{\mu r_{11}}{EI} = 2r^2, \qquad \frac{2\mu(1-\nu)}{EI(1-2\nu)} = s^4, \tag{2.10.20}$$

so that (2.10.19) becomes

$$v^{(4)}(x) - 2r^2 v^{(2)}(x) + s^4 v(x) = [q(x) + P\delta(x)]/EI. \tag{2.10.21}$$

The fundamental solutions of the homogeneous part of this equation take different forms depending on whether s is greater than, equal to, or less than r. In the first, and most important case, the solution is

$$v(x) = \exp(-\alpha x)(A \cos \beta x + B \sin \beta x) +$$
$$+ \exp(\alpha x)(C \cos \beta x + D \sin \beta x), \tag{2.10.22}$$

where

$$\alpha = \{(s^2+r^2)/2\}^{1/2}, \qquad \beta = \{(s^2-r^2)/2\}^{1/2}. \tag{2.10.23}$$

The analysis of the problem now proceeds very much in the same way as that for the Winkler foundation. Full details concerning infinite and finite beams, in complete contact, on various foundations are given in Vlasov and Leont'ev (1966) together with a variety of graphical results; the analysis of partial contact situations may be made along the lines described earlier in this section; see Ting (1973b).

Examples 2.10

1. Show that the dimensionless deformation u at the centre of a beam of length 2ℓ, supported by a Winkler foundation of stiffness k, symmetrically loaded with a concentrated load P at its centre $\xi = 0$, and with free ends ($u^{(2)}(\pm\lambda) = 0 = u^{(3)}(\pm\lambda)$, $\lambda = \gamma\ell$) is

$$u(0) = P\left\{\frac{4+2\cos 2\lambda+2\cosh 2\lambda-\sin 2\lambda}{2(\sin 2\lambda+\sinh 2\lambda)}\right\}.$$

2. Consider a beam supported by a Winkler foundation and subject to an (upward) load $P < -2q/\beta$ at the origin. Assume that there is a single region $|\xi| < \bar{\xi}$ of separation. Write down

the equations governing u in the two regions $|\xi| < \bar{\xi}$ and $|\xi| > \bar{\xi}$ (since the deflection will be even in ξ one may consider only $\xi > 0$) and the boundary conditions at $\xi = \bar{\xi}$. Find the lower bound of P so that there is no further separation region for $|\xi| > \bar{\xi}$.

3. Show that the constants r_{11}, s_{11} for the Vlasov-Leont'ev model defined by $\psi_1^*(y) = \sinh \gamma(H-y)/\sinh\gamma H$ are

$$r_{11} = \frac{\sinh \gamma H \cosh \gamma H - \gamma H}{2\gamma \sinh^2\gamma H}, \quad s_{11} = \frac{\gamma \sinh \gamma H \cosh \gamma H + \gamma H}{2 \sinh^2\gamma H},$$

and evaluate their limiting values as $H \to \infty$.

4. Carry out the analysis of partial contact described in (2.10.9)-(2.10.12) for the general problem (2.10.17).

CHAPTER III - FUNCTIONS OF A COMPLEX VARIABLE

"Two added to one - if that could but be done,"
It said, "with one's fingers and thumbs!"
Recollecting with tears how, in earlier years,
It had taken no pains with its sums.

- *The Hunting of the Snark*

3.1 *Introduction*

The purpose of this chapter is to derive some results which may be used in the analysis of plane contact problems. The most important relate to the theory of Cauchy integrals. Emphasis will be placed on the results rather than their derivations or the precise or most general conditions under which they hold; for such details the reader is referred to the work of Muskhelishvili (1953a),(1953b) in which references to earlier literature may be found and for which it is hoped that this chapter will be an introduction.

First we state some basic definitions referring to curves and regions lying in the complex plane. An *arc* is a continuous non-intersecting, i.e. simple, line which has a continuously varying tangent except at a finite number of points. A *contour* is a closed arc.

A *region* is an open connected set in the plane. If a region $\mathcal{D}^+$ has a boundary made up of one or more contours, the positive sense of description of each contour is taken to be that for which $\mathcal{D}^+$ lies to the left, as shown in Figure 3.1.1. The open set exterior to $\mathcal{D}^+$ and the bounding contours we denote by $\mathcal{D}^-$, so that as a boundary contour is described, $\mathcal{D}^+$ lies to the left and $\mathcal{D}^-$ to the right. The symbols S^+ and S^- will be used to denote the upper and lower half planes $y > 0$, $y < 0$ respectively, while S will be used to denote an arbitrary region or point set.

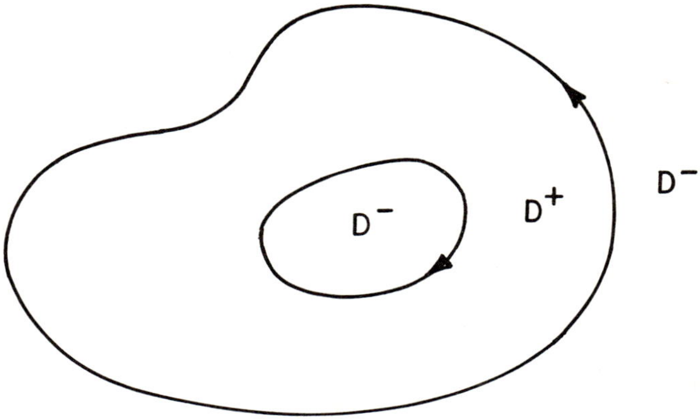

Figure 3.1.1 - The boundary is traversed so that $\mathcal{D}^+$ lies to the left.

Let S be an arbitrary point set in the complex plane. If to each point $z = x + iy$ in S there corresponds a complex number $f(z) = p(x,y) + iq(x,y)$ then $f(z)$ is said to be a complex function of the complex variable z on S. Thus z^2, $z\bar{z}$, $x + 2y$ may all be considered as complex functions of z, but the theory of functions of a complex variable is concerned with a more restricted class of functions, as will now be described.

In view of the relations

$$\left.\begin{aligned} z &= x + iy, & \bar{z} &= x - iy, \\ x &= (z+\bar{z})/2, & y &= (z-\bar{z})/(2i), \end{aligned}\right\} \tag{3.1.1}$$

we define the partial derivative operators $\partial/\partial z$, $\partial/\partial\bar{z}$ by

$$\frac{\partial}{\partial z} = \frac{1}{2}\left(\frac{\partial}{\partial x} - i\,\frac{\partial}{\partial y}\right), \qquad \frac{\partial}{\partial \bar{z}} = \frac{1}{2}\left(\frac{\partial}{\partial x} + i\,\frac{\partial}{\partial y}\right), \tag{3.1.2}$$

and then

$$\frac{\partial f}{\partial z} = \frac{1}{2}\left(\frac{\partial}{\partial x} - i\,\frac{\partial}{\partial y}\right)(p+iq) = \frac{1}{2}\left(\frac{\partial p}{\partial x} + \frac{\partial q}{\partial y}\right) + \frac{i}{2}\left(\frac{\partial q}{\partial x} - \frac{\partial p}{\partial y}\right), \tag{3.1.3}$$

$$\frac{\partial f}{\partial \bar{z}} = \frac{1}{2}\left(\frac{\partial}{\partial x} + i\frac{\partial}{\partial y}\right)(p+iq) = \frac{1}{2}\left(\frac{\partial p}{\partial x} - \frac{\partial q}{\partial y}\right) + \frac{i}{2}\left(\frac{\partial q}{\partial x} + \frac{\partial p}{\partial y}\right). \qquad (3.1.4)$$

If $\partial f/\partial \bar{z} = 0$ at a point, we see that the *Cauchy-Riemann* equations

$$\frac{\partial p}{\partial x} = \frac{\partial q}{\partial y}, \qquad \frac{\partial q}{\partial x} = -\frac{\partial p}{\partial y} \qquad (3.1.5)$$

are satisfied. It may be shown that if the first partial derivatives of p,q satisfy the Cauchy-Riemann equations and are continuous at $z = z_0$, then this is a necessary and sufficient condition for the existence of the complex derivative

$$\frac{df}{dz} \equiv f'(z) = \lim_{\delta z \to 0} \frac{f(z+\delta z)-f(z)}{\delta z} \qquad (3.1.6)$$

at $z = z_0$. Note that the limit will exist and have the same value for whichever way the complex number $z_0 + \delta z$ tends to z_0.

The function f(z) is said to be *holomorphic* (sometimes the terms *regular* or *analytic* are used) in a finite region *S* if it is single-valued in *S* and its complex derivative f'(z) exists at every point in *S*. The condition that f(z) have a complex derivative is so strong that it may be proved that if f(z) is holomorphic in *S*, then it will possess complex derivatives $f^{(n)}(z)$ of any finite order at every point in *S*, so that each such derivative will also be holomorphic in *S*. (Note the contrast with functions of a real variable where the existence of f"(x) by no means follows from the existence of f'(x)). Further, it may be expanded in a series

$$f(z) = \sum_{n=0}^{\infty} a_n (z-z_0)^n \qquad (3.1.7)$$

about any point z_0 of *S*. If the region *S* is infinite, then f(z) is said to be *holomorphic at infinity* if f(1/z) is holomorphic at the origin. This means that, for large $|z|$, f(z) may be expanded in the form

$$f(z) = \sum_{n=0}^{\infty} a_n z^{-n}. \tag{3.1.8}$$

If f(z) is holomorphic in the entire complex plane, except the point at infinity, then it must be a polynomial in z. If in addition it is holomorphic at infinity, then it must be a constant. Linked with this concept is the idea of an analytic function of the real variables x,y. A function f(x,y) is said to be analytic in a finite region S of the x-y plane if, about each point (x_0,y_0) in S, it may be expanded in a double series

$$f(x,y) = \sum_{m=0}^{\infty} \sum_{n=0}^{\infty} a_{m,n}(x-x_0)^m(y-y_0)^n. \tag{3.1.9}$$

It may be proved that if f(z) is holomorphic in S, then $f(x,y) \equiv f(z)$ is analytic in S.

It follows from equation (3.1.5) that if f(z) is holomorphic, then each of p(x,y), q(x,y) is harmonic, i.e.

$$\nabla_1^2 p(x,y) \equiv \frac{\partial^2 p}{\partial x^2} + \frac{\partial^2 p}{\partial y^2} = \frac{\partial}{\partial x}\left(\frac{\partial q}{\partial y}\right) + \frac{\partial}{\partial y}\left(-\frac{\partial q}{\partial x}\right)$$

$$= 0 = \nabla_1^2 q(x,y). \tag{3.1.10}$$

Conversely, if p(x,y) is harmonic in S, and if q(x,y) is related to p(x,y) by equations (3.1.5), then f(z) = p(x,y) + iq(x,y) is a holomorphic function of z.

3.2 *Cauchy integrals*

Let L be a simple closed contour dividing the plane into two parts, a finite part D^+ lying on the left and an infinite part D^- lying on the right, as shown in Figure 3.2.1.

Cauchy's theorem states: *If L is a simple closed contour lying wholly in a region S in which the function* f(z) *is holomorphic, then*

$$\int_L f(t)dt = 0, \tag{3.2.1}$$

where we shall use t to denote the generic point of the contour L. Apply this theorem to the function $f(z) = 1/(z - z_0)$ which is holomorphic in any region excluding z_0.

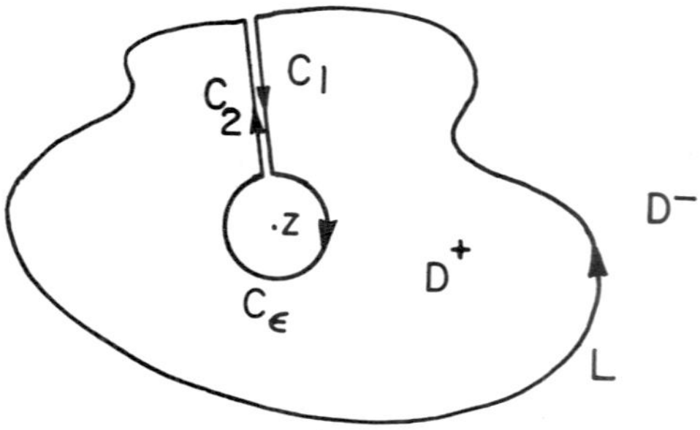

Figure 3.2.1 - The simple closed contour L *divides the plane into two parts.*

If z_0 lies outside L, i.e. in $\mathcal{D}^-$, then Cauchy's theorem gives

$$\frac{1}{2\pi i}\int_L \frac{dt}{t-z_0} = 0 \quad \text{for } z_0 \text{ in } \mathcal{D}^-. \tag{3.2.2}$$

If z_0 lies inside L, then we construct the contour $L + C_1 + C_\varepsilon + C_2$ so that z_0 again lies outside the region bounded by the contour and

$$\left\{\int_L + \int_{C_1} + \int_{C_\varepsilon} + \int_{C_2}\right\} \frac{dt}{t-z_0} = 0.$$

But the integrals along C_1, C_2 are equal and opposite and the integral around C_ε may be evaluated by writing $t = z_0 + \varepsilon \exp(i\theta)$, $dt = i\varepsilon \exp(i\theta)d\theta$ so that

$$-\int_{C_\varepsilon} \frac{dt}{t-z_0} = \int_0^{2\pi} \frac{i\varepsilon \exp(i\theta)d\theta}{\varepsilon \exp(i\theta)} = 2\pi i,$$

and therefore

$$\frac{1}{2\pi i}\int_L \frac{dt}{t-z_0} = 1 \quad \text{for } z_0 \text{ in } \mathcal{D}^+. \tag{3.2.3}$$

Now write

$$\int_L \frac{f(t)}{t-z_0}\, dt = \int_L \frac{f(t)-f(z_0)}{t-z_0}\, dt + f(z_0)\int_L \frac{dt}{t-z_0}\ . \tag{3.2.4}$$

If L is a closed contour lying in a region in which f(z) is holomorphic, then $(f(z) - f(z_0))/(z - z_0)$ will also be holomorphic so that the first integral in (3.2.4) will be zero, giving

$$\frac{1}{2\pi i}\int_L \frac{f(t)}{t-z}\, dt = \begin{cases} f(z) & \text{if } z \text{ is in } \mathcal{D}^+ \\ 0 & \text{if } z \text{ is in } \mathcal{D}^-. \end{cases} \tag{3.2.5}$$

Now let L again be a simple closed contour and let f(t) be a function given and continuous on L. The equation

$$F(z) = \frac{1}{2\pi i}\int_L \frac{f(t)}{t-z}\, dt \tag{3.2.6}$$

defines a function which may easily be shown to be holomorphic everywhere except on L. Such a function is called a *Cauchy integral*. If f(t) happens to be the boundary value of a function f(z) holomorphic in $\mathcal{D}^+$, then

$$F(z) = \begin{cases} f(z) & \text{if } z \text{ is in } \mathcal{D}^+ \\ 0 & \text{if } z \text{ is in } \mathcal{D}^-. \end{cases} \tag{3.2.7}$$

However, F(z) may be defined by (3.2.6) provided only that f(t) is continuous on L. If f(t) is the boundary value of a function holomorphic in the (infinite) region $\mathcal{D}^-$ exterior to L, then by integrating around the contour shown in Figure 3.2.2 and noting that, for large t, $f(t)/(t - z) = a/t + 0(t^{-1})$, where $a = f(\infty)$, we find

$$F(z) = \begin{cases} a & \text{if } z \text{ is in } \mathcal{D}^+ \\ a - f(z) & \text{if } z \text{ is in } \mathcal{D}^-. \end{cases} \tag{3.2.8}$$

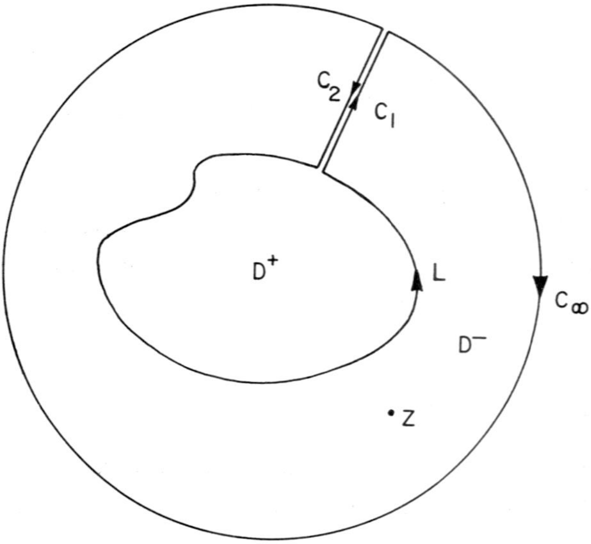

Figure 3.2.2 - The point z lies outside the contour L.

This result will be needed later in an extended form. Suppose that f(z) is holomorphic in $\mathcal{D}^-$ except at infinity where it has the form

$$f(z) = a_0 + a_1 z + \dots. + a_p z^p + 0(z^{-1})$$

$$= f_0(z) + 0(z^{-1}) \; ; \tag{3.2.9}$$

then

$$\frac{1}{2\pi i}\int_L \frac{f(t)}{t-z}\,dt + \frac{1}{2\pi i}\int_{C_\infty} \frac{f(t)}{t-z}\,dt = \begin{cases} -f(z) & \text{if } z \text{ is in } \mathcal{D}^- \\ 0 & \text{if } z \text{ is in } \mathcal{D}^+. \end{cases}$$

The negative sign arises because $\mathcal{D}^-$, the interior of the contour, lies on the right. But

$$\frac{1}{2\pi i}\int_{C_\infty} \frac{f(t)}{t-z}\,dt = \frac{1}{2\pi i}\int_{C_\infty} \frac{f_0(t)}{t-z}\,dt = -f_0(z),$$

whether z is in $\mathcal{D}^+$ or $\mathcal{D}^-$. Therefore

$$F(z) = \frac{1}{2\pi i}\int_L \frac{f(t)}{t-z}\,dt = \begin{cases} f_0(z) & \text{if } z \text{ is in } \mathcal{D}^+ \\ f_0(z) - f(z) & \text{if } z \text{ is in } \mathcal{D}^-. \end{cases} \tag{3.2.10}$$

We now suppose that in addition to being continuous on L, f(t) satisfies the so-called Hölder condition. The function f(t) is said to satisfy a Hölder condition on L if there exist parameters A,λ where $0 < \lambda < 1$ such that, for every two points t_1,t_2 of L, we have

$$|f(t_2) - f(t_1)| < A|t_2 - t_1|^\lambda.$$

The function f(t) is said to satisfy the Hölder condition in the neighbourhood of a point t_0 of L if, for all points t_1,t_2 sufficiently near t_0, we have

$$|f(t_2) - f(t_1)| < A|t_2 - t_1|^\lambda.$$

Under this condition it can be shown, and we shall sketch the proof below, that the Cauchy integral F(z) may be given a meaning when z is on L, and F(z) tends to definite limits as z tends to L from $\mathcal{D}^+$ and as z tends to L from $\mathcal{D}^-$.

Let t_1 be a point of L, in the neighbourhood of which f(t) satisfies a Hölder condition, and let t',t" be two points on L on either side of t_0, such that

$$|t_0 - t'| = |t_0 - t''| = \varepsilon,$$

as shown in Figure 3.2.3. The *Cauchy Principal Value* of the integral (3.2.6) at t_0 is defined as

$$(P)\frac{1}{2\pi i}\int_L \frac{f(t)}{t-t_0}\,dt = \lim_{\varepsilon\to 0}\frac{1}{2\pi i}\int_{L-\ell}\frac{f(t)}{t-t_0}\,dt,$$

where ℓ is the arc t't". The integral may be written

$$\frac{1}{2\pi i}\int_{L-\ell}\frac{f(t)}{t-t_0}\,dt = \frac{1}{2\pi i}\int_{L-\ell}\frac{f(t)-f(t_0)}{t-t_0}\,dt + \frac{f(t_0)}{2\pi i}\int_{L-\ell}\frac{dt}{t-t_0}$$

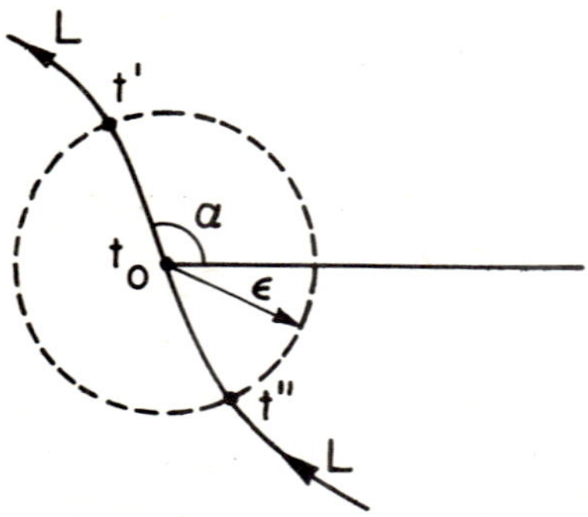

Figure 3.2.3 - The points t',t" *are equidistant from* t_0.

Since $|[f(t) - f(t_0)]/(t - t_0)| < A|t - t_0|^{\lambda-1}$, the limit of the first integral exists in the ordinary sense, i.e. provided only that t',t" tend to t_0; it is not necessary for $|t' - t_0|$, $|t'' - t_0|$ to be equal. The second integral is

$$\int_{L-\ell}\frac{dt}{t-t_0} = [\ell n(t-t_0)]_{t''}^{t'},$$

where a branch of $\ell n\, z$ has been taken which is continuous on $L-\ell$. Now

$$t' = t_0 + \varepsilon \exp i(\alpha+\pi), \quad t'' = t_0 + \varepsilon \exp(i\alpha),$$

so that

$$\ell n(t'-t_0) - \ell n(t''-t_0) = i\pi,$$

and

$$\frac{1}{2\pi i}(P)\int_L \frac{f(t)}{t-t_0}\, dt = \frac{1}{2\pi i}\int_L \frac{f(t)-f(t_0)}{t-t_0}\, dt + \frac{1}{2} f(t_0). \tag{3.2.11}$$

This is the meaning which will be attached to the integral (3.2.6) when z lies on L; thus

$$F(t_0) = \frac{1}{2} f(t_0) + \frac{1}{2\pi i}\int_L \frac{f(t)-f(t_0)}{t-t_0}\, dt. \tag{3.2.12}$$

Now return to equation (3.2.6) and write

$$F(z) = \frac{1}{2\pi i}\int_L \frac{f(t)-f(t_0)}{t-z}\, dt + \frac{f(t_0)}{2\pi i}\int_L \frac{dt}{t-z},$$

where t_0 is a point on L. Then it may be proved (Muskhelishvili (1953b, p. 39)) that the first integral tends to

$$\frac{1}{2\pi i}\int_L \frac{f(t)-f(t_0)}{t-t_0}\, dt$$

as z tends to t_0, from whichever side of L. The second integral has, by the argument used in (3.2.3), the value $f(t_0)$ if z is in $\mathcal{D}^+$, 0 if z is in $\mathcal{D}^-$. Thus the limits of F(z) as z tends to t_0 from $\mathcal{D}^+$ and $\mathcal{D}^-$, denoted by $F^+(t_0)$, $F^-(t_0)$ respectively, are

$$F^+(t_0) = \frac{1}{2\pi i}\int_L \frac{f(t)-f(t_0)}{t-t_0}\, dt + f(t_0), \tag{3.2.13}$$

$$F^-(t_0) = \frac{1}{2\pi i}\int_L \frac{f(t)-f(t_0)}{t-t_0}\, dt. \tag{3.2.14}$$

Now expressing these integrals in terms of the Cauchy Principal Value in equation (3.2.11), we find

$$F^{+}(t_0) = \frac{1}{2} f(t_0) + \frac{1}{2\pi i}(P)\int_L \frac{f(t)}{t-t_0}\,dt, \tag{3.2.15}$$

$$F^{-}(t_0) = -\frac{1}{2} f(t_0) + \frac{1}{2\pi i}(P)\int_L \frac{f(t)}{t-t_0}\,dt. \tag{3.2.16}$$

In future the (P) will be omitted and it will be understood that any singular integral must be interpreted as a Cauchy Principal Value. It is often convenient to rewrite equations (3.2.15), (3.2.16) in the form

$$F^{+}(t_0) - F^{-}(t_0) = f(t_0), \tag{3.2.17}$$

$$F^{+}(t_0) + F^{-}(t_0) = \frac{1}{\pi i}\int_L \frac{f(t)}{t-t_0}\,dt. \tag{3.2.18}$$

These equations, called the *Plemelj formulae*, have been defined for a contour - a closed arc. However if f(t) is defined and continuous on an open arc then, provided that f(t) is zero at the ends, we may close the contour and define f(t) to be zero on the closure. Then equations (3.2.15),(3.2.16) give the limiting values as z tends to t_0 from the left and right respectively. If f(t) is not zero at the ends, then if f(t) is defined to be zero on the closure, it will not be continuous, and will not satisfy the Hölder condition at the ends. Equations (3.2.15),(3.2.16) will then not hold at the ends, but will hold at other points of L.

Muskhelishvili (1953b) proves the following results concerning Cauchy integrals near the ends of an open arc L, with ends z = a, z = b, as shown in Figure 3.2.4. Let f(t) be a function defined on L and satisfying the following two conditions:

a) f(t) satisfies a Hölder condition on any closed interval of L, not including the ends a,b.

b) Near the ends a,b the function f(t) is of the form

$$f(t) = f^*(t)(t-c)^{-\gamma}, \quad \gamma = \alpha + i\beta, \quad 0 \leq \alpha < 1, \tag{3.2.19}$$

where c is either of the ends; a, b, α and β are real constants and $f^*(t)$ satisfies the Hölder condition near and at c. In (3.2.19), $(t - c)^{-\gamma}$ is any branch which varies continuously on L.

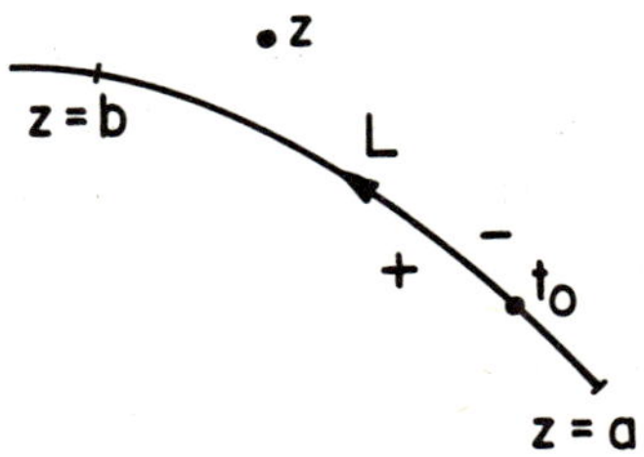

Figure 3.2.4 - The arc L *joins the points* z = a, z = b.

Let z be a point near c but not on L, and suppose F(z) is given by (3.2.6); then:

(1) If $\gamma = 0$,

$$F(z) = \mp \frac{f(c)}{2\pi i} \ln(z-c) + F_0(z), \tag{3.2.20}$$

where the upper sign is taken for c = a, the lower for c = b, and $\ln(z - c)$ is a branch, single-valued near c in the plane cut along L, and $F_0(z)$ is a bounded function tending to a definite limit as $z \to c$ along any path.

(2) If $\gamma = \alpha + i\beta \neq 0$,

$$F(z) = \pm \frac{\exp(\pm\gamma\pi i)}{2i \sin \gamma\pi} f^*(c)(z-c)^{-\gamma} + F_0(z), \tag{3.2.21}$$

where the signs are taken as in (1); $(z - c)^{-\gamma}$ is a branch single-valued near c in the plane cut along L, and taking the value $(t - c)^{-\gamma}$ on the left(+) side of L. The function $F_0(z)$ is bounded and tends to a definite limit as $z \to c$ if $\alpha = 0$, while if $\alpha > 0$ it satisfies

$$|F_0(z)| < C|z - c|^{-\alpha_0}, \tag{3.2.22}$$

where C, α_0 are real constants such that $\alpha_0 < \alpha$.

For the point $z = t_0$ lying on L the following results hold:

(3) If $\gamma = 0$

$$F(t_0) = \mp \frac{f(c)}{2\pi i} \ln(t_0 - c) + F^*(t_0), \tag{3.2.23}$$

where the signs are chosen as in (1) and $F^*(t_0)$ satisfies a Hölder condition near and at c.

(4) If $\gamma = \alpha + i\beta \neq 0$

$$F(t_0) = \pm \frac{\cot \gamma\pi}{2i} f^*(c)(t_0 - c)^{-\gamma} + F^*(t_0). \tag{3.2.24}$$

If $\alpha = 0$, $F^*(t_0)$ satisfies a Hölder condition near and at c; if $\alpha > 0$

$$F^*(t_0) = F^{**}(t_0)|t_0 - c|^{-\alpha_0}, \tag{3.2.25}$$

where $F^{**}(t_0)$ satisfies a Hölder condition near and at c and $\alpha_0 < \alpha$; the signs are chosen as in (1).

We sketch the proof of the results (2) and (4). The integral $F(z)$ may be written

$$F(z) = f^*(c)\Omega(z) + F_1(z), \tag{3.2.26}$$

where

$$\Omega(z) = \frac{1}{2\pi i}\int_L \frac{(t-c)^{-\gamma}}{t-z}\, dt, \quad F_1(z) = \frac{1}{2\pi i}\int_L \frac{f^*(t) - f^*(c)}{(t-c)^{\gamma}(t-z)}\, dt. \tag{3.2.27}$$

It may easily be shown that the second integral satisfies the condition (3.2.22) stated for $F_1(z)$. We examine the first integral, near a. Let $(z - a)^{-\gamma}$ denote that branch holomorphic in the plane cut along L and which takes the value $(t_0 - a)^{-\gamma}$ on the + side of L; then the boundary value on the - side will be $[(t_0 - a)\exp(2i\pi)]^{-\gamma}$ and

$$[(t_0-a)^{-\gamma}]^+ - [(t_0-a)^{-\gamma}]^- = (t_0-a)^{-\gamma}[1-\exp(-2i\,\gamma\pi)]. \quad (3.2.28)$$

The Plemelj formulae give

$$\Omega^+(t_0) - \Omega^-(t_0) = (t_0-a)^{-\gamma}. \quad (3.2.29)$$

Writing $[1 - \exp(-2i\,\gamma\pi)]^{-1} = \exp(i\gamma\pi)/[2i \sin \gamma\pi]$, and

$$\omega(z) = \frac{\exp(i\gamma\pi)}{2i \sin \gamma\pi}(z-a)^{-\gamma}, \quad (3.2.30)$$

we see that equation (3.2.28) gives

$$\omega^+(t_0) - \omega^-(t_0) = (t_0-a)^{-\gamma}, \quad (3.2.31)$$

so that $[\Omega(t_0) - \omega(t_0)]^+ = [\Omega(t_0) - \omega(t_0)]^-$. This means that $\Omega(z) - \omega(z)$ is holomorphic in the neighbourhood of a. Thus

$$\Omega(z) = \frac{\exp(i\gamma\pi)}{2i \sin \gamma\pi}(z-a)^{-\gamma} + \Omega_0(z), \quad (3.2.32)$$

where $\Omega_0(z)$ may be expanded in a power series near a; equations (3.2.26),(3.2.32) now yield (3.2.21) for c = a.

To prove (4), we note that $F(t_0) = 1/2(F^+(t_0) + F^-(t_0))$, so that the principal part, arising from $\Omega(z)$, is

$$\frac{f^*(c)}{2}[\Omega^+(t_0) + \Omega^-(t_0)] = \frac{f^*(a)\exp(i\gamma\pi)}{4i \sin \gamma\pi}[(t_0-a)^{-\gamma} +$$

$$+ \{(t_0-a)\exp(2i\pi)\}^{-\gamma}] + f^*(c)\Omega_0(t_0)$$

$$= \frac{\cot \gamma\pi}{2i} f^*(a)(t_0-a)^{-\gamma} + f^*(c)\Omega_0(t_0), \quad (3.2.33)$$

which leads immediately to (3.2.24).

3.3 *Cauchy integrals along the real axis*

In the sequel we shall need to know the limiting values of Cauchy integrals when L is not a closed, finite contour, but the whole real axis. Suppose that f(t) is defined on the real axis, is finite and integrable along any finite part of the axis, and satisfies at infinity the condition

$$f(t) = f(\infty) + O(|t|^{-\lambda}), \quad \lambda > 0. \tag{3.3.1}$$

We define the Cauchy integral of f(t) along the real axis as

$$F(z) = \lim_{N\to\infty} \frac{1}{2\pi i}\int_{-N}^{N} \frac{f(t)}{t-z}\, dt. \tag{3.3.2}$$

This integral, in which the end points are taken at equal distances along the real axis, is called the Cauchy Principal Value of this infinite integral. Now

$$\frac{1}{2\pi i}\int_{-N}^{N} \frac{f(t)}{t-z}\, dt = \frac{1}{2\pi i}\int_{-N}^{N} \frac{f(t)-f(\infty)}{t-z}\, dt + \frac{f(\infty)}{2\pi i}\int_{-N}^{N} \frac{dt}{t-z}\, .$$

Under the condition (3.3.1) the limit of the first integral exists in the ordinary sense - it is not necessary for the two points to be equidistant from the origin. The second integral is

$$\int_{-N}^{N} \frac{dt}{t-z} = [\ell n(t-z)]_{-N}^{N},$$

where again one branch of ℓn must be taken. If z is in S^+ then Figure 3.3.1 shows that

$$N - z = r_1 \exp(i\alpha_1), \quad -N - z = r_2 \exp(i\alpha_2),$$

so that

$$[\ell n(t-z)]_{-N}^{N} = \ell n(r_1/r_2) + i(\alpha_1 - \alpha_2).$$

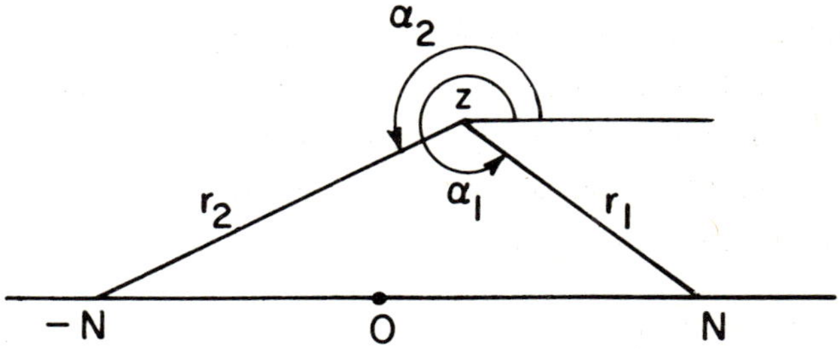

Figure 3.3.1 - The lines join z *to the points* ±N.

Now

$$\lim_{N\to\infty}(r_1/r_2) = 1, \quad \lim_{N\to\infty}(\alpha_1-\alpha_2) = \pi.$$

If z is in S^- then a similar argument may be used; the result is

$$\frac{1}{2\pi i}\int_{-\infty}^{\infty} \frac{f(t)}{t-z}\,dt = \pm\frac{1}{2}f(\infty) + \frac{1}{2\pi i}\int_{-\infty}^{\infty}\frac{f(t)-f(\infty)}{t-z}\,dt \tag{3.3.3}$$

where the sign is ± according to whether z is in $S^{\pm}$. This equation may be taken to give the meaning to be attached to, i.e., the Cauchy Principal Value of, the integral on the left.

When z lies on the real axis the integral will be interpreted as a Cauchy Principal Value in both of the senses that have been discussed. That is

$$F(t_0) = \frac{1}{2\pi i}\lim_{\substack{N\to\infty\\ \varepsilon\to 0}}\left\{\int_{-N}^{t_0-\varepsilon}\frac{f(t)}{t-t_0}\,dt + \int_{t_0+\varepsilon}^{N}\frac{f(t)}{t-t_0}\,dt\right\}. \tag{3.3.4}$$

By applying the arguments used before we may easily prove that if f(t) satisfies a Hölder condition in the neighbourhood of t_0, and satisfies conditions (3.3.1), then the limit exists. In fact, since

$$\lim_{N\to\infty}\int_{-N}^{N}\frac{dt}{t-t_0} = 0,$$

the integral (3.3.2) may be evaluated as either

$$F(t_0) = \frac{1}{2\pi i}\lim_{N\to\infty}\int_{-N}^{N}\frac{f(t)-f(t_0)}{t-t_0}\,dt,$$

or

$$F(t_0) = \frac{1}{2\pi i}\lim_{\varepsilon\to 0}\left\{\int_{-\infty}^{t_0-\varepsilon}\frac{f(t)-f(\infty)}{t-t_0}\,dt + \int_{t_0+\varepsilon}^{\infty}\frac{f(t)-f(\infty)}{t-t_0}\,dt\right\}.$$

Finally we evaluate the limiting values of $F(z)$ as z approaches L and find

$$F^{\pm}(t_0) = \pm\frac{1}{2}f(t_0) + \frac{1}{2\pi i}\int_{-\infty}^{\infty}\frac{f(t)}{t-t_0}\,dt, \tag{3.3.5}$$

where again the integral must be interpreted as a Cauchy Principal Value in both senses. Equations (3.3.5) may be combined to give the Plemelj formulae

$$F^{+}(t_0) - F^{-}(t_0) = f(t_0), \tag{3.3.6}$$

$$F^{+}(t_0) + F^{-}(t_0) = \frac{1}{\pi i}\int_{-\infty}^{\infty}\frac{f(t)}{t-t_0}\,dt. \tag{3.3.7}$$

Equation (3.3.5) gives the limiting values of $F(z)$ as z approaches a finite point t_0 from the upper or lower half-planes. In order to investigate the boundary values as z *tends to infinity* it is necessary to introduce further restrictions on $f(t)$. Consider the coordinate transformation $z = -1/\zeta$. The point at infinity in the z-plane is transformed into $\zeta = 0$, and, *vice versa*, $z = 0$ is transformed into $\zeta = \infty$. The upper and lower halves of the z-plane are transformed into the corresponding halves of the

ζ-plane. Figure 3.3.2 shows more details of the transformation.

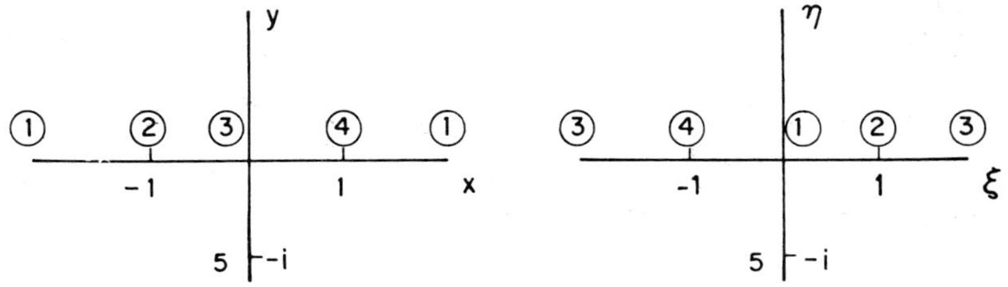

Figure 3.3.2 - The transformation $z = -1/\zeta$.

Now put

$$F(z) = F(-1/\zeta) = F^*(\zeta), \qquad f(t) = f(-1/\sigma) = f^*(\sigma),$$

and introduce the further condition that $f^*(\sigma)$ satisfies a Hölder condition in the neighbourhood of $\sigma = 0$. This means that, for all sufficiently small σ_1, σ_2,

$$|f^*(\sigma_2) - f^*(\sigma_1)| < A|\sigma_2 - \sigma_1|^\lambda.$$

This leads to the condition

$$|f(t_2) - f(t_1)| < A|1/t_2 - 1/t_1|^\lambda, \tag{3.3.8}$$

for all sufficiently large $|t_1|, |t_2|$. This condition, which is called the Hölder condition for the neighbourhood of the point at infinity, implies (3.3.1), but is stronger than it. Now

$$F^*(\zeta) = \frac{\zeta}{2\pi i}\int_{-\infty}^{\infty} \frac{f^*(\sigma)}{\sigma(\sigma-\zeta)}\, d\sigma,$$

or

$$F^*(\zeta) = \frac{1}{2\pi i}\int_{-\infty}^{\infty} \frac{f^*(\sigma)}{\sigma-\zeta}\, d\sigma - \frac{1}{2\pi i}\int_{-\infty}^{\infty} \frac{f^*(\sigma)}{\sigma}\, d\sigma.$$

These integrals will exist, in the sense of Cauchy Principal Values, if f(t) satisfies a Hölder condition for finite t and condition (3.3.8) at infinity. The behaviour of F(z) at infinity is reduced to the behaviour of $F^*(\zeta)$ at $\zeta = 0$, and this can be deduced from equation (3.3.5). In fact if z tends to infinity along a path lying wholly in $S^{\pm}$ then

$$F^{\pm}(\infty) = F^{*\pm}(0) = \pm \frac{1}{2} f^*(0) + \frac{1}{2\pi i}\int_{-\infty}^{\infty} \frac{f^*(\sigma)}{\sigma}\, d\sigma - \frac{1}{2\pi i}\int_{-\infty}^{\infty} \frac{f^*(\sigma)}{\sigma}\, d\sigma,$$

or

$$F^{\pm}(\infty) = \pm \frac{1}{2} f(\infty). \tag{3.3.9}$$

Under these conditions F(z) will be holomorphic in each half-plane, including the point at infinity.

3.4 *The problem of linear relationship*

It will be found in the following chapter that the contact problems for the half-plane may be reduced to the following equations

$$\kappa\Phi^{+}(t) + \Phi^{-}(t) = F(t) \quad \text{on } L', \tag{3.4.1}$$

$$\Phi^{+}(t) - \Phi^{-}(t) = 0 \quad \text{on } L''. \tag{3.4.2}$$

Here L' is the contact region, made up of a number of arcs of the real axis, L'' is the remainder of the real axis, which is unloaded, κ is a complex constant and F(t) is assumed to satisfy a Hölder condition. Near an end point t_0 of L' it is required that $\Phi(t)$ satisfy the condition

$$|\Phi(t)| = O(|t - t_0|^{-\lambda}), \quad 0 \leq \lambda < 1, \tag{3.4.3}$$

or equivalently

$$\lim_{t\to t_0}(t-t_0)\Phi(t) = 0. \tag{3.4.4}$$

The solution $\Phi(z)$ must also be holomorphic at infinity.

In order to solve equations (3.4.1),(3.4.2) we first find functions X(z) holomorphic in the plane cut along L' and satisfying the homogeneous equations

$$\kappa X^{+}(t) + X^{-}(t) = 0 \quad \text{on } L', \tag{3.4.5}$$

$$X^{+}(t) - X^{-}(t) = 0 \quad \text{on } L''. \tag{3.4.6}$$

First consider the case in which L' is the single arc $-a < t < a$ shown in Figure 3.4.1; let α be an arbitrary complex number, and consider the branch of the function

$$X_0(z) = (z-a)^{-1/2+i\alpha}(z+a)^{-1/2-i\alpha} \tag{3.4.7}$$

which is holomorphic in the plane cut along L' and at infinity satisfies

$$X_0(z) = z^{-1} + 0(z^{-2}). \tag{3.4.8}$$

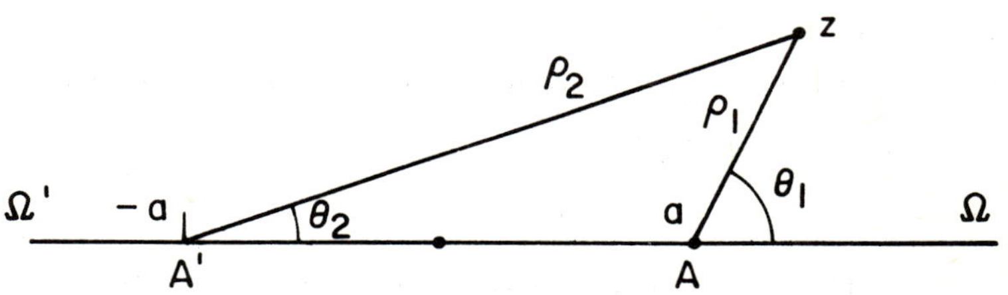

Figure 3.4.1 - Notation for defining the boundary values of $X_0(z)$.

In order that $X_0(z)$ satisfy (3.4.3) it is necessary that, if $\alpha = \alpha_1 + i\alpha_2$, then

$$-1/2 < \alpha_2 < 1/2. \tag{3.4.9}$$

Consider $X_0(z)$ more closely. The branches of the two factors making up $X_0(z)$ are

$$(z-a)^{-1/2+i\alpha} = [\rho_1 \exp(i\theta_1)]^{-1/2+i\alpha},$$

$$(z+a)^{-1/2-i\alpha} = [\rho_2 \exp(i\theta_2)]^{-1/2-i\alpha}.$$

The boundary value of $X_0(z)$ on the upper side of $A\Omega$ is

$$X_0^+(t) = (t-a)^{-1/2+i\alpha}(t+a)^{-1/2-i\alpha}, \tag{3.4.10}$$

where this is defined precisely as

$$(t^2-a^2)^{-1/2}[(t+a)/(t-a)]^{\alpha_2}[(t-a)/(t+a)]^{i\alpha_1},$$

and the last factor is

$$[(t-a)/(t+a)]^{i\alpha_1} = \cos\theta + i\sin\theta,$$

$$\theta = \alpha_1 \ln[(t-a)/(t+a)]. \tag{3.4.11}$$

On the lower side of $A\Omega$

$$X_0^-(t) = [(t-a)\exp(2i\pi)^{-1/2+i\alpha}[(t+a)\exp(2i\pi)]^{-1/2-i\alpha}$$

$$= X_0^+(t) \quad \text{on } L''. \tag{3.4.12}$$

The boundary values on A'A are

$$X_0^+(t) = [(a-t)\exp(i\pi)]^{-1/2+i\alpha}(a+t)^{-1/2-i\alpha}, \tag{3.4.13}$$

$$X_0^-(t) = [(a-t)\exp(i\pi)]^{-1/2+i\alpha}[(a+t)\exp(2i\pi)]^{-1/2-i\alpha}, \tag{3.4.14}$$

so that

$$X_0^-(t) = -\exp(2\pi\alpha)X_0^+(t) \quad \text{on } L'. \tag{3.4.15}$$

Thus $X_0(z)$ will satisfy equation (3.4.5) provided that

$$\exp(2\pi\alpha) = \kappa, \tag{3.4.16}$$

and, together with (3.4.9), this determines α uniquely.

If $X(z)$ is any other solution of equations (3.4.5), (3.4.6) then the function $Q(z) = X(z)/X_0(z)$, which is clearly continuous across L'', satisfies on L' the condition

$$Q^-(t) = X^-(t)/X_0^-(t) = -\kappa X^+(t)/[-\kappa X_0^+(t)] = Q^+(t).$$

Thus $Q(z)$ is holomorphic in the whole plane excluding the point of infinity and must therefore be a polynomial.

Thus

$$X(z) = Q(z)X_0(z). \tag{3.4.17}$$

If $X(z)$ is to be holomorphic at infinity, then $Q(z)$ must be linear. Thus the general solution of equations (3.4.5),(3.4.6), which is holomorphic in the cut plane and at infinity, is

$$X(z) = (Az+B)X_0(z). \tag{3.4.18}$$

If $X(z)$ is to vanish at infinity, then A must be zero.

Now return to equations (3.4.1) and (3.4.2) and divide through by $X_0^+(t) \equiv -\kappa^{-1}X_0^-(t)$. Then, on L',

$$\kappa\Phi^+(t)/X_0^+(t) - \kappa\Phi^-(t)/X_0^-(t) = F(t)/X_0^+(t). \tag{3.4.19}$$

Thus, writing $\Omega(z) = \Phi(z)/X_0(z)$, we deduce that

$$\Omega^+(t) - \Omega^-(t) = F(t)/(\kappa X_0^+(t)) \quad \text{on } L', \tag{3.4.20}$$

$$\Omega^+(t) - \Omega^-(t) = 0 \qquad \text{on } L''. \tag{3.4.21}$$

Since this is a pair of linear non-homogeneous equations for $\Omega^+(t)$, $\Omega^-(t)$, its general solution may be expressed as the sum of one particular solution and the general solution of the homogeneous equations (3.4.5),(3.4.6). Thus the general solution which is holomorphic in the plane cut along L', and including the point at infinity, is

$$\Phi(z) \equiv X_0(z)\Omega(z) = \frac{X_0(z)}{2\pi i\kappa}\int_{L'} \frac{F(t)dt}{X_0^+(t)(t-z)} + (Az+B)X_0(z). \tag{3.4.22}$$

Sometimes it is required to find a solution $\Phi(z)$ which is *zero* at one or more ends of L'. Thus if $\Phi(z)$ is required to be zero at $z = a$, we introduce

$$X_1(z) = (z-a)X_0(z), \tag{3.4.23}$$

and divide both sides of (3.4.1),(3.4.2) by $X_1^+(t)$, write $\Omega(z) = \Phi(z)/X_1(z)$ and obtain

$$\Omega^+(t) - \Omega^-(t) = F(t)/(\kappa X_1^+(t)) \quad \text{on } L', \tag{3.4.24}$$

$$\Omega^+(t) - \Omega^-(t) = 0 \quad \text{on } L'', \tag{3.4.25}$$

and

$$\Phi(z) \equiv X_1(z)\Omega(z) = \frac{X_1(z)}{2\pi i\kappa}\int_{L'} \frac{F(t)dt}{X_1^+(t)(t-z)} + CX_1(z). \tag{3.4.26}$$

Notice that, since $X_1(z) = O(1)$ for large z, only a constant term C appears in the complementary solution. If $\Phi(z)$ is required to be zero at *both* ends, then we take

$$X_2(z) = (z-a)(z+a)X_0(z), \tag{3.4.27}$$

and obtain

$$\Phi(z) = \frac{X_2(z)}{2\pi i\kappa}\int_{L'} \frac{F(t)dt}{X_2^+(t)(t-z)}; \tag{3.4.28}$$

but now this will not be holomorphic at infinity. For by expanding $(t - z)^{-1} = -z^{-1}(1 - t/z)^{-1} = -z^{-1} - tz^{-2} \ldots$ we find that, for large $|z|$,

$$\Phi(z) = -\frac{X_2(z)z^{-1}}{2\pi i\kappa}\int_{L'} \frac{F(t)dt}{X_2^+(t)} + O(1),$$

so that $\Phi(z)$ will be holomorphic at infinity only if

$$\int_{L'} \frac{F(t)dt}{X_2^+(t)} = 0. \tag{3.4.29}$$

The general case in which L' is the union of n arcs (a_k, b_k) $k = 1,2,\ldots n$ on the real axis and a solution is sought which is zero at some of these ends is discussed by Muskhelishvili. The corresponding results relating to one or more arcs, not necessarily on the real axis, are generally straightforward generalizations of the particular results above.

If F(t) is a polynomial, then the integrals in (3.4.22), (3.4.26) and (3.4.28) may be evaluated explicitly by integrating around the contour Λ in Figure 3.4.2. Provided that α_2 satisfies equation (3.4.9), the integrals around the small circles at each of the ends of the arcs will be negligible and

$$I(z) = \frac{1}{2\pi i}\int_{\Lambda} \frac{F(\zeta)d\zeta}{X(\zeta)(\zeta - z)} = \frac{1}{2\pi i}\int_{L'} \frac{F(t)dt}{X^{+}(t)(t-z)} -$$

$$- \frac{1}{2\pi i}\int_{L'} \frac{F(t)dt}{X^{-}(t)(t-z)}, \tag{3.4.30}$$

and since $X^{-}(t) = -\kappa X^{+}(t)$, we have

$$\frac{1}{2\pi i\kappa}\int_{L'} \frac{F(t)dt}{X^{+}(t)(t-z)} = \frac{I(z)}{\kappa+1}. \tag{3.4.31}$$

Figure 3.4.2 - The contour Λ *used in evaluating integrals along* -a, a.

But $F(\zeta)/X(\zeta)$ is holomorphic outside Λ, except possibly at infinity. Suppose that, at infinity,

$$F(z)/X(z) = S(z) + O(z^{-1}), \tag{3.4.32}$$

where $S(z)$ is a polynomial (the polynomial *conjugate* to F); then, by equation (3.2.10), noting that Λ is traversed clockwise, we deduce that

$$I(z) = F(z)/X(z) - S(z), \tag{3.4.33}$$

and therefore

$$\Phi(z) = [F(z) - S(z)X(z)]/(\kappa+1) + CX(z). \tag{3.4.34}$$

The results derived in this chapter will be applied in Chapter IV to the complex variable solution of plane problems and, in Chapter VI, to provide solutions to various integral equations which arise in contact problems.

Examples 3.4

1. Use the identity

$$T_n(z) - (z^2-1)^{1/2}U_{n-1}(z) = [z + (z^2-1)^{1/2}]^{-n}$$

given by equation (2.3.18) and Ex. 2.3.7 to deduce that if $X(z) = (z^2 - 1)^{1/2}$, $F(z) = T_n(z)$ then the polynomial $S(z)$ is $U_{n-1}(z)$. Deduce that

$$\frac{X(z)}{\pi i}\int_{-1}^{1} \frac{T_n(t)dt}{X^+(t)(t-z)} = T_n(z) - X(z)U_{n-1}(z) = [z + (z^2-1)^{1/2}]^{-n}.$$

This recovers the results obtained in §2.3.

2. Derive the corresponding result for $X_0(z) = (z^2 - 1)^{-1/2}$, namely

$$\frac{X_0(z)}{\pi i}\int_{-1}^{1} \frac{U_{n-1}(t)dt}{X_0^+(t)(t-z)} = U_{n-1}(z) - X_0(z)T_n(z)$$

$$= -(z^2-1)^{-1/2}[z + (z^2-1)^{1/2}]^{-n}.$$

3.5 *Integral equations with Cauchy kernels*

One of the most important applications of Cauchy integrals is to the solution of the singular integral equation

$$\varphi\phi(x) + \frac{1}{\pi i}\int_{-1}^{1} \frac{\phi(t)}{t-x}\,dt = g(x), \quad |x| < 1. \tag{3.5.1}$$

Here the function $g(x)$ is assumed to satisfy certain Hölder conditions and $\phi(t)$ is assumed to satisfy certain conditions at the end points ± 1. From the analysis of the preceding sections of this chapter we may deduce the following. Put

$$\Phi(z) = \frac{1}{2\pi i}\int_{-1}^{1} \frac{\phi(t)}{t-z}\,dt. \tag{3.5.2}$$

Provided that $\phi(t)$ satisfies a Hölder condition on $(-1,1)$, i.e. provided that there exist parameters A,λ where $0 < \lambda < 1$ such that, for every two points t_1,t_2 of $(-1,1)$, we have

$$|\phi(t_2) - \phi(t_1)| < A|t_2 - t_1|^{\lambda},$$

then the function $\Phi(z)$ will be holomorphic in the plane cut along $[-1,1]$, and across this line will satisfy

$$\Phi^+(x) - \Phi^-(x) = \phi(x), \tag{3.5.3}$$

$$\Phi^+(x) + \Phi^-(x) = \frac{1}{\pi i}\int_{-1}^{1} \frac{\phi(t)}{t-x}\,dt. \tag{3.5.4}$$

The function $\phi(t)$ will satisfy (3.5.1) provided that

$$\varphi(\Phi^+(x)-\Phi^-(x)) + (\Phi^+(x)+\Phi^-(x)) = g(x), \qquad |x| < 1, \tag{3.5.5}$$

$$\Phi^+(x) - \Phi^-(x) = 0, \qquad |x| > 1. \tag{3.5.6}$$

Equation (3.5.5) may be rewritten

$$\kappa\Phi^+(x) + \Phi^-(x) = F(x), \qquad |x| < 1, \tag{3.5.7}$$

where

$$\kappa = (1+\varphi)/(1-\varphi), \qquad F(x) = g(x)/(1-\varphi), \tag{3.5.8}$$

and now equations (3.5.7),(3.5.6) are an example of equations (3.4.1),(3.4.2). According to §3.4 the general solution of (3.5.6), (3.5.7) is

$$\Phi(z) = \frac{X_0(z)}{2\pi i\kappa}\int_{-1}^{1} \frac{F(t)dt}{X_0^+(t)(t-z)} + (Az+B)X_0(z), \qquad (3.5.9)$$

where $X_0(z) = (z-1)^{-1/2+i\alpha}(z+1)^{-1/2-i\alpha}$, $\exp(2\pi\alpha) = \kappa$, and $\alpha = \alpha_1 + i\alpha_2$, where $-1/2 < \alpha_2 < 1/2$. Equation (3.5.9) gives the following form for $\phi(x)$ in $|x| < 1$:

$$\phi(x) = -\sinh \pi\alpha \cosh \pi\alpha g(x) + \frac{\cosh^2\pi\alpha}{\pi i} X_0^+(x)\int_{-1}^{1} \frac{g(t)dt}{X_0^+(t)(t-x)} +$$

$$+ 2\exp(\pi\alpha)\cosh \pi\alpha(Ax+B)X_0^+(x). \qquad (3.5.10)$$

Note that if φ lies in the range $-1 < \varphi < 1$, so that κ is real and positive, as is often the case in applications, then $\alpha_2 = 0$; if $\varphi = 0$, then $\kappa = 1$ and $\alpha = 0$.

If

$$\int_{-1}^{1} \frac{F(t)dt}{X_2^+(t)} = 0, \qquad (3.5.11)$$

then the unique solution of (3.5.6),(3.5.7) which is bounded, actually zero, at the ends ± 1 is

$$\Phi(z) = \frac{X_2(z)}{2\pi i\kappa}\int_{-1}^{1} \frac{F(t)dt}{X_2^+(t)(t-z)}, \qquad (3.5.12)$$

where $X_2(z) = (z-1)^{1/2+i\alpha}(z+1)^{1/2-i\alpha}$, giving

$$\phi(x) = -\sinh \pi\alpha \cosh \pi\alpha g(x) + \frac{\cosh^2\pi\alpha}{\pi i} X_2^+(x)\int_{-1}^{1} \frac{g(t)dt}{X_2^+(t)(t-x)}. \qquad (3.5.13)$$

In applications it is important to be able to evaluate the integrals when $F(x)$ is a polynomial. Here we use the results of §3.4 which state that if $F(t)$ is a polynomial, and $S_0(z)$ is the polynomial with the property that

$$\frac{F(z)}{X_0(z)} - S_0(z) = 0(z^{-1}) \quad \text{for large } (z), \tag{3.5.14}$$

then the $\Phi(z)$ of (3.5.9) may be written

$$\Phi(z) = (\kappa+1)^{-1}\{F(z) - X_0(z)S_0(z)\} + (Az+B)X_0(z). \tag{3.5.15}$$

Similarly if

$$\frac{F(z)}{X_2(z)} - S_2(z) = 0(|z|^{-1}) \quad \text{for large } |z|, \tag{3.5.16}$$

then the $\Phi(z)$ of (3.5.12) may be written

$$\Phi(z) = (\kappa+1)^{-1}\{F(z) - X_2(z)S_2(z)\}. \tag{3.5.17}$$

These results mean in particular that the general solution (3.5.10) may be written

$$\phi(t) = -X_0^+(t)S_0(t) + (\kappa+1)X_0^+(t)(At+B), \tag{3.5.18}$$

while the solution (3.5.13) is

$$\phi(t) = -X_2^+(t)S_2(t). \tag{3.5.19}$$

These expressions show that the behaviour of $\phi(t)$ near the ends ± 1 is the same as that of the corresponding fundamental solution $X^+(t)$.

Pairs of polynomials $F(z), S(z)$ are provided by the Jacobi polynomials $P_n^{(\gamma,\delta)}(z)$ (Abramowitz and Stegun (1972, Ch. 22)). These have the property

$$V_n(z) \equiv (z-1)^{\gamma}(z+1)^{\delta}P_n^{(\gamma,\delta)}(z) - 2^{\gamma+\delta}P_{n+\gamma+\delta}^{(-\gamma,-\delta)}(z)$$

$$= -2\pi^{-1}\sin(\gamma\pi)(z-1)^{\gamma}(z+1)^{\delta}Q_n^{(\gamma,\delta)}(z), \tag{3.5.20}$$

where $Q_n^{(\gamma,\delta)}$ is the Jacobi function of the second kind. This property may be deduced from the results given in Erdelyi (1955, Vol. 3, 170-172). It is valid in the plane cut along $|z| \leq 1$, provided that $\gamma + \delta$ is an integer and $m \equiv \gamma + \delta + n + 1 > 0$.

$V_n(z)$ has an integral representation

$$V_n(z) = -\frac{2^{-n}}{\pi}\sin\gamma\pi\int_{-1}^{1}(z-t)^{-n-1}(1-t)^{n+\gamma}(1+t)^{n+\delta}dt. \qquad (3.5.21)$$

which shows that $V_n(z) = O(z^{-n-1})$ for large $|z|$. $V_n(z)$ has a hypergeometric expansion

$$V_n(z) = -\frac{2^m}{\pi}\frac{\sin\gamma\pi}{(z-1)^{n+1}}B(m-\gamma,\ m-\delta)\ \cdot$$

$$\cdot\ {}_2F_1\left(n+1,\ m-\delta;\ 2m-\gamma-\delta;\ \frac{2}{1-z}\right), \qquad (3.5.22)$$

valid for $|1 - z| > 2$. This expansion may be continued analytically by using the known transformations of the hypergeometric functions (Abramowitz and Stegun (1972, Ch. 15)); B is the Beta function. By putting $\gamma = 1/2 - i\alpha$, $\delta = 1/2 + i\alpha$ in (3.5.20), we note that $(z - 1)^{\gamma}(z + 1)^{\delta} = 1/X_0(z)$, so that

$$V_n(z) = \frac{P_n^{(\gamma,\delta)}}{X_0(z)} - 2P_{n+1}^{(-\gamma,-\delta)}(z) = O(z^{-n-1}). \qquad (3.5.23)$$

This means that we may take

$$f(z) = P_n^{(\gamma,\delta)}(z), \qquad S_0(z) = 2P_{n+1}^{(-\gamma,-\delta)}(z). \qquad (3.5.24)$$

The particular part of the solution (3.5.18) is

$$\phi(t) = -X_0^{+}(t)S_0(t). \qquad (3.5.25)$$

The value of the integral $\Phi(z)$, given in (3.5.2), is

$$\Phi(z) = (\kappa+1)^{-1}X_0(z)V_n(z)$$

$$= (\kappa+1)^{-1}\left[P_n^{(\gamma,\delta)}(z) - 2X_0(z)P_{n+1}^{(-\gamma,-\delta)}(z)\right]$$

$$= \pi^{-1}\exp(-\pi\alpha)Q_n^{(\gamma,\delta)}(z), \qquad (3.5.26)$$

and in particular the value of $\Phi(x)$, interpreted as a Cauchy Principal Value on $-1 \leq x \leq 1$, is

$$\frac{1}{2\pi i}\int_{-1}^{1} \frac{\phi(t)}{t-x}\,dt = \frac{1}{2}(\Phi^+(x)+\Phi^-(x)) = \frac{1}{\kappa+1}\left\{P_n^{(\gamma,\delta)}(x) - \right.$$
$$\left. - (1-\kappa)X_0^+(x)P_{n+1}^{(-\gamma,-\delta)}(x)\right. . \qquad (3.5.27)$$

The case $n = -1$ may be considered to give the 'complementary' solution $\phi(t) = -2X_0^+(t)$, $f(t) = 0$, if $P_{-1}^{(\gamma,\delta)}(z)$ is taken to be zero.

Corresponding results for the $X_2(z)$ solution may be found by taking $\gamma = -1/2 - i\alpha$, $\delta = -1/2 + i\alpha$, as in Ex. 3.5.3.

The particular case $\alpha = 0$ leads to the known results for the Chebyshev polynomials, since

$$P_n^{(1/2,1/2)}(z) = 2C_nU_n(z), \quad P_n^{(-1/2,-1/2)}(z) = C_{n-1}T_n(z), \qquad (3.5.28)$$

where $C_n = \Gamma(n + 3/2)\pi^{-1/2}/(n + 1)!$. Thus for $\gamma = \delta = -1/2$, $V_n(z)$ becomes

$$V_n(z) = C_{n-1}(z^2-1)^{-1/2}[T_n(z) - (z^2-1)^{1/2}U_{n-1}(z)]$$
$$= C_{n-1}(z^2-1)^{-1/2}[z + (z^2-1)^{1/2}]^{-n}. \qquad (3.5.29)$$

Examples 3.5

1. Carry out an analysis of the real integral equation

$$\phi(x) + \frac{\cot \pi\alpha}{\pi}\int_{-1}^{1} \frac{\phi(y)}{y-x}\,dy = \psi(x), \quad -1 < x < 1,$$

similar to that given above for equation (3.5.1); assume $0 < \alpha < 1/2$. Show that the general solution with integrable singularities at ± 1 and finite at infinity is

$$\phi(x) = \sin^2\pi\alpha\psi(x) - \frac{\sin\pi\alpha\cos\pi\alpha}{\pi} X_0^+(x)\int_{-1}^{1} \frac{\psi(y)dy}{X_0^+(y)(y-x)} +$$

$$+ (Ax+B)X_0^+(x),$$

where $X_0^+(x) = (1 - x)^{-1/2+\alpha}(1 + x)^{-1/2-\alpha}$.

2. Use the integral representation

$$Q_n^{(\gamma,\delta)}(z) = \frac{(z-1)^{-\gamma}(z+1)^{-\delta}}{2}\int_{-1}^{1} \frac{(1-t)^{\gamma}(1+t)^{\delta}}{t-z} P_n^{(\gamma,\delta)}(t)dt$$

to show that

$$Q_n^{(1/2,1/2)}(z) = -\pi C_n G_n(z), \quad Q_n^{(-1/2,-1/2)}(z) = (\pi/2)C_{n-1}R_n(z),$$

where $G_n(z)$, $R_n(z)$ are the functions defined in §2.3.

3. By taking $\gamma = -1/2 - i\alpha$, $\delta = -1/2 + i\alpha$ show that the solution of equation (3.5.1) for $f(t) = P_{n-1}^{(-\gamma,-\delta)}(t)$ is

$$\phi(t) = -1/2\ X_2^+(t)P_{n-1}^{(-\gamma,-\delta)}(t),$$

where the case $n = 0$ is excluded by (3.5.11). Show that

$$\Phi(z) = \frac{\exp(-\pi\alpha)}{\pi} Q_n^{(\gamma,\delta)}(z),$$

and

$$\frac{1}{\pi i}\int_{-1}^{1} \frac{\phi(t)}{t-x}\,dt = \frac{\exp(-\pi\alpha)}{\pi} Q_n^{(\gamma,\delta)}(x), \quad |x| > 1.$$

4. Carry out an analysis of the equation

$$\phi(x) + \frac{\varphi}{\pi i}\int_{-1}^{1} \frac{\phi(t)}{t-x}\,dt = g(x), \quad |x| < 1,$$

where $-1 < \varphi < 1$. Show that $\Phi(z)$ satisfies

$$\kappa\Phi^+(x) - \Phi^-(x) = F(x), \quad |x| < 1,$$

where the notation is as in (3.5.8). Now the fundamental solution $X_0(z)$ is $X_0(z) = (z - 1)^{i\alpha}(z + 1)^{-i\alpha}$ and

$$\phi(x) = \cosh^2\pi\alpha g(x) - \frac{\sinh 2\pi\alpha}{2\pi i} X_0^+(x)\int_{-1}^{1} \frac{g(t)dt}{X_0^+(t)(t-x)} + AX_0^+(x).$$

Show, moreover, that if $F(z)$, $S(z)$ are polynomials such that

$$\frac{F(z)}{X_0(z)} - S(z) = 0(z^{-1}),$$

then

$$\Phi(z) = (\kappa-1)^{-1}\{F(z) - X_0(z)S(z) - AX_0(z)\},$$

$$\phi(x) = X_0^+(x)S(x) + AX_0^+(x), \qquad |x| < 1.$$

CHAPTER IV - PLANE CONTACT PROBLEMS

> "I'm afraid I don't quite understand," said Alice. "It gets easier further on," Humpty Dumpty replied.
>
> - *Through the Looking Glass*

4.1 *Introduction*

At first sight it might appear that to assume that two contacting bodies are *perfectly elastic*, *homogeneous* and *isotropic*, and are in a *plane* state, i.e. in plane strain, plane stress or generalized plane stress, would be a sufficient idealization to allow effective solutions to be obtained for a significant number of problems. Unfortunately this is not so, and although it is possible to make some general statements about two such contacting bodies (§4.10) it is usual to make further simplifying assumptions. The two most common are that the curvature of one of the bodies, in the contact region, is so small that it may be modelled as an elastic *half-plane*, and that the other body is so much more stiff than the first that it may be treated as a *rigid body*, or *punch*. Most of the analysis in this chapter is based on these assumptions. In §4.10 it will be shown that certain problems regarding the contact of *two elastic bodies* are governed by the same equations as those for the contact between a rigid body and an elastic half-plane. Other simplifications that are made to the problem concern the presence or absence of *friction* between the contacting bodies. Here there are two extreme assumptions, viz. that there is *no friction*, and there is *complete adhesion*, between the bodies; and a number of less drastic assumptions in which some friction is allowed. Frictionless contact problems were discussed in §2.4, and will be considered again in §4.5; contact problems with adhesion or friction will be studied in §§4.6, 4.7, respectively.

Contact problems for the elastic half-plane, i.e. problems in which one body is a punch and the other an elastic half-plane, fall into the class of problems treated by the *classical theory of elasticity*. This theory was largely developed in the 19th century and is fully described with historical references in Love's (1927) treatise, the first edition of which appeared in 1893. This treatise contains much of the mathematical apparatus necessary for the solution of contact problems (in 2 and 3 dimensions) but the only theory contained in it which relates specifically to contact problems is that developed by Hertz, which has been described in part in §2.9 above. Chaplygin's collected works (1950) contain an unpublished manuscript, dated 1900, on the pressure of a punch with plane face in contact with a half-plane. Sadowsky (1928) considered several frictionless contact problems, including those for one punch with a plane base, and an infinite number of such punches periodically spaced, in contact with an elastic half-plane.

Modern research into contact problems for the half-plane began in the USSR in 1935 and, for the period 1935-50, was centred around N. I. Muskhelishvili and his colleagues at Tbilisi. This work, which was part of a general study of complex variable methods in the plane theory of elasticity, was based not only on the labours of the early elasticians but also on that of researchers in complex variable analysis, including Cauchy, Riemann, Hilbert, Kolosov, Carleman, Morera, Plemelj, Privalov and Tricomi. References to the early work on contact problems (1935-50) may be found in Muskhelishvili's treatise (1953a) and are summarized in Chapter 13 of Muskhelishvili (1953b). Two other important sources for Soviet work are the books on *Contact Problems in the Theory of Elasticity* by Shtaerman (1949) and Galin (1961); the latter is available in an English translation. References to Soviet work will be given only if they occur in one of the Russian journals in the list given just before the Bibliography; for Muskhelishvili's work, reference will be made to his treatises, rather than to his original papers.

4.2 *The basic equations of the complex variable theory of elasticity*

The general complex formulation of elasticity theory has been the subject of numerous books; besides the treatises of Muskhelishvili already cited, the reader may consult Green and Zerna (1954), Sokolnikoff (1956), Milne-Thomson (1960) and England (1971 a). Teodorescu (1964) has made a survey of 100 years of the plane theory of elasticity.

The equations of plane elastostatics are:

$$e_{xx} = \frac{\partial u}{\partial x}, \quad e_{xy} = \frac{1}{2}\left(\frac{\partial u}{\partial y} + \frac{\partial v}{\partial x}\right), \quad e_{yy} = \frac{\partial v}{\partial y}, \tag{4.2.1}$$

$$\tau_{xx} = \frac{2\mu}{1-2\nu}[(1-\nu)e_{xx} + \nu e_{yy}], \quad \tau_{xy} = 2\mu e_{xy}, \tag{4.2.2}$$

$$\tau_{yy} = \frac{2\mu}{1-2\nu}[(1-\nu)e_{yy} + \nu e_{xx}], \tag{4.2.3}$$

$$\frac{\partial \tau_{xx}}{\partial x} + \frac{\partial \tau_{xy}}{\partial y} + F_x = 0 = \frac{\partial \tau_{xy}}{\partial x} + \frac{\partial \tau_{yy}}{\partial y} + F_y \tag{4.2.4}$$

where, for generalized plane stress, ν must be replaced by $\nu^* = \nu/(1 + \nu)$ and all displacements, strains, stresses and body forces must be interpreted as mean values, as described in §1.11.

The solution of these equations was obtained in complex form in equation (1.11.18); it is

$$2\mu D = \kappa\phi(z) - z\overline{\phi'(z)} - \overline{\psi(z)} + \varphi W. \tag{4.2.5}$$

Corresponding to this complex displacement we define complex strain and stress components. Using equations (1.11.15) we note that

$$\frac{\partial D}{\partial z} = \frac{1}{2}\left(\frac{\partial D}{\partial x} - i\,\frac{\partial D}{\partial y}\right) = \frac{1}{2}\left(\frac{\partial u}{\partial x} + i\,\frac{\partial v}{\partial x} - i\,\frac{\partial u}{\partial y} + \frac{\partial v}{\partial y}\right), \tag{4.2.6}$$

$$\frac{\partial \overline{D}}{\partial z} = \frac{1}{2}\left(\frac{\partial \overline{D}}{\partial x} - i\,\frac{\partial \overline{D}}{\partial y}\right) = \frac{1}{2}\left(\frac{\partial u}{\partial x} - i\,\frac{\partial v}{\partial x} - i\,\frac{\partial u}{\partial y} - \frac{\partial v}{\partial y}\right), \tag{4.2.7}$$

which may be combined to give

$$\varepsilon = e_{xx} + e_{yy} = \frac{\partial D}{\partial z} + \frac{\partial \overline{D}}{\partial \overline{z}} \tag{4.2.8}$$

$$e_{yy} - e_{xx} + 2ie_{xy} = -2\,\frac{\partial \overline{D}}{\partial z}\,. \tag{4.2.9}$$

This suggests that we should introduce the stress combinations

$$\Theta = \tau_{xx} + \tau_{yy}, \qquad \Xi = \tau_{yy} - \tau_{xx} + 2i\tau_{xy}. \tag{4.2.10}$$

In terms of the strains, these are given by

$$(1-2\nu)\Theta = 2\mu\varepsilon, \qquad \Xi = 2\mu(e_{yy}-e_{xx}+2ie_{xy}), \tag{4.2.11}$$

so that

$$\Theta = 2[\phi'(z) + \overline{\phi}'(\overline{z})] + 2(1-\varphi)\partial W/\partial z, \tag{4.2.12}$$

$$\Xi = 2[\overline{z}\phi''(z) + \psi'(z)] - 2\varphi\,\partial\overline{W}/\partial z. \tag{4.2.13}$$

It is convenient to introduce functions Φ,Ψ such that

$$\Phi(z) = \phi'(z), \quad \Psi(z) = \psi'(z); \tag{4.2.14}$$

then

$$\Theta = 2[\Phi(z) + \overline{\Phi}(\overline{z})] + 2(1-\varphi)\partial W/\partial z, \tag{4.2.15}$$

$$\Xi = 2[\overline{z}\Phi'(z) + \Psi(z)] - 2\varphi\,\partial\overline{W}/\partial z. \tag{4.2.16}$$

The functions $\Phi(z),\Psi(z)$ and also $\phi(z),\psi(z)$ are called complex potentials.

It is convenient to introduce the function $F(z,\overline{z})$, such that

$$F(z,\overline{z}) = 2[\phi(z) + z\overline{\phi}'(\overline{z}) + \overline{\psi}(\overline{z})] - 2\varphi W(z,\overline{z}); \tag{4.2.17}$$

then Θ,Ξ may be written

$$\Theta = \frac{\partial F}{\partial z} + 2\,\frac{\partial W}{\partial z}\,, \qquad \Xi = \frac{\partial F}{\partial \overline{z}}\,. \tag{4.2.18}$$

In the engineering problems that will be considered the effects of body forces are relatively unimportant. It will therefore be assumed, unless otherwise stated, that there are *no body forces* present. In that case F is related directly to the Airy stress function $\mathcal{U}$ of Ex. 1.11.3 by the equation

$$F = 4\,\frac{\partial \mathcal{U}}{\partial \bar{z}}\,. \tag{4.2.19}$$

Consider the stresses acting over a curve AB as shown in Figure 4.2.1. Since $\hat{n} = \sin\alpha\hat{i} - \cos\alpha\hat{j}$ equation (1.4.7) shows that

$$\left.\begin{aligned}\tau_{nx} &= \sin\alpha\tau_{xx} - \cos\alpha\tau_{xy}\\ \tau_{ny} &= \sin\alpha\tau_{xy} - \cos\alpha\tau_{yy}\end{aligned}\right\} \tag{4.2.20}$$

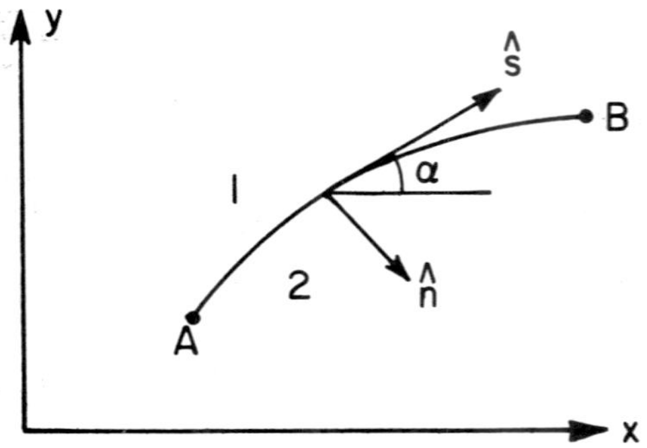

Figure 4.2.1 - The normal $\hat{n}$ *points from side 1 to side 2.*

which, when combined with Ex. 1.11.3, give

$$\left.\begin{aligned}\tau_{nx} &= \sin\alpha\,\frac{\partial^2\mathcal{U}}{\partial y^2} + \cos\alpha\,\frac{\partial^2\mathcal{U}}{\partial x\partial y} = \frac{\partial}{\partial s}\left(\frac{\partial\mathcal{U}}{\partial y}\right),\\ \tau_{ny} &= -\sin\alpha\,\frac{\partial^2\mathcal{U}}{\partial x\partial y} - \cos\alpha\,\frac{\partial^2\mathcal{U}}{\partial x^2} = -\,\frac{\partial}{\partial s}\left(\frac{\partial\mathcal{U}}{\partial x}\right).\end{aligned}\right\} \tag{4.2.21}$$

Thus

$$\tau_{nx} + i\tau_{ny} = -i\left(\frac{\partial}{\partial x} + i\,\frac{\partial}{\partial y}\right)\frac{\partial\mathcal{U}}{\partial s} = -2i\,\frac{\partial}{\partial s}\,\frac{\partial\mathcal{U}}{\partial\bar{z}} = -\,\frac{i}{2}\,\frac{\partial F}{\partial s}\,, \tag{4.2.22}$$

so that

$$dF = 2d[\phi(z) + z\overline{\phi'}(\overline{z}) + \overline{\psi}(\overline{z})] = 2i(\tau_{nx}+i\tau_{ny})ds, \tag{4.2.23}$$

and this yields an important criterion for the single-valuedness of F. The change in F in going from A to B along a particular path Γ_{AB} is

$$[F]_A^B = 2i\int_{\Gamma_{AB}} (\tau_{nx}+i\tau_{ny})ds, \tag{4.2.24}$$

so that F will be single-valued if $[F]_A^B$ is independent of the path Γ_{AB} for all points A,B. This is equivalent to the condition that

$$\int_{\Gamma}(\tau_{nx}+i\tau_{ny})ds = 0 \tag{4.2.25}$$

for all *closed* paths Γ in the body. If the body is simply-connected this criterion will be satisfied provided that the body is in equilibrium; if the body is not simply connected, equation (4.2.25) will provide an extra condition which must hold for any Γ enclosing an interior boundary. When (4.2.25) is satisfied one may deduce that the stresses in a plane elastic body due to prescribed surface tractions (and in the absence of body-forces) are independent of the elastic constants. This result, due to Michell (1899), follows from the fact that Θ,Ξ in (4.2.18) are given as derivatives of the single-valued function F.

Equations (4.2.17),(4.2.24) show that the total force acting across Γ_{AB}, and according to the convention of Figure 1.4.1 this is the force which the material on side 2 exerts on the material on side 1, is

$$X_n + iY_n = -i[F]_A^B/2 = -i[\phi(z) + z\overline{\phi'}(z) + \overline{\psi}(\overline{z})]_A^B. \tag{4.2.26}$$

The moment about 0 of the forces applied across AB is

$$M = \int_A^B (x\tau_{ny} - y\tau_{nx})\,ds$$

$$= -\int_A^B \left\{x\frac{\partial}{\partial s}\left(\frac{\partial U}{\partial x}\right) + y\frac{\partial}{\partial s}\left(\frac{\partial U}{\partial y}\right)\right\} ds,$$

and integration by parts shows that

$$M = -\left[x\frac{\partial U}{\partial x} + y\frac{\partial U}{\partial y}\right]_A^B + \int_A^B \left(\frac{\partial U}{\partial x}\frac{\partial x}{\partial s} + \frac{\partial U}{\partial y}\frac{\partial y}{\partial s}\right) ds$$

$$= \left[U - x\frac{\partial U}{\partial x} - y\frac{\partial U}{\partial y}\right]_A^B.$$

Noting that

$$x\frac{\partial U}{\partial x} + y\frac{\partial U}{\partial y} = \text{Re}\left\{(x-iy)\left(\frac{\partial U}{\partial x} + i\frac{\partial U}{\partial y}\right)\right\} = \text{Re}\left\{2\bar{z}\frac{\partial U}{\partial \bar{z}}\right\},$$

we find that

$$M = \text{Re}[\chi(z) - z\psi(z) - z\bar{z}\phi'(z)]_A^B. \tag{4.2.27}$$

Examples 4.2

1. Show that equations (4.2.4) may be combined into one complex equation

$$\frac{\partial \Theta}{\partial \bar{z}} - \frac{\partial \bar{\Xi}}{\partial z} - 2\frac{\partial V}{\partial \bar{z}} = 0.$$

2. Show that, when $W \equiv 0$, the stress combinations Θ, Ξ may be written

$$\Theta = \nabla_1^2 U \equiv 4\frac{\partial^2 U}{\partial z \partial \bar{z}}, \qquad \Xi = 4\frac{\partial^2 U}{\partial z^2}.$$

3. Show that

$$\tau_{rr} + \tau_{\theta\theta} = \Theta, \qquad \tau_{\theta\theta} - \tau_{rr} + 2i\tau_{r\theta} = \exp(2i\theta)\Xi.$$

4. In the notation of Figure 4.2.1 show that

$$2(\tau_{nx} + i\tau_{ny}) = -i[\exp(i\alpha)\Theta + \exp(-i\alpha)\bar{\Xi}].$$

5. By considering the equilibrium of a small circular region around the origin show that the complex potentials $\Phi(z), \Psi(z)$ relating to a force $X + iY$ applied at the origin are

$$\Phi(z) = -\frac{(X+iY)}{2\pi(1+\kappa)z}, \qquad \Psi(z) = \frac{\kappa(X-iY)}{2\pi(1+\kappa)z}.$$

6. Show that

$$2(\tau_{\theta\theta}-i\tau_{r\theta})\exp(i\theta) = -2i(\tau_{\theta x}+i\tau_{\theta y}) = \partial F/\partial r,$$

and hence that

$$\int_0^r (\tau_{\theta\theta}-i\tau_{r\theta})\exp(i\theta)dr = [\phi(z) + z\overline{\phi'(\overline{z})} + \overline{\psi}(\overline{z})]_0^r.$$

7. Show that if there are no body forces, then Θ is harmonic, i.e.

$$\nabla_1^2\Theta \equiv \left(\frac{\partial^2}{\partial x^2} + \frac{\partial^2}{\partial y^2}\right)\Theta \equiv 4\,\frac{\partial^2\Theta}{\partial z\partial\overline{z}} = 0.$$

4.3 *Conditions satisfied by the complex potentials*

A definitive study of the conditions satisfied by the complex potentials $\phi(z), \psi(z), \Phi(z), \Psi(z)$ may be found in Muskhelishvili (1953a). He considers an elastic body occupying a general multiply-connected region of the z-plane and discusses the questions of existence and uniqueness of solution, and the form which the potentials take, particularly at infinity. In this section we shall merely summarize, with at most intuitive verification, particular cases of some of these results, assuming only enough generality to cover the kind of problems which will be considered in later sections. We shall make the following assumptions:

1) The region $\mathcal{D}^+$ occupied by the elastic body is (in this chapter) the upper half-plane S^+, $y \geq 0$. In Chapter VIII the region will be assumed to be the interior or exterior of the unit circle, or a region which may be conformally

mapped onto one of these.

2) The stresses are single-valued continuous functions of x,y having continuous derivatives up to the second order (except perhaps at certain boundary points, as discussed below) and are bounded at infinity.

3) The displacements are single-valued, continuous functions (again except perhaps at the points mentioned in 2)) having first-order derivatives connected to the stresses by equations (4.2.1)-(4.2.4).

4) Forces are applied to the body by means of surface tractions on a finite portion of the boundary, body forces, or stresses at infinity.

First consider to what extent the complex potentials are determined if the state of stress and body forces are known. Equation (4.2.15) shows that $\mathrm{Re}[\Phi(z)]$ is known so that $\Phi(z)$ is determined apart from an imaginary constant. Thus

$$\Phi(z) = \Phi_0(z) + Ci, \qquad \phi(z) = \phi_0(z) + Ciz + E, \tag{4.3.1}$$

where $\Phi_0(z), \phi_0(z)$ are determined, C,E are arbitrary constants with C real. Now equation (4.2.16) shows that if, as we suppose, Ξ and W are known, then $\Psi(z)$ is determined so that

$$\Psi(z) = \Psi_0(z), \qquad \psi(z) = \psi_0(z) + E' \tag{4.3.2}$$

where E' is a complex constant. Equation (4.2.5) shows that the arbitrary terms C,E,E' yield a complex displacement

$$2\mu D = (\kappa+1)Ciz + (\kappa E-\overline{E}'). \tag{4.3.3}$$

The first term represents a rigid-body rotation, the second a translation. If the displacements are known in addition to (or instead of) the stresses, then $\Phi(z)$ and $\Psi(z)$ will be completely determined, i.e. C = 0 and $\phi(z),\psi(z)$ will be determined apart from arbitrary constants E,E' satisfying

$$\kappa E - \overline{E}' = 0. \tag{4.3.4}$$

We shall now show, in a non-rigorous fashion, that the complex potentials $\Phi(z),\Psi(z)$ are holomorphic functions of z in the region occupied by the body. In the absence of body forces, the stress combination Θ is harmonic (Ex. 4.2.7), i.e.

$$\nabla_1^2\Theta = 0. \tag{4.3.5}$$

But by equation (4.2.15)

$$\Theta(x,y) = 4\,\mathrm{Re}[\Phi(z)], \tag{4.3.6}$$

so that, using the analysis described in §3.1, we may show that $\Phi(z)$ is holomorphic in $\mathcal{D}^+$ except possibly at infinity. Assumption 2), that the stresses are bounded at infinity, implies that $\Phi(z)$ is holomorphic at infinity, so that

$$\Phi(z) = \Gamma + \gamma z^{-1} + O(z^{-2}) \quad \text{for large } |z|, \tag{4.3.7}$$

where $O(z^{-2})$ denotes a quantity $\varepsilon(z)$ such that $|\varepsilon(z)| < A|z|^{-2}$, for some constant A, as $|z| \to \infty$. Now by using Ex. 1.11.3, Ex. 4.2.2 and equation (4.2.14) we find

$$\nabla_1^2[\chi(z) + \overline{\chi}(\overline{z})] = \nabla_1^2[2\mathcal{U} - \overline{z}\phi(z) - z\overline{\phi}(\overline{z})]$$

$$= 2\Theta - 4\Phi(z) - 4\overline{\Phi}(\overline{z}) = 0, \tag{4.3.8}$$

so that $\chi(z)$ is holomorphic, except possibly at infinity. Now Ξ, like the stresses, is bounded at infinity, so that equation (4.2.16) implies that $\Psi(z)$, the second derivative of $\chi(z)$, must satisfy

$$\Psi(z) = \Gamma' + \gamma' z^{-1} + O(z^{-2}) \quad \text{for large } |z|. \tag{4.3.9}$$

Equations (4.3.7),(4.3.9) now lead to the results

$$\phi(z) = \Gamma z + \gamma \ln z + \phi^*(z), \tag{4.3.10}$$

$$\psi(z) = \Gamma' z + \gamma' \ln z + \psi^*(z), \tag{4.3.11}$$

where $\phi^*(z),\psi^*(z)$ are holomorphic in $\mathcal{D}^+$ including the point at infinity. The two complex constants Γ,Γ' are related to the stresses and rotation at infinity (Ex. 4.3.1), while γ,γ' are related to the total force applied to the finite part of the boundary, as will now be shown.

Consider, for the sake of example, the case when the body occupies the upper half-plane. Take the branch of $\ell n\ z$ defined in S^+ by

$$z = |z|\exp(i\theta), \qquad \ell n\ z = \ell n|z| + i\theta . \tag{4.3.12}$$

Equation (4.2.26) shows that the force applied to the upper half-plane over a segment AB, where A = (-a,0), B = (b,0), and a,b are large, as shown in Figure 4.3.1, is

$$X + iY = -i[\gamma\ \ell n\ z + \bar{\gamma}(z/\bar{z}) + \bar{\gamma}'\ \ell n\ \bar{z} + o(1) + \text{const}]_A^B.$$

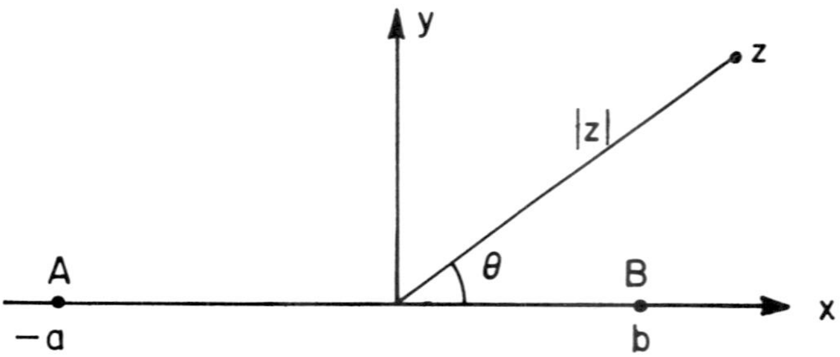

Figure 4.3.1 - The points A, B *are far from the origin.*

Now at A, $\ell n\ z = \ell n\ a + i\pi$, $\ell n\ \bar{z} = \ell n\ a - i\pi$, and at B, $\ell n\ z = \ell n\ b = \ell n\ \bar{z}$, so that

$$X + iY = -i[(\gamma+\bar{\gamma}')\ell n(b/a) - i\pi(\gamma-\bar{\gamma}') + o(1)]. \tag{4.3.13}$$

This must be finite for all b,a so that

$$\gamma + \bar{\gamma}' = 0, \tag{4.3.14}$$

and then the force applied to the x-axis is

$$X + iY = -\pi(\gamma-\bar{\gamma}') = -2\pi\gamma. \tag{4.3.15}$$

Returning to equations (4.3.7),(4.3.9) we see that the complex potentials Φ,Ψ satisfy

$$\Phi(z) = \Gamma - \frac{(X+iY)}{2\pi z} + 0(z^{-2}),$$

$$\Psi(z) = \Gamma' + \frac{(X-iY)}{2\pi z} + 0(z^{-2}), \qquad (4.3.16)$$

for large $|z|$.

Although the stresses are bounded at infinity, the displacements are not. In fact equation (4.2.5) shows that for large $|z|$

$$4\pi\mu D = 2\pi[(\kappa\Gamma-\overline{\Gamma})z - \overline{\Gamma}'\overline{z}] - \kappa(X+iY)\ell n\ z - (X+iY)\ell n\ \overline{z} +$$

$$+ (X-iY)(z/\overline{z}) + 0(1). \qquad (4.3.17)$$

This shows that even if Γ,Γ' are zero, i.e. there is no rotation and no stress at infinity, the displacements will have a logarithmic singularity at infinity unless $X = 0 = Y$. This is one of the unfortunate unrealities of plane elasticity. One of the consequences of the unboundedness of the displacements is that if a punch is pressed into the half-plane, it is not possible to find the actual penetration or 'normal approach' of the punch. Nor is it possible to find the actual displacement of the half-plane, only the relative displacement of two points. Note the arbitrary constant C of equation (2.4.9); such an arbitrary constant does not appear in the axisymmetric problems of §2.6. Various attempts have been made to remove this arbitrariness in plane problems; these are discussed in Chapter VIII.

Example 4.3

1. Show that the stresses corresponding to $\Phi(z) = \Gamma = \Gamma_1 + i\Gamma_2$, $\Psi(z) = \Gamma' = \Gamma_1' + i\Gamma_2'$ are $\tau_{xx} = 2\Gamma_1 - \Gamma_1'$, $\tau_{yy} = 2\Gamma_1 + \Gamma_1'$, $\tau_{xy} = \Gamma_2'$ and that the rotation ω_{xy} of equation (1.2.8) is given by $2\mu\omega_{xy} = (\kappa + 1)\Gamma_2$.

4.4 *The stress boundary-value problem for the half-plane*

In contrast to Muskhelishvili (1953a), who solved problems of great generality, we shall consider only some very simple problems. First let us recapitulate some of the results already obtained. We suppose that the elastic body occupies the upper half-plane S^+ and that the conditions 1)-4) mentioned at the beginning of §4.3 are satisfied. The stress combinations Θ, Ξ and the displacement D are then given by

$$\Theta = 2[\Phi(z) + \overline{\Phi}(\overline{z})], \tag{4.4.1}$$

$$\Xi = 2[\overline{z}\Phi'(z) + \Psi(z)], \tag{4.4.2}$$

$$2\mu D = \kappa\phi(z) - z\overline{\phi}'(\overline{z}) - \overline{\psi}(\overline{z}), \tag{4.4.3}$$

where

$$\phi'(z) = \Phi(z), \qquad \psi'(z) = \Psi(z), \tag{4.4.4}$$

and $\Phi(z), \Psi(z)$ are holomorphic in S^+ including the point at infinity where they satisfy equations (4.3.7),(4.3.9). The functions $\phi(z)$, $\psi(z)$ are holomorphic in S^+ except at infinity where equations (4.3.10),(4.3.11) hold.

Equations (4.4.1),(4.4.2) may be combined to give

$$\tau_{yy} - i\tau_{xy} = \frac{1}{2}\{\Theta + \overline{\Xi}\} = \Phi(z) + \overline{\Phi}(\overline{z}) + z\overline{\Phi}'(\overline{z}) + \overline{\Psi}(\overline{z}), \tag{4.4.5}$$

while equation (4.2.5) may be differentiated with respect to x to give

$$2\mu D' = 2\mu(u'+iv') = \kappa\Phi(z) - \overline{\Phi}(\overline{z}) - z\overline{\Phi}'(\overline{z}) - \overline{\Psi}(\overline{z}), \tag{4.4.6}$$

where

$$u' = \frac{\partial u}{\partial x}, \qquad v' = \frac{\partial v}{\partial x}. \tag{4.4.7}$$

Although our primary interest lies in mixed boundary-value problems, we shall consider first the simpler stress boundary-value problem defined in §1.9. Suppose that the elastic body is acted on by specified stresses distributed along the real axis, L and that these stresses satisfy an H-condition on L including the point at infinity, and that they vanish at infinity. Let

$$\tau_{yy}(x,0) = -p(x), \qquad \tau_{xy}(x,0) = -s(x); \tag{4.4.8}$$

then, on the real axis

$$\Phi(x) + \overline{\Phi(x)} + x\overline{\Phi'(x)} + \overline{\Psi(x)} = -(p(x) - is(x)), \tag{4.4.9}$$

where these values are the boundary values as z tends to x from S^+. The functions $\Phi(z),\Psi(z)$ are defined only in S^+, but $\Phi(z)$ will now be defined in S^-. The function $\Phi(z)$ is defined and holomorphic when $\mathrm{Im}(z) > 0$, so that $\overline{\Phi}(z)$ is defined and holomorphic when $\mathrm{Im}(z) < 0$, i.e. when z is in S^-. (Here $\overline{\Phi}(z)$ has the following meaning. Divide $\Phi(z)$ into its real and imaginary parts, i.e. $\Phi(z) = \Phi(x,y) = U(x,y) + iV(x,y)$; then $\overline{\Phi}(z) \equiv \{\overline{\Phi(\overline{z})}\} = U(x,y) - iV(x,y)$). So also the function $-\overline{\Phi}(z) - z\overline{\Phi}'(z) - \overline{\Psi}(z)$ is defined and holomorphic in S^-. This is taken to be the definition of $\Phi(z)$ in S^-, i.e.

$$\Phi(z) = -\overline{\Phi}(z) - z\overline{\Phi}'(z) - \overline{\Psi}(z), \qquad z \text{ in } S^-. \tag{4.4.10}$$

This equation may be inverted to give $\Psi(z)$ for z in S^+ in terms of $\Phi(z)$ and $\overline{\Phi}(z)$; in fact

$$\Psi(z) = -\Phi(z) - \overline{\Phi}(z) - z\Phi'(z), \qquad z \text{ in } S^+. \tag{4.4.11}$$

Thus equations (4.4.5),(4.4.6) become, for z in S^+,

$$\tau_{yy} - i\tau_{xy} = \Phi(z) - \Phi(\overline{z}) + (z-\overline{z})\overline{\Phi}'(\overline{z}), \tag{4.4.12}$$

$$2\mu D' = \kappa\Phi(z) + \Phi(\overline{z}) - (z-\overline{z})\overline{\Phi}'(\overline{z}). \tag{4.4.13}$$

Two conditions will now be introduced regarding the degree of singularity that $\Phi(z)$ may possess on the axis. These are

$$\lim_{y\to 0} y\Phi'(x + iy) = 0, \tag{4.4.14}$$

except possibly at a finite number of points x_0 on the axis --

ends of a loaded portion, for instance. Near these points however

$$\lim_{x \to 0}(x - x_0)\Phi(x) = 0. \tag{4.4.15}$$

This has the effect of excluding point loads. Under condition (4.4.14), equations (4.4.12),(4.4.13) yield the important relations

$$-p(x) + is(x) = \Phi^+(x) - \Phi^-(x), \qquad x \text{ on } L, \tag{4.4.16}$$

$$2\mu(u'(x) + iv'(x)) = \kappa\Phi^+(x) + \Phi^-(x), \qquad x \text{ on } L. \tag{4.4.17}$$

Notice that if z is in S^+ then $\bar{z}$ is in S^- so that the limiting value of $\Phi(\bar{z})$ as $z \to x + i0$ is $\Phi^-(x)$. Equation (4.4.16) shows that if part of the x-axis is unloaded then $\Phi^+(x)$ and $\Phi^-(x)$ are equal; this situation is described by saying that the $\Phi(z)$ which was defined in S^- by means of (4.4.10) is an *analytic continuation* through the unloaded part of the boundary of the $\Phi(z)$ defined in S^+.

The solution of the first fundamental problem follows immediately. For equation (4.4.16) has the same form as equation (3.3.6) and since, by assumption, $-p(x) + is(x)$ satisfies the conditions prescribed for f(x) in §3.3, the solution of equation (4.4.16) is

$$\Phi(z) = \frac{1}{2\pi i}\int_{-\infty}^{\infty} \frac{-p(t)+is(t)dt}{t-z} . \tag{4.4.18}$$

The simplest example of the foregoing is provided by a constant pressure p acting over a portion [-a,a] of the axis. Then

$$\Phi(z) = \frac{-p}{2\pi i}\int_{-a}^{a} \frac{dt}{t-z} = \frac{-p}{2\pi i}\, \ell n\left(\frac{z-a}{z+a}\right), \tag{4.4.19}$$

and $\bar{\Phi}(z) = -\Phi(z)$ so that, by equation (4.4.11),

$$\Psi(z) = -z\Phi'(z) = paz/[\pi i(z^2-a^2)]. \tag{4.4.20}$$

We note that the complex potential $\Phi(z)$ given by (4.4.19) satisfies condition (4.4.14) except at the ends of the loaded portion. This is to be expected since equation (3.3.6) does not hold at the end of a finite segment at which f(t) is not zero -- or alternatively at a point at which f(t) is discontinuous. The stresses may be calculated by writing $z - a = \rho_1 \exp(i\theta_1)$, $z + a = \rho_2 \exp(i\theta_2)$ as shown in Figure 4.4.1; then

$$2\pi i\Phi(z) = -p\,\ln(\rho_1/\rho_2) - ip(\theta_1-\theta_2), \tag{4.4.21}$$

$$\pi p^{-1}\tau_{xx} = -(\theta_1-\theta_2) - 2ay\,\cos(\theta_1+\theta_2)/(\rho_1\rho_2), \tag{4.4.22}$$

$$\pi p^{-1}\tau_{yy} = -(\theta_1-\theta_2) + 2ay\,\cos(\theta_1+\theta_2)/(\rho_1\rho_2), \tag{4.4.23}$$

$$\pi p^{-1}\tau_{xy} = -2ay\,\sin(\theta_1+\theta_2)/(\rho_1\rho_2). \tag{4.4.24}$$

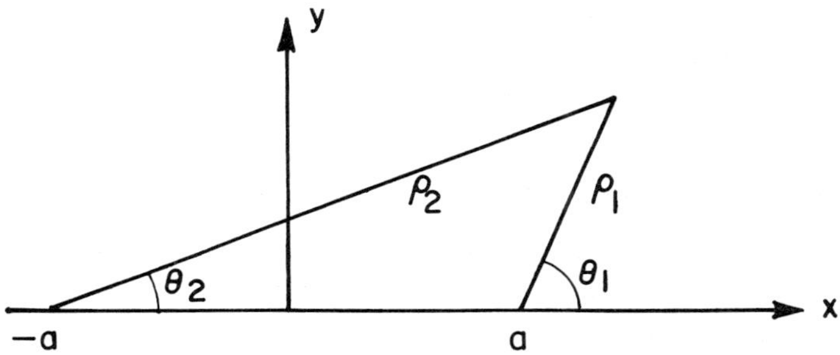

Figure 4.4.1 - The constant pressure p *acts along* -a, a.

Thus, since $y = \rho_1 \sin\theta_1 = \rho_2 \sin\theta_2$ the stresses are continuous up to the boundary with the exclusion of $x = \pm a$; the limiting values at the end points depend on the direction of approach to the points. The displacements are continuous up to the boundary including the end points $\pm a$, but with the exclusion of the point at infinity, near which they behave like $\ln|z|$, as stated in §4.3.

Example 4.4

1. Express the displacement derivative D' given by (4.4.6) in terms of $\rho_1, \rho_2, \theta_1, \theta_2$, for the problem of Figure 4.4.1.

4.5 *Frictionless contact problems for the half-plane*

The simplest contact problem for the half-plane relates to the deformation produced by a single rigid punch pressed in frictionless contact against the half-plane. The solution for a punch with a plane base was given by Sadowsky (1928). The analysis given here is based on 1935 work by Muskhelishvili, as described in Muskhelishvili (1953b), and it provides an alternative to that given in §2.2. Various solutions for punches of particular shapes, e.g. flat punches with rounded edges, appear in Shtaerman (1949).

Consider a rigid punch with a profile $y = f(x)$ which is pressed vertically into frictionless contact with the boundary of the half-plane. The assumption of infinitesimal elasticity implies that the boundary conditions must be applied on the undeformed surface, i.e. the x-axis. This will be realistic for a given punch only if its radius of curvature is large compared to the extent of the contact region. This will hereafter be assumed to be true for all punch problems considered. After contact, the vertical displacement of the region of contact will be given by

$$v^+(x) = f(x) + e \quad \text{on } L', \tag{4.5.1}$$

where e is a constant. The condition of frictionless contact implies

$$\tau_{xy} = 0 \quad \text{on } L', \tag{4.5.2}$$

while on the remainder L'' of the axis

$$\tau_{yy} - i\tau_{xy} = 0 \quad \text{on } L''. \tag{4.5.3}$$

Again define $\Phi(z)$ in S^- by means of equation (4.4.10); then on L,

$$\tau_{yy} - i\tau_{xy} = -p(x) + is(x) = \Phi^+(x) - \Phi^-(x), \tag{4.5.4}$$

so that

$$-2i\tau_{xy} = \Phi^{+}(x) - \Phi^{-}(x) - \overline{\Phi^{+}(x)} + \overline{\overline{\Phi}(x)} = 0 \quad \text{on } L. \tag{4.5.5}$$

But $\overline{\Phi^{+}(x)} = \overline{\Phi}^{-}(x)$, $\overline{\overline{\Phi}(x)} = \overline{\Phi}^{+}(x)$ so that

$$-2i\tau_{xy} = \Phi^{+}(x) + \overline{\Phi}^{+}(x) - \{\Phi^{-}(x) + \overline{\Phi}^{-}(x)\} = 0 \quad \text{on } L. \tag{4.5.6}$$

This means that the function $\Phi(z) + \overline{\Phi}(z)$ is holomorphic in the entire plane; since it vanishes at infinity it must vanish everywhere. Therefore

$$\overline{\Phi}(z) = -\Phi(z), \tag{4.5.7}$$

and

$$2\mu D'(z) = \kappa\Phi(z) + \Phi(\overline{z}) - (z - \overline{z})\,\overline{\Phi}'(\overline{z}) \tag{4.5.8}$$

gives

$$\begin{aligned} 4\mu i v'(x) &= \kappa\Phi^{+}(x) + \Phi^{-}(x) - \overline{\kappa\Phi^{+}(x)} - \overline{\Phi^{-}(x)} \\ &= \kappa\Phi^{+}(x) + \Phi^{-}(x) + \kappa\Phi^{-}(x) + \Phi^{+}(x), \end{aligned}$$

so that equations (4.5.1),(4.5.4) may be written

$$\Phi^{+}(x) + \Phi^{-}(x) = \vartheta^{-1} i f'(x) \quad \text{on } L', \tag{4.5.9}$$

$$\Phi^{+}(x) - \Phi^{-}(x) \equiv -p(x) = 0 \quad \text{on } L'', \tag{4.5.10}$$

where $\vartheta = (1 - \nu)/\mu$. Equation (4.5.9) has the form of equation (3.4.1) with $\kappa = 1$, so that $\alpha = 0$ and the solution is

$$\vartheta\Phi(z) = \frac{X_0(z)}{2\pi}\int_{L'} \frac{f'(t)dt}{X_0^{+}(t)(t-z)} + CX_0(z). \tag{4.5.11}$$

If L' is $[-a,a]$ then $X_0 = (z^2 - a^2)^{-1/2}$. The $\Phi(z)$ of (4.5.11) still has to satisfy equation (4.5.7); but (3.4.12) shows that, on L', $X_0^{+}(t) = -i(a^2 - t^2)^{-1/2} = -\overline{X}_0^{+}(t)$, while $\overline{X}_0(z) = X_0(z)$. Thus, since $f'(t)$ is real, the first term in (4.5.11) satisfies (4.5.7); the second term will satisfy it only if C is imaginary, i.e. $C = iB$. The constant B may be linked to the total force iP_0 applied by the punch since $\Phi(z)$ given by (4.5.11) satisfies $\vartheta\Phi(z) = iBz^{-1} + O(z^{-2})$ so that equation (4.3.15) gives $B = -\vartheta P_0/(2\pi)$.

For a punch with a straight horizontal base $f'(t) = 0$ and

$$2\pi\Phi(z) = -iP_0(z^2-a^2)^{-1/2} \tag{4.5.12}$$

and the stress under the punch is given by

$$p(x) = \pi^{-1}P_0(a^2-x^2)^{-1/2}. \tag{4.5.13}$$

For a punch with a straight inclined base, as in Figure 2.4.1, $f'(x) = \alpha$ so that

$$\vartheta\Phi(z) = \frac{X_0(z)}{2\pi}\alpha\int_{L'}\frac{dt}{X_0^+(t)(t-z)} - \frac{i\vartheta}{2\pi}P_0X_0(z). \tag{4.5.14}$$

The first integral may be calculated by the method of §3.4. The result is

$$2\vartheta\Phi(z) = \alpha i(1-zX_0(z)) - i\vartheta P_0X_0(z)/\pi. \tag{4.5.15}$$

The pressure under the punch is

$$p(x) = \Phi^-(x) - \Phi^+(x) = (\vartheta P_0+\pi\alpha x)(a^2-x^2)^{-1/2}/(\pi\vartheta). \tag{4.5.16}$$

For a punch of general profile there are two possible boundary conditions under the punch: *complete* contact, in which the contact region is specified and there will be (square root) singularities for $p(x)$ at the ends of the contact region, and *incomplete* contact in which the extent of the contact region is determined from the condition that the contact pressure is zero at the ends, as in Ex. 4.5.2, 4.5.3.

Examples 4.5

1. If $L' = [-a,a]$ and $f(x) = a\sum_{n=0}^{N} b_nT_n(x/a)$ show that $f'(x) = \sum_{n=0}^{N} nb_nU_{n-1}(x/a)$. Now use the result of Ex. 3.4.2 to deduce that if the punch is in complete contact then

$$2\vartheta\Phi(z) = i\sum_{n=1}^{N} nb_n\{U_{n-1}(z/a) - a(z^2-a^2)^{-1/2}T_n(z/a)\} -$$

$$- i\vartheta P_0(z^2-a^2)^{-1/2}/\pi.$$

Now use equation (4.5.4) to deduce

$$\vartheta p(x) = a(a^2-x^2)^{-1/2}\sum_{n=1}^{N} nb_nT_n(x/a) + \vartheta P_0(a^2-x^2)^{-1/2}/\pi.$$

2. Use the result of Ex. 4.5.1 to find the complex potential $\Phi(z)$ and the contact pressure distribution under a parabolic punch. Note that the contact radius a must be found so that $p(\pm a) = 0$.

3. If there is to be no singularity in the pressure at the ends of the contact region for the punch of Ex. 4.5.1 then $p(\pm a) = 0$. This means

$$\pi a\sum_{n=1}^{N} nb_n + \vartheta P_0 = 0, \text{ n even} \qquad \sum_{n=1}^{N} nb_n = 0. \text{ n odd}$$

For a punch which is symmetrical about $x = 0$ the odd b_n will be zero and the second condition will be satisfied automatically. The first condition then gives the relation between contact force P_0 and contact length a. For an unsymmetrical punch, it is necessary to make such a shift in the origin in the x-direction so that, when the punch profile is expressed in terms of the new coordinates, the contact region is $-a < x < a$, and both conditions have to be satisfied.

4. Show that equation (4.5.11) may be rearranged in the form

$$\vartheta\Phi(z) = \frac{X_2(z)}{2\pi}\int_{L'}\frac{f'(t)dt}{X_2^+(t)(t-z)} + iX_0(z)(A+Bz),$$

where the real constants A,B must be determined so that $\Phi(z)$ is holomorphic at infinity.

5. Show that when $\Phi(z)$ satisfies (4.5.7) the rotation ω_{xy} of (1.2.8) is $2\vartheta\ \text{Im}\{\Phi(z)\}$.

4.6 *Adhesive contact problems for the half-plane*

The assumptions of complete adhesion and complete absence of friction represent two extremes regarding the conditions in the contact zone between two bodies. This section relates to adhesive contact between a single rigid punch and a half-plane.

Suppose that $L' = [-a,a]$ and follow the procedure of §4.4; then for z in S^+

$$\Psi(z) = -\Phi(z) - \overline{\Phi}(z) - z\Phi'(z), \tag{4.6.1}$$

$$2\mu D'(z) = \kappa\Phi(z) + \Phi(\overline{z}) - (z-\overline{z})\overline{\Phi}'(\overline{z}), \tag{4.6.2}$$

$$\tau_{yy} - i\tau_{xy} = \Phi(z) - \Phi(\overline{z}) + (z-\overline{z})\overline{\Phi}'(\overline{z}). \tag{4.6.3}$$

The displacements and stresses on the +ve side of the boundary are

$$2\mu D'(x) = \kappa\Phi^+(x) + \Phi^-(x), \tag{4.6.4}$$

$$-p(x) + is(x) = \Phi^+(x) - \Phi^-(x). \tag{4.6.5}$$

For adhesive contact it is assumed that when a point on the boundary $y = 0$ of the half-plane comes into contact with a point of the punch the two remain in contact. This means, in particular, that if the punch is pressed normally into the half-plane then, *after a point of the half-plane comes into contact with the punch*, it will displace normally, i.e. $u'(t) = 0$, $v'(t) = f'(t)$, where $y = f(t) + c$ is the punch profile. Note however that this condition will hold only over the (unknown) contact region.

In the simplest case of a punch with a flat base the whole base $-a < x < a$ will be in contact and the boundary conditions will be

$$2\mu[u'(x) + iv'(x)] \equiv \kappa\Phi^+(x) + \Phi^-(x) = 0 \quad \text{on } L', \tag{4.6.6}$$

$$-p(x) + is(x) \equiv \Phi^+(x) - \Phi^-(x) = 0 \quad \text{on } L''. \tag{4.6.7}$$

This is an example of the fundamental mixed problem of the half-plane. When L' is a single segment the solution was obtained by Abramov (1937), using Mellin transforms. Muskhelishvili (1953a) found a solution in 1941 for an arbitrary number of segments; a similar solution was found independently by Glagolev (1942),(1943). The analysis of §3.4 shows that the solution satisfying (4.4.14) at the ends and zero at infinity is

$$\Phi(z) = CX_0(z) = iP_0X_0(z)/(2\pi), \tag{4.6.8}$$

where $X_0(z)$ and α (which is real) are given by (3.4.7),(3.4.16) and P_0 is the (vertical) force applied by the punch.

The contact stresses are given by (4.6.5) and are

$$\begin{aligned} -p(x) + is(x) &= \Phi^+(x) - \Phi^-(x) \\ &= -iP_0[X_0^+(x) - X_0^-(x)]/(2\pi). \end{aligned} \tag{4.6.9}$$

Using equations (3.4.13),(3.4.14) we may write

$$\begin{aligned} -p(x) + is(x) = \frac{-iP_0}{2\pi} \cdot (\kappa+1)\exp\left[-\frac{i\pi}{2} - \pi\alpha\right] \cdot \\ \cdot (a^2-x^2)^{-1/2}\left(\frac{a+x}{a-x}\right)^{-i\alpha}, \end{aligned} \tag{4.6.10}$$

and

$$\left.\begin{matrix} p(x) \\ \\ s(x) \end{matrix}\right\} = \frac{P_0}{2\pi} \cdot \frac{\kappa+1}{\kappa^{1/2}} \cdot (a^2-x^2)^{-1/2}\left\{\begin{matrix} \cos\gamma \\ \\ \sin\gamma, \end{matrix}\right. \tag{4.6.11}$$

where $\gamma = \alpha\,\ell n[(a + x)/(a - x)]$. Note that both $p(x)$ and $s(x)$ change sign an infinity of times as x approaches $\pm a$ so that if the punch is to remain in contact with the half-space it must be physically bonded to it. This bonding can be considered as providing an infinite coefficient of friction between the bodies. However, as Muskhelishvili (1953a, p. 467) shows, the smallest possible value of $|x|$ for which there is a negative $p(x)$ occurs when $\nu = 0$, $\kappa = 3$ and then the change of signs occurs when $|x| = 0.9997a$.

The displacement on the unstressed portion of the boundary is given by

$$2\mu D'(x) = \kappa\Phi^+(x) + \Phi^-(x) = (\kappa+1)\Phi^+(x) \quad \text{on } L'',$$

$$= -(\kappa+1)iP_0X_0^+(x)/(2\pi), \tag{4.6.12}$$

so that, when t > a, equation (3.4.10) shows that

$$D'(x) = -(\kappa+1)iP_0(x^2-a^2)^{-1/2}[(x-a)/(x+a)]^{i\alpha}/(4\pi\mu). \tag{4.6.13}$$

The solution for a punch of more general shape may be obtained by using a similarity argument described in §10.4 for axisymmetric punch problems, and which is due to Spence (1968a). We consider the special case of a parabolic indenter $y = -x^2/(2R)$ and leave the reader to formulate the general case. Consider such an indenter pressed into an elastic half-plane and suppose that, at some stage, its contact length is 2a. The contact situation is *similar* in the sense that the values of

$$\frac{u}{a}, \quad \frac{v}{a}, \quad \frac{\tau_{xy}}{2\mu}, \quad \frac{\tau_{yy}}{2\mu}$$

on $y = 0$ may be expressed as functions of x/a only multiplied by a scale factor, $\delta(a)$ say, the same for all four, in such a way that the boundary conditions are satisfied. The appropriate boundary conditions in the contact region may be formulated by considering the displacements on $y = 0$ to be functions of x and the contact length a. Since any lateral displacement must occur before contact, the displacement $u(a,x)$ must be constant (i.e. independent of a) once x lies in the contact region. Thus

$$\frac{\partial u}{\partial a}(a,x) = 0, \quad v(a,x) = v_0(a) - \frac{x^2}{2R}, \quad |x| < a; \tag{4.6.14}$$

$$\tau_{xx} = 0 = \tau_{xy}, \quad |x| > a. \tag{4.6.15}$$

In order that $v(a,x)$ should have the stated similar form, we must have $\delta(a) = a$ and then if $u(a,x)$ is to be independent of a, u/a must be proportional to $(x/a)^2$, and since u must be an odd function of x

$$\frac{u}{a} = aB\left(\frac{x}{a}\right)^2 \text{sign}(x), \qquad \frac{v}{a} = a\left\{A - \frac{1}{2R}\left(\frac{x}{a}\right)^2\right\},$$

or
$$\frac{u}{a} = aBt|t|, \qquad \frac{v}{a} = a\left\{A - \frac{t^2}{2R}\right\},$$

where $t = x/a$, and A,B are independent of a, and of dimension $(\text{length})^{-1}$. Since R is the only other length appearing in the problem, A,B must be proportional to R^{-1} so that

$$\frac{u}{a} = \frac{Ba}{R}\, t|t|, \qquad \frac{v}{a} = \frac{a}{R}\left(A - \frac{t^2}{2}\right), \tag{4.6.16}$$

where now A,B are dimensionless constants. Since the problem is self-similar the complex potential $\Phi(z)$ will have the form $\Phi(z) = a\Omega(z/a) = a\Omega(\zeta)$ so that the boundary conditions for $\Omega^{\pm}(t)$ are

$$\kappa\Omega^{+}(t) + \Omega^{-}(t) = \frac{2\mu}{a}\frac{dD}{dt} = \frac{2\mu}{R}\{2B|t| - it\}, \qquad |t| < 1, \tag{4.6.17}$$

$$\Omega^{+}(t) - \Omega^{-}(t) \equiv \frac{1}{a}\{-p(t) + is(t)\} = 0, \qquad |t| > 1. \tag{4.6.18}$$

Since the stresses are to be bounded at $|t| = 1$ we must use the solution of (4.6.17),(4.6.18) which is zero at the ends. There is such a solution

$$\Omega(\zeta) = \frac{2\mu}{R} \cdot \frac{X_2(\zeta)}{2\pi i}\int_{-1}^{1} \frac{(2B|t|-it)dt}{X_2^{+}(t)(t-\zeta)} \tag{4.6.19}$$

only if B satisfies

$$\int_{-1}^{1} \frac{(2B|t|-it)dt}{X_2^{+}(t)} = 0, \tag{4.6.20}$$

where

$$X_2(\zeta) = (\zeta-1)^{1/2+i\alpha}(\zeta+1)^{1/2-i\alpha}. \tag{4.6.21}$$

By using (3.4.13) we may write (4.6.20) as

$$\int_0^1 \frac{tdt}{(1-t^2)^{1/2}}\left[(2B-i)\left(\frac{1+t}{1-t}\right)^{i\alpha} + (2B+i)\left(\frac{1-t}{1+t}\right)^{i\alpha}\right] = 0,$$

and with the substitution

$$\frac{1+t}{1-t} = \exp(u), \qquad t = \tanh u, \tag{4.6.22}$$

this becomes

$$\int_0^\infty \tanh u \operatorname{sech} u(B \cos \alpha u + \sin \alpha u)du = 0, \tag{4.6.23}$$

which may be integrated by parts to give

$$[\operatorname{sech} \mathbf{u}[B \cos \alpha u + \sin \alpha u)]_0^\infty + \alpha\int_0^\infty \operatorname{sech} u(B \sin \alpha u - \cos \alpha u)du = 0.$$

But Gradshteyn and Ryzhik (1965, p. 503) gives

$$\int_0^\infty \operatorname{sech} u \cos \alpha u du = \frac{\pi}{2} \operatorname{sech} \frac{\pi\alpha}{2},$$

$$\int_0^\infty \operatorname{sech} u \sin \alpha u du = -\frac{\pi}{2} \tanh \frac{\pi\alpha}{2} + 4\alpha \sum_{k=0}^\infty \{\alpha^2 + (1+4k)^2\}^{-1},$$

so that

$$B\left[4\alpha^2 \sum_{k=0}^\infty \{\alpha^2 + (1+4k)^2\}^{-1} - 1 - \frac{\pi}{2}\alpha \tanh \frac{\pi\alpha}{2}\right] = \frac{\pi\alpha}{2} \operatorname{sech} \frac{\pi\alpha}{2}. \tag{4.6.24}$$

Numerical solution of this equation gives the values

ν	0.1	0.2	0.3	0.4	0.5
B	-0.242	-0.199	-0.148	-0.084	0.

Note that Muskhelishvili (1953a, §114) takes the displacement boundary condition under a completely adhesive punch to be simply

$$u' = 0, \qquad v'(x) = f'(x), \tag{4.6.25}$$

so that the governing equations are

$$\kappa\Phi^{+}(x) + \Phi^{-}(x) = 2\mu i f'(x), \qquad x \text{ on } L', \tag{4.6.26}$$

$$\Phi^{+}(x) - \Phi^{-}(x) = 0, \qquad x \text{ on } L''. \tag{4.6.27}$$

These equations lead to a stress distribution which is singular at the end of the contact region, as shown in Ex. 4.6.1. Similar assumptions were made in the earlier and very interesting paper by Okubo (1940) who used curvilinear coordinates and found solutions corresponding to $u = 0$, $v = \sum a_n T_{2n}(x)$ on $[-1,1]$.

Examples 4.6

1. Suppose that the boundary conditions for the parabolic punch in complete adhesion are

   $$u(x) = 0, \qquad v(x) = x^2/(2R), \qquad |x| < a;$$

   $$\tau_{xy} = 0 = \tau_{yy}, \qquad |x| > a.$$

   Solve the equations (4.6.6),(4.6.7) in this case and find the contact stress distribution.

2. Find the contact stresses for $\Omega(\zeta)$ given by (4.6.19) when B satisfies (4.6.24).

4.7 *Contact problems with limiting friction*

The analysis of §§4.5 and 4.6 represent the two extreme cases of no friction and complete adhesion. Numerous attempts have been made to produce more realistic analyses corresponding to a finite coefficient of friction. An account of these attempts is given in §6.5 as it is best phrased in terms of integral equations rather than Cauchy integrals.

In this section we describe the relatively simple analysis for the case in which the contact region is assumed to be all in a state of *limiting friction*. This means that if k is the coefficient of friction then

$$s(x) = kp(x) \quad \text{on } L', \tag{4.7.1}$$

$$v'(x) = f'(x) \quad \text{on } L', \tag{4.7.2}$$

$$p(x) = 0 = s(x) \quad \text{on } L''. \tag{4.7.3}$$

Equation (4.7.1) implies that, in addition to being pressed into the half-plane, the punch is being pushed in the direction of the positive x-axis. This problem was solved by Glagolev (1942) for the particular case of a straight base, by using Carleman's (1922) analysis of integral equations; by Muskhelishvili, also in 1942, see (1953a or b) using Cauchy integrals, and in 1953 by Galin (1961); the latter gives a detailed analysis, with tabulated results, for various problems. Here we present Muskhelishvili's analysis, for a single punch.

Equation (4.7.1) may be transformed into a condition on $\Phi(x)$, namely

$$\{\Phi^{+}(x) - \Phi^{-}(x)\} + \{\overline{\Phi}^{+}(x) - \overline{\Phi}^{-}(x)\} + ki\{\Phi^{+}(x) - \Phi^{-}(x)\} - \\ - ki\{\overline{\Phi}^{+}(x) - \overline{\Phi}^{-}(x)\} = 0,$$

or

$$(1+ki)\Phi^{+}(x) + (1-ki)\overline{\Phi}^{+}(x) = (1+ki)\Phi^{-}(x) + (1-ki)\overline{\Phi}^{-}(x). \tag{4.7.4}$$

This shows that the function

$$\Omega(z) = (1+ki)\Phi(z) + (1-ki)\overline{\Phi}(z) \tag{4.7.5}$$

is continuous across L'. Since it is also continuous across L'', it is holomorphic in the whole plane. It vanishes at infinity and is therefore identically zero. Therefore

$$4\mu i f'(x) = \kappa\Phi^{+}(x) + \Phi^{-}(x) - \overline{\kappa\Phi^{+}(x)} - \overline{\Phi^{-}(x)}$$

$$= \kappa\Phi^+(x) + \Phi^-(x) - \kappa\overline{\Phi}^-(x) - \overline{\Phi}^+(x)$$

$$= \kappa\Phi^+(x) + \Phi^-(x) + (1+ki)(\kappa\Phi^-(x) + \Phi^+(x))/(1-ki),$$

or $$4\mu i f'(x)/\delta = \gamma\Phi^+(x) + \Phi^-(x), \tag{4.7.6}$$

where

$$\gamma = \frac{\kappa+1-(\kappa-1)ki}{\kappa+1+(\kappa-1)ki}, \qquad \delta = \frac{\kappa+1+(\kappa-1)ki}{1-ki}. \tag{4.7.7}$$

It is convenient to introduce β such that

$$\tan \pi\beta = k(\kappa-1)/(\kappa+1), \quad 0 < \beta < 1/2, \tag{4.7.8}$$

and then

$$\gamma = \exp(-2\pi i\beta), \quad \delta = (\kappa+1)\exp(\pi i\beta)/\{(1-ki)\cos \pi\beta\}. \tag{4.7.9}$$

When L' is $(-a,a)$ the function $X_0(z)$ of (3.4.7) will satisfy the homogeneous equation corresponding to (4.7.6) provided that

$$\exp(2\pi\alpha) = \gamma = \exp(-2\pi i\beta),$$

i.e. $\alpha = -i\beta + in$, where n is integral. Thus

$$X_0(z) = (z-a)^{-1/2+\beta-n}(z+a)^{-1/2-\beta+n}, \tag{4.7.10}$$

and n must be zero if $X_0(z)$ is to have integrable singularities at the ends. Thus

$$\Phi(z) = X_0(z)\cdot\frac{4\mu i}{2\pi\gamma\delta}\int_{L'}\frac{f'(t)dt}{X_0^+(t)(t-z)} + CX_0(z). \tag{4.7.11}$$

If P_0 is the total normal force applied then equations (4.3.9), (4.3.10),(4.3.15) show that

$$C = \frac{-(X+iY)}{2\pi} = \frac{-kP_0-iP_0}{2\pi} = \frac{-i(1-ik)P_0}{2\pi}. \tag{4.7.12}$$

For a punch with a straight horizontal base $f'(t) = 0$ and

$$2\pi\Phi(z) = -i(1-ik)P_0X_0(z). \tag{4.7.13}$$

The contact problem for a punch of general profile may be solved by using the solutions described in §3.5; an alternative, but rather more complicated solution, is given by Popov (1966) who considers also the effect of thermal stresses. See also Popov (1963b), A. S. Solov'ev (1969) and V. M. Aleksandrov (1970).

Examples 4.7

1. Show that the normal pressure $p(t)$ corresponding to (4.7.13) is $p(t) = P_0 \cos \pi\beta(a - t)^{-1/2+\beta}(a + t)^{-1/2-\beta}$.
2. Show that for a punch of parabolic profile $f(t) = -t^2/2R$ the complex potential is

$$\Phi(z) = \frac{-4\mu i X_2(z)}{2\pi\gamma\delta}\int_{L'} \frac{t\,dt}{X_2^+(t)(t-z)},$$

where $X_2(z) = (z^2 - a^2)X_0(z)$. Hence show that

$$\Phi(z) = -4\mu i(z-X_2(z))/\{\delta(\gamma+1)\},$$

and that the length $2a$ of the contact region is connected to the applied force P_0 by $P_0 = 2\pi a^2(1 - 4\beta^2)\mu/[(\kappa + 1)R]$.
3. Solve Ex. 4.7.2 by using the solution (3.4.22) and demanding that $p(t)$ be zero at the ends of the contact region.
4. Use the results of §3.5 to find the displacements on the line $y = 0$, outside the punch of Ex. 4.7.2.

4.8 *Problems involving two or more punches*

The theory described in this chapter has been generalized by Begiashvili (1940) and Muskhelishvili; see Muskhelishvili (1953a), (1953b), for application to contact problems involving more than one punch. The following section gives a partial summary of this analysis as far as it concerns contact problems. An alternative treatment of many of the problems may be found in Shtaerman (1949).

Let $L' = L_1 + L_2 + \ldots + L_n$ be the union of a finite number of segments (a_k,b_k) of the real axis, as shown in Figure 4.8.1.

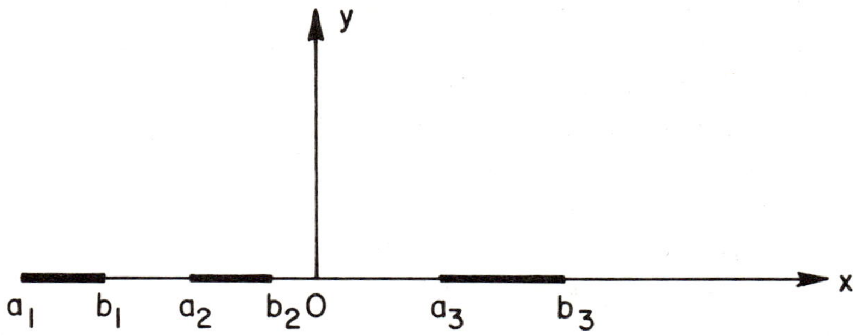

Figure 4.8.1 - The arcs a_k, b_k *lie on the* x-*axis*.

The elastic body is assumed to occupy the region S^+, and its boundary, the real axis, is denoted by $L = L' + L''$. There are two classes of problems relating to the application of punches to the segments L_k. In the first, denoted class A by Muskhelishvili, the punches are rigidly interconnected, while in the second they act independently. In each case the part of the boundary L'' is stress free, while the conditions on L' may be those of frictionless or adhesive contact, the latter in the sense of (4.6.25). In class A there will be an extra set of conditions on the displacements to express the fact that the punches are rigidly interconnected; in this case only the total force applied to the set of punches, and not the individual forces applied to each punch, will be known *a priori*. In class B the forces applied to each punch will be known.

The analysis of these problems follows very much the same lines as that already described for one punch, and need not be repeated in detail. The question of uniqueness of solution of the problems under the conditions (4.4.14),(4.4.15) is demonstrated by Muskhelishvili.

The problem may be reduced, as in §4.5, to

$$iv'(t) \equiv \vartheta\{\Phi^+(t) + \Phi^-(t)\} = if'(t) \quad \text{on } L', \tag{4.8.1}$$

$$-p(t) \equiv \Phi^{+}(t) - \Phi^{-}(t) = 0 \quad \text{on } L'', \tag{4.8.2}$$

for frictionless punches, and as in §4.6 - see Ex. 4.6.1 - to

$$2\mu i v'(t) \equiv \kappa\Phi^{+}(t) + \Phi^{-}(t) = 2\mu i f'(t) \quad \text{on } L', \tag{4.8.3}$$

$$-p(t) + is(t) \equiv \Phi^{+}(t) - \Phi^{-}(t) = 0 \quad \text{on } L'', \tag{4.8.4}$$

for adhesive contact. In both cases f(t) for t in (a_k, b_k) specifies the profile of the kth punch, apart from an arbitrary constant. First suppose that the punches are frictionless; then the general solution of (4.8.1),(4.8.2) may be obtained, as in §4.5, as

$$\vartheta\Phi(z) = \frac{X_0(z)}{2\pi i}\int_{L'} \frac{if'(t)dt}{X_0^{+}(t)(t-z)} + Q_{n-1}(z)X_0(z) \equiv \Psi_0(z) + $$
$$+ Q_{n-1}(z)X_0(z) \tag{4.8.5}$$

where

$$X_0(z) = \prod_{k=1}^{n} [(z-a_k)(z-b_k)]^{-1/2} \tag{4.8.6}$$

is the fundamental solution of the homogeneous equations corresponding to (4.8.1),(4.8.2) and

$$Q_{n-1}(z) = C_0 z^{n-1} + C_1 z^{n-2} + \dots C_{n-1} \tag{4.8.7}$$

is an arbitrary polynomial of degree n-1 or less -- notice that now $X_0(z) = 0(|z|^{-n})$ for large $|z|$ -- the coefficients of which must be determined by the additional conditions of the problem. For problems in class A there is the condition of rigid interconnection. Since condition (4.8.1) has been satisfied, we have

$$v(t) = f(t) + d_k \quad \text{on } L_k, \tag{4.8.8}$$

and we must now impose the condition

$$d_1 = d_2 = \dots = d_n, \tag{4.8.9}$$

and this is equivalent to the (n-1) conditions

$$f(a_{k+1}) - f(b_k) = \int_{b_k}^{a_{k+1}} v'(t)dt\ , \qquad k = 1,\ldots n-1. \tag{4.8.10}$$

Equations (4.8.1),(4.8.5) show that the value of $v'(t)$ on (b_k, a_{k+1}) -- which is part of the unloaded boundary L'' -- is given by

$$iv'(t) = \Psi_0(t) + Q_{n-1}(t)X_0(t), \tag{4.8.11}$$

and when substituted into (4.8.10) this yields (n-1) equations for C_0, $C_1,\ldots.C_{n-1}$. The nth equation is obtained by linking C_0 to the total force P_0 applied to the set of punches.

$$C_0 = -i\vartheta P_0/(2\pi). \tag{4.8.12}$$

For problems in class B the conditions relate to the forces applied to each punch. Now

$$-p(t) = \Phi^+(t) - \Phi^-(t) = \frac{X_0^+(t)}{\pi i}\int_{L'} \frac{if'(\tau)d\tau}{X_0^+(\tau)(\tau-t)} + $$

$$+ 2Q_{n-1}(t)X_0^+(t), \tag{4.8.13}$$

and the equations

$$\int_{a_k}^{b_k} p(t)dt = P_k, \quad k = 1,2,\ldots n, \tag{4.8.14}$$

yield n equations for C_0, $C_1,\ldots C_{n-1}$.

Equations (4.8.3),(4.8.4) may be treated in a similar fashion. The fundamental solution $X_0(z)$ of the homogeneous equations corresponding to (4.8.3),(4.8.4) is

$$X_0(z) = \prod_{k=1}^{n} (t-a_k)^{-1/2-i\alpha}(t-b_k)^{-1/2+i\alpha}, \tag{4.8.15}$$

where α is given by (3.4.16). The general solution of (4.8.3),

(4.8.4) is

$$\Phi(z) = 2\mu \frac{X_0(z)}{2\pi i}\int_{L'} \frac{if'(t)dt}{X_0^+(t)(t-z)} + Q_{n-1}(z)X_0(z), \tag{4.8.16}$$

where the coefficients of $Q_{n-1}(z)$ may be found as before.

In the particular case of two equal segments and frictionless contact the analysis of (3.5.22)-(3.5.23), or equivalently that given in Ex. 3.4.1, 3.4.2, may be extended to give a set of solutions corresponding to polynomial indenters. Thus if $L_1 = (-a,-b)$, $L_2 = (b,a)$ then we may take

$$X_0(z) = \frac{a^2-b^2}{2(z^2-a^2)^{1/2}(z^2-b^2)^{1/2}} . \tag{4.8.17}$$

Start from the identity

$$\begin{aligned}[(\zeta+1)^{1/2} - (\zeta-1)^{1/2}]^2 &= 2[\zeta - (\zeta^2-1)^{1/2}] \\ &= 2[\zeta + (\zeta^2-1)^{1/2}]^{-1},\end{aligned} \tag{4.8.18}$$

and obtain

$$\begin{aligned}[(\zeta+1)^{1/2} - (\zeta-1)^{1/2}]^{2n} &= 2^n[\zeta + (\zeta^2-1)^{1/2}]^{-n} \\ &= 2^n[T_n(\zeta) - (\zeta^2-1)^{1/2}U_{n-1}(\zeta)].\end{aligned} \tag{4.8.19}$$

Now write

$$\zeta = (2z^2-a^2-b^2)/(a^2-b^2), \tag{4.8.20}$$

so that

$$\zeta+1 = \frac{2(z^2-b^2)}{a^2-b^2}, \quad \zeta-1 = \frac{2(z^2-a^2)}{a^2-b^2}, \quad X_0(z) = (\zeta^2-1)^{-1/2}, \tag{4.8.21}$$

$$\begin{aligned}[(\zeta+1)^{1/2} - (\zeta-1)^{1/2}]^{2n} &= 2^{2n}[(\zeta+1)^{1/2} + (\zeta-1)^{1/2}]^{-2n} \\ &= [2(a^2-b^2)]^n[(z^2-a^2)^{1/2} + (z^2-b^2)^{1/2}]^{-2n}.\end{aligned} \tag{4.8.22}$$

This means that equation (4.8.19) may be rewritten

$$\frac{U_{n-1}(\zeta)}{X_0(z)} - T_n(\zeta) = \frac{-(a^2-b^2)^n}{[(z^2-a^2)^{1/2}+(z^2-b^2)^{1/2}]^{2n}}, \tag{4.8.23}$$

which shows that $U_n(\zeta)$ and $T_n(\zeta)$ form a pair $F(z), S(z)$ in the sense of (3.5.14). Indeed, since the R.H.S. of (4.8.23) is $O(|z|^{-2n})$ for large $|z|$, $z^m U_{n-1}(\zeta)$ and $z^m T_n(\zeta)$ will also be a pair for $m \leq 2n-1$. From this we may conclude that

$$\frac{1}{2\pi i}\int_{L'} \frac{t^m U_{n-1}(\tau)dt}{X_0^+(t)(t-z)} = -\frac{(a^2-b^2)^n z^m}{2\{(z^2-a^2)^{1/2}+(z^2-b^2)^{1/2}\}^{2n}}, \tag{4.8.24}$$

where $n = 1,2,\ldots;\ 0 \leq m \leq 2n-1$ and

$$\tau = (2t^2-a^2-b^2)/(a^2-b^2). \tag{4.8.25}$$

It may be proved in a similar way that if

$$X_2(z) = 2(z^2-a^2)^{1/2}(z^2-b^2)^{1/2}/(a^2-b^2), \tag{4.8.26}$$

then

$$\frac{1}{2\pi i}\int_{L'} \frac{t^m T_n(\tau)dt}{X_2^+(t)(t-z)} = \frac{1}{2}\frac{(a^2-b^2)^n z^m}{X_2(z)\{(z^2-a^2)^{1/2}+(z^2-b^2)^{1/2}\}^{2n}}, \tag{4.8.27}$$

for $n = 0,1,\ldots$ and $m \leq n \leq 2n+1$.

We may apply this result to the problem of two identical parabolic punches, distance $2c$ apart, as shown in Figure 4.8.2. The polynomial

$$y = x^4/(8Rc^2) - x^2/(4R) + \delta \tag{4.8.28}$$

fits the parabolas in the sense that

$$y'(\pm c) = 0, \qquad y''(\pm c) = 1/R, \tag{4.8.29}$$

so that

$$f'(t) = t^3/(2Rc^2) - t/(2R), \tag{4.8.30}$$

or, in terms of τ,

$$f'(t) = \frac{t}{2Rc^2}\left[\frac{(a^2-b^2)}{2}T_1(\tau) + \frac{a^2+b^2}{2} - c^2\right]. \tag{4.8.31}$$

Since there can be no singularity in the contact pressure at the

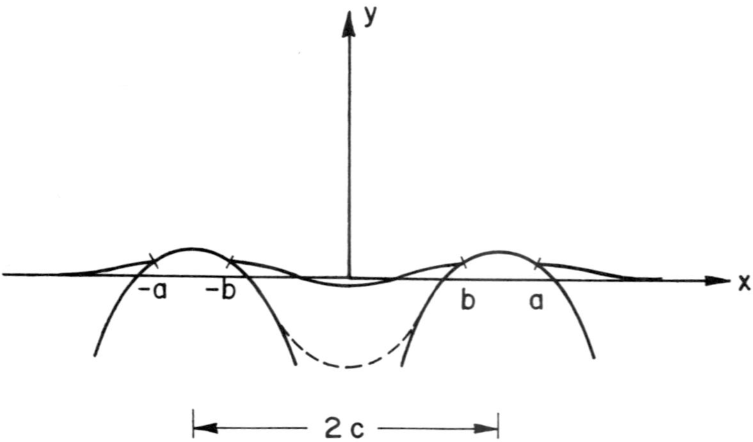

Figure 4.8.2 - Two identical interconnected parabolic punches indent the upper half-plane.

points $\pm a, \pm b$ we must use the solution involving $X_2(z)$. Thus

$$\vartheta\Phi(z) = \frac{X_2(z)}{2\pi i}\int_{L'} \frac{if'(t)dt}{X_2^+(t)(t-z)}, \tag{4.8.32}$$

and equations (4.8.27),(4.8.31) now give

$$\vartheta\Phi(z) = X_2(z)\cdot\frac{iz}{8Rc^2}\left[\frac{(a^2-b^2)^2}{X_2(z)\{(z^2-a^2)^{1/2}+(z^2-b^2)^{1/2}\}^2} + \right.$$

$$\left. + \frac{(a^2+b^2-2c^2)}{X_2(z)}\right],$$

$$= \frac{iz}{8Rc^2}\left[\frac{(a^2-b^2)^2}{\{(z^2-a^2)^{1/2}+(z^2-b^2)^{1/2}\}^2} + (a^2+b^2-2c^2)\right]. \tag{4.8.33}$$

For large $|z|$, $\Phi(z)$ must be $O(|z|^{-1})$. The first term in the brackets is $O(|z|^{-2})$ so that the condition will be satisfied if and only if

$$2c^2 = a^2 + b^2. \tag{4.8.34}$$

The total force on the punches and the form of the surface deflection between them is the subject of Ex. 4.8.2 and 4.8.3. The interested reader may explore the connection between the analysis just presented and those given by Sadowsky (1956), Tonoian (1964), Srivastava and Lowengrub (1970) and Kucherov (1974).

Examples 4.8

1. Show that the boundary values of the $X_2(z)$ of (4.8.26) are given by

   $$X_2^+(t) = A(t)\psi(t), \qquad X_2^-(t) = A(t)\overline{\psi(t)},$$

   where

   $$A(t) = 2|a^2-t^2|^{1/2}|b^2-t^2|^{1/2}/(a^2-b^2),$$

   and $\psi(t)$ is given by the table

	$t < -a$	$-a < t < -b$	$-b < t < b$	$b < t < a$	$a < t$
$\psi(t)$	1	$-i$	-1	i	1

2. Show that the contact pressure under the punches of Figure 4.8.2 is given by

   $$p(t) = t(a^2-t^2)^{1/2}(t^2-b^2)^{1/2}/[2\vartheta Rc^2],$$

   and that the total force on one punch is

   $$P_0 = \pi(a^2-b^2)^2/[32\vartheta Rc^2].$$

   This equation and (4.8.34) give a,b in terms of P_0 and c. Show that as c increases the result of Ex. 4.5.2 is reproduced.

3. Show that the deflection between the punches of Figure 4.8.2 is given by

   $$v'(t) = t\{2t^2-(a^2+b^2)+2(a^2-t^2)^{1/2}(b^2-t^2)^{1/2}\}/[4Rc^2].$$

4. Show that the use of the Plemelj formulae together with (4.8.22) leads to the result

$$-\frac{1}{\pi}\int_{L'}\frac{\operatorname{sign}(t)t^m U_{n-1}(\tau)dt}{\{(a^2-t^2)(t^2-b^2)\}^{-1/2}(t-x)} =$$

$$= \frac{(a^2-b^2)}{2}x^m\begin{cases}[\chi(x)]^{-2n}, & |x| \geq a\\ T_n(\xi), & b < |x| < a\\ (-)^n[\chi(x)]^{-2n}, & |x| \leq b\end{cases}$$

where $n = 1,2,\ldots,\ 0 \leq m \leq 2n-1,\ \xi = (2x^2-a^2-b^2)/(a^2-b^2)$,

$\chi(x) = \{|a^2-x^2|^{1/2} + |b^2-x^2|^{1/2}\}/(a^2-b^2)^{1/2}$,

and the remaining notation is as in this section. Note that the integral is a continuous function of x.

5. Show similarly that if $n = 0,1\ldots$ and $0 \leq m \leq 2n+1$

$$\frac{1}{\pi}\int_{L'}\frac{\operatorname{sign}(t)t^m T_n(\tau)dt}{\{(a^2-t^2)(t^2-b^2)\}^{1/2}(t-x)} =$$

$$= \begin{cases}\theta_{m,n}(x), & |x| > a\\ -2x^m U_{n-1}(\xi)/(a^2-b^2), & b < |x| < a\\ (-)^{n+1}\theta_{m,n}(x), & |x| < b\end{cases}$$

where $\theta_{m,n}(x) = \{(x^2-a^2)(x^2-b^2)\}^{-1/2}x^m[\chi(x)]^{-2n}$, and U_{-1} is taken to be zero.

6. Show that the sequences of polynomials $\{A_n\},\{B_n\}$ defined below are orthogonal with weight functions $|t|\{(a^2-t^2)(t^2-b^2)\}^{-1/2}$ and $|t|\{(a^2-t^2)(t^2-b^2)\}^{1/2}$ respectively, over L'.

$$A_{2n}(t) = T_n(\tau), \quad A_{2n+1}(t) = \operatorname{sign}(t)T_n(\tau)$$

$$B_{2n}(t) = U_n(\tau), \quad B_{2n+1}(t) = \operatorname{sign}(t)U_n(\tau)$$

where $n = 0,1,2,\ldots.$ Show that

$$\frac{2}{\pi}\int_{L'}|t|\{(a^2-t^2)(t^2-b^2)\}^{-1/2}A_m(t)A_n(t)dt = \begin{cases}0, & n \neq m\\ 1, & n = m \neq 0\\ 2, & n = m = 0,\end{cases}$$

$$\frac{2}{\pi}\int_{L'} |t|\{(a^2-t^2)(t^2-b^2)\}^{1/2} B_m(t)B_n(t)dt = \delta_{n,m}(a^2-b^2)^2/4.$$

4.9 *Problems involving an infinite number of contact regions*

Suppose that the contact region L' consists of an infinity of regions of length 2a, with their centres distance d(>2a) apart, as shown in Figure 4.9.1, and that identical frictionless punches are pressed into the half-plane $y > 0$ on these regions. As in §4.8 the problem may be reduced to

$$iv'(t) \equiv \vartheta\{\Phi^+(t) + \Phi^-(t)\} = if'(t) \quad \text{on } L', \tag{4.9.1}$$

$$-p(t) \equiv \Phi^+(t) - \Phi^-(t) = 0 \quad \text{on } L'', \tag{4.9.2}$$

where all displacements and stresses are periodic with period d. First it is necessary to find the fundamental solution $X_0(t)$. Using (4.8.6) and rearranging the terms we may take

$$X_0(z) = \left\{2\pi^2 \frac{(z^2-a^2)}{d^2} \prod_{n=-\infty}^{\infty} \left[1 - \frac{(z-a)}{nd}\right]\left[1 - \frac{(z+a)}{nd}\right]\right\}^{-1/2}. \tag{4.9.3}$$

The infinite product expansion for sin x is

$$\sin x = x \prod_{n=1}^{\infty} [1 - (x/n\pi)^2], \tag{4.9.4}$$

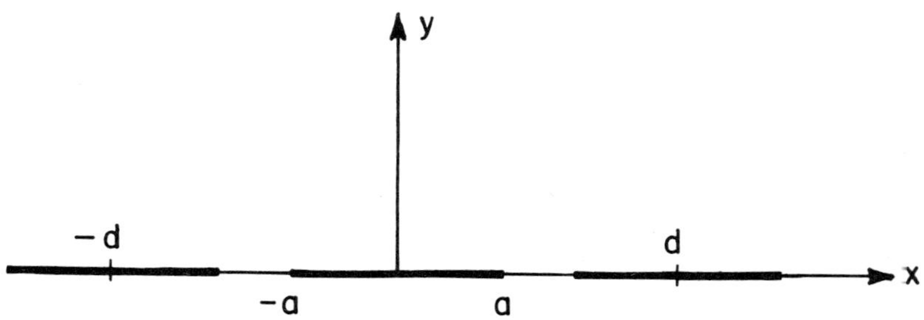

Figure 4.9.1 – Infinite array of identical contact regions.

so that

$$X_0(z) = \left\{ 2\sin\left[\frac{\pi(z-a)}{d}\right]\sin\left[\frac{\pi(z+a)}{d}\right]\right\}^{-1/2},$$

$$= \left\{\cos\left(\frac{2\pi a}{d}\right) - \cos\left(\frac{2\pi z}{d}\right)\right\}^{-1/2}. \tag{4.9.5}$$

It may easily be verified that the boundary values of $X_0(z)$ are given by

$$X_0^+(t) = M(t)\phi(t), \quad X_0^-(t) = M(t)\overline{\phi}(t), \tag{4.9.6}$$

where

$$M(t) = \left|\cos\left(\frac{2\pi a}{d}\right) - \cos\left(\frac{2\pi t}{d}\right)\right|^{-1/2}, \tag{4.9.7}$$

and $\phi(t)$ is as shown in Figure 4.9.2. Note that $X_0(z)$ is not periodic with period d, as (4.9.5) would suggest; instead $X_0(z)$ satisfies the same functional equation

$$F(z+d) = -F(z) \tag{4.9.8}$$

which governs $\sin(\pi z/d)$ and $\cos(\pi z/d)$. Now dividing equations (4.9.1),(4.9.2) by $X_0^+(t)$ we find

$$\frac{iv'(t)}{X_0^+(t)} = \vartheta\left\{\frac{\Phi^+(t)}{X_0^+(t)} - \frac{\Phi^-(t)}{X_0^-(t)}\right\} = \frac{if'(t)}{X_0^+(t)} \quad \text{on } L', \tag{4.9.9}$$

$$\frac{-p(t)}{X_0^+(t)} = \frac{\Phi^+(t)}{X_0^+(t)} - \frac{\Phi^-(t)}{X_0^-(t)} = 0 \quad \text{on } L'', \tag{4.9.10}$$

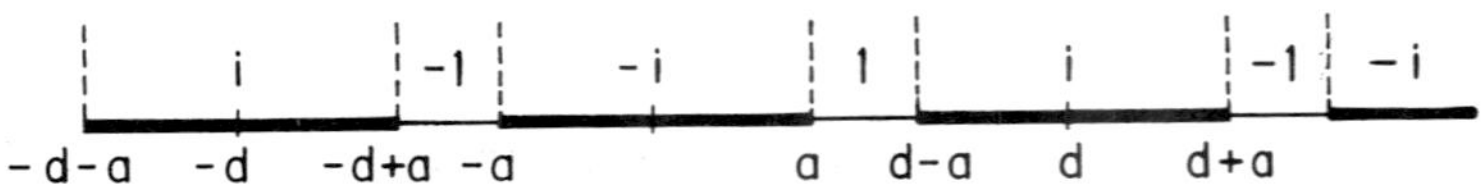

Figure 4.9.2 - The value of $\phi(t)$ *on the x-axis*

from which we deduce that

$$\vartheta \frac{\Phi(z)}{X_0(z)} = \frac{1}{2\pi}\int_{L'} \frac{f'(t)dt}{X_0^+(t)(t-z)} + \Psi(z), \tag{4.9.11}$$

where $\Psi(z)$ is yet to be determined. Now Figure 4.9.2 shows that $X_0^+(t)$ is purely imaginary on L' so that the first part of (4.9.11) gives a function $\Phi(z)$ satisfying (4.5.7). The remaining part, $\Psi(z)$, can have no singularities in the finite part of the plane, must be such that $X_0(z)\Psi(z)$ is periodic with period d, and must satisfy (4.5.7); it may be deduced that $\Psi(z)$ must have the form

$$\Psi(z) = i[A\cos(\pi z/d) + B\sin(\pi z/d)] \tag{4.9.12}$$

where A,B are real constants. Now $f'(t)$ takes identical values on each region [-a+nd, a+nd] while the value of $X_0^+(t)$ on the nth region is $(-)^n$ times the value of $X_0^+(t)$ on the central region (n = 0). Thus, collecting terms, we find

$$\frac{1}{2\pi}\int_{L'} \frac{f'(t)dt}{X_0^+(t)(t-z)} = \frac{1}{2\pi}\int_{-a}^{a} \frac{f'(t)}{X_0^+(t)}\left\{\frac{1}{t-z} + \right.$$

$$\left. + 2(t-z)\sum_{n=1}^{\infty} \frac{(-)^n}{(t-z)^2-n^2d^2}\right\}dt \tag{4.9.13}$$

and since

$$\operatorname{cosec} \pi x = \frac{1}{\pi x} + \frac{2x}{\pi}\sum_{n=1}^{\infty} \frac{(-)^n}{x^2-k^2}, \tag{4.9.14}$$

equations (4.9.11)-(4.9.14) give

$$\vartheta\Phi(z) = \frac{X_0(z)}{2d}\int_{-a}^{a} \frac{f'(t)dt}{X_0^+(t)\sin[\pi(t-z)/d]} +$$

$$+ iX_0(z)\{A\cos(\pi z/d) + B\sin(\pi z/d)\}. \tag{4.9.15}$$

The contact pressure p(t) is given by

$$-\vartheta p(x) = 2iX_0^+(x)\{A\cos(\pi x/d) + B\sin(\pi x/d)\} +$$

$$+ \frac{X_0^+(x)}{d}\int_{-a}^{a} \frac{f'(t)dt}{X_0^+(t)\sin[\pi(t-x)/d]} . \tag{4.9.16}$$

For application to particular examples it is convenient to proceed as in Ex. 4.5.4 and rewrite (4.9.15) in the form

$$\vartheta\Phi(z) = \frac{X_2(z)}{2d}\int_{-a}^{a} \frac{f'(t)dt}{X_2^+(t)\sin[\pi(t-z)/d]} +$$

$$+ iX_0(z)\{A' \cos(\pi z/d) + B' \sin(\pi z/d), \tag{4.9.17}$$

where

$$X_2(z) = \left\{ \cos\left(\frac{2\pi a}{d}\right) - \cos\left(\frac{2\pi z}{d}\right)\right\}^{1/2}, \tag{4.9.18}$$

$$X_2^+(t) = i\left\{\cos\left(\frac{2\pi t}{d}\right) - \cos\left(\frac{2\pi a}{d}\right)\right\}^{1/2}, \quad |t| < a. \tag{4.9.19}$$

In order to evaluate integrals such as (4.9.17) we recall a result from §2.3, namely

$$G_{n-1}(\zeta) = \frac{1}{\pi}\int_{-1}^{1} \frac{T_n(\tau)d\tau}{(1-\tau^2)^{1/2}(\tau-\zeta)}, \quad n = 0,1,2,\ldots. \tag{4.9.20}$$

Following Morar' and Popov (1971) write

$$\frac{\pi z}{d} = \xi, \quad \frac{\pi t}{d} = \eta, \quad \frac{\pi a}{d} = \alpha, \quad \tau = \tan\eta \cot\alpha,$$

$$\zeta = \tan\xi \cot\alpha; \tag{4.9.21}$$

then

$$X_2^+(t) = i\{\cos 2\eta - \cos 2\alpha\}^{1/2},$$

$$X_2(z) = \{\cos 2\xi - \cos 2\alpha\}^{1/2}, \tag{4.9.22}$$

$$G_{n-1}(\zeta) = \frac{2^{1/2} i \cos\xi \sin\alpha}{\pi}\int_{-\infty}^{\infty} \frac{T_n(\tan\eta \cot\alpha)d\eta}{X_2^+(t)\sin(\eta-\xi)} .$$

But

$$G_{n-1}(\zeta) = -(\zeta^2-1)^{-1/2}[\zeta - (\zeta^2-1)^{1/2}]^n$$

$$= -2^{-n}(\zeta^2-1)^{-1/2}[(\zeta+1)^{1/2} - (\zeta-1)^{1/2}]^{2n}$$

$$= -\frac{2^{1/2}\cos\xi\sin\alpha}{X_2(z)}[g(\xi)]^n, \tag{4.9.23}$$

where

$$g(\xi) = \frac{[\sin(\xi+\alpha)]^{1/2}-[\sin(\xi-\alpha)]^{1/2}}{[\sin(\xi+\alpha)]^{1/2}+[\sin(\xi-\alpha)]^{1/2}}. \tag{4.9.24}$$

This means that if

$$f'(t) = \sum_{n=0}^{N} a_n T_n(\tan\eta\cot\alpha), \tag{4.9.25}$$

then

$$\vartheta\Phi(z) = (1/2)i\sum_{n=0}^{N} a_n[g(\xi)]^n +$$

$$+ iX_0(z)\{A'\cos\xi + B'\sin\xi\}. \tag{4.9.26}$$

First consider the case when the punches are flat, so that $f'(t) = 0$. The contact stress is $p(t)$, where

$$-\vartheta p(t) = 2\{A'\cos\eta + B'\sin\eta\}\{\cos 2\eta - \cos 2\alpha\}^{-1/2}, \tag{4.9.27}$$

and the force on the punch is P_0, where

$$\vartheta P_0 = -2^{1/2}A'd. \tag{4.9.28}$$

The rotation of the plane is given by Ex. 4.5.5 as

$$\omega = 2\vartheta\,\mathrm{Im}\{\Phi(z)\}, \tag{4.9.29}$$

and for large y

$$\vartheta\Phi(z) = 2^{-1/2}A' + iB'. \tag{4.9.30}$$

If the rotation is to be zero for large y then $B' = 0$. The total pressure on one contact region is

$$P_0 = \int_{-a}^{a} p(t)dt = \frac{-2A'd}{\vartheta}\int_{-\alpha}^{\alpha} \frac{\cos \eta d\eta}{\{\cos 2\eta - \cos 2\alpha\}^{1/2}} . \tag{4.9.31}$$

The integral may be calculated by using the substitution (4.9.21);

$$A' = -(2^{1/2}d)^{-1}P_0\vartheta, \tag{4.9.32}$$

$$\Phi(z) = -i(2^{1/2}d)^{-1}P_0X_0(z)\cos \xi, \tag{4.9.33}$$

$$p(t) = \frac{P_0 \cos \eta}{d\{\sin^2\alpha - \sin^2\eta\}^{1/2}} . \tag{4.9.34}$$

The pressure agrees with Sadowsky's (1928) result. Equilibrium considerations show that the plane must be constrained by stresses at infinity; in fact the stress is hydrostatic and

$$\tau_{xx} + \tau_{yy} = 2^{3/2}A'/\vartheta = -4P_0/d. \tag{4.9.35}$$

If the punches have inclined bases, so that $f'(t) = \varepsilon$, then equation (4.9.26) shows that

$$\vartheta\Phi(z) = i\varepsilon/2 + iX_0(z)\{A' \cos \xi + B' \sin \xi\}; \tag{4.9.36}$$

the constant A' is still given by (4.9.32), while the condition that $\omega = 0$ for large y gives

$$B' = -\varepsilon/2. \tag{4.9.37}$$

This agrees with the result obtained by England and Green (1963) using a slightly different, complex variable, method. For further analysis and references see Dundurs, Tsai and Keer (1973), Krenk (1976) and Kuznetzov (1976).

Example 4.9

1. Use the substitution $\tau = \tan(\pi t/d)\cot(\pi a/d)$ to calculate the integral (4.9.31).

4.10 *Contact between two elastic bodies*

The analysis of this chapter has so far been concerned with the contact between a rigid punch (or punches) and an elastic body occupying the upper half-plane. Muskhelishvili (1953a),(1953b) considered the equations governing the *frictionless* contact between two bodies which in the neighbourhood of their contact region are so flat that i) the boundary conditions for contact may be applied on the x-axis, ii) the bodies themselves may be approximated by the upper and lower half-planes $y > 0$, $y < 0$, and iii) the departures of the boundaries of the bodies from the x-axis, both before and after contact, are infinitesimal. (These assumptions are the 2-dimensional analogues of the assumptions made by Hertz (see §2.9) who found the force-deflection relations for the two contacting bodies by assuming each occupied a half-space.) Muskhelishvili showed that the equations governing the contact have exactly the same *form* as the equations governing the contact between a rigid punch and an elastic half-plane, the only difference being the appearance of parameters involving the elastic constants of both bodies. Such analyses also appear in Shtaerman (1949). Galin (1961), using an unneccessarily complicated analysis, showed that a similar result holds for the contact between two such elastic bodies in the presence of (limiting) friction. The second line of research in the area started with Michell's (1899) paper, discussed in §4.2. Michell showed that provided equation (4.2.25) is satisfied and there are no body forces, the stresses produced in a plane elastic body by specified surface tractions are independent of the elastic constants. Dundurs and Stippes (1970), building on earlier papers of Dundurs (1967),(1969), showed that Michell's result carries over to the contact between two plane isotropic elastic bodies, having the same material properties, provided that the contact is *receding*; this term and the related *advancing* are defined below. Dundurs (1975) extended this analysis to the contact between plane isotropic elastic bodies of different materials.

It is convenient to reverse the chronological order and consider the more general results obtained by Dundurs and Stippes first, and then turn to Muskhelishvili's and Galin's results later. Let C denote the extent of the contact between two (plane or 3-dimensional) elastic bodies in the deformed configuration and let $C°$ denote the initial contact region in the load-free state. Each of $C, C°$ will consist of one or more areas, curves and points, and $C°$ may be empty if the bodies are simply-supported and start to touch only after the loads have reached a certain level. The *classification* of contacts is based on the criterion of whether or not new surface points come into contact as the bodies deform. A contact is called *receding* if C is contained in $C°$, i.e. $C \subseteq C°$, and *advancing* if C is not contained in $C°$, as shown in Figure 4.10.1. Note that the term *stationary*, which was used by Dundurs and Stippes to denote the case $C = C°$, was later used by Dundurs for the case in which C was *independent of the level of loading*, provided that there *was* loading; the latter meaning will be used here.

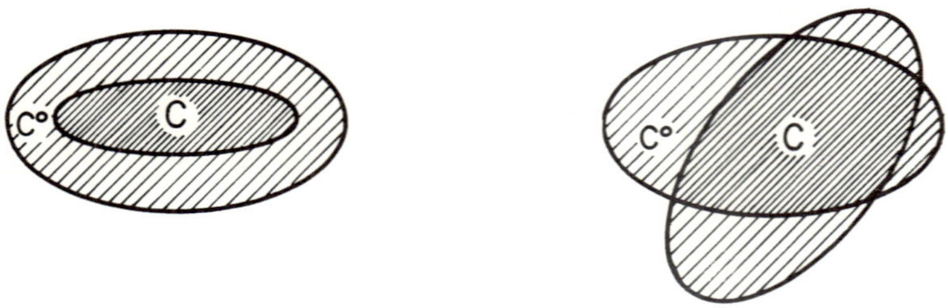

Figure 4.10.1 - Receding and advancing contact.

The first set of results apply equally to 3-dimensional and plane problems; they will be phrased in terms of three dimensions. Contact is being considered within the framework of linear elasticity so that stresses, rotations and displacements are small and the boundary conditions are imposed on the original geometry. The contact C can extend only over those parts of the surfaces that initially have (nearly) parallel tangent planes and are separated by an initial gap g° which is small in comparison to the overall dimensions involved. Let $\hat{n}$ be the common normal to the tangent planes, taken from body 1 to body 2. The contact regions C, C° may be made up from more than one region; consider one such region and choose a local coordinate frame there with the z-axis pointing along $\hat{n}$. In the neighbourhood of $C \cap C^\circ$ the equations of the surfaces will be $z = f_1(x,y)$, $z = f_2(x,y)$ respectively in the undeformed state, so that the gap between them is $g^\circ = f_2(x,y) - f_1(x,y)$, as shown in Figure 4.10.2. As the bodies deform, the point $P_i \equiv (x,y,f_i(x,y))$ goes to $(x,y,f_i(x,y) + w_i)$ so that the gap between P_1 and P_2 becomes

$$g = f_2(x,y) + w_2 - f_1(x,y) - w_1 = g^\circ + w_2 - w_1. \qquad (4.10.1)$$

Since the bodies cannot penetrate each other, this gap must be positive or zero, so that $g \geq 0$; the contact region is given by

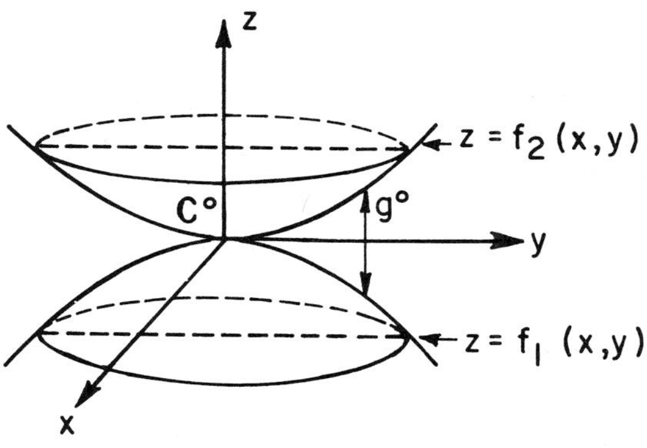

Figure 4.10.2 - The gap between the two bodies in their undeformed state is g°.

the condition $g = 0$, i.e.

$$w_1 - w_2 = g^\circ \quad \text{on } C.$$

We may write this boundary condition in a form which applies to a contact region composed of more than one part by replacing w by $\vec{d}\cdot\hat{n}$; thus

$$(\vec{d}\cdot\hat{n})_1 - (\vec{d}\cdot\hat{n})_2 = g^\circ \quad \text{on } C.$$

Now we note that C is made up of two portions -- the portion (if any) of C in C°, namely $C \cap C^\circ$, and in this part g° is zero since, by definition, g° is the initial gap; and the remainder $C - C \cap C^\circ$. Thus

$$(\vec{d}\cdot\hat{n})_1 - (\vec{d}\cdot\hat{n})_2 = \begin{cases} 0 & \text{on } C \cap C^\circ \\ g^\circ & \text{on } C - C \cap C^\circ. \end{cases} \tag{4.10.2}$$

It is this condition which yields the fundamental distinction between receding and advancing contacts. For receding contacts $C = C \cap C^\circ$ so that $C - C \cap C^\circ$ is empty and equation (4.10.2) is *homogeneous*, i.e.

$$(\vec{d}\cdot\hat{n})_1 - (\vec{d}\cdot\hat{n})_2 = 0 \quad \text{on } C. \tag{4.10.3}$$

For general, advancing contacts, it is non-homogeneous.

Equation (4.10.3) has far-reaching implications. The equations governing the deformation of the two bodies will be the equations of equilibrium *within* the two bodies, involving the body forces, if any; the boundary conditions on the parts of the surfaces outside C, where displacements or surface tractions may be specified; and boundary conditions on C. The latter will include (4.10.3) and will also include others depending on the nature of the contact, i.e. whether it is frictionless, adhesive or in some intermediate state with friction. We assume that these equations represent a well-posed problem with a unique solution.

The crux of the issue is that, when the contact is receding, all these equations and boundary conditions are *homogeneous*, except for those relating to the body forces and the applied tractions and/or displacements; these may be collectively termed *loads*. If the loads are all changed in the *same proportion* this is termed a change in the *level of the loading*. The conclusions for receding contact are immediate:

1) The displacements, strain and stress are proportional to the level of loading.
2) The contact region C is independent of the level of loading, i.e. it remains stationary.
3) Except for the special case $C = C°$, the change from $C°$ to C is discontinuous.

Note that although the induced elastic fields are proportional to the applied loads, the law of superposition does not apply in general. Thus the elastic field due to the sum of two systems of loads will be equal to the sum of the elastic fields due to each system separately only if the contact region C is the same for each system.

The second set of results relates to contact between two *isotropic* elastic bodies in plane strain or plane stress. For such it is convenient to use the complex variable formulation of §4.2. In the absence of body forces, equations (4.2.5),(4.2.12), (4.2.13) give

$$2\mu D = 2\mu(u+iv) = \kappa\phi(z) - z\overline{\phi}'(\overline{z}) - \overline{\psi}(\overline{z}), \tag{4.10.4}$$

$$\Theta = \tau_{xx} + \tau_{yy} = 2[\phi'(z) + \overline{\phi}'(\overline{z})], \tag{4.10.5}$$

$$\Xi = \tau_{yy} - \tau_{xx} + 2i\tau_{xy} = 2[\overline{z}\phi''(z) + \psi'(z)]. \tag{4.10.6}$$

Note that equations (4.10.5),(4.10.6) giving the stresses in terms of the complex potentials contain no elastic constants. Consequently the dependence of the stresses on the elastic constants is the same as that for the complex potentials. Provided that equation (4.2.25) is satisfied, the elastic constants can enter the complex potentials only through the boundary conditions. If the bodies are loaded by prescribed surface tractions, and not by prescribed displacements, the elastic constants can appear only through the boundary conditions on C. We therefore examine these in detail for the case when C is receding.

Let θ be the angle between $\hat{n}$ and the x-axis, as shown in Figure 4.10.3. The contact may be frictionless, adhesive or in some intermediate state. The two boundary conditions which hold in all these cases are (4.10.3) and the equilibrium equation

$$(\tau_{nx}+i\tau_{ny})_1 = (\tau_{nx}+i\tau_{ny})_2. \qquad (4.10.7)$$

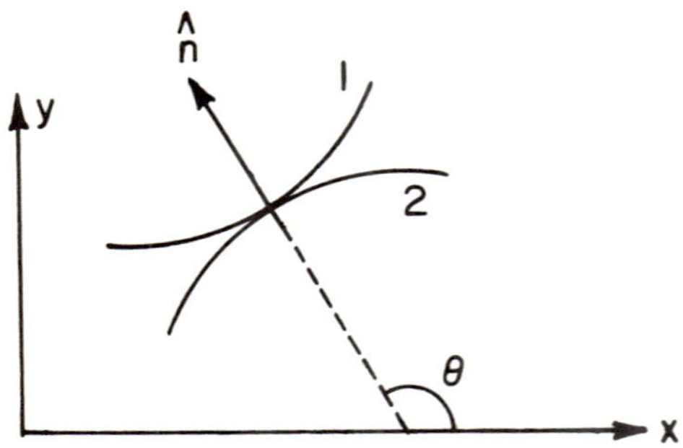

Figure 4.10.3 - θ is the angle which the normal makes with the x*-axis.*

Now if the shear modulus and parameter κ of body j are μ_j,κ_j (j = 1,2) respectively, then the normal displacements of the two bodies are

$$4\mu_j u_{n,j} = (\kappa_j\phi_j - z\overline{\phi_j'} - \overline{\psi_j})\exp(-i\theta) + (\kappa_j\overline{\phi_j} - \overline{z}\phi_j' - \psi_j)\exp(i\theta), \qquad (4.10.8)$$

where the repeated suffice does *not* imply summation. Equation (4.10.3) becomes

$$\Gamma[(\kappa_1\phi_1 - z\overline{\phi}_1' - \overline{\psi}_1)\exp(-i\theta) + (\kappa_1\overline{\phi}_1 - \overline{z}\phi_1' - \psi_1)\exp(i\theta)]$$

$$= (\kappa_2\phi_2 - z\overline{\phi}_2' - \overline{\psi}_2)\exp(-i\theta) + (\kappa_2\overline{\phi}_2 - \overline{z}\phi_2' - \psi_2)\exp(i\theta), \qquad (4.10.9)$$

where $\Gamma = \mu_2/\mu_1$. Equations (4.2.23),(4.10.7) yield

$$d(\phi_1 + z\overline{\phi}_1' + \overline{\psi}_1) = d(\phi_2 + z\overline{\phi}_2' + \overline{\psi}_2), \qquad (4.10.10)$$

and if (4.2.25) is satisfied for all closed contours lying in the region occupied by *both* bodies, then F_1, F_2 will each be single-valued and will be equal on C so that

$$A \equiv \phi_1 + z\overline{\phi}_1' + \overline{\psi}_1 = \phi_2 + z\overline{\phi}_2' + \overline{\psi}_2 \quad \text{on } C. \qquad (4.10.11)$$

Equation (4.10.9) involves Γ, κ_1 and κ_2, but by using (4.10.11), which involves no elastic constants, we may replaced (4.10.9) by a new equation involving just two constants. Form (4.10.9) + + $1/2[(\Gamma+1)[A\exp(-i\theta) + \overline{A}\exp(i\theta)]$ and divide the sum by $\Gamma(\kappa_1+1) + \kappa_2 + 1$; the result is

$$(2+\alpha+\beta)[\exp(-i\theta)\phi_1 + \exp(i\theta)\overline{\phi}_1] - (\alpha-\beta)[\exp(-i\theta)(z\overline{\phi}_1' + \overline{\psi}_1) +$$

$$+ \exp(i\theta)(\overline{z}\phi_1' + \psi_1)] - (2-\alpha-\beta)[\exp(-i\theta)\phi_2 + \exp(i\theta)\overline{\phi}_2] -$$

$$- (\alpha-\beta)[\exp(-i\theta)(z\overline{\phi}_2' + \overline{\psi}_2) + \exp(i\theta)(\overline{z}\phi_2' + \psi_2)] = 0, \qquad (4.10.12)$$

where

$$\alpha = \frac{\Gamma(\kappa_1+1) - (\kappa_2+1)}{\Gamma(\kappa_1+1) + (\kappa_2+1)}, \qquad \beta = \frac{\Gamma(\kappa_1-1) - (\kappa_2-1)}{\Gamma(\kappa_1+1) + (\kappa_2+1)}. \qquad (4.10.13)$$

Dundurs (1969),(1970) notes that α, β are measures of the mismatches in the uniaxial and voluminal compliances of the two elastic materials, and that $-1 \leq \alpha \leq 1$, $-1 \leq \beta \leq 1$, with $\alpha = 0 = \beta$ for identical materials. This last statement implies that when the two materials are identical the two boundary conditions are *independent* of the elastic constants.

For limiting friction on C, the final boundary condition will be that $\tau_{ns} = k\tau_{nn}$; this equation will not involve the elastic constants: the case of the frictionless contact corresponds to $k = 0$. For adhesive receding contact in the sense of (4.6.25),

$$\vec{d}_1 = \vec{d}_2, \tag{4.10.14}$$

so that equation (4.10.4) gives

$$\Gamma(\kappa_1\phi_1 - z\overline{\phi}_1' - \overline{\psi}_1) = \kappa_2\phi_2 - z\overline{\phi}_2' - \overline{\psi}_2, \tag{4.10.15}$$

Proceeding as before, adding ΓA to (4.10.15) and dividing by $\Gamma(\kappa_1+1) + (\kappa_2+1)$, we find

$$(1+\alpha)\phi_1 = (1-\beta)\phi_2 + (\alpha-\beta)(z\overline{\phi}_2'+\overline{\psi}_2), \tag{4.10.16}$$

so that the contact equations (4.10.11),(4.10.16) again involve only α, β.

We conclude that, in the contact between an assemblage of two or more bodies in frictionless or adhesive contact, *if*

a) there are just two sets of elastic constants involved,

b) the bodies are loaded by specified surface traction,

c) equation (4.2.25) holds in the union of the regions occupied by the bodies,

d) the contacts are receding,

then the extent of the contact and the stresses depend only on α, β.

It should be noted that no attempt has been made to discuss the detailed solution of contact problems involving two bodies of general shape (Blackketter and Christensen (1969)); rather, attention has been confined to certain properties of the solution, particularly the dependence on the elastic constants.

Further reductions in the dependence on the elastic constants can be obtained by specialization. Thus consider receding frictionless contact *along a straight boundary*. Placing the x-axis along this boundary we have

$$v_1 = v_2. \tag{4.10.17}$$

This condition can be replaced, for a wide class of problems, by

$$\partial v_1/\partial x = \partial v_2/\partial x, \tag{4.10.18}$$

and equation (4.10.4) shows that

$$4i\mu\partial v/\partial x = (\kappa+1)(\phi'-\overline{\phi}') + x(\phi''-\overline{\phi}'') + \psi' - \overline{\psi}'. \tag{4.10.19}$$

But equation (4.10.6) shows that

$$2i\tau_{xy} = x(\phi''-\overline{\phi}'') + \psi' - \overline{\psi}', \tag{4.10.20}$$

so that

$$\frac{\kappa_1+1}{\mu_1}(\phi_1'- \overline{\phi}_1') = \frac{\kappa_2+1}{\mu_2}(\phi_2'-\overline{\phi}_2'), \tag{4.10.21}$$

and equations (4.10.13) show that this is equivalent to

$$(1+\alpha)(\phi_1'-\overline{\phi}_1') = (1-\alpha)(\phi_2'-\overline{\phi}_2'). \tag{4.10.22}$$

In this case therefore the stresses and the extent of the contact depend only on α, and note that there is no dependence on the elastic constants if the materials are identical ($\alpha = 0$) or if one of the materials is rigid ($\alpha = \pm 1$). Equation (4.10.22) depends on equation (4.10.18) being equivalent to (4.10.17). This will certainly be true if one of the bodies is not constrained against rigid-body motion and the contact is over a single interval; if however contact is over more than one interval, the two displacements cannot generally be matched merely by matching the derivatives, so that equation (4.10.18) is not generally valid.

Muskhelishvili and Galin considered the contact between two bodies 1,2 occupying the lower and upper half-planes respectively, and assumed that the contact takes place over a union of segments denoted by L'. If each body is analyzed as in §4.4, body j will have complex potentials Φ_j, Ψ_j ($j = 1,2$). The values of $\Psi_1(z)$ in S^+, and of $\Psi_2(z)$ in S^- will be given by equations corresponding to (4.4.11), namely

$$\Psi_1(z) = -\Phi_1(z) - \overline{\Phi}_1(z) - z\Phi_1'(z), \quad z \text{ in } S^+, \tag{4.10.23}$$

$$\Psi_2(z) = -\Phi_2(z) - \overline{\Phi}_2(z) - z\Phi_2'(z), \quad z \text{ in } S^-, \tag{4.10.24}$$

so that, on the x-axis,

$$\tau_{yy}^{(1)} - i\tau_{xy}^{(1)} = -p_1(x) + is_1(x) = \Phi_1^-(x) - \Phi_1^+(x), \qquad (4.10.25)$$

$$\tau_{yy}^{(2)} - i\tau_{xy}^{(2)} = -p_2(x) + is_2(x) = \Phi_2^+(x) - \Phi_2^-(x), \qquad (4.10.26)$$

$$2\mu_1(u_1'+iv_1') = \kappa_1\Phi_1^-(x) + \Phi_1^+(x), \qquad (4.10.27)$$

$$2\mu_2(u_2'+iv_2') = \kappa_2\Phi_2^+(x) + \Phi_2^-(x). \qquad (4.10.28)$$

Muskhelishvili and Galin did not make the distinction between advancing and receding contact, but considered a general contact, receding or advancing, occurring over a region L'.

We suppose that the contact is either a) frictionless, b) completely adhesive in the sense of (4.6.26) or c) in limiting friction. Without loss of generality we may confine ourselves to the case when the remainder, L'', of the x-axis is free from traction; then $\Phi_1(z),\Phi_2(z)$ will be holomorphic in the plane cut along L'. Whichever contact condition applies, equilibrium on L' gives $p_1(x) = p_2(x) = p(x)$, $s_1(x) = s_2(x) = s(x)$, so that

$$\Phi_1^-(x) - \Phi_1^+(x) = \Phi_2^+(x) - \Phi_2^-(x), \qquad (4.10.29)$$

which implies that $\Phi_1(z) + \Phi_2(z)$ is continuous across L'. But since $\Phi_1(z) + \Phi_2(z)$ is continuous across L'', it is holomorphic in the entire plane; and since it vanishes at infinity it is identically zero; thus

$$\Phi_2(z) = -\Phi_1(z) \quad \text{for all } z. \qquad (4.10.30)$$

If $y = f_1(x)$, $y = f_2(x)$ are the equations of the boundaries of the lower and upper half-planes before contact ($f_1(x)$, $f_2(x)$ being small quantities) then by using the argument leading to (4.10.2) we deduce that

$$v_1'(x) - v_2'(x) = f_2'(x) - f_1'(x) = -f'(x), \quad x \text{ on } L'. \qquad (4.10.31)$$

Equations (4.10.27),(4.10.28),(4.10.30) now give

$$\left(\frac{1}{4\mu_1}+\frac{\kappa_2}{4\mu_2}\right)(\Phi_2^+(x)-\overline{\Phi_2^+(x)})+\left(\frac{\kappa_1}{4\mu_1}+\frac{1}{4\mu_2}\right)\cdot$$

$$\cdot(\Phi_2^-(x)-\overline{\Phi_2^-(x)}) = if'(x), \quad x \text{ on } L'. \qquad (4.10.32)$$

Now consider the second condition on L'; this will depend on the contact condition:

a) frictionless contact. Here $s(x) = 0$ on L' and the argument of §4.5 shows that $\overline{\Phi}_1(z) = -\Phi_1(z)$, $\overline{\Phi}_2(z) = -\Phi_2(z)$ so that $\overline{\Phi_2^+(x)} = \overline{\Phi}_2^-(x) = -\Phi_2^-(x)$, $\overline{\Phi_2^-(x)} = \overline{\Phi}_2^+(x) = -\Phi_2^+(x)$ and equations (4.10.26),(4.10.32) reduce to

$$\Phi_2^+(x) + \Phi_2^-(x) = \vartheta^{-1}if'(x), \quad x \text{ on } L', \qquad (4.10.33)$$

$$-p(x) \equiv \Phi_2^+(x) - \Phi_2^-(x) = 0, \quad x \text{ on } L'', \qquad (4.10.34)$$

where

$$\vartheta = \frac{\kappa_1+1}{4\mu_1} + \frac{\kappa_2+1}{4\mu_2} = \vartheta_1 + \vartheta_2. \qquad (4.10.35)$$

Equations (4.10.33),(4.10.34) are exactly the same as equations (4.5.9),(4.5.10), which refer to the contact between a rigid punch and an elastic half-space. As far as the discussion on dependence on elastic constants is concerned equations (4.10.33),(4.10.34) show that if the two bodies may be approximated by half-spaces, and the contact is receding, so that $f_1'(x) = f_2'(x)$ on L', i.e. $f'(x) = 0$ on L', then the extent of the contact and the stresses will be independent of the elastic constants; they will be determined by the other conditions in the problem, e.g. the total force applied to produce contact, and the conditions on the contact stress at the ends of the contact region.

b) complete adhesion. Here $u_1' = u_2'$ on L' so that equation (4.10.32) may be replaced by

$$\frac{\kappa_2\Phi^+ + \Phi_2^-}{2\mu_2} - \frac{(\kappa_1\Phi_1^- + \Phi_1^+)}{2\mu_1} = if'(x), \quad x \text{ on } L', \qquad (4.10.36)$$

which when combined with (4.10.30) yields

$$(1-\beta)\Phi_2^+(x) + (1+\beta)\Phi_2^-(x) = \vartheta^{-1}if'(x), \qquad x \text{ on } L', \qquad (4.10.37)$$

$$-p(x) + is(x) \equiv \Phi_2^+(x) - \Phi_2^-(x) = 0, \qquad x \text{ on } L'', \qquad (4.10.38)$$

which have exactly the same form as equations (4.6.26),(4.6.27). Note that the assumption that the contacting bodies may be approximated by half-planes reduces the dependency of a receding contact problem from the dependence on α,β in the general adhesive problem to dependence only on β.

c) limiting friction. Here $s(x) = kp(x)$ for x on L' so that the argument of §4.7 yields

$$(1+ik)\Phi_2(z) + (1-ik)\overline{\Phi}_2(z) = 0 \quad \text{for all } z, \qquad (4.10.39)$$

which, when combined with equation (4.10.32), yields equation (4.7.6).

Example 4.10

1. Show that if the two bodies in contact have 'thicknesses' h_1,h_2 so that equation (4.10.7) has to be replaced by

   $$h_1(\tau_{nx}+i\tau_{ny})_1 = h_2(\tau_{nx}+i\tau_{ny})_2$$

   then the appropriate mismatch parameters α,β are obtained by replacing Γ everywhere by $\Gamma h_2/h_1$.

CHAPTER V - INTEGRAL TRANSFORMS

> "She began thinking over other children she knew, who might do very well as pigs, and was just saying to herself, "if one only knew the right way to change them -"
>
> \- *Alice in Wonderland*

5.1 *Introduction*

The use of integral transforms for the solution of the partial differential equations of mathematical physics is based on the method of separation of variables introduced by d'Alembert, Daniel Bernoulli, and Euler in the middle of the eighteenth century. To illustrate the method we shall consider its use in the solution of the Dirichlet problem for a half-plane which may be formulated in the form: Find the function $\phi(x,y)$ which satisfies Laplace's equation

$$\frac{\partial^2\phi}{\partial x^2} + \frac{\partial^2\phi}{\partial y^2} = 0 \tag{5.1.1}$$

in the half-plane $-\infty < x < \infty$, $y > 0$ and which satisfies the boundary condition

$$\phi(x,0) = f(x) \ , \tag{5.1.2}$$

where the function $f(x)$ is prescribed for all real values of x and is such that

$$\int_{-\infty}^{\infty} |f(x)|dx < \infty \ ;$$

further, ϕ satisfies the limiting condition $\phi(x,y) \to 0$ as $(x^2+y^2)^{1/2} \to \infty$.

By the method of separation of variables it is easily shown that the function

$$M(x,y;\xi) = (2\pi)^{-1/2}\exp(-i\xi x - |\xi|y) \qquad (5.1.3)$$

satisfies (5.1.1) for all real values of ξ. Since Laplace's equation (5.1.1) is linear we deduce from the principle of linear superposition that it is *probable* that the function

$$\phi(x,y) = \int_{-\infty}^{\infty} \Phi(\xi)M(x,y;\xi)d\xi \qquad (5.1.4)$$

will also be a solution of (5.1.1) for any arbitrary real-valued function Φ defined on the whole real line, at least if Φ is such that the integral on the right-hand side of this last equation converges uniformly for all real values of x and all positive real values of y. The function $\phi(x,y)$ defined by equation (5.1.4) will satisfy the boundary condition (5.1.2) if we are able to solve the integral equation

$$f(x) = (2\pi)^{-1/2} \int_{-\infty}^{\infty} \Phi(\xi)\exp(-i\xi x)d\xi \ , \qquad (5.1.5)$$

$(-\infty < x < \infty)$, for the unknown function Φ, and the condition at infinity will also be satisfied if Φ is such that the integral on the right of equation (5.1.5) tends to zero as $|x| \to \infty$.

The method can be generalized in an obvious way. Suppose that we have to determine the solution in a region $\mathcal{D}$ of a homogeneous partial differential equation of the form

$$L\phi(x,y) = 0 \ , \qquad (5.1.6)$$

in which L is a *linear* differential operator in the variables x and y. It may be possible to find a solution $M(x,y;\xi)$ of (5.1.6) involving an arbitrary constant ξ whose values, real or complex, are taken from a set Ω. Since the operator L is linear, it is probable that the function

$$\phi(x,y) = \int_\Omega M(x,y;\xi)F(\xi)d\xi \qquad (5.1.7)$$

will also be a solution for a wide class of functions F.

If, in addition,

$$u(x,0) = f(x) ,$$

then we must choose $F(\xi)$ in such a way that

$$f(x) = \int_\Omega K(x,\xi)F(\xi)d\xi , \qquad x \in \mathcal{D}_1 , \qquad (5.1.8)$$

where $K(x,\xi) = M(x,0;\xi)$ and $\mathcal{D}_1$ is the intersection of the region $\mathcal{D}$ and the line $y = 0$.

When a function f is defined in terms of a function F by means of an integral relation of the type (5.1.8) we say that f is the *integral transform* of the function F by the *kernel* $K(x,\xi)$ and we employ a notation such as

$$f(x) = \mathcal{T}[F(\xi);x] . \qquad (5.1.9)$$

For a particular choice of kernel we do not use the general symbol $\mathcal{T}$ but some symbol which reflects the nature of the kernel. For example, the kernel

$$K(x,\xi) = \exp(-x\xi) , \qquad Re(x) > 0 ,$$

defines the *Laplace transform*

$$\int_0^\infty F(\xi)\exp(-x\xi)d\xi$$

of F and we denote this transform by $L[F(\xi);x]$.

We shall now list the main properties of the Fourier transforms which will be used for the formulation and solution of plane problems. Only the briefest outline is given here; for details the reader may consult Sneddon (1951) or Sneddon (1972).

5.2 *Fourier transforms of functions of one variable*

If f(t) is such that the integral

$$\int_{-\infty}^{\infty} f(t)\exp(i\xi t)dt$$

exists for all real values of ξ, we define the *Fourier transform* of f by the equation

$$F(\xi) = (2\pi)^{-1/2} \int_{-\infty}^{\infty} f(t)\exp(i\xi t)dt \ . \tag{5.2.1}$$

We also write this equation in the form

$$F(\xi) = F[f(t);\xi] \ ,$$

and call F the operator of the Fourier transform. We define F^*, the *conjugate* of F, by the equation

$$F^*[G(\xi);x] = (2\pi)^{-1/2} \int_{-\infty}^{\infty} G(\xi)\exp(-i\xi x)d\xi \ . \tag{5.2.2}$$

It should be observed that

$$F^*[G(\xi);x] = F[G(\xi);-x] \ . \tag{5.2.3}$$

For a wide class of functions, which includes most of the functions occurring in physical applications,

$$\lim_{|\xi| \to \infty} F[f(t);\xi] = 0 ; \tag{5.2.4}$$

this result is known as the *Riemann-Lebesque lemma.*

Similarly, the *Fourier inversion theorem* states that if

$$F(\xi) = F[f(t);\xi] , \tag{5.2.5}$$

then

$$f(t) = F^*[F(\xi);t] . \tag{5.2.6}$$

In terms of the operators F, F^* and the identity operator I we can write this result in the form

$$F^*F = I , \tag{5.2.7}$$

or in the alternative form

$$F^{-1} = F^* . \tag{5.2.8}$$

The Fourier transform is used for functions whose domain is the whole real line. To treat functions whose domain is the positive real line we introduce the Fourier cosine transform defined by

$$F_c(\xi) \equiv F_c[f(t);\xi] = (2/\pi)^{1/2} \int_0^\infty f(t)\cos(\xi t)dt , \tag{5.2.9}$$

and the Fourier sine transform

$$F_s(\xi) \equiv F_s[f(t);\xi] = (2/\pi)^{1/2} \int_0^\infty f(t)\sin(\xi t)dt . \tag{5.2.10}$$

The *Fourier cosine inversion formula* is

$$F_c[F_c(\xi);t] = f(t) , \tag{5.2.11}$$

i.e.

$$\mathcal{F}_c^{-1} = \mathcal{F}_c , \tag{5.2.12}$$

and the Fourier sine inversion formula is

$$\mathcal{F}_s[F_s(\xi);t] = f(t) , \tag{5.2.13}$$

i.e.

$$\mathcal{F}_s^{-1} = \mathcal{F}_s . \tag{5.2.14}$$

The reader unacquainted with Fourier transforms is urged to work the following examples before proceeding further.

Precise conditions which a function $f(t)$ must satisfy in order that its Fourier transform will exist and, more importantly, the Fourier inversion theorem will hold, will be found, for example, in Titchmarsh (1937), Sneddon (1951) or Sneddon (1972). There are in fact a number of alternative sets of sufficient conditions. Thus Sneddon (1951) shows that if $f(x)$ satisfies *Dirichlet's conditions* in $-\infty < x < \infty$, viz. $f(x)$ has only a finite number of maxima and minima in any finite interval, and has no infinite discontinuities, and if $f(x)$ is absolutely integrable on $-\infty < x < \infty$, i.e.

$$\int_{-\infty}^{\infty} |f(x)|dx$$

exists, then

$$\mathcal{F}^*[\mathcal{F}\{f(t);x\};\xi] = \frac{1}{2}[f(\xi+0) + f(\xi-0)] , \tag{5.2.15}$$

where $f(\xi+0)$, $f(\xi-0)$ denote the limits of $f(x)$ as x tends to ξ from above and below. At a point where $f(x)$ is continuous the right-hand side of (5.2.15) is $f(\xi)$. Sneddon (1972) replaces Dirichlet's conditions by the condition that $f(x)$ shall be piecewise-continuously differentiable and, in §2.11, discusses much more general conditions.

Examples 5.2

1. Show that the Fourier transform of $f(t/a)$ is $aF(a\xi)$, and that the corresponding results hold for $\mathcal{F}_c$ and $\mathcal{F}_s$.
2. Show that if $f(t) = (2/\pi)^{1/2}t^{-1}\sin at$ then

$$F_c(\xi) = \begin{cases} 1 & 0 < \xi < a \\ 0 & \xi > a \end{cases}, \qquad F_s(\xi) = \pi^{-1}\ln|(\xi+a)/(\xi-a)| \ .$$

   Hint: establish the first part by using the inverse Fourier cosine transform.
3. Show that if $f(t)$ is an even function then $F(\xi) = F_c(\xi)$ while, if $f(t)$ is an odd function, $F(\xi) = iF_s(\xi)$.
4. Show that $\mathcal{F}^*[\exp(-|\xi|y);x] = (2/\pi)^{1/2}y(x^2+y^2)^{-1}$.
5. Show that $\mathcal{F}[f(x-a);\xi] = \exp(ia\xi)\mathcal{F}[f(x);\xi]$.

5.3 *Relations between Fourier transforms*

In the calculation of Fourier transforms use is often made of the convolution theorem

$$\mathcal{F}^*[F(\xi)G(\xi);t] = (2\pi)^{-1/2}\int_{-\infty}^{\infty} f(s)g(t-s)ds \ , \qquad (5.3.1)$$

in which $F = \mathcal{F}f$ and $G = \mathcal{F}g$. The integral on the right-hand side of this equation is called the *convolution* of the functions f and g and is written $(f\circ g)(t)$. It will be observed that $f\circ g = g\circ f$ and that the convolution theorem may be written in the form

$$\mathcal{F}^*[F(\xi)G(\xi);t] = (f\circ g)(t) \ , \qquad (5.3.2)$$

or in the alternative form

$$\mathcal{F}[(f\circ g)(t);\xi] = F(\xi)G(\xi) \ . \qquad (5.3.3)$$

Relations between Fourier transforms

Fourier transforms are used, as noted in the introduction, in the solution of differential equations; their use in such situations depends on certain simple relations which exist between the transforms of a function and of its derivatives. Thus, if f(t) is differentiable, integration by parts shows that

$$F[f'(t);\xi] = (2\pi)^{-1/2}\int_{-\infty}^{\infty} f'(t)\exp(i\xi t)dt$$

$$= (2\pi)^{-1/2}[f(t)\exp(i\xi t)]_{-\infty}^{\infty} - (2\pi)^{-1/2}i\xi\int_{-\infty}^{\infty} f(t)\exp(i\xi t)dt \ , \tag{5.3.4}$$

provided that the integrals exist. This shows that provided $f(t) \to 0$ as $t \to \pm\infty$, and this is so for most functions appearing in physical problems,

$$F[f'(t);\xi] = -i\xi F[f(t);\xi] \ . \tag{5.3.5}$$

This is the basic relation; the corresponding results for the F_c and F_s transforms are the subject of Example 5.3.1.

The notation used above has to be modified when f is a function of more than one real variable. Suppose, for instance, that we are concerned with a function f of two real variables x and y. Then we define $F[f(x,y);x \to \xi]$ and $F[f(x,y);y \to \eta]$ respectively by the equations

$$F[f(x,y);x \to \xi] = (2\pi)^{-1/2}\int_{-\infty}^{\infty} f(x,y)\exp(i\xi x)dx \ , \tag{5.3.6}$$

$$F[f(x,y);y \to \eta] = (2\pi)^{-1/2}\int_{-\infty}^{\infty} f(x,y)\exp(i\eta y)dy \ . \tag{5.3.7}$$

Equation (5.3.5) is correspondingly modified to

$$F[D_x f(x,y);x \to \xi] = -i\xi F[f(x,y);x \to \xi] \ , \tag{5.3.8}$$

where D_x denotes the differential operator $\partial/\partial x$. We also have the additional relation

$$\mathcal{F}[D_y f(x,y);x \to \xi] = D_y \mathcal{F}[f(x,y);x \to \xi] \ , \tag{5.3.9}$$

where D_y denotes the differential operator $\partial/\partial y$, and it is assumed that all the integrals exist.

In an exactly similar way we define $\mathcal{F}_c[f(x,y);x \to \xi]$ and $\mathcal{F}_s[f(x,y);x \to \xi]$. The analogues of equation (5.3.8) are

$$\mathcal{F}_c[D_x f(x,y);x \to \xi] = \xi \mathcal{F}_s[f(x,y);x \to \xi] - (2/\pi)^{1/2} f(0,y) \ , \tag{5.3.10}$$

$$\mathcal{F}_s[D_x f(x,y);x \to \xi] = -\xi \mathcal{F}_c[f(x,y);x \to \xi] \ . \tag{5.3.11}$$

These relations can often be used to derive unknown transforms from known ones. The results involving second derivatives can be deduced immediately; they are

$$\mathcal{F}_c[D_x^2 f(x,y);x \to \xi] = -\xi^2 \mathcal{F}_c[f(x,y);x \to \xi] - (2/\pi)^{1/2} f_x(0,y) \ , \tag{5.3.12}$$

$$\mathcal{F}_s[D_x^2 f(x,y);x \to \xi] = -\xi^2 \mathcal{F}_s[f(x,y);x \to \xi] + (2/\pi)^{1/2} \xi f(0,y) \ , \tag{5.3.13}$$

where in the first of these equations $f_x(0,y)$ denotes $[\partial f/\partial x]_{x=0}$. From these results we may deduce the transforms of $\nabla_1^2 f(x,y)$, where ∇_1^2 denotes the two-dimensional Laplacian operator $\nabla_1^2 = D_x^2 + D_y^2$; that for $\mathcal{F}$ is given by

$$\mathcal{F}[\nabla_1^2 f(x,y);x \to \xi] = (D_y^2 - \xi^2) F(\xi,y) \ . \tag{5.3.14}$$

Similarly for the biharmonic operator $\nabla_1^4 = \nabla_1^2 \nabla_1^2$ we have

$$\mathcal{F}[\nabla_1^4 f(x,y);x \to \xi] = (D_y^2 - \xi^2)^2 F(\xi,y) \ . \tag{5.3.15}$$

Other results may be found in the Examples.

Examples 5.3

1. By integrating the appropriate transform of f'(t) by parts show that

$$F_c[f'(t);\xi] = \xi F_s[f(t);\xi] - (2/\pi)^{1/2}f(0) ,$$

$$F_s[f'(t);\xi] = -\xi F_c[f(t);\xi] .$$

2. Use the result corresponding to (5.3.9) for F_c and Example 5.2.2 to show that if

$$f(x) = \frac{1-\cos ax}{x^2} , \quad F_c(\xi) = \begin{cases} (\pi/2)^{1/2}\xi , & 0 < \xi < a \\ (\pi/2)^{1/2}a , & \xi \geq a . \end{cases}$$

3. Show that

$$D_x F_s[\xi^{-1}A(\xi);x] = F_c[A(\xi);x] .$$

4. Show that

$$F_c[\nabla_1^2 f(x,y);x \to \xi] = (D_y^2 - \xi^2)F_c(\xi,y) - (2/\pi)^{1/2}f_x(0,y) ,$$

$$F_s[\nabla_1^2 f(x,y);x \to \xi] = (D_y^2 - \xi^2)F_s(\xi,y) - (2/\pi)^{1/2}\xi f(0,y) ,$$

$$F_c[\nabla_1^4 f(x,y);x \to \xi] = (D_y^2 - \xi^2)^2 F_c(\xi,y) + (\xi^2-2D_y^2)f_1(y) - f_3(y) ,$$

$$F_s[\nabla_1^4 f(x,y);x \to \xi] = (D_y^2 - \xi^2)^2 F_s(\xi,y) - \xi(\xi^2+2D_y^2)f_0(y) + \xi f_2(y),$$

where

$$f_r(y) = (2/\pi)^{1/2}[\partial^r f/\partial x^r]_{x=0} .$$

5. Use (5.3.8) and Example 5.2.4 to deduce that

$$F^*[i\xi y\exp(-|\xi|y);\xi \to x] = 2(2/\pi)^{1/2}xy^2(x^2+y^2)^{-2} .$$

5.4 *Boussinesq's solution for the half-plane*

We shall now show how integral transform methods may be used to derive Boussinesq's (or Flamant's) solution for the half-plane. We saw in §5.1 that the function $F^*[\Psi(\xi)\exp(-|\xi|y);\xi \to x]$ is harmonic in the half-plane $y \geq 0$. If therefore in the Papkovich-Neuber solution (1.10.11) we take the components of $\vec{\psi}$ to be

$$\psi_1 = 0 , \quad \psi_2(x,y) = F^*[|\xi|\xi^{-2}A(\xi)\exp(-|\xi|y);\xi \to x] , \quad \psi_3 = 0 , \tag{5.4.1}$$

and

$$\phi(x,y) = F^*[\xi^{-2}C(\xi)\exp(-|\xi|y);\xi \to x] , \tag{5.4.2}$$

we obtain the displacements and stresses

$$u(x,y) = F^*[U(\xi,y);\xi \to x] , \quad v(x,y) = F^*[V(\xi,y);\xi \to x] , \tag{5.4.3}$$

$$\tau_{xx}(x,y) = F^*[T_{11}(\xi,y);\xi \to x] , \quad \tau_{xy}(x,y) = F^*[T_{12}(\xi,t);\xi \to x] , \tag{5.4.4}$$

$$\tau_{yy}(x,y) = F^*[T_{22}(\xi,y);\xi \to x] , \tag{5.4.5}$$

where

$$\begin{bmatrix} 2\mu U \\ 2\mu V \\ -T_{11} \\ -T_{12} \\ -T_{22} \end{bmatrix} = \begin{bmatrix} \{|\xi|yA + C\}i\xi^{-1} \\ \{(3-4\nu)A + |\xi|yA + C\}|\xi|^{-1} \\ 2\nu A - |\xi|yA - C \\ \{(1-2\nu)A + |\xi|yA + C\}i\,\mathrm{sign}(\xi) \\ 2(1-\nu)A + |\xi|yA + C \end{bmatrix} \exp(-|\xi|y) . \tag{5.4.6}$$

Suppose now that $A(\xi)$, $C(\xi)$ have to be found to satisfy the boundary conditions

$$\tau_{xy}(x,0) = 0 , \quad \tau_{yy}(x,0) = -p(x) , \quad -\infty < x < \infty . \tag{5.4.7}$$

Equation (5.4.6) shows that $A(\xi)$, $C(\xi)$ must satisfy

$$(1-2\nu)A(\xi) + C(\xi) = 0 , \qquad (5.4.8)$$

and then

$$\tau_{yy}(x,0) = -F^*[A(\xi);x] , \qquad (5.4.9)$$

so that

$$A(\xi) = F[p(x);\xi] = P(\xi) ,$$

and the matrix of (5.4.6) is given by

$$\begin{bmatrix} 2\mu U \\ 2\mu V \\ -T_{11} \\ -T_{12} \\ -T_{22} \end{bmatrix} = \begin{bmatrix} \{|\xi|y - (1-2\nu)\}i\xi^{-1} \\ \{|\xi|y + 2(1-\nu)\}|\xi|^{-1} \\ -|\xi|y + 1 \\ \{|\xi|y\}i\,\mathrm{sign}(\xi) \\ |\xi|y + 1 \end{bmatrix} P(\xi)\exp(-|\xi|y) . \qquad (5.4.10)$$

The expressions for the displacements and stresses may be expressed directly in terms of $p(x)$ by using the convolution theorem. Thus since (Example 5.3.5)

$$F^*[i\xi y\exp(-|\xi|y);\xi \to x] = 2(2/\pi)^{1/2}xy^2(x^2+y^2)^{-2} ,$$

$$F^*[(1-|\xi|y)\exp(-|\xi|y);\xi \to x] = 2(2/\pi)^{1/2}x^2y(x^2+y^2)^{-2} ,$$

$$F^*[(1+|\xi|y)\exp(-|\xi|y);\xi \to x] = 2(2/\pi)^{1/2}y^3(x^2+y^2)^{-2} ,$$

we may write

$$\tau_{xx}(x,y) = -\int_{-\infty}^{\infty} T_{xx}(x-x',y)p(x')dx' , \quad \tau_{xy}(x,y) = -\int_{-\infty}^{\infty} T_{xy}(x-x',y)p(x')dx' , \qquad (5.4.11)$$

$$\tau_{yy}(x,y) = -\int_{-\infty}^{\infty} T_{yy}(x-x',y)p(x')dx' , \qquad (5.4.12)$$

where the kernels are defined by the equations

$$T_{xx}(x,y) = 2\pi^{-1}xy^2(x^2+y^2)^{-2} , \quad T_{xy}(x,y) = 2\pi^{-1}x^2y(x^2+y^2)^{-2} , \qquad (5.4.13)$$

$$T_{yy}(x,y) = 2\pi^{-1}y^3(x^2+y^2)^{-2} . \qquad (5.4.14)$$

These results, which agree with those obtained in §2.1, were obtained by Sneddon (1951).

Integrals for the displacements in terms of $p(x)$ may be obtained in a similar way, but slight complications arise. We shall first carry out the analysis formally and then comment on it. For the displacement $u(x,y)$ we may use the result

$$F^*[-i\xi^{-1}(1-2\nu-|\xi|y)\exp(-|\xi|y);\xi \to x] = (2/\pi)^{1/2}[xy(x^2+y^2)^{-1} + $$

$$+ (1-2\nu)\mathrm{Tan}^{-1}(y/x)] ,$$

where $\mathrm{Tan}^{-1}(y/x)$ is defined in (2.1.15), to give

$$u(x,y) = \int_{-\infty}^{\infty} G_{12}(x-x',y)p(x')dx', \qquad (5.4.15)$$

where $G_{12}(x,y)$ is given in (2.1.18). For the displacement $v(x,y)$ we must first differentiate (5.4.3) with respect to x to eliminate the $|\xi|^{-1}$ singularity and obtain

$$2\mu\partial v/\partial x = -iF^*[\mathrm{sign}(\xi)(2-2\nu+|\xi|y)P(\xi)\exp(-|\xi|y);\xi \to x] , \qquad (5.4.16)$$

and then use

$$iF^*[\mathrm{sign}(\xi)(2-2\nu+|\xi|y)\exp(-|\xi|y);\xi \to x] = 2(2/\pi)^{1/2}[xy^2(x^2+y^2)^{-2} + $$

$$+ (1-\nu)x(x^2+y^2)^{-1}]$$

to obtain

$$\partial v(x,y)/\partial x = \int_{-\infty}^{\infty} G_{22,x}(x-x',y)p(x')dx' , \tag{5.4.17}$$

where

$$G_{22,x}(x,y) = -(\pi\mu)^{-1}[xy^2(x^2+y^2)^{-2} + (1-\nu)x(x^2+y^2)^{-1}] . \tag{5.4.18}$$

Equation (5.4.18) may now be formally integrated to yield

$$v(x,y) = \int_{-\infty}^{\infty} G_{22}(x-x',y)p(x')dx' , \tag{5.4.19}$$

where $G_{22}(x,y)$ is given by equation (2.1.19).

To comment on this formal analysis we note that if

$$P(0) \equiv (2\pi)^{-1/2} \int_{-\infty}^{\infty} p(t)dt$$

is not zero, i.e., if the total force acting on the boundary is not zero, then the integrand in the expression for $u(x,y)$ has a ξ^{-1} singularity so that the integral must be interpreted as a Cauchy Principal Value. This affects the validity of equation (5.4.15) only in that $u(x,y)$ does not tend to zero at infinity, whereas the Riemann-Lebesque Lemma (5.2.4) states that 'most' functions expressible as Fourier transforms, and equation (5.4.3) does so express $u(x,y)$, do tend to zero at infinity. The situation regarding $v(x,y)$ is similar, but more acute. If $P(0) \neq 0$, the integrand in (5.4.10) has a $|\xi|^{-1}$ singularity, which is not integrable even as a Cauchy Principal Value. This shows itself in the fact that $v(x,y)$ tends to infinity at infinity, as discussed in §2.1. There is no difficulty in either integral if $P(0) = 0$. An interesting discussion of this matter may be found in Golecki (1975).

Examples 5.4

1. Show that the state-vector transform corresponding to the boundary stress

$$\tau_{xy}(x,0) = -s(x) \quad , \quad \tau_{yy}(x,0) = 0 \ ,$$

is given by

$$\begin{bmatrix} 2\mu U \\ 2\mu V \\ -T_{11} \\ -T_{12} \\ -T_{22} \end{bmatrix} = \begin{bmatrix} \{-|\xi|y + 2(1-\nu)\}|\xi|^{-1} \\ \{|\xi|y + (1-2\nu)\}i\xi^{-1} \\ (-|\xi|y + 2)i\,\mathrm{sign}(\xi) \\ -|\xi|y + 1 \\ |\xi|y\,i\,\mathrm{sign}(\xi) \end{bmatrix} \exp(-|\xi|y)S(\xi) \ ,$$

where $S(\xi) = \mathcal{F}[s(x);\xi]$. Use this result to derive the equations given in Example 2.1.2.

2. Show that if $p(x)$ is an even (odd) function and if $P(\xi) = \mathcal{F}_c(\mathcal{F}_s)p$, then the operators $\mathcal{F}^*$ in equations (5.4.3) - (5.4.5) in front of U, T_{12} must be replaced by $\mathcal{F}_s(-\mathcal{F}_c)$ and those in front of V, T_{11}, T_{22} by $\mathcal{F}_c(\mathcal{F}_s)$. The quantities U, V, T_{11}, T_{12}, T_{22} are related to $P(\xi)$ by equation (5.4.10) except that $|\xi|$ is replaced by ξ and i, sign (ξ) are omitted.
Show similarly that if $s(x)$ is an even (odd) function and if $S(\xi) = \mathcal{F}_c(\mathcal{F}_s)s$, then the operators $\mathcal{F}^*$ in equations (5.4.3) - (5.4.5) in front of U, T_{12} must be replaced by $\mathcal{F}_c(\mathcal{F}_s)$ and those in front of V, T_{11}, T_{22} by $\mathcal{F}_s(-\mathcal{F}_c)$. Again U, V, T_{11}, T_{12}, T_{22} are related to $S(\xi)$ by the equations in Example 5.4.1 except that $|\xi|$ is replaced by ξ and i, $\mathrm{sign}(\xi)$ are omitted.

3. Show that the formal transforms of the stress-strain relations (4.2.2), (4.2.3) for plane strain are

$$(1-2\nu)T_{11} = 2\mu\{-i(1-\nu)\xi U + \nu\partial V/\partial y\} \ ,$$
$$(1-2\nu)T_{22} = 2\mu\{-i\nu\xi U + (1-\nu)\partial V/\partial y\} \ , \quad T_{12} = \mu\{\partial U/\partial y - i\xi V\}$$

4. If the corresponding capital letters are used to denote Fourier transforms show that the formal transforms $x \rightarrow \xi$ of the Papkovich-Neuber equations (1.11.5), (1.11.6) for plane

strain are

$2\mu U = 4(1-\nu)\Psi_1 + i\xi(-i\partial\Psi_1/\partial\xi + y\Psi_2 + \Phi)$,

$2\mu V = 4(1-\nu)\Psi_2 - \partial/\partial y(-i\partial\Psi_1/\partial\xi + y\Psi_2 + \Phi)$,

$2(1-\nu)(\partial^2/\partial y^2-\xi^2)\Psi_i = -F_i$, $2(1-\nu)(\partial^2/\partial y^2-\xi^2)\Phi = -i\partial F_1/\partial\xi + yF_2$,

where F_1, F_2 are the transforms of F_x, F_y respectively.

5. Show that if $F_y \equiv 0$ and F_x is independent of y the particular solutions for Ψ_1, Φ of Example 5.4.4 are given by

$$2(1-\nu)\Psi_1 = \xi^{-2}F_1 , \quad 2(1-\nu)\Phi = i\xi^{-2}\partial F_1/\partial\xi ,$$

and that the corresponding displacement and stress transforms are given by $V \equiv 0 \equiv T_{12}$ and

$$2\mu(1-\nu)U = (1-2\nu)\xi^{-2}F_1 , \quad T_{11} = -i\xi^{-1}F_1 , \quad (1-\nu)T_{22} = -i\nu\xi^{-1}F_1 .$$

5.5 *Hilbert Transforms*

In the analysis of certain contact problems (see e.g., (5.4.9)) we shall encounter the situation in which

$$F^*[F(\xi);x] = 0 , \quad |x| > a . \tag{5.5.1}$$

If we let

$$F^*[F(\xi);x] = f(x) , \quad |x| \leq a , \tag{5.5.2}$$

then, by making use of the Fourier inversion theorem, we see that

$$F(\xi) = (2\pi)^{-1/2}\int_{-a}^{a} f(x)\exp(i\xi x)dx . \tag{5.5.3}$$

Let us now evaluate the transform

$$g(x,y) = F^*[-i\,\mathrm{sign}(\xi)F(\xi)\exp(-|\xi|y);x] , \tag{5.5.4}$$

where $y > 0$; the reason for choosing this transform will become clear below. Interchanging the order of integration, we have

$$g(x,y) = (2\pi)^{-1}\int_{-a}^{a} f(t)\left\{\int_{-\infty}^{\infty} -i\,\mathrm{sign}(\xi)\exp[-|\xi|y+i\xi(t-x)]d\xi\right\}dt\ .$$

The inner integral is

$$\int_{0}^{\infty}\left\{-i\exp\xi[-y+i(t-x)] + i\ \exp\xi[-y-i(t-x)]\right\}d\xi = \frac{2(t-x)}{(t-x)^2+y^2}\ ,$$

so that

$$g(x,y) = \frac{1}{\pi}\int_{-a}^{a}\frac{(t-x)f(t)dt}{(t-x)^2+y^2}\ . \qquad (5.5.5)$$

This is the type of integral that appeared in §§2.1 and 5.4. For the moment we are interested more in the limit of the integral as $y \to 0+$; this will be defined as the *Hilbert transform* of f(x). We have

$$F^*[-i\,\mathrm{sign}(\xi)F(\xi);x] = h[f(t);x] = \frac{1}{\pi}\int_{-a}^{a}\frac{f(t)dt}{t-x}\ . \qquad (5.5.6)$$

Note that the Fourier transform is defined as $\lim_{y\to 0+} g(x,y)$ and the Hilbert transform will have to be interpreted as the Cauchy Principal Value

$$\lim_{\varepsilon\to 0}\frac{1}{\pi}\left\{\int_{-a}^{x-\varepsilon}\frac{f(t)dt}{t-x} + \int_{x+\varepsilon}^{a}\frac{f(t)dt}{t-x}\right\} \qquad (5.5.7)$$

when x lies in [-a,a].

The analysis described above may be modified for functions defined only on the positive real axis. Thus if

$$F_s[F(\xi);x] = f(x)H(a-x) \quad \text{for} \quad x \geq 0\ , \qquad (5.5.8)$$

where H(x) is the Heaviside function

$$H(x) = \begin{cases} 1 \ , \ x \geq 0 \ , \\ 0 \ , \ x < 0 \ , \end{cases} \tag{5.5.9}$$

then

$$F(\xi) = (2/\pi)^{1/2} \int_0^a f(x)\sin(\xi x)dx \ , \tag{5.5.10}$$

and this may be written in the form (5.5.3), i.e.

$$F(\xi) = (2\pi)^{-1/2} \int_{-a}^a -if(x)\exp(i\xi x)dx \ , \tag{5.5.11}$$

where $f(x)$ is defined for negative x by the statement $f(-x) = -f(x)$. If equation (5.5.9) is used to define $F(\xi)$ for negative ξ then $F(-\xi) = -F(\xi)$, and equation (5.5.6), applied to $-if(x)$, gives

$$\mathcal{F}^*\{-i\,\mathrm{sign}(\xi)F(\xi);x\} = -i\mathcal{F}_c\{F(\xi);x\} = h[-if(t);x] \ ,$$

or

$$\mathcal{F}_c\{F(\xi);x\} = h[f(t);x] = \frac{1}{\pi}\int_{-a}^a \frac{f(t)dt}{t-x} \ . \tag{5.5.12}$$

The corresponding result for $\mathcal{F}_c$, $\mathcal{F}_s$ interchanged is the subject of Example 5.5.1.

The *inversion* of the Hilbert transform may be obtained using only real analysis, but it is easier to use the analysis of §3.5 based on the Plemelj formulae (3.3.6)-(3.3.7). Thus the analysis of §3.5 shows that if

$$\frac{1}{\pi}\int_{-a}^a \frac{f(t)}{t-x}\,dt = g(x) \ , \quad |x| < a \ , \tag{5.5.13}$$

then the general solution for $f(x)$ in the class of functions having integrable singularities at $\pm$ a is

$$f(t) = \frac{(a^2-t^2)^{-1/2}}{\pi} \int_{-a}^{a} \frac{(a^2-x^2)^{1/2} g(x)dx}{x-t} + C(a^2-t^2)^{-1/2} , \qquad (5.5.14)$$

where C is arbitrary. If f(t) is to be finite at $\pm$ a, then it will in fact be zero there and

$$f(x) = \frac{(a^2-t^2)^{1/2}}{\pi} \int_{-a}^{a} \frac{(a^2-x^2)^{-1/2} g(x)dx}{x-t} , \qquad (5.5.15)$$

provided that g(x) satisfies

$$\int_{-a}^{a} (a^2-x^2)^{-1/2} g(x)dx = 0 . \qquad (5.5.16)$$

Examples 5.5

1. Show that if $F_c[G(\xi);x] = g(x)H(a-x)$ for $x \geq 0$, then $F_s[G(\xi);x] = -h[g(t);x]$, where $g(-t) = g(t)$ for $-a \leq t \leq a$.
2. Show that

$$F^*[-i\,\mathrm{sign}(\xi)F(\xi);x] = \begin{cases} -F_s[F(\xi);x], & \text{if } F(\xi) \text{ even ,} \\ -iF_c[F(\xi);x], & \text{if } F(\xi) \text{ odd.} \end{cases}$$

3. Extend the result of Example 5.5.1 to show that if $F_c[P(\xi);x] = p(x)H(1-x)$ for $x \geq 0$, and $p(x) = p(-x)$ for $-1 \leq x \leq 0$, then

$$F_s[P(\xi)G(\xi);x] = -\pi^{-1} \int_{-1}^{1} p(t)g(t-x)dt \text{ , where}$$

$g(x) = (\pi/2)^{1/2} F_s[G(\xi);x]$.
Similarly if $F_s[P(\xi);x] = p(x)H(1-x)$ for $x \geq 0$, and $p(x) = -p(-x)$ for $-1 \leq x \leq 0$, then

$$F_c[P(\xi)G(\xi);x] = \pi^{-1} \int_{-1}^{1} p(t)g(t-x)dt .$$

These results may be considered as special cases of the convolution theorem (5.3.1).

5.6 *Hankel transforms*

Hankel transforms have important applications both in plane and in axisymmetric problems. The Hankel transform of order ν is defined in terms of $J_\nu(\xi x)$, the Bessel function of the first kind of order ν, through the equation

$$H_\nu[f(\xi);x] = \int_0^\infty xf(x)J_\nu(\xi x)dx \ . \tag{5.6.1}$$

Hankel transforms may be defined for very general, complex values of ξ and ν; this general theory is beyond the scope of this book and the interested reader may consult the standard texts, e.g., Erdelyi (1954), Sneddon (1951) or Watson (1958). For the purposes of this book it suffices to consider ν to be a positive integer, almost always 0, 1 or 2, or of the form $\nu = \pm n + \frac{1}{2}$ where n is a positive integer. Any analysis involving Hankel transforms involves the calculation of various integrals of Bessel functions. There are literally scores of different types of such integrals; some are listed in the texts cited above; others may be found in Luke (1962). The following are some of the most important properties of Bessel functions.

The Bessel function $y \equiv J_\nu(\xi x)$ is a solution of the equation

$$x^2y'' + xy' + (\xi^2x^2 - \nu^2)y = 0 \ . \tag{5.6.2}$$

When $\nu \geq 0$, $J_\nu(\xi x)$ is bounded as $x \to 0$, and has the series representation

$$J_\nu(\xi x) = (\xi x/2)^\nu \sum_{k=0}^\infty (-\xi^2x^2/4)^k[k!\Gamma(\nu+k+1)]^{-1} \ , \tag{5.6.3}$$

where Γ is the Gamma function which, for all integer and half-integer values, may be determined from the initial values

$$\Gamma(1) = 1 \ , \quad \Gamma(1/2) = \pi^{1/2} \ , \tag{5.6.4}$$

and the recurrence relations

$$\Gamma(x+1) = x\Gamma(x) , \quad \Gamma(x)\Gamma(1-x) = \pi \operatorname{cosec} \pi x . \tag{5.6.5}$$

In particular

$$\Gamma(n+1) = n! \tag{5.6.6}$$

The Bessel functions satisfy the recurrence relations

$$J_{\nu-1}(\xi x) + J_{\nu+1}(\xi x) = 2\nu(\xi x)^{-1}J_\nu(\xi x) , \tag{5.6.7}$$

$$J'_\nu(\xi x) = \xi J_{\nu-1}(\xi x) - \nu x^{-1}J_\nu(\xi x) , \tag{5.6.8}$$

$$= -\xi J_{\nu+1}(\xi x) + \nu x^{-1}J_\nu(\xi x) , \tag{5.6.9}$$

where ' denotes differentiation with respect to x. They also satisfy the more general equations

$$[(\xi x)^{-1}d/dx]^k\{x^\nu J_\nu(\xi x)\} = x^{\nu-k}J_{\nu-k}(\xi x) , \tag{5.6.10}$$

$$[(\xi x)^{-1}d/dx]^k\{x^{-\nu} J_\nu(\xi x)\} = (-)^k x^{-\nu-k}J_{\nu+k}(\xi x) , \tag{5.6.11}$$

for $k = 0,1,2,\ldots$, of which the most important cases are

$$J'_0(\xi x) = -\xi J_1(\xi x) , \quad [xJ_1(\xi x)]' = \xi x J_0(\xi x) . \tag{5.6.12}$$

When ν is half an odd integer, J_ν may be expressed in terms of trigonometric functions. It is usual to define the spherical Bessel functions of the first kind

$$j_n(x) = [\pi/(2x)]^{1/2}J_{n+1/2}(x) , \tag{5.6.13}$$

and then

$$j_{-1}(x) = x^{-1}\cos x , \quad j_0(x) = x^{-1}\sin x , \tag{5.6.14}$$

and other $j_n(x)$ may be calculated from the recurrence

$$j_{n-1}(x)+j_{n+1}(x) = (2n+1)x^{-1}j_n(x); \quad n = 0, \pm 1, \pm 2, \ldots . \tag{5.6.15}$$

The equation (5.6.2) satisfied by $J_\nu(\xi x)$ is intimately related to the form which the operator ∇^2 takes in cylindrical polar co-ordinates, namely

$$\nabla^2 = \frac{\partial^2}{\partial r^2} + \frac{1}{r}\frac{\partial}{\partial r} + \frac{1}{r^2}\frac{\partial^2}{\partial \theta^2} + \frac{\partial^2}{\partial z^2} . \tag{5.6.16}$$

Thus we notice that equation (5.6.2) may be written

$$(D_r^2 + r^{-1}D_r - \nu^2 r^{-2} + \xi^2)J_\nu(\xi r) = 0 , \tag{5.6.17}$$

so that if $\mathcal{B}_\nu = D_r^2 + r^{-1}D_r - \nu^2 r^{-2}$, then

$$\mathcal{H}_\nu[\mathcal{B}_\nu f(r);\xi] = -\xi^2 \mathcal{H}_\nu[f(r);\xi] . \tag{5.6.18}$$

If the function f(r) is now replaced by $f(r,z)\exp(i\nu\theta)$, then

$$\nabla^2 f(r,z)\exp(i\nu\theta) = \exp(i\nu\theta)(\mathcal{B}_\nu + D_z^2)f(r,z) , \tag{5.6.19}$$

so that

$$\mathcal{H}_\nu[\nabla^2 f(r,z)\exp(i\nu\theta); r \to \xi] = \exp(i\nu\theta)(D_z^2 - \xi^2)\overline{f}_\nu(\xi,z) , \tag{5.6.20}$$

where $D_z = \partial/\partial z$ and

$$\overline{f}_\nu(\xi,z) = \mathcal{H}_\nu[f(r,z); r \to \xi] , \tag{5.6.21}$$

and it is assumed that all the operations are valid. This has an important consequence; the function $f(r,z)\exp(i\nu\theta)$ will be *harmonic* if its $\mathcal{H}_\nu$ transform $\overline{f}_\nu(\xi,z)$ satisfies

$$(D_z^2 - \xi^2)\overline{f}_\nu(\xi,z) = 0 \;, \tag{5.6.22}$$

of which the general solution is

$$\overline{f}_\nu(\xi,z) = A(\xi)\exp(-\xi z) + C(\xi)\exp(\xi z) \;. \tag{5.6.23}$$

This is the basis for the Hankel transform solutions for the layer and half-space.

The use of Hankel transforms is based on the *Hankel inversion theorem*; this states that if

$$H_\nu[f(x);\xi] = \overline{f}_\nu(\xi) \;, \tag{5.6.24}$$

then

$$f(x) = H_\nu[\overline{f}_\nu(\xi);x] \;, \tag{5.6.25}$$

i.e., that

$$H_\nu^{-1} = H_\nu \;. \tag{5.6.26}$$

We need to be able to express the Hankel transform of the derivative of a function in terms of a Hankel transform of the function itself; this we can do by using equations (5.6.10), (5.6.11). We shall use D_x to denote d/dx; then taking k = 1 in (5.6.10) we deduce that

$$D_x\{x^\nu J_\nu(\xi x)\} = \xi x^\nu J_{\nu-1}(\xi x) \;, \tag{5.6.27}$$

so that

$$H_\nu[x^{\nu-1}D_x\{x^{1-\nu}f(x)\};\xi] = \int_0^\infty D_x\{x^{1-\nu}f(x)\}x^\nu J_\nu(\xi x)dx =$$

$$= [xf(x)J_\nu(\xi x)]_0^\infty - \int_0^\infty x^{1-\nu}f(x)\xi x^\nu J_{\nu-1}(\xi x)dx \;,$$

and, provided the function is such that the integrated terms vanish, we deduce that

$$H_\nu[x^{\nu-1}D_x\{x^{1-\nu}f(x)\};\xi] = -\xi H_{\nu-1}[f(x);\xi] \ . \tag{5.6.28}$$

By using equation (5.6.11) with k = 1 we may show that

$$H_\nu[x^{-\nu-1}D_x\{x^{1+\nu}f(x)\};\xi] = \xi H_{\nu+1}[f(x);\xi] \ , \tag{5.6.29}$$

again provided that the integrated terms vanish.

Many important Hankel transforms may be derived by manipulating the integral representation

$$J_\nu(\xi x) = \frac{2(\xi x/2)^\nu}{\pi^{1/2}\Gamma(\nu+1/2)} \int_0^1 (1-t^2)^{\nu-1/2} \cos(\xi x t)dt \ , \tag{5.6.30}$$

which is valid for $\nu > -1/2$ (actually for $\mathrm{Re}(\nu) > -1/2$). This may be rewritten to show $J_\nu(\xi x)$ as a Fourier cosine transform; thus

$$2^{-1/2}\Gamma(\nu+1/2)\ (\xi x/2)^{-\nu}J_\nu(\xi x) = F_c[q_\nu(t);t \to (\xi x)] \ , \tag{5.6.31}$$

where

$$q_\nu(t) = \begin{cases} (1-t^2)^{\nu-1/2} \ , & 0 \le t < 1 \ , \\ 0 \ , & t > 1 \ . \end{cases} \tag{5.6.32}$$

Now, by using the F_c inversion theorem, we obtain

$$2^{-1/2}\Gamma(\nu+1/2)\ F_c[(\xi x/2)^{-\nu}J_\nu(\xi x);x \to t] = \xi^{-1}q_\nu(t/\xi) \ . \tag{5.6.33}$$

This equation opens the way to many new relations, for the left-hand side may be expressed as an H_ν transform and this in turn may be inverted using the H_ν inversion theorem. The equation may also be differentiated and integrated to obtain other relations. Some of the more important ones will now be derived, while others appear in the Examples and later applications.

Taking $\nu = 0$ in (5.6.33) we obtain

$$(2/\pi)^{-1/2}\mathcal{F}_c[J_0(\xi x); x \to t] = \xi^{-1}q_0(t/\xi) , \tag{5.6.34}$$

which may be rewritten

$$\int_0^\infty J_0(\xi x)\cos(xt)dx = \begin{cases} (\xi^2-t^2)^{-1/2} , & 0 \leq t < \xi , \\ 0 , & t > \xi , \end{cases} \tag{5.6.35}$$

or in other words

$$\mathcal{H}_0[x^{-1}\cos xt; x \to \xi] = \xi^{-1}q_0(t/\xi) . \tag{5.6.36}$$

Now integrate this with respect to t, using $\int_0^\tau \cos xt dt = x^{-1}\sin x\tau$, to obtain

$$\mathcal{H}_0[x^{-2}\sin xt; x \to \xi] = \begin{cases} \sin^{-1}(t/\xi) , & 0 \leq t < \xi , \\ \frac{\pi}{2} , & t \geq \xi . \end{cases} \tag{5.6.37}$$

Another transform may be found by applying (5.6.29) with $\nu = 0$, $f(x) = x^{-1}\sin xt$; this gives

$$\mathcal{H}_1[x^{-1}\sin xt; x \to \xi] = \xi^{-1}\mathcal{H}_0[x^{-1}(t\cos xt); x \to \xi] = t\xi^{-2}q_0(t/\xi) , \tag{5.6.38}$$

and three other equivalent relations may be found from this equation. The $\mathcal{H}_0$ transform of $x^{-1}\sin(xt)$ is often used, and is (Example 5.6.1)

$$\mathcal{H}_0[x^{-1}\sin(xt); x \to \xi] = t^{-1}q_0(\xi/t) . \tag{5.6.39}$$

This may be rewritten as a sine transform, namely

$$(2/\pi)^{-1/2}\mathcal{F}_s[J_0(x\xi); x \to t] = t^{-1}q_0(\xi/t) . \tag{5.6.40}$$

Examples 5.6

1. Use the integral representation

$$J_0(x) = (2/\pi)\int_1^\infty (t^2-1)^{-1/2}\sin(xt)dt$$

to obtain equation (5.6.39).

2. Invert equations (5.6.39), (5.6.38), (5.6.35) to obtain respectively

$$\int_0^t \xi(t^2-\xi^2)^{-1/2}J_0(\xi x)d\xi = x^{-1}\sin xt ,$$

$$\int_t^\infty (\xi^2-t^2)^{-1/2}J_1(\xi x)d\xi = (xt)^{-1}\sin xt ,$$

$$\int_t^\infty \xi(\xi^2-t^2)^{-1/2}J_0(\xi x)d\xi = x^{-1}\cos xt .$$

3. Show that $H_\nu[\nabla^4 f(r,z)\exp(i\nu\theta);r \to \xi] = (D_z^2-\xi^2)^2\exp(i\nu\theta)\overline{f}_\nu(\xi,z)$ and deduce that $f(r,z)\exp(i\nu\theta)$ will be *biharmonic*, i.e., $\nabla^4 f(r,z)\exp(i\nu\theta) = 0$, if $\overline{f}_\nu(\xi,z) = [A(\xi) + zB(\xi)]\exp(-\xi z) + [C(\xi) + zD(\xi)]\exp(\xi z)$.

5.7 *Relations between Fourier, Hankel and Abel transforms*

We begin by defining a generalization to n-dimensions of the Fourier transform. If f is a real-valued function of the n real variables $x_1, x_2, \ldots, x_n$, we define its *n-dimensional Fourier transform* by

$$F_n(\xi) = F_{(n)}[f(\underset{\sim}{x});\underset{\sim}{\xi}] = (2\pi)^{-n/2}\int_{E_n} f(\underset{\sim}{x})\ \exp[i(\underset{\sim}{\xi}\cdot\underset{\sim}{x})]d\underset{\sim}{x} , \qquad (5.7.1)$$

where $\underset{\sim}{x}$ and $\underset{\sim}{\xi}$ respectively denote the n-vectors with components $(x_1, x_2, \ldots, x_n)$ and $(\xi_1, \xi_2, \ldots, \xi_n)$ and

$(\underset{\sim}{\xi}\cdot\underset{\sim}{x}) = \xi_1 x_1 + \xi_2 x_2 + \cdots + \xi_n x_n$ denotes the inner product of the vectors $\underset{\sim}{\xi}$ and $\underset{\sim}{x}$ in E_n, the Euclidean space of n-dimensions.

The appropriate inversion theorem is

$$f(\underset{\sim}{x}) = F^*_{(n)}[F_{(n)}(\xi);\underset{\sim}{x}] , \tag{5.7.2}$$

where the operator $F^*_{(n)}$ is defined by the equation

$$F^*_{(n)}[g(\underset{\sim}{x});\underset{\sim}{\xi}] = (2\pi)^{-n/2}\int_{E_n} g(\underset{\sim}{x})\exp[-i(\underset{\sim}{\xi}\cdot\underset{\sim}{x})]d\underset{\sim}{x} ,$$

so that

$$F^*_{(n)}[g(\underset{\sim}{x});\underset{\sim}{\xi}] = F_{(n)}[g(\underset{\sim}{x});-\underset{\sim}{\xi}] . \tag{5.7.3}$$

If we define the *convolution* of two functions of this kind by the equation

$$(f\circ g)(\underset{\sim}{x}) = (2\pi)^{-n/2}\int_{E_n} f(\underset{\sim}{x}-\underset{\sim}{y})g(\underset{\sim}{y})d\underset{\sim}{y} , \tag{5.7.4}$$

we find the *Faltung* theorem

$$F_{(n)}[(f\circ g)(\underset{\sim}{x});\underset{\sim}{x} \to \underset{\sim}{\xi}] = F_n(\underset{\sim}{\xi})G_n(\underset{\sim}{\xi}) , \tag{5.7.5}$$

a result which is frequently used in its inverse form

$$F^*_{(n)}[F_n(\underset{\sim}{\xi})G_n(\underset{\sim}{\xi});\underset{\sim}{\xi} \to \underset{\sim}{x}] = (f\circ g)(\underset{\sim}{x}) . \tag{5.7.6}$$

The first relation between Hankel and Fourier transforms appears when we have to find the Fourier transform $F_n(\underset{\sim}{\xi})$ of a function f of the radial distance $r = (x_1^2 + x_2^2 + \cdots + x_n^2)^{1/2}$. Consider the case n = 2; then

$$F_{(2)}[f(r);\underset{\sim}{\xi}] = (2\pi)^{-1}\int_{-\infty}^{\infty}\int_{-\infty}^{\infty} f(r)\exp[i(\xi_1 x_1 + \xi_2 x_2)]dx_1 dx_2 .$$

The transformation to polar coordinates

$$x_1 = r\cos\theta \ , \quad x_2 = r\sin\theta \ , \quad \xi_1 = \xi\cos\phi \ , \quad \xi_2 = \xi\sin\phi \tag{5.7.7}$$

yields

$$F_{(2)}[f(r);\underset{\sim}{\xi}] = (2\pi)^{-1}\int_0^{\infty} r\,dr\,f(r)\int_0^{2\pi} \exp[i\xi r\cos(\theta-\phi)]d\theta \tag{5.7.8}$$

and

$$\int_0^{2\pi} \exp[i\xi r\cos(\theta-\phi)]d\theta = 4\int_0^{\pi/2} \cos(\xi r\cos\theta)d\theta = 2\pi J_0(\xi r) \ , \tag{5.7.9}$$

where we have used the integral representation (5.6.30) for J_0, and put $t = \cos\theta$ in the integral. Substituting (5.7.9) into (5.7.8) we obtain

$$F_{(2)}[f(r);\underset{\sim}{\xi}] = H_0[f(r);\xi] \ . \tag{5.7.10}$$

The more general result is

$$F_{(n)}[f(r);\underset{\sim}{\xi}] = \xi^{-\nu}H_\nu[r^\nu f(r);\xi] \ , \tag{5.7.11}$$

where $\xi = (\xi_1^2 + \xi_2^2 + \cdots + \xi_n^2)^{1/2}$, $\nu = (n-2)/2$. (5.7.12)

Now taking $\underset{\sim}{\xi} = (\xi_1,\xi_2)$ and differentiating (5.7.10) with respect to ξ_1 we find

$$F_{(2)}[ix_1 f(r);(\xi_1,\xi_2)] = \xi_1\xi^{-1}\partial[H_0[f(r);\xi]/\partial\xi \ , \tag{5.7.13}$$

and making use of the relation

$$\partial J_0(\xi r)/\partial\xi = -rJ_1(\xi r) \ , \tag{5.7.14}$$

we see that

$$F_{(2)}[x_1 f(r);(\xi_1,\xi_2)] = i\xi_1\xi^{-1}H_1[rf(r);\xi] \ . \qquad (5.7.15)$$

The general equation (5.7.11), and the particular equations (5.7.10), (5.7.15) provide a convenient way of evaluating Hankel transforms by reducing them to integrals containing exponentials; these integrals may often be evaluated in closed form by using a transformation to polar coordinates and subsequent use of contour integration around the unit circle (see Example 5.7.1).

The second relation between Fourier and Hankel transforms derives from equations (5.6.13), (5.6.14) which show that sinx and cosx are simply related to $J_{1/2}(x)$ and $J_{-1/2}(x)$ respectively (see Example 5.7.3).

The third and most important set of relations may be derived by using the Hankel transforms of the trigonometric functions obtained in §5.6. Since these transforms involve $q_0(t/\xi)$ or $q_0(\xi/t)$, the final results may be conveniently stated in terms of *Abel transforms*. We define the Abel operators

$$A_1[f(t);x] = (2/\pi)^{1/2}\int_0^x (x^2-t^2)^{-1/2}f(t)dt \ , \qquad x > 0 \ , \qquad (5.7.16)$$

$$A_2[f(t);x] = (2/\pi)^{1/2}\int_x^\infty (t^2-x^2)^{-1/2}f(t)dt \ , \qquad x > 0 \ . \qquad (5.7.17)$$

Now consider the Hankel transform of $\xi^{-1}F_c(\xi)$, where $F_c(\xi)$ is the Fourier cosine transform of $f(t)$; by using (5.6.35) we have

$$H_0[\xi^{-1}F_c(\xi);x] = \int_0^\infty F_c(\xi)J_0(\xi x)d\xi =$$

$$= (2/\pi)^{1/2}\int_0^\infty f(t)dt\int_0^\infty J_0(\xi x)\cos(\xi t)d\xi = (2/\pi)^{1/2}\int_0^x (x^2-t^2)^{-1/2}f(t)dt \ ,$$

i.e.,

$$H_0[\xi^{-1}F_c(\xi);x] = A_1[f(t);x] \ . \qquad (5.7.18)$$

By using equation (5.6.39) we may prove that

$$H_0[\xi^{-1}F_s(\xi);x] = A_2[f(t);x] \ , \qquad (5.7.19)$$

and by inverting the order in which the transforms are taken we may prove that if $\overline{f}_0(\xi) = H_0[f(t);\xi]$, then

$$F_c[\overline{f}_0(\xi);x] = (2/\pi)^{1/2}\int_0^\infty \overline{f}_0(\xi)\cos(x\xi)d\xi$$

$$= (2/\pi)^{1/2}\int_0^\infty tf(t)dt\int_0^\infty J_0(\xi t)\cos(x\xi)d\xi = (2/\pi)^{1/2}\int_x^\infty tf(t)(t^2-x^2)^{-1/2}dt \ .$$

Thus

$$F_c[\overline{f}_0(\xi);x] = A_2[tf(t);x] \ , \qquad (5.7.20)$$

$$F_s[\overline{f}_0(\xi);x] = A_1[tf(t);x] \ . \qquad (5.7.21)$$

Equations (5.7.18) - (5.7.21) all have the general form $ab = c$ where a, b, c are all operators of Hankel, Fourier or Abel type. From this relation one may obtain the equivalent relations $a = cb^{-1}$, $b = a^{-1}c$, $a^{-1} = bc^{-1}$, $b^{-1} = c^{-1}a$, $c^{-1} = b^{-1}a^{-1}$. It is often difficult to recognize the equivalence of two apparently different relations when they are written in terms of H_ν, F_c or F_s, A_1 or A_2, and there are in any case so many of them; we have therefore listed the more important ones, in a standard form, in Tables 5.7.1 and 5.7.2.

Table 5.7.1 - The conversion of Hankel transforms into Fourier transforms

1.	$A_1^{-1}\{H_0[\xi^{-1}A(\xi);t];x\} = F_c\{A(\xi);x\}$
2.	$x^{-1}A_1^{-1}\{tH_1[\xi^{-1}A(\xi);t];x\} = F_s\{A(\xi);x\}$
3.	$xA_1t^{-2}D_tt^2H_2[\xi^{-1}A(\xi);t];x\} = -F_c\{A(\xi);x\} + B(0)$ $[B(0) = F_c\{A(\xi);x\}]$
4.	$A_1\{tH_0[A(\xi);t];x\} = F_s\{A(\xi);x\}$
5.	$xA_1\{H_1[A(\xi);t];x\} = -F_c\{A(\xi);x\} + B(0)$
6.	$\int_o^x uA_1\{t^{-2}D_tt^2H_2[A(\xi);t];u\}\,du = -F_s\{A(\xi);x\} + C(0)x$ $[C(0) = F_c\{\xi A(\xi);0\}]$
7.	$A_2^{-1}\{H_0[\xi^{-1}A(\xi);t];x\} = F_s\{A(\xi);x\}$
8.	$D_xxA_2\{H_1[\xi^{-1}A(\xi);t];x\} = F_c\{A(\xi);x\}$
9.	$xA_2\{t^{-2}D_tt^2H_2[\xi^{-1}A(\xi);t];x\} = F_s\{A(\xi);x\}$
10.	$A_2\{tH_0[A(\xi);t];x\} = F_c\{A(\xi);x\}$
11.	$xA_2\{H_1[A(\xi);t];x\} = F_s\{A(\xi);x\}$
12.	$\int_x^\infty uA_2\{t^{-2}D_tt^2H_2[A(\xi);t];u\}du = F_c\{A(\xi);x\}$

Table 5.7.2 - The conversion of Fourier transforms into Hankel transforms

1.	$A_1\{F_c[A(\xi);t];x\} = H_0\{\xi^{-1}A(\xi);x\}$
2.	$A_1\{tF_s[A(\xi);t];x\} = xH_1\{\xi^{-1}A(\xi);x\}$
3.	$A_1^{-1}\{F_s[A(\xi);t];x\} = xH_0\{A(\xi);x\}$
4.	$A_1^{-1}\{t^{-1}F_c[A(\xi);t];x\} = -H_1\{A(\xi);x\}$
5.	$A_2\{F_s[A(\xi);t];x\} = H_0\{\xi^{-1}A(\xi);x\}$
6.	$D_xA_2\left\{\int_u^\infty F_c[A(\xi);t]du;x\right\} = H_1\{\xi^{-1}A(\xi);x\}$
7.	$A_2^{-1}\{F_c[A(\xi);t];x\} = xH_0\{A(\xi);x\}$
8.	$A_2^{-1}\{t^{-1}F_s[A(\xi);t];x\} = H_1\{A(\xi);x\}$

Relations between Fourier, Hankel and Abel transforms

The *inverse* Abel operators A_1^{-1}, A_2^{-1} may be obtained as follows. Consider the equation

$$g(t) = D_t A_1[xf(x);t] \ . \qquad (5.7.22)$$

The transform may be written

$$A_1[xf(x);t] = (2/\pi)^{1/2} \int_0^t x(t^2-x^2)^{-1/2} f(x)dx \ .$$

Integration by parts now yields

$$A_1[xf(x);t] = (2/\pi)^{1/2} \left\{ [-(t^2-x^2)^{1/2} f(x)]_0^t + \int_0^t (t^2-x^2)^{1/2} f'(x)dx \right\} =$$

$$= (2/\pi)^{1/2} \left\{ -tf(0) + \int_0^t (t^2-x^2)^{1/2} f'(x)dx \right\} ,$$

so that differentiation with respect to t yields

$$g(t) = -(2/\pi)^{1/2} f(0) + tA_1[f'(x);t] \ .$$

Now find the A_1 transform of g(t); we have

$$A_1[g(t);\xi] = (2/\pi) \left\{ -f(0) \int_0^\xi (\xi^2-t^2)^{-1/2} dt \ + \right.$$

$$\left. + \int_0^\xi t(\xi^2-t^2)^{-1/2} dt \int_0^t f'(x)(t^2-x^2)^{-1/2} dx \right\} . \qquad (5.7.23)$$

The domain of the double integral is the triangle shown in Figure 5.7.1; interchanging the order of integration we have

$$\int_0^\xi f'(x)dx \int_x^\xi t(\xi^2-t^2)^{-1/2}(t^2-x^2)^{-1/2} dt = (\pi/2) \int_0^\xi f'(x)dx \ , \qquad (5.7.24)$$

where the inner integral may be evaluated as in Example 5.7.4.

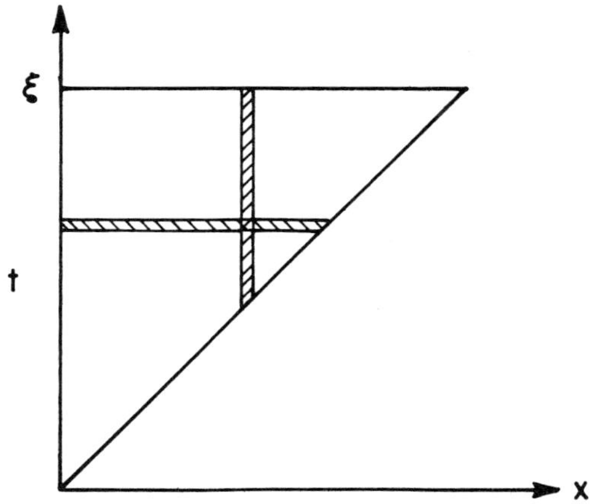

Figure 5.7.1 - The domain of integration appearing in (5.7.21).

Collecting the terms in (5.7.23) together we have

$$A_1[g(t);\xi] = -f(0) + \int_0^{\xi} f'(\xi)d\xi = f(\xi) \ . \tag{5.7.25}$$

This shows that g(t) is *an* inverse of f(ξ) and we may define the operator A_1^{-1} by the equation

$$A_1^{-1}[f(\xi);t] = D_t A_1[xf(x);t] \ . \tag{5.7.26}$$

We must note however that the equation

$$A_1^{-1}[\xi^{-1}f(\xi);t] = g(t) \tag{5.7.27}$$

has the general solution

$$f(\xi) = \xi A_1[g(t);\xi] + f(0) \; , \tag{5.7.28}$$

since $A_1^{-1}[\xi^{-1}c;t] = 0$ when c is a constant. This explains the presence of the terms involving B(0) in Table 5.7.1. We may prove in a similar way that

$$A_2^{-1}[f(\xi);t] = -D_t A_2[xf(x);t] \; , \tag{5.7.29}$$

but note that the equation $A_2^{-1}[\xi^{-1}f(\xi);t] \equiv 0$ has only the solution $f(\xi) \equiv 0$.

These inverse operators may now be used to convert the Hankel transforms into Fourier transforms. Thus by applying A_1^{-1} to equation (5.7.18), with $f(\xi)$ and $F_c(t)$ interchanged, we find

$$D_t A_1[x\mathcal{H}_0\{\xi^{-1}f(\xi);x\};t] = F_c(t) \; , \tag{5.7.30}$$

while use of (5.7.19) and (5.7.29) yields

$$D_t A_2[x\mathcal{H}_0\{\xi^{-1}f(\xi);x\};t] = -F_s(t) \; . \tag{5.7.31}$$

Suppose that it is required to find a function $f(\xi)$ such that

$$F_c(x) \equiv \mathcal{F}_c[f(\xi);x] = 0 \quad \text{for} \quad x > a \; . \tag{5.7.32}$$

One way to express $f(\xi)$ derives from the $\mathcal{F}_c$ transform, namely

$$f(\xi) = \mathcal{F}_c[F_c(x);\xi] = (2/\pi)^{1/2} \int_0^a F_c(x)\cos(\xi x)dx \; . \tag{5.7.33}$$

But another way, which is found to be more convenient, is to use the appropriate entry, (7) in this case, in Table 5.7.2. Entry (7) involves A_2^{-1} which is an *outer* operator, acting on the range (x,∞); this means that if $x > a$, so that $F_c(x)$ is zero, then the left-hand side of entry (7) is identically zero, yielding

$$A_2^{-1}[F_c(t);x] \equiv x\mathcal{H}_0[f(\xi);x] \equiv x\bar{f}_0(x) = 0 , \quad x > a ,$$

which allows us to write $f(\xi)$ as

$$f(\xi) = \mathcal{H}_0[\bar{f}_0(x);\xi] = \int_0^a x\bar{f}_0(x)J_0(\xi x)dx , \tag{5.7.34}$$

and $F_c(t)$ in the form

$$F_c(t) = A_2[x\bar{f}_0(x);t] = (2/\pi)^{1/2}\int_t^a x(t^2-x^2)^{-1/2}\bar{f}_0(x)dx \quad \text{if} \quad 0 \le t < a \tag{5.7.35}$$

The other entries in Tables 5.7.1 and 5.7.2 may be used similarly.

Examples 5.7

1. Use (5.7.10) to show that if $k > 0$, then $\mathcal{H}_0[r^{-1}e^{-kr};\xi] = (k^2+\xi^2)^{-1/2}$. Use polar coordinates and contour integration to evaluate the Fourier integral.
2. Use the result of Example 5.7.1 to show that $\mathcal{H}_1[r^{-1}e^{-kr};\xi] = \xi^{-1}[1-k(k^2+\xi^2)^{-1/2}]$.
3. Use equations (5.6.13), (5.6.14) to show that $F_c[f(x);\xi] = \xi^{1/2}\mathcal{H}_{1/2}[x^{-1/2}f(x);\xi]$, $F_s[f(x);\xi] = \xi^{1/2}\mathcal{H}_{-1/2}[x^{-1/2}f(x);\xi]$
4. Evaluate the inner integral in equation (5.7.24) by using the substitution $t^2 = x^2\cos^2\theta + \xi^2\sin^2\theta$.
5. Show that $F_s[\bar{f}_1(\xi);x] = xA_2[f(t);x]$.
6. Show that $x^{-1}A^{-1}[t\mathcal{H}_1\{\xi^{-1}f(\xi);t\};x] = F_s(x)$.
7. Use equation (5.6.37) to show that if $\bar{f}_0(\tau) = \mathcal{H}_0[f(t);\tau]$ and $f(t) = 0$ for $t > 1$, then if $\xi < 1$

$$F_s[\tau^{-1}\bar{f}_0(\tau);\xi] = (2/\pi)^{1/2}\left\{\pi/2\int_0^\xi tf(t)dt + \int_\xi^1 t\sin^{-1}(\xi/t)f(t)dt\right\}$$

while for $\xi \ge 1$

$$F_s[\tau^{-1}\bar{f}_0(\tau);\xi] = (2/\pi)^{-1/2}\int_0^1 tf(t)dt .$$

In addition show that $F_s[\bar{f}_0(\tau);\xi] = A_1[tf(t);\xi]$.

8. By differentiating (5.6.30) show that
$tJ_1(\tau t) = (2/\pi)^{1/2}A_1[\xi\sin(\tau\xi);\xi \to t]$
and hence show that $(2/\pi)^{1/2}A_1^{-1}[\sin(\tau\xi);\xi \to t] = \tau t J_0(\tau t)$.

9. Show that $F_{(2)}[\xi(\xi^2+\eta^2)^{-1};(\xi,\eta) \to (x,y)] = ix(x^2+y^2)^{-1}$.

10. Abel transforms of polynomials may be computed as follows. The addition theorem for the Legendre function (Gradshteyn and Ryzhik (1965, 8.814)) gives

$$P_n(\sin\phi\cos\theta) = P_n(0)P_n(\cos\phi) + 2\sum_{k=1}^{n}\frac{(n-k)!}{(n+k)!}P_n^k(0)P_n^k(\cos\phi)\cos k\theta .$$

Let

$$I(\rho) = \int_0^\rho (\rho^2-\tau^2)^{-1/2}P_{2m}(\tau)d\tau = \int_0^{\pi/2} P_{2m}(\rho\cos\theta)d\theta \ ;$$

then if $\rho = \sin\phi$, $n = 2m$, we find that

$$I(\rho) = \frac{\pi}{2}P_{2m}(0)P_{2m}[(1-\rho^2)^{1/2}] \ .$$

Thus $A_1[P_{2m}(\tau);\rho] = (\pi/2)^{1/2}P_{2m}(0)P_{2m}[(1-\rho^2)^{1/2}]$, and by inverting we find

$$D_\tau A_1[\rho P_{2m}[(1-\rho^2)^{1/2}];\tau] = (2/\pi)^{1/2}P_{2m}(\tau)/P_{2m}(0).$$

These formulae may be used to construct the transforms of even and odd polynomials; the transforms also are polynomials.

11. If $f(x_1,x_2)$ and $g(x_1,x_2)$ are functions of $(x_1^2+x_2^2)^{1/2} = r$ only show that $f\circ g = 2\pi H_0[F(\rho)G(\rho);r]$ where $F(\rho) = H_0[f(r);\rho]$, $G(\rho) = H_0[g(r);\rho]$.

12. Extend equation (5.7.10) to show that if $f(x_1,x_2)$ is a function of $\rho = (x_1^2/a^2+x_2^2/b^2)^{1/2}$ and if $\lambda = (a^2\xi_1^2+b^2\xi_2^2)^{1/2}$, then $F_{(2)}[f(\rho);\tilde{\xi}] = H_0[f(\rho);\lambda]$. Sneddon (1979a), (1979b) has used this result and a corresponding inversion to find an

alternative form of solution for the flat elliptic punch indenting a half-space (see §12.2).

5.8 *Applications of the two-dimensional Fourier transform*

It was shown in §1.11 that the elasticity equations admit a solution in terms of the Papkovich-Neuber solution for plane strain or generalized plane stress which is

$$2\mu u = 4(1-\nu)\psi_1 - \frac{\partial}{\partial x}(x\psi_1 + y\psi_2 + \phi) , \tag{5.8.1}$$

$$2\mu v = 4(1-\nu)\psi_2 - \frac{\partial}{\partial y}(x\psi_1 + y\psi_2 + \phi) , \tag{5.8.2}$$

where ψ_1, ψ_2 and ϕ satisfy

$$2(1-\nu)\nabla^2\psi_1 = -F_x , \quad 2(1-\nu)\nabla^2\psi_2 = -F_y , \tag{5.8.3}$$

$$2(1-\nu)\nabla^2\phi = xF_x + yF_y . \tag{5.8.4}$$

Here $\nabla^2 \equiv \partial^2/\partial x^2 + \partial^2/\partial y^2$, and in generalized plane stress the displacements and body forces (i.e., loads applied in the plane of the plate) are averaged through the plate thickness and ν must be replaced by $\nu^* = \nu/(1+\nu)$.

Suppose that

$$\psi_k(x,y) = F^*_{(2)}[\Psi_k(\xi,\eta);(\xi,\eta) \to (x,y)] , \tag{5.8.5}$$

$$\phi(x,y) = F^*_{(2)}[\Phi(\xi,\eta);(\xi,\eta) \to (x,y)] . \tag{5.8.6}$$

The formal equivalent of (5.3.8) for $F_{(2)}$ transforms is

$$F_{(2)}[D_x f(x,y);(x,y) \to (\xi,\eta)] = -i\xi F_{(2)}[f(x,y);(x,y) \to (\xi,\eta)] , \tag{5.8.7}$$

which may be written in the alternative form

$$D_\xi F(\xi,\eta) = i\mathcal{F}_{(2)}[x\mathcal{F}^*_{(2)}[F(s,t);(x,y)];(\xi,\eta)] \ . \tag{5.8.8}$$

The application of $\mathcal{F}_{(2)}$ to equations (5.8.3), (5.8.4) and the use of (5.8.7), (5.8.8) now yields

$$2(1-\nu)(\xi^2+\eta^2)\Psi_i = \overline{F}_i \ , \quad 2(1-\nu)(\xi^2+\eta^2)\Phi = iD_\xi\overline{F}_1+iD_\eta\overline{F}_2 \ , \tag{5.8.9}$$

where $\overline{F}_1$, $\overline{F}_2$ are the transforms of F_x, F_y respectively. Taking the $\mathcal{F}_{(2)}$ transforms of (5.8.1), (5.8.2) we find

$$2\mu U = 4(1-\nu)\Psi_1 + \xi[D_\xi\Psi_1 + D_\eta\Psi_2 + i\Phi] \ , \tag{5.8.10}$$

$$2\mu V = 4(1-\nu)\Psi_2 + \eta[D_\xi\Psi_1 + D_\eta\Psi_2 + i\Phi] \ , \tag{5.8.11}$$

and equations (5.8.9) show that

$$D_\xi\Psi_1 + D_\eta\Psi_2 + i\Phi = \frac{1}{2(1-\nu)}\,[\overline{F}_1 D_\xi(\xi^2+\eta^2)^{-1}+\overline{F}_2 D_\eta(\xi^2+\eta^2)^{-1}] \ , \tag{5.8.12}$$

so that

$$2\mu U = \overline{F}_1\left\{2(\xi^2+\eta^2)^{-1} + \frac{\xi}{2(1-\nu)}\,D_\xi(\xi^2+\eta^2)^{-1}\right\} + \frac{\overline{F}_2\xi}{2(1-\nu)}\,D_\eta(\xi^2+\eta^2)^{-1} \ , \tag{5.8.13}$$

$$2\eta V = \frac{\overline{F}_1\eta}{2(1-\nu)}\,D_\xi(\xi^2+\eta^2)^{-1} + \overline{F}_2\left\{2(\xi^2+\eta^2)^{-1} + \frac{\eta}{2(1-\nu)}\,D_\eta(\xi^2+\eta^2)^{-1}\right\} \ . \tag{5.8.14}$$

To recover u, v we apply the Faltung theorem (5.7.6). To find the transforms of the bracketed terms in (5.8.13), (5.8.14) we use the result (Example 5.7.9) that

$$\mathcal{F}^*_{(2)}[i\xi(\xi^2+\eta^2)^{-1};(\xi,\eta) \to (x,y)] = x(x^2+y^2)^{-1} \ . \tag{5.8.15}$$

In conjunction with (5.8.8), this gives *formally*

$$F^*_{(2)}[(\xi^2+\eta^2)^{-1};(\xi,\eta)\to(x,y)] = -\frac{1}{2}\ln(x^2+y^2) \ . \qquad (5.8.16)$$

Similarly

$$F^*_{(2)}[\xi D_\xi(\xi^2+\eta^2)^{-1}] = iD_x F^*_2[D_\xi(\xi^2+\eta^2)^{-1}] = -D_x[xF^*_{(2)}(\xi^2+\eta^2)^{-1}]$$

$$= \frac{1}{2}D_x[x\ln(x^2+y^2)] = \frac{1}{2}\ln(x^2+y^2) + x^2(x^2+y^2)^{-1} \ , \qquad (5.8.17)$$

so that the final result is

$$u(x,y) = \int_{-\infty}^{\infty}\int_{-\infty}^{\infty}\{K_{11}(x-t,y-t')F_x(t,t')+K_{12}(x-t,y-t')F_y(t,t')\}dtdt' \ , \qquad (5.8.18)$$

$$v(x,y) = \int_{-\infty}^{\infty}\int_{-\infty}^{\infty}\{K_{21}(x-t,y-t')F_x(t,t')+K_{22}(x-t,y-t')F_y(t,t')\}dtdt' \ , \qquad (5.8.19)$$

where $K_{12}(x,y) = K_{21}(x,y)$, $r^2 = x^2 + y^2$ and

$$-8\pi\mu(1-\nu)K_{11}(x,y) = (3-4\nu)\ln r + y^2r^{-2} \ , \qquad (5.8.20)$$

$$-8\pi\mu(1-\nu)K_{12}(x,y) = -xyr^{-2} \ , \qquad (5.8.21)$$

$$-8\pi\mu(1-\nu)K_{22}(x,y) = (3-4\nu)\ln r + x^2r^{-2} \ . \qquad (5.8.22)$$

The stress distribution corresponding to this problem appears in Love (1927), where is was deduced by using the Airy stress function; the displacements may be obtained by integration.

Equations (5.8.20) - (5.8.21) may be used to find the displacements in the upper half-plane produced by interior forces when the edge of the half-plane is either stress-free or rigidly supported. In the former case consider first the application of a concentrated unit load in the y-direction at (t,t'). The solution for the complete x-y plane corresponding to concentrated loads $\pm$ 1 at (t,t') and the image point $(t,-t')$, respectively, as shown in Figure 5.8.1, is

$$u(x,y) = K_{12}(x-t,y-t') - K_{12}(x-t,y+t') \ , \qquad (5.8.23)$$

$$v(x,y) = K_{22}(x-t,y-t') - K_{22}(x-t,y+t') \ . \qquad (5.8.24)$$

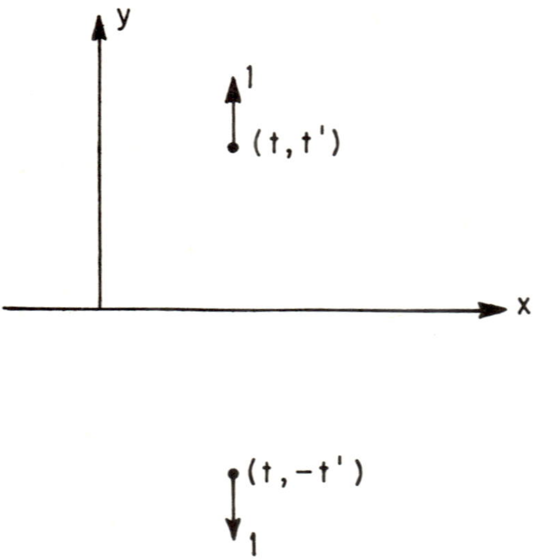

Figure 5.8.1 - Image distribution used for a stress-free x-axis.

Symmetry considerations show that these displacements satisfy the conditions $\partial u/\partial y \equiv 0 \equiv \partial v/\partial x$ on $y = 0$, so that $\tau_{xy}(x,0) \equiv 0$; the normal stress corresponding to these displacements is found to be given by

$$2\pi(1-\nu)\tau_{yy}(x,0) = t'[(1-2\nu)r^{-2} + 2t'^2r^{-4}] \ , \qquad (5.8.25)$$

where $r^2 = (x-t)^2 + t'^2$. To maintain the free-edge condition on $y = 0$, an equal and opposite normal stress must be applied to the boundary $y = 0$ of the upper half-plane, i.e., $\tau_{yy}(x,0) = -p(x)$, where

$$2\pi(1-\nu)p(x) = t'[(1-2\nu)r^{-2} + 2t'^{2}r^{-4}] \;. \qquad (5.8.26)$$

The displacements produced in the upper half-plane by this normal stress may be deduced from §5.4. Since $p(x)$ is an even function of $(x-t)$, Example 5.2.3 and Example 5.2.5 give

$$4(1-\nu)P(\xi) = (2/\pi)^{1/2}\{2(1-\nu) + t'|\xi|\}\exp(it\xi - t'|\xi|) \;. \qquad (5.8.27)$$

Equation (5.4.10) gives the displacements corresponding to this pressure; after inverting the transforms we find the total displacements, i.e., those produced by the two concentrated loads and the pressure (5.8.26), to be

$$u(x,y) = K^{*}_{12}(x-t,y,t') \quad , \quad v(x,y) = K^{*}_{22}(x-t,y,t') \;, \qquad (5.8.28)$$

where $r_1^2 = (x-t)^2 + (y-t')^2$, $r_2^2 = (x-t)^2 + (y+t')^2$,

$$8\pi\mu(1-\nu)K^{*}_{12}(x,y,t') = x(y-t')r_1^{-2} - x(y+t')r_2^{-2} + 4xyt'(y+t')r_2^{-4} -$$

$$-4(1-\nu)(1-2\nu)\tan^{-1}[x/(t'+y)] + 2[2(1-\nu)y - (1-2\nu)t']xr_2^{-2} \;, \qquad (5.8.29)$$

$$8\pi\mu(1-\nu)K^{*}_{22}(x,y,t') = -(3-4\nu)\ln(r_1/r_2) - 8(1-\nu)^2\ln r_2 -$$

$$-x^2r_1^{-2} + x^2r_2^{-2} + 4(1-\nu)(t'+y)^2r_2^{-2} + 2t'y[2(t'+y)^2r_2^{-4} - r_2^{-2}] \;. \qquad (5.8.30)$$

Notice that the derivation of K^{*}_{22} involves a divergent integral, as discussed at the end of §5.4; K^{*}_{22} has therefore to be found by finding a derivative of v and then formally integrating it.

It may be shown similarly (Example 5.8.1) that the displacement $u(x,y)$ due to a unit horizontal force at (t,t') in the upper half-plane is $K^{*}_{11}(x-t,y,t')$ where

$$8\pi\mu(1-\nu)K^*_{11}(x,y,t') = -(3-4\nu)\ell n(r_1 r_2) - y^2 r_1^{-2} - y^2 r_2^{-2} +$$

$$+ 4\{(1-2\nu)^2 \ell n r_2 - (1-2\nu)(t'+y)^2 r_2^{-2} + t'y[2(t'+y)^2 r_2^{-4} - r_2^{-2}]\} . \quad (5.8.31)$$

The reciprocal theorem yields the corresponding displacement $v(x,y)$ to be $K^*_{21}(x-t,y,t')$, where

$$K^*_{21}(x-t,y,t') = K^*_{12}(t-x,t',y) . \quad (5.8.32)$$

This problem appears to have been studied first by Riedel (1927), but his solution was incorrect. Melan (1932a) started with the known Airy stress function for a point force in an infinite plate. When the force acted normal to the boundary he superimposed the solutions corresponding to two equal (not equal and opposite) forces at (t,t') and $(t,-t')$; this provides a solution giving $\tau_{yy}(x,0) = 0$, $\tau_{xy}(x,0) \neq 0$. He then used the known Airy stress function corresponding to a given shearing stress on $y = 0$ to obtain the full stress distribution in the half-plane. He employed a similar technique when the force is parallel to the boundary. Sneddon (1944), (1951) used the image system shown in Figure 5.8.1 and used a Fourier transform representation of the Airy stress function to eliminate the normal stress. He found the displacements by integrating the strain-stress relations. Frasier and Rongved (1957) obtained the stress distribution for two half-planes of differing elastic properties, joined along $y = 0$, when a concentrated in-plane load acts inside one half-plane.

Examples 5.8.1

1. Show that the normal stress $\tau_{yy}(x,0) = p(x)$ due to equal horizontal unit forces at (t,t'), $(t,-t')$ is given by
   $2\pi(1-\nu)p(x) = \{(1-2\nu)r^{-2} - 2t'^2 r^{-4}\}(x-t)$,
   and that the shear stress on $y = 0$ is zero. Apply an equal and opposite pressure $\tau_{yy}(x,0) = -p(x)$ to the upper

half-plane, find its transform, making use of Example 5.2.3 and Example 5.2.5, and hence find $u(x,y)$, $v(x,y)$ for $y \geq 0$ due to the unit horizontal load at (t,t'), when the boundary $y = 0$ is stress-free.

2. Suppose that the force $F_x(x,y)$ is a line load on the y-axis, so that $F_x(x,y) = s(x)\delta(y)$. Show that
$\overline{F}_1(\xi,\eta) = (2\pi)^{-1/2}S(\xi)$, where $S(\xi) = \mathcal{F}[s(x);\xi]$.
By carrying out the η-integration in $\mathcal{F}^*_{(2)}[U(\xi,\eta);(x,y)]$ show that
$$8\mu(1-\nu)u(x,y) = \mathcal{F}^*[|\xi|^{-1}\{3-4\nu-|\xi y|\}S(\xi)\exp(-|\xi y|);x] .$$
Show also that
$$8\mu(1-\nu)v(x,y) = |y|\mathcal{F}^*[i\,\mathrm{sign}(\xi)S(\xi)\exp(-|\xi y|);x] ,$$
and compare these with the results of Example 5.4.1.

CHAPTER VI - INTEGRAL EQUATIONS FOR CONTACT PROBLEMS

"Just the place for a Snark!
I have said it twice:
That alone should encourage the crew.
Just the place for a Snark!
I have said it thrice:
What I tell you three times is true."

- *The Hunting of the Snark*

6.1 *Introduction*

This chapter is devoted to a development of the theory of plane contact problems based on the integral transforms of the previous chapter, rather than on the complex variable theory of Chapter III. It is shown, when appropriate, that both transform and complex variable methods lead to identical equations. Problems involving finite friction - neither frictionless nor completely adhesive contact - and strips rather than half-planes cannot be solved easily by using complex variable methods. These problems may be formulated in terms of integral equations; some of the theory of such equations is presented and numerical methods for solving these equations are discussed.

A note on nomenclature is appropriate at this stage. Two bodies will be said to be in *frictionless* contact if there is no surface tangential stress on the bodies in the contact region. If two bodies are in frictionless contact there are two possible situations that may occur. In the first, exemplified by a rigid punch with a flat face indenting an elastic half-space, the whole of the face will be in contact with the half-space and there will be a (square root) singularity in the contact pressure at the boundary of the contact region. Such contact is here called *complete*; other authors, including myself, have sometimes called this *bonded* contact. In the second, exemplified by a punch with a convex curved surface indenting a half-space, the extent of the contact region depends on the force pressing the bodies together. In this case the contact pressure will be zero at the boundary of the contact region. This contact is called *incomplete*; others, including myself (Gladwell (1976)), have sometimes called this *unbonded*. When two bodies are in frictionless contact there is, in general, tangential slippage between the two bodies in the contact region. When this slippage is completely restrained, the contact is called *adhesive*; the word *bonded* is sometimes used, and will be used here, for this situation.

A distinction must also be made between the words *strip* and *layer*. The former will be used for plane elasticity problems, the latter for three-dimensional problems.

6.2 *Integral equations for the half-plane*

The results of Chapter V lead to some alternative formulations and solutions for plane-strain problems for the half-plane.

The basic plane-strain frictionless contact problem for a rigid punch indenting an elastic half-plane over [-a,a] is specified by the boundary conditions

$$v(x,0) = f(x) + d \quad , \quad |x| < a \; , \tag{6.2.1}$$

$$\tau_{xy}(x,0) = 0 \text{ for all } x \quad ; \quad \tau_{yy}(x,0) = 0 \; , \quad |x| > a \; , \tag{6.2.2}$$

where f(x) is the profile of the punch and d is an arbitrary constant. The analysis of §4.10 shows that the contact problem for two elastic bodies which may be approximated by half-planes also leads to equations of the same type as (6.2.1), (6.2.2). If the displacement field in the half-plane is sought in the class of displacements which have no rotation at infinity, i.e.

$$\lim_{r\to\infty} r^{-1}|u(x,y) + iv(x,y)| = 0 , \qquad (6.2.3)$$

and if the stresses have integrable singularities at $(\pm a,0)$, then it can be shown that the solution is unique apart from an arbitrary translation. To obviate the complication raised by the presence of d we may express the boundary conditions (6.2.1) in the alternative form

$$v_x(x,0) = f'(x) \quad , \quad |x| < a . \qquad (6.2.4)$$

The expressions in (5.4.3), (5.4.5) yield

$$v(x,0) \equiv \vartheta F^*[|\xi|^{-1}P(\xi);x] = f(x)+d , \quad |x| < a , \qquad (6.2.5)$$

$$\tau_{yy}(x,0) \equiv -F^*[P(\xi);x] = -p(x)H(a-|x|) , \qquad (6.2.6)$$

so that the boundary conditions (6.2.1), (6.2.2) yield the pair of *dual integral equations*

$$\vartheta F^*[-i\,\mathrm{sign}(\xi)P(\xi);x] = f'(x) , \quad |x| < a , \qquad (6.2.7)$$

$$F^*[P(\xi);x] = 0 , \quad |x| > a . \qquad (6.2.8)$$

The analysis of §5.5 shows that the equations may be written in the form of *Cauchy's integral equation*

$$\vartheta h[f(t);x] \equiv \frac{\vartheta}{\pi}\int_{-a}^{a}\frac{p(t)dt}{t-x} = f'(x) , \quad |x| < a . \qquad (6.2.9)$$

The inverse of this equation is given by equation (5.5.14), i.e.

$$\vartheta p(x) = \frac{(a^2-x^2)^{-1/2}}{\pi}\int_{-a}^{a}\frac{(a^2-t^2)^{1/2}f'(t)dt}{t-x} + C(a^2-x^2)^{-1/2} , \qquad (6.2.10)$$

when the contact is complete, so that p(x) has square root singularities at $x = \pm a$. When the contact is incomplete, so that p(x) is zero at $\pm a$, then

$$\vartheta p(x) = \frac{(a^2-x^2)^{1/2}}{\pi} \int_{-a}^{a} \frac{(a^2-t^2)^{-1/2} f'(t)}{t-x}\, dt , \qquad (6.2.11)$$

provided that f'(t) satisfies the condition

$$\int_{-a}^{a} (a^2-t^2)^{-1/2} f'(t)dt = 0 . \qquad (6.2.12)$$

The Chebyshev polynomials presented in §2.3 correspond to the two particular solutions

$$\vartheta p(x) = a_n a(a^2-x^2)^{-1/2} T_n(x/a) , \quad f'(x) = a_n U_{n-1}(x/a) , \qquad (6.2.13)$$

$$\vartheta p(x) = -a_n a^{-1}(a^2-x^2)^{1/2} U_{n-1}(x/a) , \quad f'(x) = a_n T_n(x/a) , \qquad (6.2.14)$$

of equation (6.2.9). The term $C(a^2-x^2)^{-1/2}$ in (6.2.10) corresponds to (6.2.13) with n = 0, since $U_{-1} \equiv 0$. The condition (6.2.12) for incomplete contact corresponds to the condition $a_0 = 0$ in (6.2.14). It was shown in §2.3 that the displacement derivative $v_x(x,0)$ for $|x| > a$ corresponding to these solutions may be expressed in terms of the functions R_n and G_{n-1}; the displacements and stresses off the x-axis were obtained in §2.4.

Another solution of equations (6.2.1), (6.2.2) may be obtained by using the influence function G_{22} found in equation (2.1.19) or (5.4.19). Thus the boundary conditions will be satisfied if

$$\int_{-a}^{a} G_{22}(x-x',0)p(x')dx' = f(x) + d , \quad |x| < a , \qquad (6.2.15)$$

and Example (2.1.4) shows that this equation may be written

$$-\frac{\vartheta}{\pi}\int_{-a}^{a} \ln|x-x'|p(x')dx' = f(x) + d \,. \qquad (6.2.16)$$

This equation, which was the subject of Example 2.3.8, has solutions

$$\vartheta p(x) = a_n a(a^2-x^2)^{-1/2}T_n(x/a) \,, \quad f(x) = \begin{cases} a_n an^{-1}T_n(x/a) \,, & n \neq 0 \\ a_0 a\ln 2 \,, & n = 0 \,. \end{cases} \qquad (6.2.17)$$

Integral equations which have polynomial solutions of this type are termed *integral equations with π-kernels* by Popov (1969), and are discussed further in §6.11.

A third formulation of equations (6.2.1), (6.2.2) may be found by using Hankel transforms and for this it is convenient to consider separately the cases in which f(x) is an even or an odd function (any f(x) may be written as the sum of an even and an odd function). In the former case equations (6.2.5), (6.2.6) may be written

$$v(x,0) = \vartheta F_c[\xi^{-1}P(\xi);x] = f(x) + d \,, \quad 0 < x < a \,, \qquad (6.2.18)$$

$$\tau_{yy}(x,0) = -F_c[P(\xi);x] = -p(x)H(a-x) \,. \qquad (6.2.19)$$

In dimensionless variables the equations to be considered are

$$F_c[\xi^{-1}A(\xi);x] = f(x) + c \,, \quad 0 < x < 1 \,, \qquad (6.2.20)$$

$$F_c[A(\xi);x] = 0 \,, \qquad x > 1 \,. \qquad (6.2.21)$$

These equations are a particular case (see Example 5.7.3) of the equations

$$H_\nu\{\xi^\alpha A(\xi);x\} = f(x) \;, \quad 0 < x < 1 \;, \tag{6.2.22}$$

$$H_\nu\{A(\xi);x\} = g(x) \;, \quad x > 1 \;, \tag{6.2.23}$$

which have been extensively studied (see §10.2). Equations (6.2.20), (6.2.21) may be solved by first rewriting equation (6.2.20) in the form

$$F_c[\xi^{-1}A(\xi);x - 0] = f(x) - f(0) \;, \quad 0 < x < 1 \;, \tag{6.2.24}$$

where the left-hand side denotes

$$(2/\pi)^{1/2} \int_0^\infty \xi^{-1}(\cos\xi x-1)A(\xi)d\xi \;.$$

This has the advantage that it eliminates the arbitrary constant and also removes a potential singularity in the integrand at $\xi = 0$. To solve equations (6.2.21), (6.2.24) we note first that the homogeneous equations with $f(x) \equiv 0$ have the general solution $A(\xi) = CJ_0(\xi)$ where C is an arbitrary constant (Example 6.2.1). A particular solution may be found by operating on equations (6.2.24), (6.2.21) with the operators in entries 4, 6 respectively of Table 5.7.2. This gives

$$H_1\{\xi^{-1}A(\xi);x\} = -A_1^{-1}\{t^{-1}f(t);x\} \;, \quad 0 < x < 1 \;, \tag{6.2.25}$$

$$H_1\{\xi^{-1}A(\xi);x\} = 0 \;, \quad x > 1 \;. \tag{6.2.26}$$

Note that operator (4) is an inner operator acting on the range $(0,x)$ while (6) is an outer operator acting on (x,∞). Equations (6.2.25), (6.2.26) can be solved by using H_1 inversion and, when combined with the general solution, this gives

$$A(\xi) = C_1 J_0(\xi) - \xi H_1\{A_1^{-1}[t^{-1}f(t);x]H(1-x);\xi\} , \qquad (6.2.27)$$

which may be rearranged to give

$$A(\xi) = (2/\pi)^{1/2}[CJ_0(\xi) - \xi J_1(\xi)\int_0^1 (1-t^2)^{-1/2}f(t)dt +$$

$$+ \xi^2 \int_0^1 (1-t^2)^{-1/2}dt \int_0^1 xf(xt)J_0(\xi x)dx] . \qquad (6.2.28)$$

The solution in the dimensional variables is

$$P(\xi) = (2/\pi)^{1/2}[CJ_0(a\xi) - a\xi J_1(a\xi)\int_0^a (a^2-t^2)^{-1/2}f(t)dt +$$

$$+ \xi^2 \int_0^1 (1-t^2)^{-1/2}dt \int_0^a xf(xt)J_0(\xi x)dx] . \qquad (6.2.29)$$

As the first example consider the normal indentation by a rigid indenter with a plane face. Here $f(t) = 0$ so that $\vartheta P(\xi) = (2/\pi)^{1/2}CJ_0(a\xi)$. The contact pressure is given by

$$\vartheta p(x) = \vartheta F_c[P(\xi);x] = (2/\pi)C(a^2-x^2)^{-1/2} , \quad 0 < x < a , \qquad (6.2.30)$$

so that the total contact force is $P_0 = \vartheta^{-1}C$. The displacement $v(x,0) - v(0,0)$ outside the contact region is given by

$$v(x,0) - v(0,0) = (2/\pi)\vartheta P_0 \int_0^\infty \xi^{-1}(\cos\xi x - 1)J_0(a\xi)d\xi ,$$

$$= (2/\pi)\vartheta P_0(1-\cosh^{-1}(x/a)) , \qquad x > a , \qquad (6.2.31)$$

which agrees with Example 2.2.2.

Now consider indentation by the rigid wedge shown in Figure 6.2.1. Here $f(x) = -\varepsilon x/a$ so that after some manipulation we find

$$\vartheta P(\xi) = (2/\pi)^{1/2}[CJ_0(a\xi)-\varepsilon J_0(a\xi) + (\varepsilon/a)\int_0^a J_0(\xi x)dx] , \qquad (6.2.32)$$

and therefore

$$\vartheta p(x) = (2/\pi)[(C-\varepsilon)(a^2-x^2)^{-1/2}+(\varepsilon/a)\cosh^{-1}(a/x)] , \qquad 0 < x < a . \qquad (6.2.33)$$

If the contact is to be incomplete so that there is no singularity at $x = \pm a$ then $C = \varepsilon$; note however that $p(x)$ still has a logarithmic singularity at $x = 0$. The normal displacement outside the contact region is found to be

$$v'(x,0) = -2\varepsilon/(\pi a)\sin^{-1}(a/x) . \qquad (6.2.34)$$

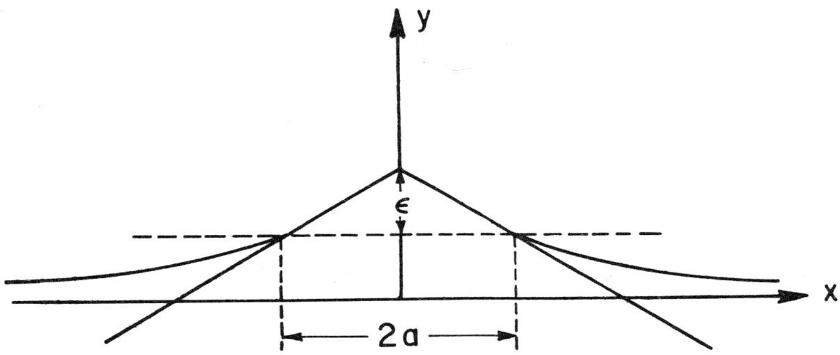

Figure 6.2.1 - Indentation of a half-plane by a rigid wedge.

Integral equations for the half-plane

In order to obtain an alternative formulation of the adhesive contact problems considered in §4.6 we consider the case in which both surface displacements are specified in $|x| < a$ and both stresses are zero for $|x| > a$ i.e.

$$u(x,0) = f_1(x) + d_1 , \quad v(x,0) = f_2(x) + d_2 , \qquad |x| < a , \tag{6.2.35}$$

$$\tau_{xy}(x,0) = 0 = \tau_{yy}(x,0) , \qquad |x| > a . \tag{6.2.36}$$

The analysis of §5.4 shows that these conditions yield the two pairs of dual integral equations

$$\left.\begin{aligned} u(x,0) &\equiv \vartheta\mathcal{F}^*[|\xi|^{-1}\{S(\xi)-i\varphi\,\mathrm{sign}(\xi)P(\xi)\};x] = f_1(x)+d , \quad |x| < a , \\ v(x,0) &\equiv \vartheta\mathcal{F}^*[|\xi|^{-1}\{P(\xi)+i\varphi\,\mathrm{sign}(\xi)S(\xi)\};x] = f_2(x)+d , \quad |x| < a , \end{aligned}\right\} \tag{6.2.37}$$

$$\left.\begin{aligned} \tau_{xy}(x,0) &\equiv -\mathcal{F}^*\{S(\xi);x\} = -s(x)H(a-|x|) , \\ \tau_{yy}(x,0) &\equiv -\mathcal{F}^*\{P(\xi);x\} \equiv -p(x)H(a-|x|) , \end{aligned}\right\} \tag{6.2.38}$$

which may be written as a complex pair

$$D(x,0) \equiv -i\vartheta\mathcal{F}^*[|\xi|^{-1}\{1-\varphi\,\mathrm{sign}(\xi)\}Q(\xi);x] = f(x)+d , \qquad |x| < a , \tag{6.2.39}$$

$$\tau_{yy}(x,0)-i\tau_{xy}(x,0) \equiv \mathcal{F}^*\{Q(\xi);x\} = q(x)H(a-|x|) , \tag{6.2.40}$$

where $q(x) = -p(x)+is(x)$, $Q(\xi) = -P(\xi)+iS(\xi)$, $f(x) = f_1(x)+if_2(x)$, etc. These equations may best be solved by writing the displacement conditions in the derivative form. Then

$$D'(x,0) = \vartheta\mathcal{F}^*[(-\mathrm{sign}(\xi)+\varphi)Q(\xi);x] = f'(x) , \qquad |x| < a , \tag{6.2.41}$$

$$\mathcal{F}^*\{Q(\xi);x\} = q(x)H(a-|x|) , \tag{6.2.42}$$

and the analysis of §5.5 shows that equation (6.2.41) may be written

$$\vartheta\left\{\varphi q(x)+\frac{1}{\pi i}\int_{-a}^{a}\frac{q(t)dt}{t-x}\right\}=f'(x)\ . \tag{6.2.43}$$

This equation is of type (3.5.1) and may be solved by using the Plemelj formulae just as the adhesive problems were solved in §4.6. In fact, writing

$$\Phi(z)=\frac{1}{2\pi i}\int_{-a}^{a}\frac{q(t)dt}{t-z}\ , \tag{6.2.44}$$

we immediately find that equation (6.2.43) is equivalent to

$$\kappa\Phi^{+}(x)+\Phi^{-}(x)=2\mu f'(x)\ , \tag{6.2.45}$$

while

$$\Phi^{+}(x)-\Phi^{-}(x)=q(x)\ . \tag{6.2.45}$$

These equations were obtained earlier in (4.6.4), (4.6.5). It should be emphasized that, except for the flat punch, for which $f'(x)$ is known, viz. zero, the function $f'(x)$ must first be found before the problem can be solved; a similarity argument leading to $f'(x)$ was described in §4.6 for one special case.

Examples 6.2

1. Show that $F_c[\xi^{-1}J_0(\xi);x]=0,\ 0<x<1$ and $F_c[J_0(\xi);x]=0,\quad x>1$.
2. Derive the displacement derivative (6.2.34) outside the contact area from the expression for $P(\xi)$ in (6.2.32).

6.3 *Integral equations for the plane strip*

The analysis of §5.4 was based on the Papkovich-Neuber solution and the use of certain simple transforms that are harmonic in the half-plane $y \geq 0$. For the plane strip $0 \leq y \leq h$ the basic harmonic functions are

$$\exp\xi(y-ix)\ , \quad \exp\xi(-y-ix)\ .$$

Thus the function $\phi(x,y)$, harmonic in $0 \leq y \leq h$, which satisfies the conditions

$$\phi(x,0) = f(x)\ , \quad \phi(x,h) = 0\ , \quad -\infty < x < \infty\ , \tag{6.3.1}$$

may be found by assuming that

$$\phi(x,y) = F^*[A(\xi)\exp(-\xi y) + B(\xi)\exp(\xi y);\xi \to x]\ . \tag{6.3.2}$$

Conditions (6.3.1) imply

$$A(\xi)\exp(-\xi h) + B(\xi)\exp(\xi h) = 0\ ,$$

$$F^*[A(\xi) + B(\xi);x] = f(x)\ ,$$

which yield $B(\xi) = -A(\xi)\exp(-2\xi h)$, $A(\xi) = F(\xi)\exp(\xi h)/(2\sinh(\xi h))$, and

$$\phi(x,y) = F^*[F(\xi)\sinh\xi(h-y)/\sinh\xi h;\xi \to x]\ , \tag{6.3.3}$$

where, as usual, $F(\xi) = F[f(x);\xi]$.

Now apply this method to the problem of one or more frictionless punches indenting one face of a strip when the other rests against a frictionless rigid foundation. The boundary conditions are

$$\tau_{xy}(x,0) = 0 = \tau_{xy}(x,h), \quad v(x,h) = 0, \quad -\infty < x < \infty\ , \tag{6.3.4}$$

$$v(x,0) = f(x), \quad x \text{ in } L'; \quad \tau_{yy}(x,0) = 0, \quad x \text{ in } L''\ , \tag{6.3.5}$$

where L' denotes the contact region, and L'' the remainder of the x-axis. In addition, let

$$\tau_{yy}(x,0) = -p(x) \;, \quad x \text{ in } L' \;. \tag{6.3.6}$$

For the Papkovich-Neuber potentials take $\psi_1 = 0 = \psi_3$ and

$$\psi_2(x,y) = F^*[\xi^{-1}\{A(\xi)\exp(-\xi y) + B(\xi)\exp(\xi y)\};\xi \to x] \;, \tag{6.3.7}$$

$$\phi(x,y) = F^*[\xi^{-2}\{C(\xi)\exp(-\xi y) + D(\xi)\exp(\xi y)\};\xi \to x] \;. \tag{6.3.8}$$

These yield the displacements and stresses

$$u(x,y) = F^*[U(\xi,y);\xi \to x] \;, \quad v(x,y) = F^*[V(\xi,y);\xi \to x] \;, \tag{6.3.9}$$

$$\tau_{xx}(x,y) = F^*[T_{11}(\xi,y);\xi \to x] \;, \quad \tau_{xy}(x,y) = F^*[T_{12}(\xi,y);\xi \to x] \;, \tag{6.3.10}$$

$$\tau_{yy}(x,y) = F^*[T_{22}(\xi,y);\xi \to x] \;, \tag{6.3.11}$$

where in matrix form

$$\begin{bmatrix} 2\mu U \\ 2\mu V \\ -T_{11} \\ -T_{12} \\ -T_{22} \end{bmatrix} = \begin{bmatrix} i\xi^{-1}(\xi yA+C) \\ \xi^{-1}\{(3-4\nu)A+\xi yA+C\} \\ 2\nu A-\xi yA-C \\ i\{(1-2\nu)A+\xi yA+C\} \\ 2(1-\nu)A+\xi yA+C \end{bmatrix} \exp(-\xi y) + \begin{bmatrix} i\xi^{-1}(\xi yB+D) \\ \xi^{-1}\{(3-4\nu)B-\xi yB-D\} \\ -2\nu B-\xi yB-D \\ i\{(1-2\nu)B-\xi yB-D\} \\ -2(1-\nu)B+\xi yB+D \end{bmatrix} \exp(\xi y) \tag{6.3.12}$$

After satisfying conditions (6.3.4) we obtain

$$\tau_{yy}(x,0) = F^*[\exp(\omega)(2\omega+\sinh 2\omega)B(\xi)/\sinh\omega;\xi \to x] \;, \tag{6.3.13}$$

$$2\mu v(x,0) = -4(1-\nu)F^*[\xi^{-1}\exp(\omega)\sinh\omega B(\xi);\xi \to x] \;, \tag{6.3.14}$$

where $\omega = h\xi$. Now $B(\xi)$ must be chosen so that conditions (6.3.5) are satisfied. Thus

$$\tau_{yy}(x,0) \equiv -F^*[P(\xi);x] = 0 \ , \quad x \text{ in } L'' \ , \tag{6.3.15}$$

$$v(x,0) \equiv \vartheta F^*[\xi^{-1}\{2\sinh^2\omega/(2\omega+\sinh 2\omega)\}P(\xi);x] = f(x) \ , \quad x \text{ in } L' \ . \tag{6.3.16}$$

The equations corresponding to some other set of boundary conditions on $y = h$, e.g., the strip is fixed to the rigid foundation, will differ from (6.3.15), (6.3.16) only in that the kernel of (6.3.16) is changed (Example 6.3.2). It is convenient to present a general form of the equations before approximate solutions are described. (An analysis of the plane strip using Muskhelishvili-type complex potentials was derived by Belen'kii (1952); it has not found general acceptance.)

Let a be a standard length, P_0 a force and introduce the dimensionless quantities

$$\xi = x/a \ , \quad s = a\xi \ , \tag{6.3.17}$$

$$\Phi(s) = P_0^{-1}P(\xi) \ , \quad \phi(\xi) = aP_0^{-1}p(x) \ , \quad \chi(\xi) = (\vartheta P_0)^{-1}f(x) \ , \tag{6.3.18}$$

where $p(x) = F^*[P(\xi);x]$. Then the equations may be written in the form

$$aP_0^{-1}\tau_{yy}(x,0) \equiv -F^*[\Phi(s);\xi] = 0 \ , \quad \xi \text{ in } \underline{L}'' \ , \tag{6.3.19}$$

$$(\vartheta P_0)^{-1}v(x,0) \equiv F^*[s^{-1}\text{sign}(s)\{1+N(s)\}\Phi(s);\xi] = \chi(\xi), \ \xi \text{ in } \underline{L}' \ , \tag{6.3.20}$$

where $\underline{L}'$, $\underline{L}''$ are the dimensionless counterparts of L', L'', and for equation (6.3.16)

$$\text{sign}(s)\{1+N(s)\} = 2\sinh^2\omega/(2\omega+\sinh 2\omega) \ . \tag{6.3.21}$$

In general $N(s)$ is an even function, and $N(s) \to 0$ as $s \to \infty$.

When there is a single punch, so that the conditions

$$v(x,0) = f(x)\ , \quad x \text{ in } L'; \qquad v_x(x,0) = f'(x)\ , \quad x \text{ in } L'; \qquad (6.3.22)$$

are equivalent (provided f(x) is differentiable), equation (6.3.20) may be replaced by

$$(2\pi\vartheta P_0)^{-1} a v_x(x,0) \equiv -iF^*[\text{sign}(s)\{1+N(s)\}\Phi(s);\xi] = \chi'(\xi)\ , \quad \xi \text{ in } \underline{L}''. \qquad (6.3.23)$$

We now present three ways of reducing the pairs (6.3.19), (6.3.20) or (6.3.19), (6.3.23) to a single integral equation when the contact region $\underline{L}'$ is [-1, 1]. First, write

$$F^*[\Phi(s);\xi] = \phi(\xi)H(1-|\xi|)\ ; \qquad (6.3.24)$$

then equation (5.5.6) shows that the first integral in (6.3.23) may be written as a Hilbert transform; the second part may be rewritten by using the convolution theorem (5.3.1). Thus

$$\frac{1}{\pi}\int_{-1}^{1} \phi(\tau)\left\{\frac{1}{\tau-\xi} + k(\xi,\tau)\right\} d\tau = \chi'(\xi)\ , \qquad |\xi| < 1\ , \qquad (6.3.25)$$

where

$$k(\xi,\tau) = \int_0^\infty N(s)\sin s(\tau-\xi)\,ds\ . \qquad (6.3.26)$$

Since $k(\xi,\tau)$ is everywhere finite the behaviour of solutions of this equation is governed by the singular term, and $\phi(\tau)$ will therefore, in general, have a square-root singularity at $\pm$ 1. Equation (6.3.25) may be reduced to a nonsingular equation by taking the term involving $k(\xi,\tau)$ to the right-hand side and using the inverse of the Hilbert equation. By writing $\phi(\tau) = (1-\tau^2)^{-1/2}\theta(\tau)$ we obtain

$$\theta(\tau) + \int_{-1}^{1} M(\tau,\xi)\theta(\xi)d\xi = w(\tau) \ , \tag{6.3.27}$$

where

$$M(\tau,\xi) = \frac{(1-\xi^2)^{-1/2}}{\pi^2} \int_{-1}^{1} \frac{(1-s^2)^{1/2}k(s,\xi)}{s-\tau} ds \ , \tag{6.3.28}$$

$$w(\tau) = -\frac{1}{\pi} \int_{-1}^{1} \frac{(1-\xi^2)^{1/2}\chi'(\xi)}{\xi-\tau} d\xi \ . \tag{6.3.29}$$

The solution of equations such as (6.3.27), called *Fredholm integral equations of the second kind*, is discussed briefly in §6.4.

The disadvantage of the method outlined above is that, by dealing with the displacement derivative, it does not immediately yield the actual amount by which the punch indents the strip. This indentation can be found by working directly with equation (6.3.20), as will now be described. First note that the part of the kernel of (6.3.20) in braces is a function of $\omega = h\xi = \gamma s$ where $\gamma = h/a$. Thus, following V.M. Aleksandrov (1962), we note that equation (6.3.20) may be written

$$\frac{1}{\pi} \int_{-1}^{1} K\left(\frac{\xi-\tau}{\gamma}\right) \phi(\tau)d\tau = \chi(\xi) \ , \tag{6.3.30}$$

where

$$K(k) = \int_{0}^{\infty} \frac{2\sinh^2\omega\cos k\omega d\omega}{\omega(2\omega+\sinh 2\omega)} \ . \tag{6.3.31}$$

In order to solve equation (6.3.30) for large γ, i.e., for layers that are deep compared to the length of the contact region, we note that $K(k)$ has a logarithmic singularity for small k. In order to isolate this singularity we write

$$C(k) \equiv \int_0^\infty \omega^{-1}(1-\exp(-q\omega)]\cos k\omega d\omega = -\ln|k|+\ln(q^2+k^2)^{1/2} , \quad q > 0 . \tag{6.3.32}$$

The behaviour of $K(k)$ for small k is dominated by the form of $2\sinh^2\omega/(2\omega+\sinh 2\omega) \simeq 1-4\omega\exp(-2\omega)$ for large ω. Thus

$$K(k) = -\ln|k|+\ln(q^2+k^2)^{1/2} + \int_0^\infty \omega^{-1}\left[\frac{2\sinh^2\omega}{2\omega+\sinh 2\omega} - 1+\exp(-q\omega)\right]\cos k\omega d\omega . \tag{6.3.33}$$

The cosine transform (with $q = 2$, say) may be computed easily since the integrand decreases exponentially, and equation (6.3.33) may be written

$$K(k) = -\ln|k| + F(k) , \tag{6.3.34}$$

so that equation (6.3.30) is

$$\frac{1}{\pi}\int_{-1}^{1}\left\{-\ln\left|\frac{\xi-\tau}{\gamma}\right| + F\left(\frac{\xi-\tau}{\gamma}\right)\right\}\phi(\tau)d\tau = \chi(\xi) , \quad |\xi| \leq 1 . \tag{6.3.35}$$

V.M. Aleksandrov (1962), following Shtaerman (1949), has shown how the principal part of this singular integral equation, i.e., the equation with $F \equiv 0$, may be inverted (Example 6.3.3) and how equation (6.3.35) may be reduced to a nonsingular integral equation of the second kind: further study of equation (6.3.35) will be deferred to §6.5.

The third method for solving equations (6.3.19), (6.3.23) is to use the operators in Table 5.7.2. It is convenient to distinguish between the cases in which the contact is symmetric, so that $\phi(\xi)$, $\chi(\xi)$ are even functions of ξ, and the contact is antisymmetric so that $\phi(\xi)$, $\chi(\xi)$ are odd. In the former case equations (6.3.19), (6.3.23) may be written

$$a(\vartheta P_0)^{-1}v_x(x,0) = -F_s[\{1+N(s)\}\Phi(s);\xi] = \chi'(\xi) , \quad 0 \leq \xi \leq 1 , \tag{6.3.36}$$

$$aP_0^{-1}\tau_{yy}(x,0) = -F_c[\Phi(s);\xi] = -\phi(\xi)H(1-\xi) , \quad \xi \geq 0 . \tag{6.3.37}$$

Operate on (6.3.37) with entry (7) to obtain

$$\tau H_0[\Phi(s);\tau] = \psi(\tau) , \tag{6.3.38}$$

where

$$\psi(\tau) = A_2^{-1}\{\phi(\xi);\tau\} \tag{6.3.39}$$

is zero if $x > 1$. Equation (6.3.38) may be inverted to give

$$\Phi(s) = \int_0^1 \psi(\tau)J_0(\tau s)d\tau , \tag{6.3.40}$$

while (6.3.39) yields

$$\phi(\xi) = A_2[\psi(\tau);\xi] = (2/\pi)^{1/2} \int_\xi^1 (\tau^2-\xi^2)^{-1/2}\psi(\tau)d\tau , \tag{6.3.41}$$

when $0 \leq \xi < 1$. Now operate on (6.3.36) with entry 3 to obtain

$$\psi(t) + tH_0\{N(s)\Phi(s);t\} = -A_1^{-1}\{\chi'(\xi);t\} , \tag{6.3.42}$$

but

$$tH_0\{N(s)\Phi(s);t\} = \int_0^\infty tsN(s)J_0(ts)\left\{\int_0^1 \psi(\tau)J_0(\tau s)d\tau\right\} ds$$

$$= \int_0^1 H(t,\tau)\psi(\tau)d\tau , \tag{6.3.43}$$

where

$$H(t,\tau) = \int_0^\infty tsN(s)J_0(ts)J_0(\tau s)ds . \tag{6.3.44}$$

Thus

$$\psi(t) + \int_0^1 H(t,\tau)\psi(\tau)d\tau = \theta(t) , \quad 0 \leq t < 1 , \tag{6.3.45}$$

where

$$\theta(t) = -A_1^{-1}\{\chi'(\xi);t\} . \qquad (6.3.45)$$

This equation is another Fredholm equation of the second kind, equivalent to (6.3.27) when ϕ, χ are even functions.

For the analysis of an *anisotropic strip lying on a rigid foundation* see Lekhnitskii (1963a), Milovic and Touzot (1970), and other papers cited there.

Examples 6.3

1. Derive equations (6.3.12) - (6.3.14).
2. Show that if the strip is rigidly attached to the plane $y = h$, so that the boundary conditions are $u(x,h) = 0 = v(x,h)$, $\tau_{xy}(x,0) = 0$, $-\infty < x < \infty$; $\tau_{yy}(x,0) = -p(x)$, then

$$v(x,0) = \vartheta F^*\left[\xi^{-1}\left\{\frac{2\kappa\sinh 2\omega - 4\omega}{\kappa^2+4\omega^2+2\kappa\cosh 2\omega}\right\}P(\xi);x\right] ,$$

$$p(x) = F^*[P(\xi);x] ,$$

where $\kappa = 3-4\nu$.

3. Show that the solution of the equation

$$-\pi^{-1}\int_{-1}^{1} \ln|\xi-t|\phi(t)dt = \chi(\xi) , \qquad |\xi| \le 1 ,$$

which has integrable singularities at $\pm$ 1, is

$$\phi(\xi) = \pi^{-1}(1-\xi^2)^{-1/2}\left[C - \int_{-1}^{1} (1-t^2)^{1/2}\chi'(t)(t-\xi)^{-1}dt\right] ,$$

where

$$C = \int_{-1}^{1} \phi(\xi)d\xi = (\ln 2)^{-1}\int_{-1}^{1} (1-t^2)^{-1/2}\chi(t)dt .$$

Show that this agrees with the result of Example 2.3.8.

4. Show that if the deformation of the strip is symmetric (antisymmetric) about the y-axis in the sense that
$v(x,y) = +(-)v(-x,y)$, $u(x,y) = -(+)u(-x,y)$
then A, B, C, D of (6.3.7), (6.3.8) must satisfy $A(\xi) = -(+)B(-\xi)$, $C(\xi) = +(-)D(-\xi)$ and that the transforms F^* in front of U, T_{12} may be replaced by $F_s(F_c)$ while those operating on V, T_{11}, T_{22} must be replaced by $F_c(-F_s)$, where equation (6.3.12) is unchanged except that i is neglected.

5. Show that if $\phi_0(\tau)$, $\phi_1(\tau)$ are the solutions of equation (6.3.30) corresponding to $\chi_0(\xi) = 1$, $\chi_1(\xi) = \xi$ respectively then the dimensionless total force $P = \int_{-1}^{1} \phi(\tau)d\tau$ and moment $M = \int_{-1}^{1} \tau\phi(\tau)d\tau$, corresponding to $\chi(\xi)$ may be written (V.M. Aleksandrov (1962))

$$P = \int_{-1}^{1} \phi_0(\xi)\chi(\xi)d\xi \ , \quad M = \int_{-1}^{1} \phi_1(\xi)\chi(\xi)d\xi \ .$$

6.4 *The solution of some integral equations*

Equations (6.3.25), (6.3.27), (6.3.30) of the previous section are all examples of *Fredholm* integral equations, the general forms of which are as follows. The equation

$$\int_a^b K(x,y)\phi(y)dy = f(x) \ , \quad a < x < b \ , \tag{6.4.1}$$

for the unknown function $\phi(y)$ is a Fredholm equation of the *first* kind, while

$$\phi(x) - \lambda \int_a^b K(x,y)\phi(y) = f(x) \ , \quad a < x < b \ , \tag{6.4.2}$$

is an equation of the *second* kind; the latter is homogeneous if $f(x) \equiv 0$, inhomogeneous otherwise. The integral equations are further categorized according to the behaviour of K(x,y). If K(x,y) is continuous in $a \leq x \leq b$, $a \leq y \leq b$, or if at least the discontinuities of K(x,y) are such that the double integral

$$\int_a^b \int_a^b |K^2(x,y)|dxdy$$

has a finite value, the equations are termed *nonsingular*. If the kernel has the form

$$K(x,y) = \frac{H(x,y)}{|x-y|^{\alpha}}, \qquad (6.4.3)$$

where H(x,y) is bounded and α is a constant satisfying $0 < \alpha < 1$, then the equations are said to have a *weak singularity*. If

$$K(x,y) = \frac{A(x,y)}{x-y}, \qquad (6.4.4)$$

where A(x,y) is a differentiable function of x and y, then the equations obtained by interpreting the integrals in (6.4.1) and (6.4.2) as Cauchy principal values (see (5.5.7)), are said to be *singular* integral equations. Equations (3.5.1), (6.3.25) are of this latter type. Equations with a $\ell n|x-y|$ singularity in the kernel behave very much like singular integral equations.

The general theory governing Fredholm integral equations, and in particular *Fredholm's alternative,* which states that *either* the nonhomogeneous equation (6.4.2) has a solution for *all* f(x), *or else* the homogeneous equation with f(x) = 0 has a nontrivial solution, may be found in many texts, e.g., Mikhlin (1957), Smirnov (1964), Smithies (1958) or Cochran (1972): the values of λ for which the homogeneous equation has a nontrivial solution are called the *characteristic values,* or *eigenvalues*, of the equation. Here we shall be content to set aside questions concerning the existence and uniqueness of solutions and concentrate on the approximate solution of the equations. Surveys of the literature on the numerical solution of integral equations have recently been made by Atkinson (1976), Baker (1977) and Delves and Walsh (1974).

When the kernel K(x,y) is nonsingular, either equation (6.4.1) or (6.4.2) may be solved by suitably approximating the integral, so that

$$\int_a^b K(x,y)\phi(y)dy = \sum_{n=1}^{n} w_n K(x,y_n)\phi(y_n) , \qquad (6.4.5)$$

where w_j, y_j are the weights and abscissae of a numerical quadrature. Now by equating both sides of the equation at N points x_m we obtain

$$\sum_{n=1}^{N} w_n K(x_m,y_n)\phi(y_n) = f(x_m) , \quad m = 1,2,\ldots,N , \qquad (6.4.6)$$

as the approximate form of (6.4.1), and

$$\phi(y_m) - \lambda \sum_{n=1}^{N} w_n K(y_m,y_n)\phi(y_n) = f(y_m) , \quad m = 1,2,\ldots,N , \qquad (6.4.7)$$

as the approximation to (6.4.2); note that in (6.4.7) the points x_n at which both sides are equated have been taken to be the quadrature points y_n. Equations (6.4.6), (6.4.7) are straightforward sets of simultaneous equations for the unknowns $\phi(y_n)$.

A somewhat similar analysis may be developed by using a set of (Legendre-like) polynomials $P_n(x)$ orthonormal over [a,b], so that

$$\int_a^b P_m(x)P_n(x)dx = \delta_{mn} , \quad m,n = 0,\ldots,N . \qquad (6.4.8)$$

If K(x,y) is continuous over $a \leq x \leq b$, $a \leq y \leq b$, the operation of finding a polynomial approximation to K(x,y) of the form

$$K(x,y) \doteq \sum_{m=0}^{N} \sum_{n=0}^{N} c_{mn} P_m(x) P_n(y) \tag{6.4.9}$$

is straightforward; this is a particular example of a so-called *separable* kernel. If therefore we assume that $\phi(x)$ and $f(x)$ have polynomial approximations

$$\phi(x) = \sum_{m=0}^{N} a_m P_m(x) , \quad f(x) = \sum_{m=0}^{N} b_m P_m(x) , \tag{6.4.10}$$

and use the orthonormality relation (6.4.8) to evaluate the integrals, we find the approximations to (6.4.1), (6.4.2) to be

$$\sum_{n=0}^{N} c_{mn} a_n = b_m , \quad m = 0,1,\ldots,N , \tag{6.4.11}$$

and

$$a_m - \lambda \sum_{n=0}^{N} c_{mn} a_n = b_m , \quad m = 0,1,\ldots,N . \tag{6.4.12}$$

A variant of this procedure which uses Chebyshev polynomials and the *discrete* orthogonality equation which they satisfy is discussed by D.A. Elliott (1967).

One of the questions which arises in approximate solution is the relation between the solution of the N+1 equations (6.4.12) and the solution of the infinite set of equations obtained by taking infinite expansions in (6.4.9), (6.4.10). The set of equations

$$a_m - \lambda \sum_{n=0}^{\infty} c_{mn} a_n = b_m , \quad m = 0,1,\ldots , \tag{6.4.13}$$

is said to be *fully regular* if there is a number θ, $0 < \theta < 1$, so that

$$s_m = \sum_{n=0}^{\infty} |\lambda c_{mn}| \leq 1 - \theta < 1 , \quad m = 0,1,2, \ldots . \tag{6.4.14}$$

In this case, provided that the $|b_m|$ are bounded, the limit as $N \to \infty$ of the solution $\{a_0, a_1, \ldots, a_N\}$ obtained by solving the first N+1 equations (6.4.12) *will be* the solution of the set of equations (6.4.13). A full discussion of the existence and convergence of the solution of equation (6.4.13) under this and other less restrictive conditions may be found in Kantorovich and Krylov (1958).

A completely different method of solving equation (6.4.2), which is discussed exhaustively in the older literature (see, e.g., Kantorovich and Krylov (1958)), is provided by the iterative solution

$$\phi(x) = \sum_{n=0}^{\infty} \lambda^n \phi_n(x) , \qquad (6.4.15)$$

where

$$\phi_0(x) = f(x) , \quad \phi_n(x) = \int_a^b K(x,y)\phi_{n-1}(y)dy , \quad n = 1,2,\ldots . \qquad (6.4.16)$$

This procedure is still useful if λ is 'small' and the integrals can be easily evaluated.

A full account of the theory of *singular* integral equations may be found in Muskhelishvili (1953b). There the integrals are not taken necessarily over a single interval [a,b] of the x-axis, but over one or more arcs or contours. We specialize our discussion to the type of equation which appeared in §6.3, namely

$$\frac{1}{\pi} \int_{-1}^{1} \phi(\tau) \left\{ \frac{1}{\tau-\xi} + K(\xi,\tau) \right\} d\tau = \chi'(\xi) , \quad |\xi| < 1 . \qquad (6.4.17)$$

(Clearly a simple linear transformation may always be used to convert the interval $a < x < b$ into $-1 < \xi < 1$.) It may be shown that if $K(\xi,\tau)$ is continuous in $-1 \le \xi \le 1$, $-1 \le \tau \le 1$ then the term $(\tau-\xi)^{-1}$ *dominates* equation (6.4.17), in the sense that the general solution of (6.4.17), like that of the equation with $K(\xi,\tau) \equiv 0$, has a square root singularity at $\xi = \pm 1$. Thus we assume that $\chi(\xi)$ may be approximated by a polynomial

$$\chi(\xi) = \sum_{m=0}^{N} a_m T_m(\xi) , \qquad (6.4.18)$$

so that

$$\chi'(\xi) = \sum_{m=1}^{N} m a_m U_{m-1}(\xi) , \qquad (6.4.19)$$

and we may seek the solution of (6.4.17) in the form

$$\phi(\tau) = (1-\tau^2)^{-1/2} \sum_{n=1}^{N} b_n T_n(\tau) . \qquad (6.4.20)$$

The first integral in (6.4.17) will then be given exactly by (2.3.26) while the second may be calculated approximately by using the Gauss-Chebyshev formula (Abramowitz and Stegun (1972, p. 889))

$$\pi^{-1} \int_{-1}^{1} (1-\tau^2)^{-1/2} f(\tau)d\tau = (N+1)^{-1} \sum_{j=1}^{N+1} f(\tau_j) , \qquad (6.4.21)$$

where the τ_n are the zeros of $T_{N+1}(\tau_j) = 0$, namely

$$\tau_j = \cos[(2j-1)\pi/(2N+2)] . \qquad (6.4.22)$$

This formula is exact if $f(\tau)$ is a polynomial of degree 2N+1 or less. Using these results we may now approximate equation (6.4.17) by

$$\sum_{n=1}^{N} b_n U_{n-1}(\xi) + \sum_{n=0}^{N} b_n c_n(\xi) = \sum_{n=1}^{N} n a_n U_{n-1}(\xi) , \quad |\xi| \leq 1 , \qquad (6.4.23)$$

where

$$c_n(\xi) = (N+1)^{-1} \sum_{j=1}^{N+1} T_n(\tau_j)K(\tau_j,\xi) . \qquad (6.4.24)$$

Equation (6.4.23) may be solved approximately by multiplying throughout by $\pi^{-1}(1-\xi^2)^{1/2}U_{m-1}(\xi)$, integrating over [-1,1], and using the orthogonality relation of Example 2.3.2 for the U_n, and the Gauss-Chebyshev formula, exact for polynomials of degree 2N-1 or less (Abramowitz and Stegun (1972, p. 889)), namely

$$\pi^{-1} \int_{-1}^{1} (1-\xi^2)^{1/2} f(\xi)d\xi = (N+1)^{-1} \sum_{k=1}^{N} w_k f(\xi_k) , \qquad (6.4.25)$$

where the ξ_k are the zeros of $U_N(\xi_k) = 0$, namely

$$\xi_k = \cos[k\pi/(N+1)] , \quad w_k = 1 - \xi_k^2 . \qquad (6.4.26)$$

Thus

$$b_m + \sum_{n=0}^{N} c_{mn} b_n = m a_m , \quad m = 1,2,\ldots,N , \qquad (6.4.27)$$

where

$$c_{mn} = 2(N+1)^{-1} \sum_{k=1}^{N} w_k U_{m-1}(\xi_k) c_n(\xi_k) . \qquad (6.4.28)$$

Typically there will be one more equation. In the physical problem to which (6.4.17) refers, $\phi(x)$ is related to the pressure p(x) by equation (6.3.18); the total force applied over the contact pressure is

$$P_0 = \int_{-a}^{a} p(x)dx = P_0 \int_{-1}^{1} \phi(\xi)d\xi = P_0 \pi b_0 , \qquad (6.4.29)$$

so that

$$\pi b_0 = 1 . \qquad (6.4.30)$$

Equations (6.4.27), (6.4.30) provide (N+1) equations for the determination of b_0, b_1, ..., b_N.

Erdogan (1969), Erdogan and Gupta (1972) and Erdogan, Gupta and Cook (1973) have used the identity (Example 6.4.2)

$$\frac{1}{(N+1)} \sum_{j=1}^{N+1} \frac{T_n(\tau_j)}{\tau_j - \xi_k} = \begin{cases} 0 & , \quad n = 0 , \\ U_{n-1}(\xi_j) , & 0 < n \leq N , \end{cases} \tag{6.4.31}$$

to rewrite equation (6.4.23) for each of the N points ξ_k. Thus

$$\sum_{n=1}^{N} b_n U_{n-1}(\xi_k) = \frac{1}{N+1} \sum_{n=0}^{N} b_n \sum_{j=1}^{N+1} \frac{T_n(\tau_j)}{\tau_j - \xi_k} = \frac{1}{N+1} \sum_{j=1}^{N+1} \frac{g(\tau_j)}{\tau_j - \xi_k} , \tag{6.4.32}$$

where

$$g(\tau) = \sum_{n=0}^{N} b_n T_n(\tau) .$$

Similarly

$$\sum_{n=0}^{N} b_n c_n(\xi_k) = \frac{1}{N+1} \sum_{n=0}^{N} b_n \sum_{j=1}^{N+1} T_n(\tau_j) K(\tau_j, \xi_k) = \frac{1}{N+1} \sum_{j=1}^{N+1} g(\tau_j) K(\tau_j, \xi_k) \tag{6.4.33}$$

so that equation (6.4.23) for the point $\xi = \xi_k$ is

$$\frac{1}{N+1} \sum_{j=1}^{N+1} g(\tau_j) \left\{ \frac{1}{\tau_j - \xi_k} + K(\tau_j, \xi_k) \right\} = \chi'(\xi_k) , \quad k = 1,2,\ldots,N , \tag{6.4.34}$$

and equation (6.4.30) may be written

$$\frac{\pi}{N+1} \sum_{j=1}^{N+1} g(\tau_j) = 1 . \tag{6.4.35}$$

Equations (6.4.34), (6.4.35) will in general not be so well conditioned as equations (6.4.27), (6.4.30) since, if the pressure distribution is 'smooth', the coefficients a_m will rapidly become small as m increases. Notice also, that if $c_{mn} \equiv 0$, and this occurs when the strip becomes a half-plane, the solution of equation (6.4.27) is clearly seen to be $b_m = ma_m$, agreeing with (2.4.10); this simple solution is not immediately apparent from equation (6.4.34). Erdogan and Gupta carry through a similar analysis for the cases in which $\phi(\tau)$ is zero at one or both of the ends ± 1, but this analysis is unneccessary as $\phi(\tau)$ may still be sought in the form (6.4.20), and the additional conditions

$$\phi(1) = 0\ , \quad \text{i.e.,} \quad \sum_{n=0}^{N} b_n = 0\ ; \quad \phi(-1) = 0\ , \quad \text{i.e.,} \quad \sum_{n=0}^{N} b_n(-)^n = 0\ , \tag{6.4.36}$$

may be simply combined with equations (6.4.27). The equations so found will have a solution only under certain circumstances, which physically will yield some relation between total applied force and contact length. This will be discussed further in the examples. Further work on the numerical solution of singular integral equations may be found in Kalandia (1959), Karpenko (1966), Margetson and Morland (1970b), Morland (1970), Krenk (1975a), (1975b), (1978), and Theocaris and Ioakimidis (1977b).

Examples 6.4

1. Derive the recurrence relation

   $$U_{N-n}(x) = U_N(x)T_n(x) - T_{N+1}(x)U_{n-1}(x)\ .$$

2. Use Example 6.4.1 and (2.3.6) to show that

   $$U_{N-n}(x)/T_{N+1}(x) = \sum_{j=1}^{N+1} a_j/(\tau_j - x)\ ,$$

   where the τ_j are the roots (6.4.22) of $T_{N+1}(x)$. Show that

   $$a_j = -U_{N-n}(\tau_j)/T'_{N+1}(\tau_j) = -U_{N-n}(\tau_j)/[(N+1)U_N(\tau_j)]$$

   $$= -T_n(\tau_j)/(N+1)\ ,$$

   and now use

$$\frac{U_{N-n}(x)}{T_{N+1}(x)} = -U_{n-1} + \frac{T_n(x)U_N(x)}{T_{N+1}(x)} = -\frac{1}{N+1}\sum_{j=1}^{n}\frac{T_n(\tau_j)}{\tau_j - x}$$

to deduce (6.4.31).

3. Use equation (3.2.24) to show that the solution $\phi(\tau)$ of equation (6.4.17) will have square root singularities at $\tau = \pm 1$.

6.5 *Indentation of a strip resting on a rigid foundation*

The equations governing the contact problem for a frictionless rigid punch indenting one face of a strip when the other face is either (a) lying in frictionless contact with, or (b) fixed to, a rigid foundation, were formulated in §6.3. In this section we describe research on a general form of the integral equation governing such problems and examine the solution of these equations in detail, for the particular cases (1) complete contact of a punch with a straight base, (2) incomplete contact of a punch with a circular (actually parabolic) profile.

The analysis has a long history; see Marguerre (1931). Hannah (1951) formulated problems a2, b2 effectively as in §6.2 and presented approximate solutions, which to-day provide little of interest. V.M. Aleksandrov (1962) used the formulation (6.3.30), namely

$$\frac{1}{\pi}\int_{-1}^{1} \phi(\tau)K\left(\frac{\xi-\tau}{\gamma}\right)d\tau = \chi(\xi)\ , \qquad |\xi| \leq 1\ , \tag{6.5.1}$$

where $\gamma = h/a$, assumed that $K(k)$ could be written

$$K(k) = -\ell n|k| + F(k)\ , \tag{6.5.2}$$

where $F(k)$ is an even function of k, and wrote equation (6.5.1) in the form

$$-\frac{1}{\pi}\int_{-1}^{1}\ln\left|\frac{\xi-\tau}{\gamma}\right|\phi(\tau)d\tau = \chi(\xi) - \frac{1}{\pi}\int_{-1}^{1}F\left(\frac{\xi-\tau}{\gamma}\right)\phi(\tau)d\tau \equiv \psi(\xi) . \tag{6.5.3}$$

If the function $\psi(\xi)$ is assumed to be known, the solution of this equation is (Example 6.3.3)

$$\phi(\xi) = \frac{1}{\pi(1-\xi^2)^{1/2}}\left[P - \int_{-1}^{1}\frac{(1-\tau^2)^{1/2}\psi'(\tau)d\tau}{\tau-\xi}\right], \tag{6.5.4}$$

where

$$P = \int_{-1}^{1}\phi(\xi)d\xi = \frac{1}{\ln(2\gamma)}\int_{-1}^{1}\frac{\psi(\tau)d\tau}{(1-\tau^2)^{1/2}} . \tag{6.5.5}$$

For large γ the function F may be approximated by its value F(0) and this yields the first approximation

$$\phi(\xi) = \frac{1}{\pi(1-\xi^2)^{1/2}}\left[P^* - \int_{-1}^{1}\frac{(1-\tau^2)^{1/2}\chi'(\tau)d\tau}{\tau-\xi}\right], \tag{6.5.6}$$

$$P^* = \frac{1}{\ln(2\gamma)+F(0)}\int_{-1}^{1}\frac{\chi(\tau)d\tau}{(1-\tau^2)^{1/2}} . \tag{6.5.7}$$

Aleksandrov states the conditions on F and χ for which these results are valid, and obtained higher order approximations, valid for large γ, by assuming that F(k) may be written

$$F(k) = \sum_{n=0}^{\infty} a_n k^{2n} , \qquad |k| < \rho . \tag{6.5.8}$$

A similar, somewhat neater, solution for problem b2, again for large γ, was given by Meijers (1968). In his comprehensive paper he presented results for the pressure, displacement, penetration depth and contact length for different values of Poisson's ratio.

V.M. Aleksandrov (1964a) considered the solution of (6.5.1) by expanding ϕ, F, χ as infinite Chebyshev series; V.M. Aleksandrov

(1968), V.M. Aleksandrov and Kucherov (1970) modified this analysis and proved the quasi-regularity of the infinite system of equations obtained, for large γ. With the regularity of the system proved, it is sufficient to take χ, ϕ in the forms (6.4.18), (6.4.20) respectively, evaluate the integral involving $\ell n|(\xi-\tau)/\gamma|$ by using Example 2.3.8 and the one involving F by using Gauss-Chebyshev quadrature: this is basically a refinement of the procedures used by V.M. Aleksandrov (1962) and Meijers (1968). Thus for a cylinder of radius R (case 2), penetrating the strip to a depth v_0, the f(x) of (6.2.5) is $f(x) = v_0 - x^2/(2R)$, so that

$$\chi(\xi) = (\vartheta P_0)^{-1}[v_0 - a^2\xi^2/(2R)] = -(\pi\lambda)^{-1}(\xi^2 - 2d) ,$$

$$= -(2\pi\lambda)^{-1}[T_2(\xi) + (1-4d)T_0(\xi)] , \tag{6.5.9}$$

where λ, d are the parameters used by Meijers, namely

$$\lambda = \frac{2\vartheta R P_0}{\pi a^2} = \frac{4(1-\nu^2)PR_0}{\pi E a^2} , \quad d = \frac{Rv_0}{a^2} . \tag{6.5.10}$$

The equation (6.5.3) may be written

$$\frac{1}{\pi}\int_{-1}^{1}\left\{-\ell n\left|\frac{\xi-\tau}{\gamma}\right| + F\left(\frac{\xi-\tau}{\gamma}\right) + \frac{1}{2\lambda}[T_2(\xi) + (1-4d)T_0(\xi)]\right\}\phi(\tau)d\tau = 0 , \tag{6.5.11}$$

where the equilibrium condition $\int_{-1}^{1}\phi(\tau)d\tau = 1$ has been used to obtain the last term. Expanding $\phi(\tau)$ in the form

$$\phi(\tau) = (1-\tau^2)^{-1/2}\sum_{n=0}^{2N} b_n T_n(\tau) , \tag{6.5.12}$$

where now only the even b_n are non-zero, we find

$$\sum_{n=0}^{2N*} n^{-1} b_n T_n(\xi) + \sum_{n=0}^{2N} b_n d_n(\xi) + (2\pi\lambda)^{-1}[T_2(\xi) + (1-4d)T_0(\xi)]b_0 = 0 , \tag{6.5.13}$$

where * denotes the fact that the first term is $b_0 \ln(2\gamma)$ and

$$d_n(\xi) = \frac{1}{\pi} \int_{-1}^{1} (1-\tau^2)^{-1/2} T_n(\tau) F\left(\frac{\xi-\tau}{\gamma}\right) d\tau \ . \tag{6.5.14}$$

Now multiplying equation (6.5.13) by $(1-\xi^2)^{-1/2} T_m(\xi)$, where $m = 0,2,\ldots,2N$, we find

$$b_0 \ln(2\gamma) + \sum_{n=0}^{2N} c_{0,n} b_n + (2\lambda)^{-1}(1-4d) b_0 = 0 \ , \tag{6.5.15}$$

$$b_2/4 + \sum_{n=0}^{2N} c_{2,n} b_n + (2\lambda)^{-1} b_0 = 0 \ , \tag{6.5.16}$$

$$b_m/(2m) + \sum_{n=0}^{2N} c_{m,n} b_n = 0 \ , \quad m = 4,6,\ldots,2N \ . \tag{6.5.17}$$

The condition for equilibrium is $\pi b_0 = 1$ and that for incomplete contact at $\xi = \pm 1$ is

$$b_0 + b_2 + \cdots + b_{2N} = 0 \ . \tag{6.5.18}$$

Equations (6.5.16) - (6.5.18) form a very simple eigenvalue problem for λ, with just one eigenvalue. Once λ and the eigenvector $b_0, b_2, \ldots, b_{2N}$, normalized so that $\pi b_0 = 1$, are known, the penetration d may be computed from (6.5.15). If these calculations are carried out, it is found that a comparatively small value of $N(<6)$ is required to give an acceptable approximate solution when $\gamma > 1$; for these values of γ the results for problem b2 agree very closely with those obtained by Meijers (1968).

If the calculations described above are performed with a decreasing sequence of γ values, from 1 to 0.05 for example, it is found that, for a given value of N, the coefficients $b_0, b_2, \ldots, b_{2N}$ decrease in magnitude less rapidly for each succeeding γ, and tend to alternate in sign. These two facts are indications that the series (6.5.12) is unable to model the behaviour of the pressure distribution. Eventually, as γ is taken smaller, the procedure breaks down altogether; the actual critical value of γ will depend on the value of N, the accuracy with which the terms in the matrix c_{mn} are computed and the stability of the procedure used to solve (6.5.16) - (6.5.18); in the calculation reported by Gladwell (1976) the critical value lay in the range 0.05 - 0.1.

Many researchers have attempted to find a solution procedure which is valid for small γ. The starting point, following V.M. Aleksandrov (1963a), is to note that the kernel K(k) may be written

$$K(k) = \int_0^\infty \omega^{-1} L(\omega) \cos k\omega \, d\omega \ , \tag{6.5.19}$$

where in case (a) equation (6.3.31) shows that

$$L(\omega) = \frac{2\sinh^2\omega}{2\omega+\sinh 2\omega} \ . \tag{6.5.20}$$

For both problems (a) and (b) $L(\omega)$ has the properties

$$L(\omega) \to A\omega \text{ as } \omega \to 0 \ ; \qquad L(\omega) \to 1 \text{ as } \omega \to \infty \ , \tag{6.5.21}$$

where for (a) $A = 1/2$ and for (b) $A = (1-2\nu)/[2(1-\nu)^2]$. When γ is small, the argument $k = (\xi-\tau)/\gamma$ will be large except when $|\xi-\tau|$ is very small. The behaviour of K(k) for large (small) k is governed by the behaviour of $L(\omega)$ for small (large) ω. The function

$$L_0(\omega) = \frac{\omega}{(\omega^2+A^{-2})^{1/2}} \tag{6.5.22}$$

behaves like $L(\omega)$ for both large and for small ω, so that, intuitively,

$$K(k) \sim \int_0^\infty \frac{L_0(\omega)\cos k\omega d\omega}{\omega} = K_0(k/A) \ , \qquad (6.5.23)$$

where K_0 is the modified Bessel function. For large k, $K_0(k/a) \to 0$, while for small k, $K_0(k/A)$ is singular. When γ is small, the kernel $K[(\xi-\tau)/\gamma]$ makes a contribution only in the neighbourhood of $\tau = \xi$; in this neighbourhood K(k) will be singular, but, since

$$\int_{-\infty}^\infty K(k)dk \sim \int_{-\infty}^\infty K_0(k/A)dk = \pi A \ , \qquad (6.5.24)$$

K(k) is approximately $\pi A\delta(k)$, where $\delta(k)$ is the Dirac delta function. Since $\delta(k) = 0$ when $k \neq 0$, the approximation to equation (6.5.1), namely

$$\frac{1}{\pi} \int_{-\infty}^\infty \pi A\delta\left(\frac{\xi-\tau}{\gamma}\right)\phi(\tau)d\tau = \chi(\xi) \ , \qquad (6.5.25)$$

has the solution

$$\phi(\xi) = \chi(\xi)/(A\gamma) \ , \qquad (6.5.26)$$

as stated earlier in V.M. Aleksandrov (1962). For problem 2, for which $\chi(\xi)$ is given by (6.5.9), this gives

$$\phi(\xi) = (\pi A\gamma\lambda)^{-1}(2d-\xi^2) \ . \qquad (6.5.27)$$

The condition of incomplete contact at $\xi = \pm 1$ now yields $d = 1/2$ and the equilibrium condition $\int_{-1}^{1} \phi(\xi)d\xi = 1$ gives $\lambda = 4/(3A\pi\gamma)$. These asymptotic values agree with those obtained by Meijers (1968).

This elastostatic problem has an electrostatic analogue. Consider a conducting strip brought near, and parallel to, a ground plane. If the distance between strip and plane is much less than the length of the strip then the electrostatic field can be approximated by that for the parallel plate condenser.

V.M. Aleksandrov (1963a) finds a better asymptotic solution for small γ as follows: for problem 1, for which $\chi(\xi) = \chi_0 = d/(\vartheta P_0)$, the solution is sought in the form

$$\phi(\xi) = \frac{\chi_0}{A\gamma}\{1 + \theta_+(\xi) + \theta_-(\xi)\} . \tag{6.5.28}$$

Here, following Krein (1955), he notes that $\theta_+(\xi)$ will have a singularity of the form $(1-\xi)^{-1/2}$ and tend rapidly to zero as $\xi < 1$, while $\theta_-(\xi)$ will have a singularity $(1+\xi)^{-1/2}$ and tend rapidly to zero as $\xi > -1$. An approximation to $\theta_-(\xi)$ may be found by changing to the variables $x = (1+\xi)/\gamma$, $t = (1+\tau)/\gamma$, for which equation (6.5.1) becomes

$$\frac{1}{\pi}\int_0^{2/\gamma} \phi(\gamma t-1)K(x-t)dt = \frac{\chi_0}{\gamma} , \tag{6.5.29}$$

and for small γ the upper limit may be taken to be infinite, giving

$$\frac{1}{\pi}\int_0^{\infty} \phi(\gamma t-1)K(x-t)dt = \frac{\chi_0}{\gamma} . \tag{6.5.30}$$

But this is the integral equation governing the indentation by a semi-infinite punch. For small γ the effect of the right-hand end of the punch on the state of stress near the left-hand end (and vice versa) can be neglected, so that for small γ the solution of (6.5.30) is

$$\phi(\xi) = \frac{\chi_0}{A\gamma}\{1 + \theta_-(\xi)\} . \tag{6.5.31}$$

Aleksandrov now solves (6.5.30) by replacing K by the approximation $K_0(k/A)$ of (6.5.23), and using the Wiener-Hopf technique. V.M. Aleksandrov and Vorovich (1964) use the symmetric change of variables

$$\xi = (1-2\gamma x)^{1/2} , \quad \tau = (1-2\gamma t)^{1/2} , \tag{6.5.32}$$

for which, and for small γ, equation (6.5.1) becomes

$$\frac{1}{\pi}\int_0^\infty \phi^*(t)K(x-t)dt = \frac{\chi^*(x)}{\gamma} . \tag{6.5.33}$$

The same Wiener-Hopf solution obtained by replacing K(k) by $K_0(k/A)$ now gives a solution which is a function of the symmetric variable $x = (1-\xi^2)/(2\gamma)$.

S.F. Smith (1964), in a long and difficult analysis, obtained an exact solution for the flat punch, valid for small γ. V.M. Aleksandrov and Babeshko (1965) sought improved approximations to $L(\omega)$ of the form (6.5.22) for which equations such as (6.5.30) could be solved. A less cumbersome analysis was given by V.M. Aleksandrov and Vorovich (1964); see also Privarnikov (1963), Koiter (1965), Babeshko (1966), V.M. Aleksandrov and Belokon (1967), Wu and Chiu (1967), Babeshko (1969), Alblas and Kuipers (1969), (1970a), (1970b), V.M. Aleksandrov and Smetanin (1971), Pao, Wu and Chiu (1971) and Potelezkho (1971). Perhaps the best overall discussion of the various asymptotics may be found in V.M. Aleksandrov (1968).

6.6 *Contact with finite friction*

A proper account of the effect of a rigid punch indenting a half-plane should take account of the fact that the contact will neither be frictionless, as assumed in §4.5, nor completely adhesive, as in §4.6, nor even in a uniform state of limiting friction, as in §4.7. Instead there will be regions in which there is complete adhesion and others in which there is slip.

Galin (1945), (1961) properly formulated the contact problem with partial slip for a rigid punch with a plane face. He was unable to find a complete solution, but he found an approximate one by using a complicated argument which to-day has little interest. It was not until 1973 that Spence (1973) provided a definitive solution. He considered not only a flat punch, but also a self-similar one with profile $y = y_0 - A|x|^n$ as shown in Figure 6.6.1(b). An outline of his analysis, which is difficult, is given below; for the details the reader may refer to the original paper.

Figure 6.6.1 (a,b) show a flat rigid punch, and one with a self-similar profile, indenting the surface of an elastic body occupying the upper half-plane. Suppose that over a central portion $|x| < c$ the friction is sufficient to prevent slip taking place as the load P is slowly increased, but that in the remaining parts $c < |x| < a$ the friction is limiting so that slip occurs under progressive application of the load. It is assumed that the load is applied sufficiently slowly for static equilibrium to apply at each stage, and is applied monotonically so that slip is always in the same direction. It is not clear at the outset that a solution of this simple form will exist, since there might be more than one adhesive region with interspersed slip regions (and this might well be the case in the corresponding 'unloading' problem) but, as Spence shows, the problem as posed has a unique solution satisfying all the physical conditions.

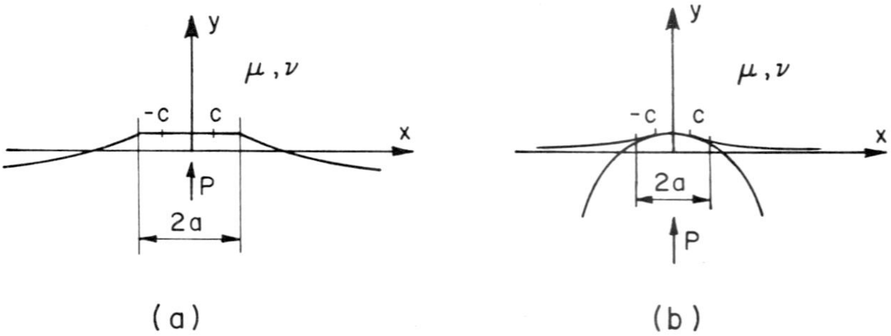

Figure 6.6.1 - Indentation with (a) a flat and (b) a power law punch pressed into an elastic half-plane .

To specify the boundary conditions for the contact we use the notation of §6.2 and write

$$\tau_{xy}(x,0) = -s(x)H(a-|x|) \;, \qquad \tau_{yy}(x,0) = -p(x)H(a-|x|) \;, \tag{6.6.1}$$

where H is the Heaviside function, s(x) is an odd function of x and p(x) an even function. The boundary conditions on the stresses may be expressed by introducing the function

$$r(x) = p(x) - \frac{s(x)}{k}\,\mathrm{sign}(x) \;, \tag{6.6.2}$$

where k is the coefficient of friction. Thus noting that the shearing stress will act outward in the slip regions we have

$$r(x) = 0 \;, \qquad c < |x| < a \;, \tag{6.6.3}$$

and since limiting friction has not been reached in the inner region

$$r(x) \geq 0 \;, \quad |x| < c \;. \tag{6.6.4}$$

The normal displacement derivative is specified throughout the contact region; thus for the polynomial indenter

$$\vartheta^{-1}\frac{dv}{dx} = -B|x|^{n-1}\operatorname{sign}(x) , \tag{6.6.5}$$

where $B = n\vartheta^{-1}A$, and in particular $B = 0$ for the flat punch. The condition governing the lateral displacement is not so straightforward. For the *flat* punch the condition is simply $u = 0$ for $|x| < c$, but for the power-law punch this does not hold as the punch comes into contact progressively and slip at a given x will occur until that point enters the non-slip region. However, by modifying the argument used for totally adhesive contact (see §4.6) and replacing a by c we may deduce that

$$\vartheta^{-1}\frac{du}{dx} = \left\{\begin{array}{ll} 0 , & \text{(flat punch)} , \\ C|x|^{n-1} , & \text{(power-law)} , \end{array}\right\} \quad |x| < c . \tag{6.6.6}$$

In the remaining intervals $c < |x| < a$, $u'(x)$ may not be specified. The values of B and C are eventually to be determined from the conditions that the stresses vanish at the ends of the contact region; for the flat punch, by contrast, the stresses become infinite at the ends.

The final equations are those linking the displacements to the stresses; equations (6.2.43) give

$$\vartheta^{-1}\frac{du}{dx} = \frac{1}{\pi}\int_{-a}^{a}\frac{s(y)dy}{y-x} - \varphi p(x) , \tag{6.6.7}$$

$$\vartheta^{-1}\frac{dv}{dx} = \frac{1}{\pi}\int_{-a}^{a}\frac{p(y)dy}{y-x} + \varphi s(x) . \tag{6.6.8}$$

Spence shows that the stress distributions p(x), s(x) may be obtained by simple quadrature operations from the corresponding functions $p_0(x)$, $s_0(x)$ relating to the flat punch (Example 6.6.1) and the the same unknown c specifying the adhesive region applies to both. This means that it is sufficient to consider the *flat* punch, for which equations (6.6.7) - (6.6.8) give

$$\frac{1}{\pi}\int_{-a}^{a}\frac{p(y)dy}{y-x}+\varphi s(x)=0 \quad , \quad |x|<a \; , \tag{6.6.9}$$

$$\frac{1}{\pi}\int_{-a}^{a}\frac{s(y)dy}{y-x}-\varphi p(x)=0 \; , \quad |x|<c \; , \tag{6.6.10}$$

which must be solved subject to equations (6.6.3), (6.6.4).

First use the fact that p(x) is even and s(x) odd to rewrite equations (6.6.9), (6.6.10) as equations for positive x. Thus, using (6.6.2) we find

$$p(x)+2\left(\frac{\cot\pi\alpha}{\pi}\right)x\int_{0}^{a}\frac{p(y)dy}{y^2-x^2}=r(x) \; , \quad 0<x<a \; , \tag{6.6.11}$$

$$-\varphi^2 p(x)+2\left(\frac{\tan\pi\alpha}{\pi}\right)\int_{0}^{a}\left(\frac{p(y)-r(y)}{y^2-x^2}\right)ydy=0 \; , \quad 0<x<c \; , \tag{6.6.12}$$

where $\tan\pi\alpha = k\varphi$. Equation (6.6.11) can be considered as an equation for p(y) in terms of r(x); it may be reduced to the standard form discussed in §3.5 by making the substitutions

$$\frac{y^2}{a^2}=\frac{1}{2}(1+\eta) \; ; \quad \frac{x^2}{a^2}=\frac{1}{2}(1+\xi) \; , \quad p(y)=y\phi(\eta) \; , \quad r(x)=x\psi(\xi) \; , \tag{6.6.13}$$

after which it becomes

$$\phi(\xi)+\frac{\cot\pi\alpha}{\pi}\int_{-1}^{1}\frac{\phi(\eta)d\eta}{\eta-\xi}=\psi(\xi) \; , \quad -1<\xi<1 \; . \tag{6.6.14}$$

This is an equation of type (3.5.1) with φ of that equation given by

$$\varphi = -i\tan\pi\alpha \; . \tag{6.6.15}$$

Using that analysis we find $\kappa = \exp(-2\pi i\alpha)$,

$$X_0(\zeta) = (\zeta-1)^{-1/2+\alpha}(\zeta+1)^{-1/2-\alpha} \tag{6.6.16}$$

and the general solution of (6.6.14) to be

$$\phi(\xi) = \sin^2\pi\alpha\psi(\xi) - \frac{\sin\pi\alpha\cos\pi\alpha}{\pi} X_0^+(\xi) \int_{-1}^{1} \frac{\psi(\eta)d\eta}{X_0^+(\eta)(\eta-\xi)} + (A\xi+B)X_0^+(\xi) \tag{6.6.17}$$

This will be singular at $\xi = -1$, i.e., $x = 0$, unless $A = 0$ and

$$B = \frac{\sin\pi\alpha\cos\pi\alpha}{\pi} \int_{-1}^{1} \frac{\psi(\eta)d\eta}{X_0^+(\eta)(\eta+1)} \; . \tag{6.6.18}$$

Then

$$\phi(\xi) = \sin^2\pi\alpha\psi(\xi) - \frac{\sin\pi\alpha\cos\pi\alpha}{\pi} \int_{-1}^{1} \frac{\omega(\eta)\psi(\eta)d\eta}{\omega(\xi)(\eta-\xi)} , \tag{6.6.19}$$

where

$$\omega(\xi) = \left(\frac{1-\xi}{1+\xi}\right)^{1/2-\alpha} \tag{6.6.20}$$

When translated into the original variables, equation (6.6.19) becomes

$$p(x) = \sin^2\pi\alpha r(x) - x\frac{\sin 2\pi\alpha}{\pi} \int_{0}^{c} \frac{\omega(y)r(y)dy}{\omega(x)(y^2-x^2)} , \qquad 0 < x < a , \tag{6.6.21}$$

where now $\omega(x) = [(a^2-x^2)/x^2]^{1/2-\alpha}$. Notice that the domain of integration $(0,a)$ has been changed to $(0,c)$ on account of (6.6.3).

Equation (6.6.21) expresses $p(x)$ in terms of $r(x)$, and this expression may be substituted for $p(x)$ in the integral appearing in equation (6.6.12). Spence shows that the integral may be simplified to give the integral equation, valid in $0 < x < c$,

$$(1-\varphi^2)p(x) - \frac{2}{\pi a} g(\alpha)\sin\pi\alpha \int_0^c G(x^2,u^2)\omega(u)r(u)du = 0 , \qquad (6.6.22)$$

where

$$G(x^2,u^2) = \{x^2A(x^2) - u^2A(u^2)\}/(x^2-u^2) , \text{ and}$$

$$A(x^2) = {}_2F_1(1,1/2;1+\alpha;x^2/a^2) , \quad g(\alpha) = \Gamma(1/2+\alpha)/[\Gamma(1+\alpha)\Gamma(1/2)]$$

Equation (6.6.22) may be written in terms of the new variables ξ, $\phi(\xi)$, $\psi(\xi)$ given by

$$x^2 = c^2\xi, \quad (1-x^2/a^2)^{1/2-\alpha}r(x) = \phi(\xi), \quad (1-x^2/a^2)^{1/2-\alpha}p(x) = \psi(\xi); \qquad (6.6.23)$$

then

$$\psi(\xi) = \frac{g(\alpha)\sin\pi\alpha}{\pi(1-\varphi^2)}\left(\frac{c}{a}\right)^{2\alpha} \int_0^1 \eta^{\alpha-1}H(\xi,\eta)\phi(\eta)d\eta , \qquad (6.6.24)$$

where

$$H(\xi,\eta) = (1-c^2\xi/a^2)^{1/2-\alpha}G(c^2\xi,c^2\eta) .$$

A second equation linking ψ and ϕ may be found by solving equation (6.6.21) for $r(x)$, i.e., $\phi(\xi)$, in terms of $p(y)$, i.e., $\psi(\eta)$. The inverse that is bounded on (0,1) is

$$\phi(\xi) = \psi(\xi) + \frac{\cot\pi\alpha}{\pi}\int_0^1 \left(\frac{\xi}{\eta}\right)^{1/2}\left(\frac{1-\xi}{1-\eta}\right)^{1/2-\alpha}\frac{\psi(\eta)d\eta}{\eta-\xi} . \qquad (6.6.25)$$

Equations (6.6.24), (6.6.25) now yield an integral equation of the form

$$\lambda\phi(\xi) = \int_0^1 \eta^{\alpha-1} K(\xi,\eta)\phi(\eta)d\eta \ , \tag{6.6.26}$$

for $\phi(\eta)$, i.e., for the pressure in the adhesive region. A detailed examination of the kernel K shows that for equation (6.6.26) there is a unique value of λ for which the equation has a solution $\phi(\xi)$ which is positive throughout (0,1), as, being related to r(x), it must be. The detailed solution of equation (6.6.26), and the determination of the eigenvalue λ which yields the relation between (c/a), k and φ, is carried out numerically. Figure 6.6.2, reproduced from Spence (1973), shows the contact pressure distribution for three typical values of the friction coefficient. Popov (1963b), (1966), A.S. Solov'ev (1969), and V.M. Aleksandrov (1970) also consider plane problems with finite friction.

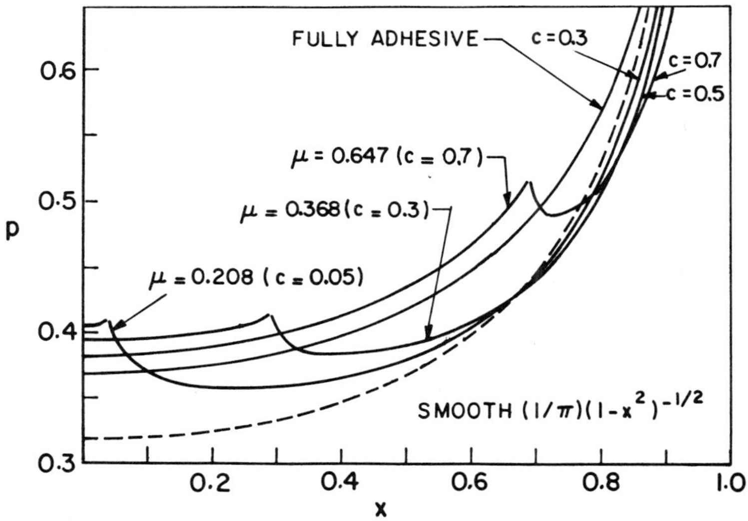

Figure 6.6.2 - Normal pressure under flat two-dimensional punch, $\nu = 0$.

Examples 6.6

1. Show that the stress distributions p(x), s(x) relating to a punch with profile $y = y_0 - A|x|^n$ are related to the corresponding quantities $p_0(x)$, $s_0(x)$ for the flat punch by the equations

$$p(x) = nx^{n-1}\int_x^1 \frac{p_0(t)dt}{t^n}, \quad s(x) = nx^{n-1}\int_x^1 \frac{s_0(t)dt}{t^n}.$$

6.7 *Beams on an elastic half-plane*

The analysis of a beam of *infinite* length in *continuous* contact with an elastic half-plane has been studied since the turn of the century. An early paper by Prager (1927) discusses plates and beams on various foundations, presents a plane analysis and gives references to earlier, mostly German, literature. Biot (1937) obtains equations (6.7.8), (6.7.9) below, gives tabular and graphical results, and considers a beam on a three-dimensional half-space. Reissner (1937) gives a very clear general presentation for a beam on a general foundation and also derives the particular result for an elastic half-plane. A more comprehensive treatment of both beams and axisymmetric plates on elastic half-spaces and Winkler foundations may be found in Marguerre (1937). Further references may be found in Hetenyi (1946) and Vlasov and Leont'ev (1966).

Consider a beam of infinite length lying on an elastic foundation. The governing equation is

$$E_1 I v^{(4)} = q(x) - p(x) , \tag{6.7.1}$$

where p(x) is the foundation pressure and q(x) is the external load, as shown in Figure 6.7.1, and (4) denotes the fourth derivative w.r.t.x. For many foundations, including the Winkler and the elastic half-plane, the foundation pressure is related to the displacement v(x) by an operator equation

$$p(x) = L\{v(t);t \to x\} , \tag{6.7.2}$$

so that the Fourier transform of p(x) is

$$F\{p(x);x \to \xi\} = f(\xi)\overline{v}(\xi) , \tag{6.7.3}$$

where $f(\xi)$ is related to the operator L. The formal Fourier transform of equation (6.7.1) is

$$E_1 I \xi^4 \overline{v}(\xi) = \overline{q}(\xi) - f(\xi)\overline{v}(\xi) , \tag{6.7.4}$$

where $\overline{q}(\xi) = F\{q(x);x \to \xi\}$. Thus

$$\overline{v}(\xi) = \overline{q}(\xi)/[E_1 I \xi^4 + f(\xi)] , \tag{6.7.5}$$

and this gives the formal solution

$$v(x) = F^*\{\overline{q}(\xi)/[E_1 I \xi^4 + f(\xi)];\xi \to x\} . \tag{6.7.6}$$

For an elastic half-plane equation (5.4.10) shows that $f(\xi) = \vartheta|\xi|^{-1}$ so that

$$v(x) = \frac{1}{E_1 I} F^*\left\{ \frac{\overline{q}(\xi)}{\xi^4 + \alpha^3|\xi|} ;\xi \to x \right\} , \tag{6.7.7}$$

where $\alpha^3 = \vartheta/(E_1 I)$. For a concentrated load P at the origin, for which $\overline{q}(\xi) = (2\pi)^{-1/2}P$, we have

$$v(x)-v(0) = \frac{P}{\pi E_1 I} \int_0^\infty \frac{\cos\xi x - 1}{\xi(\xi^3+\alpha^3)} d\xi . \tag{6.7.8}$$

The bending moment is $-E_1Iv''(x)$ and is

$$M(x) = \frac{P}{\pi}\int_0^\infty \frac{\xi\cos\xi x d\xi}{\xi^3+\alpha^3} . \tag{6.7.9}$$

Further applications and consequences of equation (6.7.5), together with graphical results, may be found in the paper by Sneddon, Gladwell and Coen (1975), which, in effect, provides a summary and uniform treatment of much of the earlier research. Other detailed results and references may be found in Vesic (1961), Hetenyi (1964).

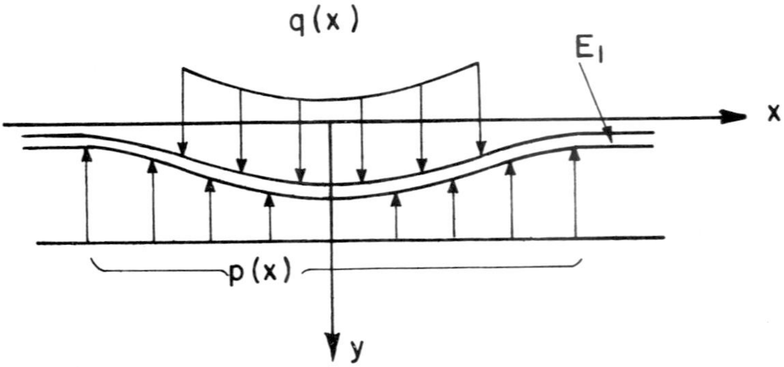

Figure 6.7.1 - Infinite beam in continuous contact with a foundation.

An elegant analysis of beams of *finite* length loaded in various ways and resting in continuous contact with an elastic half-plane was given by Klubin (1952). He proceeded as follows. Consider a uniform beam, with free ends $x = \pm L$, lying on an elastic half-plane in plane strain. The beam deflection $v(x)$ is governed by the equation

$$E_1 I v^{(4)} = q(x) - p(x) \ , \tag{6.7.10}$$

and the end conditions

$$v^{(2)} = 0 = v^{(3)} \ , \quad \text{at } x = \pm L \ , \tag{6.7.11}$$

while the relation between deflection and pressure for the half-space is

$$v(x) = \vartheta_2 \left[-\pi^{-1} \int_{-L}^{L} p(t) \ln|x-t| dt + c \right] . \tag{6.7.12}$$

The displacements for beam and half-plane must be equal in [-L,L]. Assume that the contact pressure has the form

$$p(x) = (L^2-x^2)^{-1/2} \sum_{n=0}^{\infty} A_n T_n(x/L) \ ; \tag{6.7.13}$$

find the displacement v(x) in [-L,L] for the half-plane by using Example 2.3.8 and (6.7.12); the result is

$$v(x) = \vartheta_2 \left[c + \sum_{n=1}^{\infty} n^{-1} A_n T_n(x/L) \right] . \tag{6.7.14}$$

Now integrate equation (6.7.10) with p(x) given by (6.7.13). This gives the beam deflection in the form

$$v(x) = f(x) + \sum_{n=0}^{\infty} A_n [g_n(x) \sin^{-1}(x/L) + h_n(x)(L^2-x^2)^{1/2}] \ , \tag{6.7.15}$$

where f(x) relates to the loading g(x) and will be a fourth degree polynomial when q(x) is constant, and $g_n(x)$, $h_n(x)$ are polynomials. Klubin now expands $\sin^{-1}(x/L)$ and $(L^2-x^2)^{1/2}$ in powers of x/L and equates the expressions (6.7.14) and (6.7.15). The step involving the infinite series is a major drawback for a computer solution; an effective solution may be obtained by suitably modifying the analysis given below for incomplete contact.

Weitsman (1972) correctly formulated the problem of an infinite weightless beam lying, in incomplete contact, on a half-plane, but erred in his solution by assuming that the pressure derivative $p'(x)$ would be zero at the ends $\pm$ a of the contact region; in fact it is $p(x)$ which is zero there, as is usual in incomplete contact problems. The problem of a finite beam, with weight, lying on a half-plane, was solved by Gladwell (1976) by reducing it to an integral equation with a $\ln|x-t|$ singularity, as will now be described.

The beam, shown in Figure 6.7.2, has length 2L, flexural rigidity D_1, and is pressed by its weight W_0 and an additional force P_0 distributed uniformly over $[-b,b]$. Contact occurs over $[-a,a]$. Because the contact pressure $p(x)$ satisfies the condition $p(x) = 0$ for $x = \pm a$ the equilibrium condition may be written

$$P_0 + W_0 = \int_{-a}^{a} p(x)dx = -\int_{-a}^{a} xp'(x)dx \; . \tag{6.7.16}$$

The displacement derivative for the beam may be obtained from Euler beam theory in the form

$$12D_1v'(x) = -\int_{-a}^{a} \{B(x,t)-Pf_j(x,t)t-Wf_3(x,L)t\}p'(t)dt \; , \tag{6.7.17}$$

where

$$B(x,t) = \begin{cases} t^3 + 3x^2t \; , & |t| < |x| \; , \\ (x^3+3t^2x)\text{sign}(x) \; , & |t| \geq |x| \; , \end{cases} \tag{6.7.18}$$

$$j = \begin{cases} 3 \; , \; |x| < b \; , & f_3(x,b) = (x^3+3b^2x)/b \; , \\ 4 \; , \; |x| \geq b \; , & f_4(x,b) = (b^2+3x^2)\text{sign}(x) \; , \end{cases} \tag{6.7.19}$$

$$P = P_0/(P_0+W_0) \; , \qquad W = W_0/(P_0+W_0) \; . \tag{6.7.20}$$

Notice that the terms in (6.7.17) relating to P and W have been included in the integral by using (6.7.16). The corresponding expression for $v'(x)$ for the half-space may be deduced from (6.7.12) and is

$$v'(x) = \pi^{-1}\vartheta_2 \int_{-a}^{a} \ln|x-t|\,p'(t)dt \;, \tag{6.7.21}$$

where ϑ_2 is the modulus for the foundation. Now introduce dimensionless variables

$$(\xi,\tau,\ell,\beta) = (x,t,L,b)/a \;, \tag{6.7.22}$$

$$\lambda = \pi a^3/[12\vartheta_2 D_1] \;, \qquad f(\tau) = -a^2 p'(t)/(P_0+W_0); \tag{6.7.23}$$

then $f(\tau)$ satisfies the equation

$$\int_{-1}^{1} \{\ln|\xi-\tau| + \lambda K(\xi,\tau)\} f(\tau)d\tau = 0 \tag{6.7.24}$$

and the normalizing condition

$$\int_{-1}^{1} \tau f(\tau)d\tau = 1 \;, \tag{6.7.25}$$

where

$$K(\xi,\tau) = B(\xi,\tau) - Pf_j(\xi,\beta)\xi - Wf_3(\xi,\ell)\xi \;. \tag{6.7.26}$$

Equation (6.7.24) has the form (6.3.35), and may be solved by using the approximation described in §6.4. The function $f(\tau)$, being proportional to the pressure *derivative*, will have a square root singularity at $\tau = \pm 1$, even when a is the incomplete contact length corresponding to the elastic constants and the length of the beam. The appropriate form of equations (6.4.27), (6.4.28) will be

$$b_0 \ln 2 - \lambda \sum_{n=0}^{N} d_{0,n} b_n = 0 \ , \tag{6.7.27}$$

$$(2m)^{-1} b_m - \lambda \sum_{n=0}^{N} d_{m,n} b_n = 0 \ , \quad m = 1,2,\ldots,N \ , \tag{6.7.28}$$

where

$$d_{m,n} = (N+1)^{-1} \sum_{k=1}^{N+1} T_m(\tau_k) d_n(\tau_k) \ , \tag{6.7.29}$$

$$d_n(\xi) = (N+1)^{-1} \sum_{j=1}^{N+1} K(\xi,\tau_j) T_n(\tau_j) \ , \tag{6.7.30}$$

and the τ_j are given by (6.4.22). The b's are the coefficients in the expansion

$$f(\tau) = (1-\tau^2)^{-1/2} \sum_{n=0}^{N} b_n T_n(\tau) \ ; \tag{6.7.31}$$

equation (6.7.25) shows that $b_1 = 2/\pi$, and since $p'(x)$ will be an odd function of x, all the even b's will be zero. Equations (6.7.27), (6.7.28) pose an eigenvalue problem for λ. If L,b are given, the solution procedure is as follows. (1) Choose a,W; (2) find the eigenvalues of equations (6.7.27), (6.7.28); (3) check the eigensolutions - the eigenvalue λ must be positive, the contact pressure compressive throughout, and the beam must not bend back to touch the half-space. If this procedure is carried out for a point load, b = 0, it is found that for given values of W and $L/a = \ell$ there are either just two positive eigenvalues or, if W is too large for the given ℓ, so that the contact is no longer incomplete (i.e., p(x) itself has a square root singularity at $x = \pm L$) there is no positive eigenvalue. For values of W smaller that this critical value it is found that the contact stress corresponding to one of the eigenvalues and the assumed contact is not compressive throughout, while that corresponding to the other is always compressive.

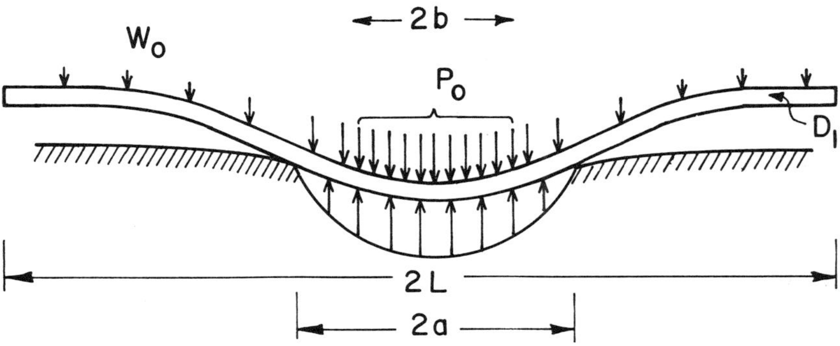

Figure 6.7.2 - Finite beam lying on an elastic half-plane.

The contact length depends only on the 'shape' of the loading - i.e., for given L,b on the ratio $W = W_0/(P_0+W_0)$ - and not on the magnitude of the loading. This agrees with the general results of §4.10. Figure 6.7.3 shows the variation of dimensionless contact length a/L with elastic constant ratio $K = \pi L^3/[12\vartheta_2 D_1]$ for different values of dimensionless weight W. The curve W = 0 is given by $a/L = (0.345K)^{-1/3}$ so that the contact length for a weightless beam under a point load is given by

$$a^3 = 12\vartheta_2 D_1/(0.345\pi) = 11.07\vartheta_2 D_1 \ . \qquad (6.7.32)$$

Weitsman (1972), by using the incorrect form of p(x) at the ends of the contact region, obtained the value 2.41.

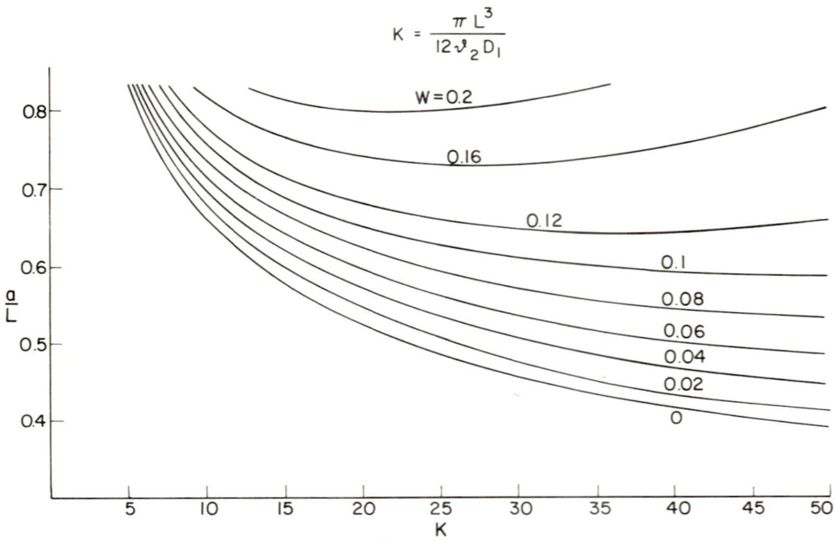

Figure 6.7.3 - The variation of contact length with elastic constant ratio.

6.8 *Receding contact between a strip and a half-plane*

The analysis of §6.7 treated the title problem by modelling the strip as a beam in flexure. Keer, Dundurs and Tsai (1972) considered the strip shown in Figure 6.8.1 as an elastic strip governed by the analysis of §6.3. The boundary conditions for the problem are straightforward, namely

$$\tau_{xy}^{(1)}(x,-h) = 0 \quad , \quad \tau_{yy}^{(1)}(x,-h) = -q(x)H(b-|x|) \ , \qquad (6.8.1)$$

$$\tau_{xy}^{(1)}(x,0) = 0 = \tau_{xy}^{(2)}(x,0) \ , \qquad (6.8.2)$$

$$\tau_{yy}^{(1)}(x,0) = \tau_{yy}^{(2)}(x,0) = -p(x)H(a-|x|) \ , \qquad (6.8.3)$$

$$v^{(1)'}(x,0) = v^{(2)'}(x,0) \quad , \quad 0 \leq |x| < a \; . \tag{6.8.4}$$

There are in addition two other conditions, that the forces p(x) and q(x) be in equilibrium, and that the contact stress be compressive; thus

$$\int_{-a}^{a} p(x)dx = Q_0 = \int_{-b}^{b} q(x)dx \; , \tag{6.8.5}$$

$$p(x) \geq 0 \; , \quad 0 \leq |x| < a \; . \tag{6.8.6}$$

The equations governing the half-plane and the strip are given in Example 5.4.2 and Example 6.3.4 respectively. Thus equations (6.8.2), (6.8.3) for the half-plane give

$$(1-2\nu_2)A_2 + C_2 = 0 \; , \tag{6.8.7}$$

$$2\mu_2[2(1-\nu_2)A_2 + C_2] = P(\xi) \; , \tag{6.8.8}$$

while equations (6.3.12) and Example 6.3.4 for the strip give

$$[(1-2\nu_1)A_1-uA_1+C_1]\exp u + [(1-2\nu_1)B_1+uB_1-D_1]\exp(-u) = 0 \; , \tag{6.8.9}$$

$$(1-2\nu_1)A_1+C_1+(1-2\nu_1)B_1-D_1 = 0 \; , \tag{6.8.10}$$

$$2\mu_1[\{2(1-\nu_1)A_1-uA_1+C_1\}\exp u + \{-2(1-\nu_1)B_1-uB_1+D_1\}\exp(-u)] = Q(\xi) \; , \tag{6.8.11}$$

$$2\mu_1[2(1-\nu_1)A_1+C_1-2(1-\nu_1)B_1+D_1] = P(\xi) \; , \tag{6.8.12}$$

where $P(\xi)$, $Q(\xi)$ are the F_c transforms of p,q respectively, and $u = \xi h$. The displacement derivatives on the line $y = 0$ are

$$v_1'(x,0) = -F_s[V_1(\xi);x] \; , \quad v_2'(x,0) = -F_s[V_2(\xi);x] \; , \tag{6.8.13}$$

where

$$V_1 = (3-4\nu_1)A_1 + C_1 + (3-4\nu_1)B_1 - D_1 \ , \qquad (6.8.14)$$

$$V_2 = (3-4\nu_2)A_2 + C_2 \ . \qquad (6.8.15)$$

Solving these equations (Example 6.8.1), we find

$$V_1 = \vartheta_1 \left\{ -P(\xi)\left[\frac{u+\sinh u\cosh u}{\sinh^2 u-u^2}\right] + Q(\xi)\left[\frac{\sinh u+u\cosh u}{\sinh^2 u-u^2}\right]\right\} \ , \qquad (6.8.16)$$

$$V_2 = \vartheta_2 P(\xi) \ , \qquad (6.8.17)$$

where $\vartheta_i = (1-\nu_i)/\mu_i$. Now equations (6.8.3), (6.8.4) give the pair of integral equations

$$p(x) \equiv \mathcal{F}_c[P(\xi);x] = 0 \ , \quad x > a \ , \qquad (6.8.18)$$

$$\mathcal{F}_s[(\vartheta_1+\vartheta_2)P(\xi) + \vartheta_1\{e(u)P(\xi)-f(u)Q(\xi)\};x] = 0 \ , \quad 0 \leq x < a \ , \qquad (6.8.19)$$

where

$$e(u) = \frac{u+u^2+\sinh u\exp(-u)}{\sinh^2 u-u^2} \ , \quad f(u) = \frac{\sinh u+u\cosh u}{\sinh^2 u-u^2} \ , \qquad (6.8.20)$$

and the kernel has been rearranged so that the second part is convergent for large u; it is also convergent at the lower limit in spite of the fact that both e(u), f(u) have the form $6u^{-3} + 0(u^{-1})$ for small u, since for small ξ

$$P(\xi) = (2/\pi)^{1/2} \int_0^a p(x)\cos\xi x dx = (2/\pi)^{1/2} \int_0^a p(x) \left\{1 - \frac{\xi^2 x^2}{2} \cdots\right\} dx$$

$$= (2/\pi)^{1/2}\{Q_0/2+0(\xi^2)\} \ , \qquad (6.8.21)$$

and because of the equilibrium condition

$$Q(\xi) = (2/\pi)^{1/2}\{Q_0/2+0(\xi^2)\} . \qquad (6.8.22)$$

Now operate on equations (6.8.18), (6.8.19) by operators (7) and (3) respectively of Table 5.7.2. We find

$$A_2^{-1}[p(x);t] = tH_0[P(\xi);t] = 0 , \quad t > a , \qquad (6.8.23)$$

giving

$$P(\xi) = H_0[t^{-1}\phi(t);\xi] = \int_0^a \phi(t)J_0(\xi t)dt , \qquad (6.8.24)$$

and

$$p(x) = F_c[P(\xi);x] = (2/\pi)^{1/2}\int_x^a (t^2-x^2)^{-1/2}\phi(t)dt , \qquad (6.8.25)$$

so that

$$Q_0 = 2\int_0^a p(x)dx = 2(2/\pi)^{1/2}\frac{\pi}{2}\int_0^a \phi(t)dt , \qquad (6.8.26)$$

and from equation (6.8.19) we find

$$(\vartheta_1+\vartheta_2)H_0[P(\xi);s] + \vartheta_1 H_0[e(u)P(\xi)-f(u)Q(\xi);s] = 0 . \qquad (6.8.27)$$

But $H_0[P(\xi);s] = s^{-1}\phi(s)$ and

$$H_0[e(u)P(\xi)-f(u)Q(\xi);s] = \int_0^\infty \xi[e(u)P(\xi)-f(u)Q(\xi)]J_0(\xi s)d\xi$$

$$= \int_0^\infty \xi[e(u)\int_0^a \phi(t)J_0(\xi t)dt-f(u)Q(\xi)\{Q^{-1}(2\pi)^{1/2}\int_0^a \phi(t)dt\}]J_0(\xi s)d\xi$$

$$= -s^{-1}\int_0^a K(s,t)\phi(t)dt , \qquad (6.8.28)$$

where

$$K(s,t) = s \int_0^\infty [f(u)\hat{Q}(\xi)-e(u)J_0(\xi t)]\xi J_0(\xi s)d\xi \qquad (6.8.29)$$

and the bracketed term multiplying $Q(\xi)$ in the second line of (6.8.28) is seen to be unity on account of (6.8.26): $\hat{Q}(\xi) = Q_0^{-1}(2\pi)^{1/2}Q(\xi)$. Equation (6.8.27) now becomes

$$\int_0^a K(s,t)\phi(t)dt = \frac{2}{1+\alpha}\,\phi(s) \ , \qquad (6.8.30)$$

where

$$\alpha = (\vartheta_1-\vartheta_2)/(\vartheta_1+\vartheta_2) \qquad (6.8.31)$$

is one of Dundurs' (1969) mismatch parameters. Equation (6.8.30), which was derived by Keer et al in a slightly different form, is a homogeneous equation for $\phi(t)$. They noted that although it is possible to take given values of ϑ_1, ϑ_2 and a given load distribution $q(x)$ and then find the contact length a, it is much easier to take $q(x)$ and a as given and find the combination of materials for which this a is the contact length. In this approach $2/(1+\alpha)$ is found as an eigenvalue of equation (6.8.30). In a numerical solution of equation (6.8.30), based upon a straightforward approximation of the integral as a weighted sum of values of $\phi(t)$, they found that only one eigenvalue satisfied the essential conditions (1) $-1 \le \alpha \le 1$ and (2) the contact pressure derived from $\phi(t)$ is everywhere compressive. Note that if $\phi(t)$ is finite in $0 < t < a$, then equation (6.8.25) shows that $p(x) \to 0$ as $x \to a^-$. Figure 6.8.2 shows a/h as a function of α for the case in which the load is applied by means of a concentrated force. On the same figure is shown the curve obtained from the equation $a^3 = 11.07\vartheta_2 D_1$ by replacing D_1 by the cylindrical plate bending value $D_1 = h^3/6\vartheta_1$, i.e.

$$(a/h)^3 = 1.65\vartheta_2/\vartheta_1 = 1.65(1-\alpha)/(1+\alpha) \quad . \tag{6.8.32}$$

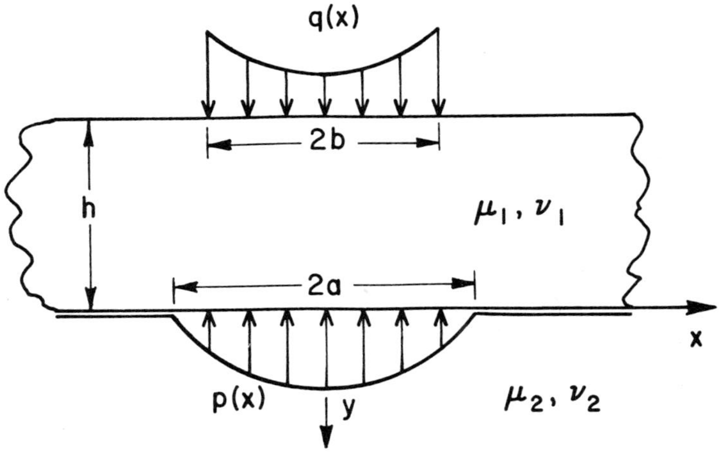

Figure 6.8.1 - Receding contact between a strip and a half-space .

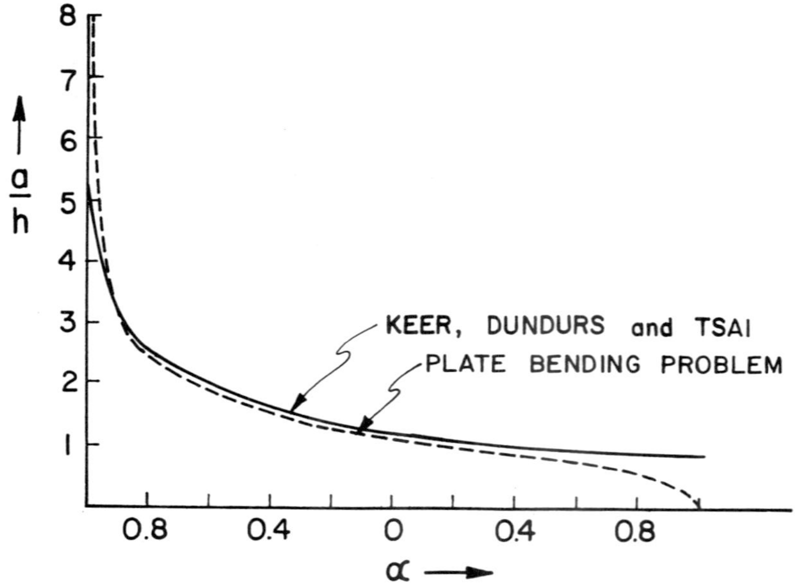

Figure 6.8.2 - Dimensionless contact length a/h varies with mismatch parameter α.

It should be noted that since the loading $q(x)$ enters equation (6.8.29) only in the form of $\hat{Q}(\xi)$, the extent of contact is independent of the level of loading Q_0, in agreement with the general result for receding contact problems (§4.10). In addition, as expected, the elastic constants μ_1, ν_1, μ_2, ν_2 enter only through the parameter α, and it then follows from the general properties of stress fields in plane elastostatics that also the stresses inside the two bodies depend only on this one parameter. When the two materials are identical, so that $\alpha = 0$, the contact length and the stresses are independent of μ, ν.

Tsai, Dundurs and Keer (1972) extended this analysis to the case in which the strip has a slightly curved bottom as shown in

Figure 6.8.3. This problem is interesting on two counts. First, the contact region *advances* with the loading level, unlike the case in which the strip is flat, in which the contact is receding and independent of the level of load. Secondly, the integral equation governing the contact is the nonhomogeneous counterpart of equation (6.8.30), as given in Example 6.8.2. They solve this equation numerically for the case of a parabolic hump which is long enough (i.e., $b/a > 1$) to ensure that contact occurs only in the central region $|x| < a$. The reader is referred to their paper for further details.

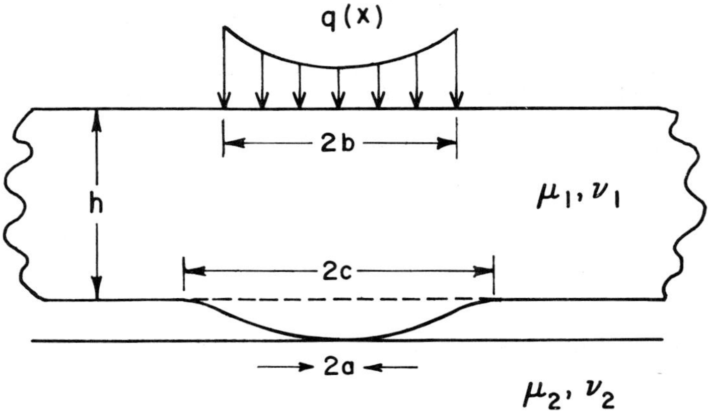

Figure 6.8.3 - Advancing contact for a strip with curved bottom.

Keer and Chantaramungkorn (1972) consider a layer loaded uniformly except for a central portion of length 2b as shown in Figure 6.8.4. They show that as c/h is increased the layer comes away from the half-space at the origin, but a/h is always less then c/h. They analyze the problem in detail and give numerical results for various values of α.

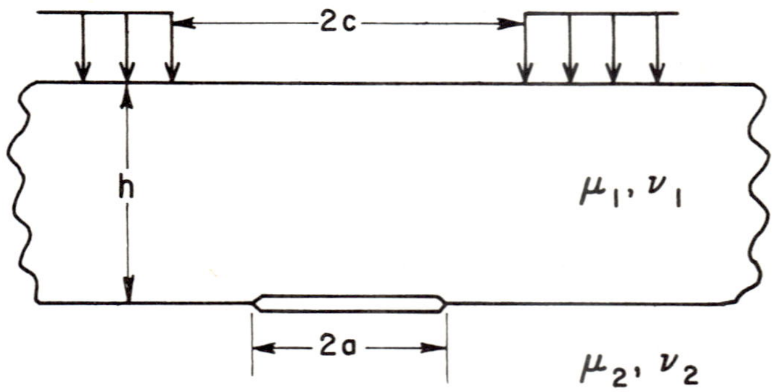

Figure 6.8.4 - The strip is loaded uniformly outside $|x| \le c$.

Tsai, Dundurs and Keer (1974) consider the general case of the problem shown in Figure 6.8.1 in which q(x) is not constrained to by symmetric, and give explicit numerical results for a number of loadings. See also Dundurs, Tsai and Keer (1975).

Examples 6.8

1. Show that the solution of equations (6.8.9)-(6.8.12) is given by

$$4\mu_1 A_1(\sinh^2 u - u^2) = Q(u\exp(-u)+\sinh u) - P\exp(-u)(u\exp u+\sinh u) \ ,$$

$$C_1 = -\frac{(3-4\nu_1)A_1}{2} + \frac{A_1}{2}\left\{\frac{u\exp u+\sinh u}{u\exp(-u)+\sinh u}\right\} + \frac{uP\exp(-u)}{4\mu_1(u\exp(-u)+\sinh u)},$$

$$B_1 = A_1\left\{\frac{u\exp u+\sinh u}{u\exp(-u)+\sinh u}\right\} - \frac{P}{2\mu_1}\left\{\frac{\sinh u}{u\exp(-u)+\sinh u}\right\} \ ,$$

$$D_1 = 4(1-\nu_1)(1-2\nu_1)A_1+(3-4\nu_1)C_1-(1-2\nu_1)P/(2\mu_1) \ .$$

2. Assuming that the hump of Figure 6.8.3 is very shallow in the sense that the gap g(x) between the two bodies satisfies $g(b) \ll b$ and $g(b) \ll h$ show that the governing equations are (6.8.1)-(6.8.6) with the exception that (6.8.4) is

changed to

$$v'^{(1)}(x,0) - v'^{(2)}(x.0) = g'(x) ,$$

and hence show that the final integral equation for $\phi(t)$ is

$$\frac{2}{1+\alpha}\phi(s) - \int_0^a K(s,t)\phi(t)dt = -\frac{1}{\vartheta_1} A_1^{-1}[g'(x);s] .$$

3. Ratwani and Erdogan (1973) consider the case in which the strip and half-plane are pressed together by means of a symmetric rigid punch, as shown in Figure 6.8.5. Show that in this case the governing equations are (6.8.7)-(6.8.15) together with

$$v'(x,-h) = -F_s[V(\xi);x] = g'(x) , \quad 0 \leq x < b ,$$

where $V(\xi) = 2(1-\nu_1)[A_1 \exp u + B_1 \exp(-u)]$. By using Example 6.8.1 show that

$$V(\xi) = \vartheta_1\{[1+e(u)]Q(\xi)-f(u)P(\xi)\} ,$$

so that the governing equations are (6.8.18), (6.8.19) and

$$F_c[Q(\xi);x] = 0 , \quad x > b ,$$

$$\vartheta_1 F_s[f(u)P(\xi)-(1+e(u))Q(\xi);x] = g'(x) , \quad 0 \leq x < b .$$

Now use Example 5.5.3 to reduce these equations to

$$\int_{-b}^{b}\frac{q(t)dt}{t-x} + \int_{-b}^{b} E(t-x)q(t)dt - \int_{-a}^{a} F(t-x)p(t)dt = \pi\vartheta_1^{-1}g'(x) ,$$

$$\frac{\vartheta_1+\vartheta_2}{\vartheta_1}\int_{-a}^{a}\frac{p(t)dt}{t-x} - \int_{-b}^{b} F(t-x)q(t)dt - \int_{-a}^{a} E(t-x)p(t)dt = 0 ,$$

where

$$E(x) = (\pi/2)^{1/2}F_s[e(u);\xi \to x] ,$$

$$F(x) = (\pi/2)^{1/2}F_s[f(u);\xi \to x] .$$

These equations are examined in detail by Ratwani and Erdogan (1973).

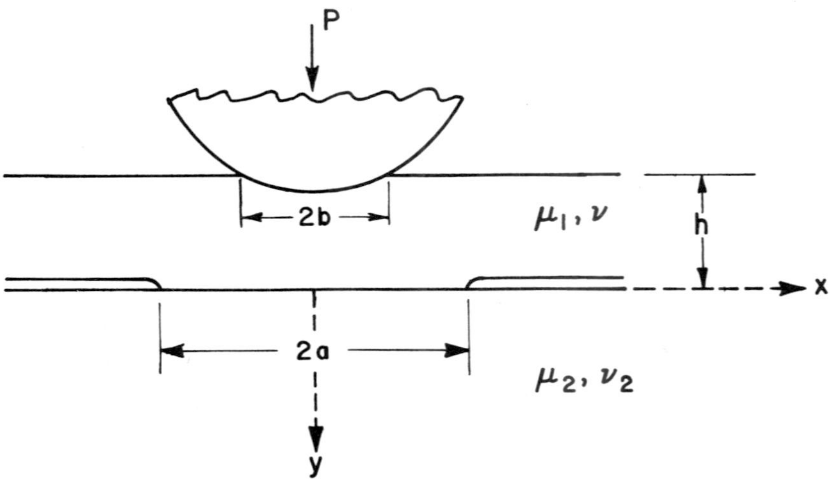

Figure 6.8.5 - A strip and a half-plane are pressed together by a symmetric rigid punch .

6.9 *Further problems concerning a plane strip and a foundation*

This section is devoted to a discussion of some particular problems relating to a plane strip. They may be viewed as examples of how the theoretical results concerning Abel, Fourier, Hankel and Hilbert transforms derived in Chapter 5 may be used to reduce the integral equations governing a typical contact situation into a form amenable to computation.

Keer and Silva (1972) analyze the plane, receding contact problem of a strip lying on a rigid foundation by using Fourier and Abel transforms. The strip, shown in Figure 6.9.1, is initially compressed by a uniform pressure $\tau_{yy}(x,h) = -p$. Then an inextensible membrane that is bonded to the end $x = 0$ lifts the strip, causing it to lose contact with the foundation in $y = 0$, $0 \le x \le \ell$; at $x = \ell$ the contact stress $\tau_{yy}(x,0)$ is zero, as is appropriate for incomplete contact.

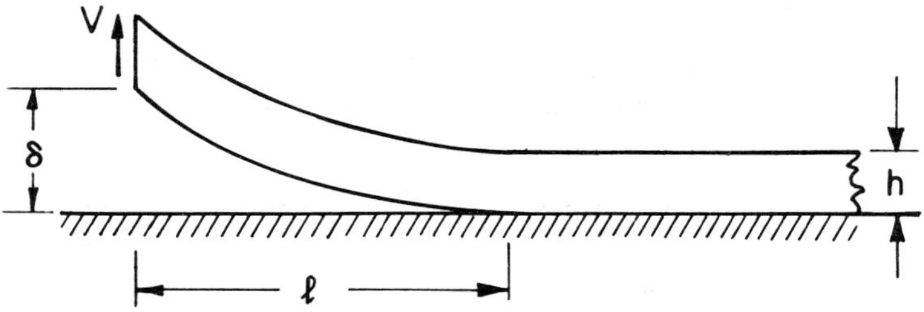

Figure 6.9.1 - Lifting a strip from a rigid foundation.

Keer and Silva seek the displacements of the strip as the superposition of two displacements; the first, given by

$$2\mu u(x,y) = \nu px \quad , \qquad 2\mu v(x,y) = -(1-\nu)py + 2\mu\delta \ , \qquad (6.9.1)$$

corresponds to uniform compression of the strip and a rigid body displacement normal to the plane. The second has the Papkovich-Neuber form used in (6.3.12). In order to pose the problem properly it is necessary to specify certain conditions on the face $x = 0$. They assume that the v displacement is as given above, while the normal stress is zero. Thus

$$2\mu v(0,y) = -(1-\nu)py+2\mu\delta \ , \quad \tau_{xx}(0,y) = 0 \ , \quad 0 \le y \le h \ . \qquad (6.9.2)$$

The remaining boundary conditions are

$$\tau_{xy}(x,h) = 0 = \tau_{xy}(x,0) \;, \quad \tau_{yy}(x,h) = -p \;, \quad 0 \le x < \infty \;, \qquad (6.9.3)$$

$$\tau_{yy}(x,0) = 0 \;, \quad 0 \le x < \ell \;; \quad v(x,0) = 0 \;, \quad \ell < x < \infty \;. \qquad (6.9.4)$$

The conditions (6.9.2) demand that the Papkovich-Neuber solution to be added to (6.9.1) be antisymmetric in the sense of Example 6.3.4 so that

$$B(-\xi) = A(\xi) \;, \quad D(-\xi) = -C(\xi) \;, \quad -\infty < \xi < \infty \;, \qquad (6.9.5)$$

and, in the notation of Example 6.3.4 and equations (6.3.9)-(6.3.11),

$$u(x,y) = F_c[U(\xi,y);\xi \to x] + \nu px/(2\mu) \;, \qquad (6.9.6)$$

$$v(x,y) = -F_s[V(\xi,y);\xi \to x] - (1-\nu)py/(2\mu) + \delta, \qquad (6.9.7)$$

$$\tau_{xy}(x,y) = F_c[T_{12}(\xi,y);\xi \to x] \;, \qquad (6.9.8)$$

$$\tau_{yy}(x,y) = -F_s[T_{22}(\xi,y);\xi \to x] - p \;, \qquad (6.9.9)$$

where U, V, T_{12}, T_{22} are related to A, B, C, D by equations obtained by deleting i everywhere in (6.3.12).

Now apply the boundary conditions on y = h and the condition $\tau_{xy}(x,0) = 0$. The resulting equations may be brought to the dimensionless form

$$\delta^{-1}v(x,0) = 1 - (2\lambda)^{-1}F_s[\tau^{-1}F(\tau);t] \;, \qquad (6.9.10)$$

$$p^{-1}\tau_{yy}(x,0) = -1 + F_s[(1-G(\omega))F(\tau);t] \;, \qquad (6.9.11)$$

where $x = \ell t$, $\ell\xi = \tau$, $\omega = \xi h$,

$$\frac{2(\sinh 2\omega+2\omega)}{-1+2\omega+\exp(2\omega)}\,A(\xi) = \ell p F(\tau)\ , \qquad \frac{\delta}{2p\ell\vartheta} = \lambda\ , \tag{6.9.12}$$

$$G(\omega) = 1 - \frac{(2\sinh^2\omega-2\omega^2)}{\sinh 2\omega+2\omega} = \frac{\omega+\omega^2+\sinh\omega\exp(-\omega)}{\omega+\sinh\omega\cosh\omega}\ . \tag{6.9.13}$$

Now $F(\tau)$ must be chosen so that $v(x,0) = 0$ for $\xi > 1$. Example 5.7.7 shows that if

$$F(\tau) = (2/\pi)^{1/2}\mathcal{H}_0[s^{-1}f(s);\tau]\ , \tag{6.9.14}$$

where $f(s) = 0$ for $s > 1$, then

$$\mathcal{F}_s[\tau^{-1}F(\tau);t] = \int_0^1 f(s)ds \quad \text{for} \quad t > 1\ . \tag{6.9.15}$$

This means that the representation (6.9.14) for $F(\tau)$ will make $v(x,0) = 0$ for $x > \ell$, i.e. $t > 1$, if

$$\int_0^1 f(s)ds = 2\lambda\ . \tag{6.9.16}$$

Now substitute the expression (6.9.14) for $F(\tau)$ in equation (6.9.11) and use Example 5.7.7 to calculate the first integral. The equation $\tau_{yy}(x,0) = 0$ for $0 \le x < \ell$ becomes

$$(2/\pi)^{1/2}A_1[f(s);t] - (2/\pi)\int_0^1 K_1(t,s)f(s)ds = 1\ , \tag{6.9.17}$$

where

$$K_1(t,s) = \int_0^\infty G(\omega)\sin\tau t J_0(\tau s)d\tau\ . \tag{6.9.18}$$

Now invert equation (6.9.17) using the Abel inverse (5.7.26) and Example 5.7.8; the result is

$$f(t) - t\int_0^1 K(s,t)f(s)ds = 1\ , \tag{6.9.19}$$

where

$$K(s,t) = \int_0^\infty \tau G(\omega) J_0(\tau s) J_0(\tau t) d\tau \ , \qquad (6.9.20)$$

may be expressed in terms of $\gamma = h/\ell$ as

$$K(s,t) = \gamma^{-2} \int_0^\infty uG(u) J_0(us/\gamma) J_0(ut/\gamma) du \ . \qquad (6.9.21)$$

The procedure described in equations (6.9.14)-(6.9.21) is followed, more or less, whenever a pair of dual integral equations is to be reduced to a single Fredholm equation. Numerous examples may be found in the literature. Variants occur in (6.3.24)-(6.3.29), (6.8.18)-(6.8.30).

Equation (6.9.19) is a standard non-singular Fredholm integral equation of the second kind which may be solved numerically (see §6.4). Once $f(t)$ has been found, λ may be found from equation (6.9.16). The displacement $\delta^{-1}v(x,0)$ in the free portion $0 \leq \xi \leq 1$ may be expressed in terms of $f(t)$ by using Example 5.7.7; it is

$$\delta^{-1}v(x,0) = 1-(2\lambda)^{-1}\left\{\pi/2 \int_0^\xi f(t)dt + \int_\xi^1 \sin^{-1}(\xi/t)f(t)dt\right\} \ , \qquad (6.9.22)$$

while the contact stress for $\xi > 1$ is given by

$$p^{-1}\tau_{yy}(x,0) = -1+(2/\pi)\int_0^1 (\xi^2-t^2)^{-1/2} f(t)dt + (2/\pi)\int_0^1 K_1(\xi,s)f(s)ds \ . \qquad (6.9.23)$$

It may easily be verified that the conditions

$$v(x,0) = 0 = v'(x,0) = \tau_{yy}(x,0) \ , \quad x = \ell \ , \qquad (6.9.24)$$

will be satisfied by the expressions (6.9.22), (6.9.23).

Keer and Silva report results of a numerical solution of equation (6.9.19) using Simpson's rule to evaluate both the finite integral and the infinite integral in the kernel. As is to be expected, the computation becomes increasingly difficult as γ becomes smaller. They present results for the displacement and contact pressure and for the ratio of the load S applied to the end x = 0 to the deflection δ (Example 6.9.1). They find that the dimensionless ratio $(1-\nu)S\ell^3/[2\mu\delta h^3]$ differs from the value 16/21 deduced from cylindrical plate bending theory by about 2% when $h/\ell = 0.1$.

Civelek and Erdogan (1975) consider the effect of the lifting of a strip under the combined action of an upward point force P_0, the downward pull of the weight of the strip, and a uniform clamping pressure p, as shown in Figure 6.9.2.

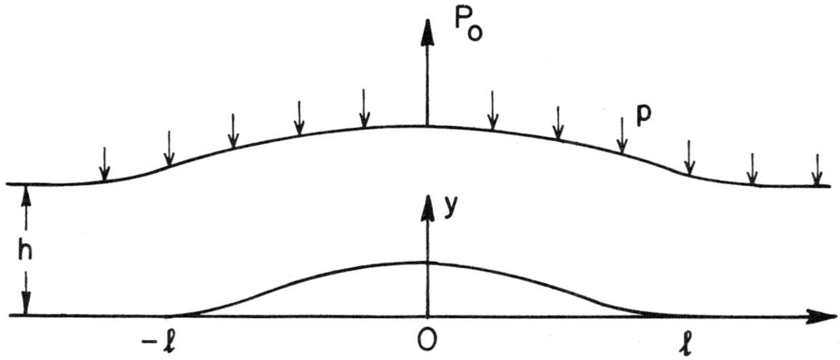

Figure 6.9.2 - Lifting a symmetrical strip.

The solution may be sought as the superposition of two parts, namely the displacements due to the gravity body force and the clamping pressure, which are given by

$$2\mu u_0(x,y) = \nu(p+\rho gh/2)x \ , \qquad (6.9.25)$$

$$2\mu v_0(x,y) = -(1-\nu)py+\rho gy[\nu(y-h)-(1-\nu)h/2] , \qquad (6.9.26)$$

and a Papkovich-Neuber solution of the form (6.3.12). The boundary conditions are

$$\tau_{xy}(x,0) = 0 = \tau_{xy}(x,h) , \quad \tau_{yy}(x,h) = P_0\delta(x)-p , \quad -\infty < x < \infty , \qquad (6.9.27)$$

$$v(x,0) = 0 , \quad |x| > a , \quad \tau_{yy}(x,0) = 0 , \quad |x| < a . \qquad (6.9.28)$$

Symmetry considerations show that u(x,y) will be odd in x, v(x,y) will be even. This means that

$$B(-\xi) = -A(\xi) , \quad D(-\xi) = C(\xi) , \quad -\infty < \xi < \infty , \qquad (6.9.29)$$

and the Papkovich-Neuber displacements may be written as in Example 6.3.4,

$$u(x,y) = F_s[U(\xi,y);\xi \to x] , \qquad (6.9.30)$$

$$v(x,y) = F_c[V(\xi,y);\xi \to x] , \qquad (6.9.31)$$

while the total stresses may be written

$$\tau_{xy}(x,y) = F_s[T_{12}(\xi,y);\xi \to x] , \qquad (6.9.32)$$

$$\tau_{yy}(x,y) = F_c[T_{22}(\xi,y);\xi \to x] - p + \rho g(y-h) , \qquad (6.9.33)$$

where U, V, T_{12} and T_{22} are given by (6.3.12) with i omitted.

Since

$$F_c[1;\xi \to x] = (2\pi)^{1/2}\delta(x) , \qquad (6.9.34)$$

the conditions on the face y = h imply that

$$2C = Q\exp(\omega) + B\exp(2\omega) - (3-4\nu)A - 2\omega A , \tag{6.9.35}$$

$$2D = Q\exp(-\omega) - A\exp(-2\omega) + (3-4\nu)B - 2\omega B , \tag{6.9.36}$$

where $(2\pi)^{1/2}P\delta = Q$. Now the condition $\tau_{xy}(x,0) = 0$ gives

$$A[\exp(-2\omega)-1-2\omega] + B[-1+2\omega+\exp(2\omega)] + Q\sinh\omega = 0 , \tag{6.9.37}$$

so that

$$v'(x,0) = -F_s[A+B;\xi \to x] , \tag{6.9.38}$$

$$\tau_{yy}(x,0) = -F_c[2(1+\nu)A+C-2(1-\nu)B+D;\xi \to x] - p - \rho gh . \tag{6.9.39}$$

There are two cases to consider. If the force P_0 is not sufficiently large, the strip will remain completely in contact with the foundation so that $a = 0$ and $v(x,0) = 0$ for all x. In this case, writing

$$x = ht , \quad \lambda = P_0/[h(p+\rho gh)] , \tag{6.9.40}$$

we find

$$(p+\rho gh)^{-1}\tau_{yy}(x,0) = -1+2\lambda\pi^{-1}\int_0^\infty H(\omega)\cos(\omega t)d\omega , \tag{6.9.41}$$

where

$$H(\omega) = (\sinh\omega+\omega\cosh\omega)/(\sinh 2\omega+2\omega) . \tag{6.9.42}$$

As λ increases from zero this contact pressure will be compressive everywhere until, for a critical value λ_{cr}, the contact pressure will become zero, at the origin. The critical value is given by

$$2\lambda_{cr}\int_0^\infty H(\omega)d\omega = \pi . \tag{6.9.43}$$

Civelek and Erdogan found $\lambda_{cr} = 1.088$.

For $\lambda > \lambda_{cr}$, the strip will come away from the foundation so that a, being nonzero, may be used as the standard length. Put

$$x = at \ , \quad a\xi = \tau \ , \quad \gamma = h/a \ , \quad p+\rho gh = p_e \ , \tag{6.9.44}$$

$$v'(x,0) = \vartheta p_e f(t) \ , \quad \tau_{yy}(x,0) = p_e g(t) \ , \tag{6.9.45}$$

$$-(p_e a)^{-1}(A(\xi) + B(\xi)) = F(\tau) \ ; \tag{6.9.46}$$

then equation (6.9.38) may be written

$$f(t) = F_s[F(\tau);t] \ , \quad 0 \leq \xi < 1 \ , \tag{6.9.47}$$

while (6.9.39) becomes

$$g(t) = F_c[(1+K(\omega))F(\tau);\tau \to t] + (2/\pi)\lambda\gamma \int_0^\infty H(\omega)\cos\tau t d\tau - 1 \ , \tag{6.9.48}$$

where

$$K(\omega) = \{\exp(-2\omega) - (1+2\omega+2\omega^2)\}/(\sinh 2\omega + 2\omega) \ . \tag{6.9.49}$$

Now the terms in $F(\tau)$ may be expressed in terms of $f(\xi)$ by using equation (5.5.12) and Example 5.5.3. Equation (6.9.48) becomes

$$g(t) = \frac{1}{\pi}\int_{-1}^{1} \frac{f(\tau)d\tau}{\tau - t} + \frac{1}{\pi}\int_{-1}^{1} f(\tau)k(\tau-t)d\tau + \frac{2\lambda\gamma}{\pi}\int_0^\infty H(u)\cos\tau t d\tau - 1 \ , \tag{6.9.50}$$

where

$$k(\xi) = \int_0^\infty K(\omega)\sin(\tau\xi)d\tau = \gamma^{-1}\int_0^\infty K(u)\sin(\omega\xi/\gamma)du \ , \tag{6.9.51}$$

and, as usual, $f(\tau)$ has been extended as an odd function into $-1 < \xi < 0$. The equation $g(t) = 0$, $|t| < 1$, now has the form

(6.3.25) and may be solved by the methods outlined in §6.4; the condition of unbonded contact means that f(x) must be sought in the form (6.4.20) under the condition (6.4.36). Figure 6.9.3, reproduced from Civelek and Erdogan (1975), shows λ plotted against a/h, with the curve $\lambda = 4a/(3h)$ obtained from a simple beam solution for comparison. Civelek and Erdogan (1976) consider the effect of a compressive normal force on a strip which lies on a rigid foundation.

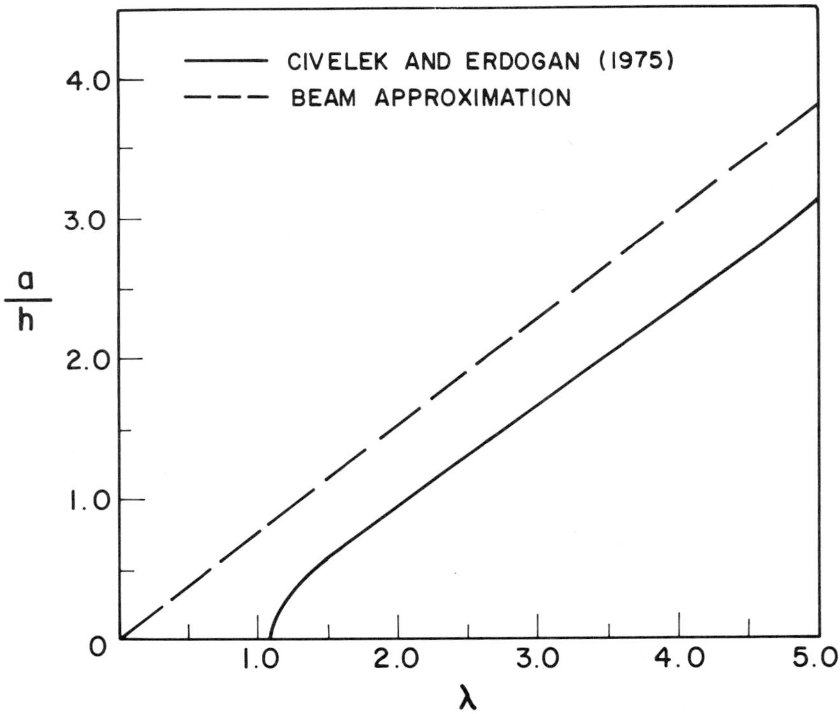

Figure 6.9.3 - a/h is a function of λ

Problems concerning stiffeners bonded to a plate are considered in Chapter 7. A number of papers consider *strips bonded or partially bonded to another strip or to a half-plane;* these include Goland and Reissner (1944), Keer (1974), Chang and Muki (1974), Chantaramungkorn and Keer (1975), and Keer and Chantaramungkorn (1975).

Examples 6.9

1. Show that the total shearing force applied to the end of the strip in Figure 6.9.1 is

$$S = \frac{2\ell p}{\pi}\int_0^1 L(t)f(t)dt$$

where

$$L(t) = \int_0^\infty \omega^{-1}\left(\frac{\sinh^2\omega-\omega^2}{\sinh\omega\cosh\omega+\omega}\right) J_0(\omega t\ell/h)d\omega \ .$$

6.10 *The inhomogeneous half-plane*

The equations governing plane strain in the absence of body forces are the displacement-strain equations (1.2.4), (1.2.5), the strain-stress equations

$$\left.\begin{aligned} &2\mu e_{xx} = (1-\nu)(\tau_{xx}+\tau_{yy})-\tau_{yy}\ , \quad 2\mu e_{yy} = (1-\nu)(\tau_{xx}+\tau_{yy})-\tau_{xx},\\ &2\mu e_{xy} = \tau_{xy}\ , \end{aligned}\right\} \qquad (6.10.1)$$

and the equilibrium equations

$$\frac{\partial\tau_{xx}}{\partial x} + \frac{\partial\tau_{xy}}{\partial y} = 0 = \frac{\partial\tau_{xy}}{\partial x} + \frac{\partial\tau_{yy}}{\partial y}\ . \qquad (6.10.2)$$

The solution of the equilibrium equations may be expressed in terms of the Airy stress function, as in Example 1.11.3. When the medium is homogeneous, so that μ,ν are constant, the compatibility condition of Example 1.11.2 shows that the Airy stress function U is biharmonic. When μ,ν are functions of x,y the compatibility equation yields

$$\frac{\partial^2}{\partial y^2}\left[\frac{1}{\mu}\left\{(1-\nu)\nabla^2 U - \frac{\partial^2 U}{\partial x^2}\right\}\right] + \frac{\partial^2}{\partial x^2}\left[\frac{1}{\mu}\left\{(1-\nu)\nabla^2 U - \frac{\partial^2 U}{\partial y^2}\right\}\right] +$$

$$+ \frac{2\partial^2}{\partial x\partial y}\left\{\frac{1}{\mu}\frac{\partial^2 U}{\partial x\partial y}\right\} = 0 . \tag{6.10.3}$$

In the practically more important case in which ν is constant this becomes (Du Ching Hua (1961), Rostovtsev (1964b))

$$(1-\nu)\nabla^2\left\{\frac{1}{\mu}\nabla^2 U\right\} = \frac{\partial^2}{\partial x^2}\left(\frac{1}{\mu}\right)\frac{\partial^2 U}{\partial y^2} + \frac{\partial^2}{\partial y^2}\left(\frac{1}{\mu}\right)\frac{\partial^2 U}{\partial x^2} - \frac{2\partial^2}{\partial x\partial y}\left(\frac{1}{\mu}\right)\frac{\partial^2 U}{\partial x\partial y} \tag{6.10.4}$$

Considerable attention has been focused on the case in which

$$\mu = my^{\lambda} . \tag{6.10.5}$$

Then

$$\nabla^2(y^{-\lambda}\nabla^2 U) = \frac{\lambda(\lambda+1)}{1-\nu}\, y^{-\lambda-2}\,\frac{\partial^2 U}{\partial x^2} . \tag{6.10.6}$$

Rostovtsev (1964b) solved this equation in terms of Whittaker functions. These functions may be used to give the solution for a half-space $y \geq c > 0$ or a strip $c \leq y \leq d$; see also Golecki (1960) Gibson et al (1971) and Plevako (1973a).

In the special case of the half-plane $y \geq 0$ and the μ-variation (6.10.5) the above-mentioned solution has meaning only when $0 \leq \lambda < 1$. In this case Rostovtsev (1964b) found the state of stress due to tractions on the boundary $y = 0$ by finding the counterparts of Flamant's solution (§2.1), i.e., by finding purely radial states of stress; Lekhnitskii (1962a), (1963b) had shown how μ,ν must vary if such a state is to exist, and $\mu = my^{\lambda}$, $\nu = \nu_0$ has the required type of variation. For a purely radial stress $\tau_{r\theta} = 0 = \tau_{\theta\theta}$ and τ_{rr} must satisfy

$$\frac{\partial}{\partial r}(r\tau_{rr}) = 0 , \tag{6.10.7}$$

giving $\tau_{rr} = r^{-1}F(\theta)$. The corresponding strains are (1.6.13), i.e.,

$$e_{rr} = \frac{(1-\nu)\tau_{rr}}{2\mu} , \quad e_{\theta\theta} = -\frac{\nu\tau_{rr}}{2\mu} , \tag{6.10.8}$$

and the compatibility equation (Example 1.11.5) is

$$\left(\frac{\partial^2}{\partial r^2} + \frac{2}{r}\frac{\partial}{\partial r}\right)e_{\theta\theta} + \left(\frac{1}{r^2}\frac{\partial^2}{\partial\theta^2} - \frac{1}{r}\frac{\partial}{\partial r}\right)e_{rr} = 0 . \tag{6.10.9}$$

Substituting for e_{rr}, $e_{\theta\theta}$ from (6.10.8) and supposing that μ is variable, ν constant, we find

$$(1-\nu)\nabla^2\left(\frac{\Psi}{r}\right) = \frac{1}{r}\frac{\partial^2\Psi}{\partial r^2} , \quad \Psi = \frac{m}{\mu} r\tau_{rr} = \frac{m}{\mu} F(\theta) . \tag{6.10.10}$$

This equation has a closed form solution with arbitrary parameters

$$\Psi(r,\theta) = r^{\alpha}(A\cos q\theta + C\sin q\theta) , \tag{6.10.11}$$

$$q = q(\alpha) = \{(1-\alpha)[1+\alpha\nu/(1-\nu)]\}^{1/2} , \tag{6.10.12}$$

and a more general solution may be found from the Steiltjes integral

$$\Psi = \int r^{\alpha}\{\cos[\theta q(\alpha)]dA(\alpha) + \sin[\theta q(\alpha)]dC(\alpha)\} . \tag{6.10.13}$$

In the case (6.10.5), $\mu = m(r\sin\theta)^{\lambda}$ and

$$F(\theta) = r^{\lambda}(\sin\theta)^{\lambda} r^{\alpha}(A\cos q\theta + C\sin q\theta) \quad , \tag{6.10.14}$$

so that $\alpha = -\lambda$, $q = q(-\lambda)$ and

$$\tau_{rr} = r^{-1}(\sin\theta)^{\lambda}(A\cos q\theta + C\sin q\theta) \ . \tag{6.10.15}$$

This solution may be identified by proceeding as in §2.1. The total force produced by the stress on the semi-circular region around the origin has components in the x- and y- directions given by

$$-P = \int_0^{\pi} \tau_{rr} r\sin\theta d\theta = \int_0^{\pi} \sin\theta(\sin\theta)^{\lambda}(A\cos q\theta + C\sin q\theta)d\theta \ , \tag{6.10.16}$$

$$-S = \int_0^{\pi} \tau_{rr} r\cos\theta d\theta = \int_0^{\pi} \cos\theta(\sin\theta)^{\lambda}(A\cos q\theta + C\sin q\theta)d\theta \ . \tag{6.10.17}$$

Using Gradshteyn and Ryzhik (1965, 3.631), we find

$$-2IP = \pi(A\cos\beta + C\sin\beta) \ , \quad -2JS = \pi(A\sin\beta - C\cos\beta) \ , \tag{6.10.18}$$

where $\beta = \pi q/2$, and

$$I = 2^{\lambda}(\lambda+2)B \ , \quad qJ = 2^{\lambda}(\lambda+1)(\lambda+2)B \ , \quad B = B\left\{\frac{\lambda+q+3}{2} \ , \ \frac{\lambda-q+3}{2}\right\} \ , \tag{6.10.19}$$

and the latter is the Beta function. Thus

$$\pi A = -2[IP\cos\beta + JS\sin\beta] \ , \quad \pi C = -2[IP\sin\beta - JS\cos\beta] \ , \tag{6.10.20}$$

$$\tau_{rr} = -\frac{2\sin^{\lambda}\theta}{\pi r}\left\{IP\cos q\left(\frac{\pi}{2} - \theta\right) + JS\sin q\left(\frac{\pi}{2} - \theta\right)\right\} \ . \tag{6.10.21}$$

In the homogeneous case $\lambda = 0$, $q = 1 = I = J$ and

$$\tau_{rr} = -\frac{2}{\pi r}\{P\sin\theta + S\cos\theta\}, \tag{6.10.22}$$

in agreement with (2.1.11) and Example 2.1.1. Thus (6.10.21) gives the stress distribution in the upper half-plane due to a force with components S, P acting at the origin. The displacements, found from (6.10.21), are

$$u_r = \frac{1-\nu}{\pi m} \cdot \frac{r^{-\lambda}}{\lambda} \left\{ IP\cos q\left(\frac{\pi}{2} - \theta\right) + JS\sin q\left(\frac{\pi}{2} - \theta\right) \right\} , \qquad (6.10.23)$$

$$u_\theta = \frac{1-\nu}{\pi m} \cdot \frac{qr^{-\lambda}}{\lambda(1+\lambda)} \left\{ IP\sin q\left(\frac{\pi}{2} - \theta\right) - JS\cos q\left(\frac{\pi}{2} - \theta\right) \right\} . \qquad (6.10.24)$$

From these we may deduce the Cartesian displacements, u,v on the axis y = 0 due to a distribution of stresses

$$\tau_{yy}(t,0) = -p(t) , \quad \tau_{xy}(t,0) = -s(t) , \qquad (6.10.25)$$

$$u(x,0) = \int_{-\infty}^{\infty} \{g_{11}s(t) + g_{12}\operatorname{sign}(x-t)p(t)\}|x-t|^{-\lambda}dt , \qquad (6.10.26)$$

$$v(x,0) = \int_{-\infty}^{\infty} \{-g_{12}\operatorname{sign}(x-t)s(t) + g_{22}p(t)\}|x-t|^{-\lambda}dt , \qquad (6.10.27)$$

$$g_{11} = \frac{(1-\nu)J\sin\beta}{\pi m\lambda} , \quad g_{12} = \frac{(1-\nu)I\cos\beta}{\pi m\lambda} , \quad g_{22} = \frac{(1-\nu)qI\sin\beta}{\pi m\lambda(1+\lambda)} . \qquad (6.10.28)$$

Rostovtsev (1964b) uses these results to find the corresponding results for a nonhomogeneous *half-space*, for which see §12.5.

Since the integrals (6.10.26), (6.10.27) have considerable interest in soil mechanics we examine them for a number of special cases. If the *loading is normal and uniform* over [-b,b] and if $0 < \lambda < 1$, then the displacement v(x,0) is

$$v(x,0) = g_{22}p_0 \int_{-b}^{b} |x-t|^{-\lambda}dt , \qquad (6.10.29)$$

so that

$$v(x,0) = \begin{cases} g_{22}p_0[(b-x)^{1-\lambda}+(b+x)^{1-\lambda}]/(1-\lambda) , & |x| \le b , \\ g_{22}p_0[(|x|+b)^{1-\lambda}-(|x|-b)^{1-\lambda}]/(1-\lambda) , & |x| > b . \end{cases} \tag{6.10.30}$$

There are three special cases:

(i) $\lambda = 0$, the *homogeneous medium*. Here the passage from (6.10.22) to (6.10.23), (6.10.24) is inadmissable, and the normal displacement is given in Example 2.1.6.

(ii) $\lambda = 1$, $\nu \neq 1/2$. This is a *compressible medium* in which the shear modulus is directly proportional to the depth. In this case, the integral (6.10.29) is singular when $|x| < b$; however $v(x,0) - v(0,0)$ may be calculated from (6.10.29) as a Cauchy principal value, i.e.,

$$v(x,0)-v(0,0) = g_{22}p_0\left\{\int_{-b}^{x-\varepsilon}\frac{dt}{x-t} + \int_{x+\varepsilon}^{b}\frac{dt}{t-x} - \int_{-b}^{-\varepsilon}\frac{dt}{-t} - \int_{\varepsilon}^{b}\frac{dt}{t}\right\} =$$

$$= g_{22}p_0\ell n(1-x^2/b^2) ; \tag{6.10.31}$$

when $|x| > b$,

$$v(x,0) = g_{22}p_0\ell n[(|x|+b)/(|x|-b)] ; \tag{6.10.32}$$

when $\lambda = 1$, the expression (6.10.28) reduces to $g_{22} = \varphi/(2m)$. Equations (6.10.31), (6.10.32) appear in Awojobi and Gibson (1973).

(iii) $\lambda = 1$, $\nu = 1/2$. This is the so-called *'Gibson' medium* In this case we may take the limit of (6.10.30) as $\lambda \to 1-$. The limit of $g_{22}/(1-\lambda)$ as $\lambda \to 1-$ is $1/(2m)$, giving

$$v(x,0) = \begin{cases} p_0/(2m) & |x| < b \\ 0 & |x| > b . \end{cases} \tag{6.10.33}$$

Gibson (1967) pointed out that, since the principle of superposition holds, equation (6.10.33) implies that:

(a) the medium behaves exactly like a Winkler foundation (§2.10), in the sense that, if an arbitrary normal load $p(x)$ is applied to the boundary $y = 0$, then it is related to the boundary displacement by the equation

$$p(x) = kv(x,0) \quad , \quad k = 2m \ . \tag{6.10.34}$$

He also found that:

(b) the stress distribution (but not the displacement field) in the Gibson medium due to normal loading on the boundary is exactly the same as that in a homogeneous medium. Calladine and Greenwood (1978), in a lucid paper, showed that (b) could be proved simply by considering equation (6.10.21). When $\lambda = 1$, $\nu = 1/2$, then $q = 0$ and $I = 1$ and the stress field due to a normal load P at the origin is

$$\tau_{rr} = -2P\sin\theta/(\pi r) \ , \tag{6.10.35}$$

which is the same as that given by (6.10.22) with $S = 0$. Calladine and Greenwood started from the purely radial displacement field

$$u_r = P/(2\pi m r) \ , \quad u_\theta = 0 \ , \tag{6.10.36}$$

which is given by (6.10.23)-(6.10.24) in this case.

The surface settlement $v(x,0)$ for the general case $\nu = \nu_0$, $\mu = \mu_0 + my$ was given by Gibson, Brown and Andrews (1971), P.T. Brown and Gibson (1972) and in greater depth by Awojobi and Gibson (1973). Awojobi (1976) analyzed a layer of depth h in which $\mu = \mu_0 h/(h-y)$. The behaviour of a transversely isotropic Gibson medium in plane strain was discussed by Gibson and Kalsi

(1974), Gibson and Sills (1975) and by Calladine and Greenwood (1978). The axis of transverse isotropy is taken to be normal to the surface (the y-axis), the plane strain to be in the x,y-plane. Now there are two shear moduli μ, μ' both of which are proportional to y, i.e., $\mu = my$, $\mu = n\mu'$. Such a medium has property (a) above, but now k depends on n as well as m; property (b) no longer holds.

6.11 *The method of orthogonal polynomials*

In many cases the solution of a contact problem may be reduced to the solution of one or more integral equations. The unknown(s) in the integral equation(s) is (are) usually the contact stress(es) required to give the known values of the displacements in the contact region. One of the methods that has already been used in preceding chapters, and will be used below, to obtain a solution of the integral equation, is to find basic stress distributions for which the integrals (i.e., the displacements) may be explicitly computed. The method has been intensively developed, especially by G. Ia. Popov.

In the simplest case, the frictionless isotropic half-plane, the integral equation is given in Example 2.1.4, namely

$$v(x) = -\frac{\vartheta}{\pi}\int_{-a}^{a} \ln\left|\frac{x-x'}{a}\right| p(x')dx' \ , \tag{6.11.1}$$

for which the appropriate integral relation is

$$-\frac{1}{\pi}\int_{-1}^{1}\frac{\ln|x-t|T_n(t)dt}{(1-t^2)^{1/2}} = \begin{cases}\ln 2 & , \quad n = 0 \ , \\ n^{-1}T_n(x) \ , & n \neq 0 \ , \end{cases} \tag{6.11.2}$$

as given by Popov (1963a) and discussed in Chapter 2. The kernel $\ln|x-t|$ is an example of what Popov (1963a), (1969) calls a polynomial-kernel, or in short, a π-kernel. Popov gives a number of definitions, but for our purposes it is sufficient to say that $\Pi(x,t)$ is a π-kernel if there is a set of polynomials $p_n(x)$ such that

$$\int_a^b \Pi(t,x)g(t)p_n(t)dt = \sigma_n h(x)p_n(x) \quad , \quad n = 1,2,\ldots \; . \tag{6.11.3}$$

Here $p_n(x)$ has degree n, and g(x), h(x) are functions independent of n. For the frictionless contact problem in the nonhomogeneous half-plane, equation (6.10.27) gives

$$v(x,0) = g_{22}\int_{-a}^{a} \frac{p(t)dt}{|x-t|^{\lambda}} \; , \tag{6.11.4}$$

for which Popov (1963a) found the integral relation

$$\frac{1}{\pi}\int_{-1}^{1} \frac{C_n^{\lambda/2}(t)dt}{|x-t|^{\lambda}(1-t^2)^{(1-\lambda)/2}} = \frac{\Gamma(n+\lambda)}{\cos(\pi\lambda/2)\Gamma(\lambda)n!}\, C_n^{\lambda/2}(x) \quad , \quad |x| \le 1 \; , \tag{6.11.5}$$

valid for $0 \le \lambda < 1$. Here $C_n^{\lambda/2}(x)$ is a Gegenbauer polynomial. The relation (6.11.2) may be obtained from (6.11.5) by using the limits

$$\lim_{\lambda\to 0}\left[\left\{\frac{1}{|x-t|^{\lambda}} - 1\right\}\frac{1}{\lambda}\right] = \ln\left\{\frac{1}{|x-t|}\right\} , \quad \lim_{\lambda\to 0}\Gamma(\lambda/2)C_n^{\lambda/2}(x) = \frac{2}{n}\,T_n(x) \; . \tag{6.11.6}$$

Popov (1963a) exhibits (6.11.5) as a special case of a more general relation, valid for $0 \le \lambda < 1$, $\mu > -1$,

$$\int_0^1 (1-t^2)^{-\omega}t^{\mu+1}W_{\mu}^{\lambda}(t,x)P_n^{(\mu,-\omega)}(1-2t^2)dt = \mu_n x^n P_n^{(\mu,-\omega)}(1-2x^2) \; , \tag{6.11.7}$$

where $\omega = (1-\lambda)/2$, $P_n^{(\mu,-\omega)}$ is the Jacobi polynomial, and

$$W_\mu^\lambda(t,x) = \int_0^\infty \xi^\lambda J_\mu(\xi t)J_\mu(\xi x)d\xi \ , \qquad (6.11.8)$$

$$\mu_n = \frac{(-)^n 2^{\lambda-1}\pi\Gamma(1+n+\mu-\omega)}{\sin\pi\omega\Gamma(1+n+\mu)\Gamma(\omega-n)n!} \ . \qquad (6.11.9)$$

The case $\lambda = 0 = \mu$ gives the result (2.6.12), (2.6.13), since

$$P_m^0(t) = P_m^{(0,1/2)}(1-2t^2) = P_{2m}[(1-t^2)^{1/2}] \ , \qquad (6.11.10)$$

while the case $\lambda = 0$, $\mu = m$ appears in §10.5 (see Example 10.3.4 and equations (10.5.25)-(10.5.27); the polynomial $P_n^{(m,-1/2)}$ is there expressed in terms of the Legendre function P_n^m).

The equation governing contact with limiting friction on the isotropic half-plane (Popov (1963b)) can be obtained from Example 2.1.4 by putting $s(x) = kp(x)$; it is

$$\frac{2\mu}{(1-2\nu)k} v(x) = \int_{-\infty}^{\infty} \left\{\frac{1}{2}\,\text{sign}(x-x') - \frac{1}{\pi k\varphi}\,\ell n\left|\frac{x-x'}{a}\right|\right\} p(x')dx' \ , \qquad (6.11.11)$$

for which Popov (1966), (1968) found the appropriate integral relation, for $|x| \le 1$,

$$\int_{-1}^{1} \left\{\frac{1}{2}\,\text{sign}(x-t) - \frac{\cot\pi\alpha}{\pi}\,\ell n|x-t|\right\} \frac{P_{n,\alpha}(t)dt}{\phi_\alpha(t)} = \mu_n P_{n,-\alpha}(x) \ , \qquad (6.11.12)$$

where now $P_{n,\alpha}(t) = P_n^{(-\gamma,-\delta)}(t)$, $\phi_\alpha(t) = (1-t)^\gamma(1+t)^\delta$ and $\gamma = 1/2 - \alpha$, $\delta = 1/2 + \alpha$, $|\text{Re}(\alpha)| < 1/2$, $n = 0,1,2,\ldots$,

$$2\mu_0 = \pi\sec\pi\alpha - 2\,\text{cosec}\,\pi\alpha[\ell n2 + \psi(\delta) - \psi(1)] \ , \ \mu_n = (n\sin\pi\alpha)^{-1} \ , \qquad (6.11.13)$$

where ψ is Euler's psi function. The equation governing the corresponding problem in the nonhomogeneous half-plane of §6.10 is

$$v(x) = \int_{-\infty}^{\infty} \left\{ g_{22} - kg_{12}\,\text{sign}(x-t) \right\} \frac{p(t)dt}{|x-t|^{\lambda}} , \qquad (6.11.4)$$

for which Popov (1967a), (1967b) obtained a generalization of his earlier results; see also Popov (1964), (1967c), (1970), (1972a), (1973a).

CHAPTER VII - STIFFENED PLATES

"The method employed I would gladly explain,
While I have it so clear in my head,
If I had but the time and you had but the brain -
But much yet remains to be said."
The Hunting of the Snark

7.1 *Introduction*

Problems concerning the diffusion of load from a stiffener or stringer to an elastic plate have been the subject of numerous researches because of their importance to the design of aircraft structures.

Research on the subject effectively began with two papers by Melan (1932a), (1932b). In the first, discussed in §5.8, he derived the stress distribution in the upper half-plane when a unit concentrated in-plane force acts at an interior point and the boundary $y = 0$ is stress-free. This result was later repeatedly used to obtain the effect on a plate of the shearing stresses arising from the attachment of a stiffener, as described below. In the second, he solved two basic problems for a plate with an infinite, thin straight stiffener. The first problem concerns the response of a plate occupying the upper half-plane when the complete boundary $y = 0$ is stiffened and an in-plane force is applied at the origin in the x-direction, as shown in Figure 7.2.1. The second, shown in Figure 7.2.2, concerns a plate occupying the whole x-y plane and stiffened along the x-axis. The analyses for the two problems are formally identical and either has since been termed 'Melan's problem'. In both problems the stiffeners act only in tension.

Reissner (1940) was the first to formulate the third basic problem for a stiffened plate, shown in Figure 7.7.1. Here the load due to a force applied normal to the boundary is diffused through the plate by means of a transverse stringer, i.e., one which is normal to the boundary. Reissner assumed that the stringer was thin, so that the loading was applied over a line, and he recognized the similarity between the resulting singular integral equation and that appearing in Prandtl's lifting line theory in aerodynamics. He did not solve the equation, nor did he conclusively determine the boundary conditions on the shearing stress to be applied at the end of the stringer and at the point where it is first attached; instead he argued on intuitive grounds that no concentrated loads can be applied at these points.

Subsequent papers can be divided generally into two groups: those concerned primarily with the mathematical analysis of some generalization of one of the three basic problems, and those attempting to elucidate the mechanics of the load diffusion, and in particular, to answer the questions raised by Reissner. In a certain sense it is, as often, difficult to separate the mathematics from the mechanics; a thorough analysis of the governing integral equations for the different configurations is the only proper way to clarify the mechanics.

The first of the primarily mathematical papers is that by E.L. Buell (1948); he considered Melan's first problem for a semi-infinite stiffener, i.e., one occupying the half-line $x > 0$, as shown in Figure 7.4.1. Using a potential solution of the governing biharmonic equation of plane elasticity and a conformal mapping of the entire x-y plane onto the interior of the unit circle, he obtained an approximate Fourier series solution. A greatly simplified version of his analysis appears in §7.4. Buell's results were extended by E.H. Brown (1957) to other possible loadings of the half-plane. Koiter (1955) noted that although Buell's analysis was entirely adequate for practical purposes, it is not rigorous as he does not establish the convergence of the solution of his governing infinite system of equations. Koiter therefore proceeds to obtain a rigorous solution of the semi-infinite versions of Melan's two problems, basing his analysis on the Mellin transform method of §9.2. Kalandia (1969b) gave an elegant solution of the problem based on Muskhelishvili's methods, as discussed in §7.4. Vorob'ev and Popov (1970) solved the mathematically more difficult problem of a unit load applied not at the *end* of the semi-infinite stiffener, but at some arbitrary distance from the end.

Pflüger (1947) was the first to formulate Melan's (first) problem for a thin stiffener of finite length (and he allowed the stiffener thickness to vary). He manipulated its governing integral equation so that it had exactly the same form as that appearing in Prandtl's lifting line theory, and, although he did not solve the equation, he considered the effect of different shearing stress distributions on the plate. Benscoter (1949) independently formulated Melan's (second) problem and used a numerical method devised for the aerodynamic equation to solve it. His formulation appears in §7.3; there the equation is solved by using Chebyshev polynomials, and Benscoter's numerical results are verified. Glover (1950), in commenting on Benscoter's paper noted that if it is assumed that the shearing forces communicated to the plate are distributed uniformly across the width of the stiffener, rather than applied along the centre line, then the kernel of the integral equation has a simple jump discontinuity at $x = t$, rather than a Cauchy singularity $(x-t)^{-1}$. In this case there are concentrated loads transmitted to the plate at the ends of the stiffener.

During the period 1938-1948 a detailed and essentially separate literature on load diffusion was built up in the Reports and Memoranda of the British Aeronautical Research Council. In this work, by H.L. Cox, E.H. Mansfield, J.H. Argyris, and others, Melan's and Reissner's problems were formulated for various configurations of plates, both when the plates were modelled by using the plane stress solution, and by using a simplified 'sheet' model. In particular the basic response functions for in-plane loads applied to infinite and semi-infinite plates were derived, and several 'inverse' or design solutions, i.e., the shape of stiffener cross-section corresponding to given stress distributions, were found. Reference to this literature may be found in the review paper by Mansfield (1948).

Goodier and Hsu (1954) returned to Reissner's problem. They first considered two plates in (generalized) plane stress bonded together over some region D. They showed that if the Poisson's ratios ν_1, ν_2 of the two plates are equal then the bond forces will be distributed only on the boundary of the contact region D, while if $\nu_1 \neq \nu_2$ there will be a distribution throughout D in addition to the line load on the boundary of D. They then obtained explicit expressions for the (in-plane) strains and stresses just inside D, and the line distribution on D, in terms of the stresses on the part of the plate just outside D. (This kind of problem was discussed in definitive form by Muki and Sternberg (1968b)). They then applied this analysis to Reissner's problem, but made simplifying assumptions. They assumed that the stringer applies a line-load to the plate, although, as they note in their closure of discussion raised by Krahn (1955), when the stringer and plate have identical Poisson's ratios, as they assume, the load will be applied as two parallel line loads along the two edges of the stiffener. They set up an integral equation for the stringer tension $p(y)$, use their general theory to specify the concentrated load applied at the point where plate and stiffener meet, but use an extraneous assumption, $p'(a) = 0$, to provide a needed condition at the end of the stringer. In spite of these assumptions they obtain fair agreement with their experimental results. Krahn (1955), in commenting upon the paper, suggests that the shear loading should be distributed uniformly across the width of the stringer, and he exhibits the integral equation corresponding to this assumption. Goodier and Hsu, however, state that such a distribution was considered but that the results obtained did not agree well with experiment. Lefevre, Mudge and Dickie (1966) returned effectively to Krahn's model, set up the integral equation and found the loads applied to the plate at the two ends, or what is equivalent, the discontinuities in tension $p(y)$ at the ends $y = 0$, $y = \ell$. However, their derivation of the appropriate end conditions is cumbersome, and their analysis is marred by an elementary algebraic error which affects the numerical results. The definitive study of Reissner's problem, including a comparison between results obtained on the basis of the various loading assumptions and from experiment, was carried out by Muki and Sternberg (1968a); these results will be discussed in detail below.

After Benscoter (1949), Melan's problems were taken up by Büfler (1959), (1961), (1964), Budiansky and Wu (1961), Tamate and Kamada (1963), and Muki and Sternberg (1967). Büfler generally discussed a variety of stiffener and load configurations while Muki and Sternberg gave an analysis of the various possible assumptions concerning the way in which the load is diffused in the basic configuration of infinite edge stiffener. These papers will be discussed below.

More complex three-dimensional problems of load diffusion have been analyzed by Muki and Sternberg in a series of papers (1968b), (1969), (1970); these and others are summarized in Sternberg (1970).

7.2 *Melan's original problems*

The stiffener, of Young's modulus E_1 and area of cross-section A, is attached to the boundary $y = 0$ of a semi-infinite plate of thickness h in generalized plane stress. Let $q(x)$ be the longitudinal force in the stiffener, $s(x)$ the shearing stress between stiffener and plate, and $f(x)$ the force per unit length applied to the stiffener, as shown in Figure 7.2.1. It is assumed that the bending stiffness of the stiffener, and hence the normal pressure applied to the stiffener, may be neglected, so that the stiffener is in a uniaxial state of stress. Figure 7.2.1 shows that

$$\frac{dq}{dx} + f(x) = hs(x) \; ; \tag{7.2.1}$$

the longitudinal strain in the stiffener is

$$\varepsilon^{(1)}(x) = q(x)/(AE_1) \; . \tag{7.2.2}$$

The displacements in the upper half-plane due to the shear stress $\tau_{xy}(x,0) = -s(x)$ may be obtained from Example 5.4.1, replacing ν by $\nu^* = \nu/(1+\nu)$ for generalized plane stress; thus

$$E_2 u(x,0) = 2F^*[|\xi|^{-1}S(\xi);x] \quad , \quad S(\xi) = F[s(x);\xi] \; , \tag{7.2.3}$$

so that

$$E_2\varepsilon^{(2)}(x,0) = -2iF^*[S(\xi);x] \; . \tag{7.2.4}$$

But since stiffener and plate are joined, $\varepsilon^{(1)}(x) = \varepsilon^{(2)}(x,0)$ so that, substituting $q(x) = AE_1\varepsilon^{(2)}(x,0)$ in (7.2.1), we obtain

$$f(x) = F^*[\{h+2AE_1|\xi|/E_2)\}S(\xi);x] \; , \tag{7.2.5}$$

which may be inverted to give $S(\xi)$. If the total force P is applied by means of a uniform force per unit length $f = P/(2a)$ applied over $[-a,a]$ then

$$\{h+2AE_1|\xi|/E_2\}S(\xi) = F[f(x);\xi] = (2\pi)^{-1/2}P\sin a\xi/(a\xi) \; . \tag{7.2.6}$$

In particular if $a \to 0$ so that the force becomes a point force at the origin then

$$\{h + 2AE_1|\xi|/E_2\}S(\xi) = (2\pi)^{-1/2}P \,. \tag{7.2.7}$$

Introduce the length $\ell = 2AE_1/E_2h$, and put $x = \ell\bar{x}$, $\bar{\xi} = \ell\xi$; the shear stress is

$$s(x) = F_c[S(\xi);x] = \frac{PE_2}{2\pi AE_1}\int_0^\infty \frac{\cos\bar{\xi}\bar{x}}{1+\bar{\xi}}\,d\bar{\xi} \,. \tag{7.2.8}$$

This integral may be expressed in terms of

$$si(\bar{x}) = -\int_x^\infty \frac{\sin t}{t}\,dt \,, \qquad Ci(\bar{x}) = -\int_x^\infty \frac{\cos t}{t}\,dt \,, \tag{7.2.9}$$

for which, see, for example, Abramowitz and Stegun (1972). Then

$$s(x) = \frac{-PE_2}{2\pi AE_1}\left\{\cos\bar{x}\,Ci(\bar{x}) + \sin\bar{x}\,si(\bar{x})\right\}, \tag{7.2.10}$$

and the stress $\tau_{xx}(x,0)$ at the surface, deduced from Example 5.4.1, is

$$\tau_{xx}(x,0) = \frac{PE_2}{\pi AE_1}\left\{\cos\bar{x}\,si(\bar{x}) - \sin\bar{x}\,Ci(\bar{x})\right\} \,. \tag{7.2.11}$$

Example 7.2.3 shows how the stresses decay away from the edge. Figure 7.2.2 shows the dimensionless quantities

$$-AE_1 s(x)/(PE_2) \,, \qquad AE_1\tau_{xx}/(PE_2) \,.$$

Clearly $s(x)$ is negligible for $|x| > 8\ell$, whereas at $|x| = 8\ell$ τ_{xx} is about 10% of its maximum value.

Melan's original problem

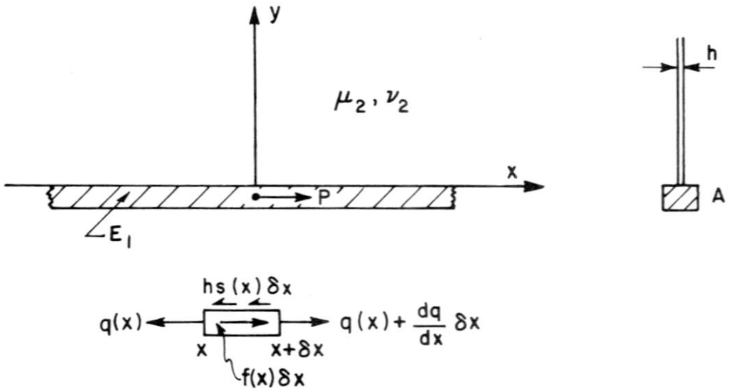

Figure 7.2.1 - Melan's first problem

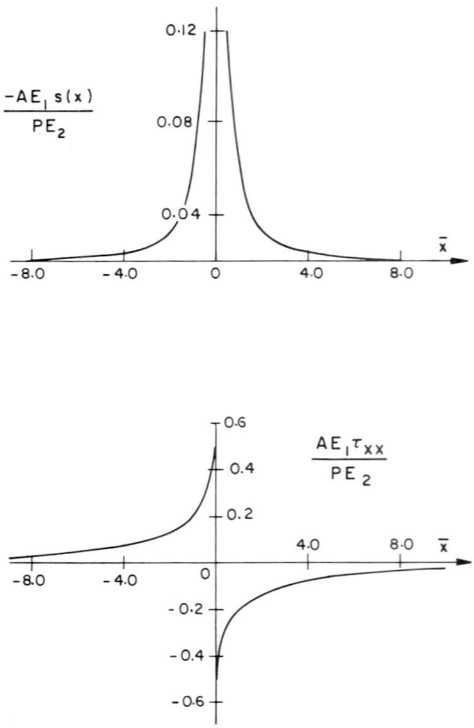

Figure 7.2.2 - Dimensionless stresses in Melan's problem

Figure 7.2.3 shows the configuration for Melan's second problem. A stiffener of area A and width h is bonded to the complete x-y plane, assumed to be in a state of generalized plane stress, along the x-axis. Since bending effects are being neglected it is preferable to suppose that, instead of one stiffener of area A, there are two, each of area A/2, on opposite sides of the plate. The equation linking q(x), f(x) and s(x) is again (7.2.1); the strain is given by (7.2.2), but now equation (7.2.3) has to be replaced by the equation giving the displacement due to a distribution of shearing stress in the whole x-y plane. Following Melan's original paper, we assume that the loading is applied on the line $y = 0$. The displacements in the plane may be obtained from Example 5.8.2 by replacing ν by ν^*. Thus

$$E_2^* u(x,0) = 2F^*[|\xi|^{-1}S(\xi);x] \tag{7.2.12}$$

where

$$E_2^* = 8E_2/[(3-\nu)(1+\nu)] \ . \tag{7.2.13}$$

Now, by comparing (7.2.3) with (7.2.12), we see that the problem is equivalent to Melan's first problem with E_2 replaced by E_2^*, so that the remainder of the analysis follows as before.

Melan's original problem

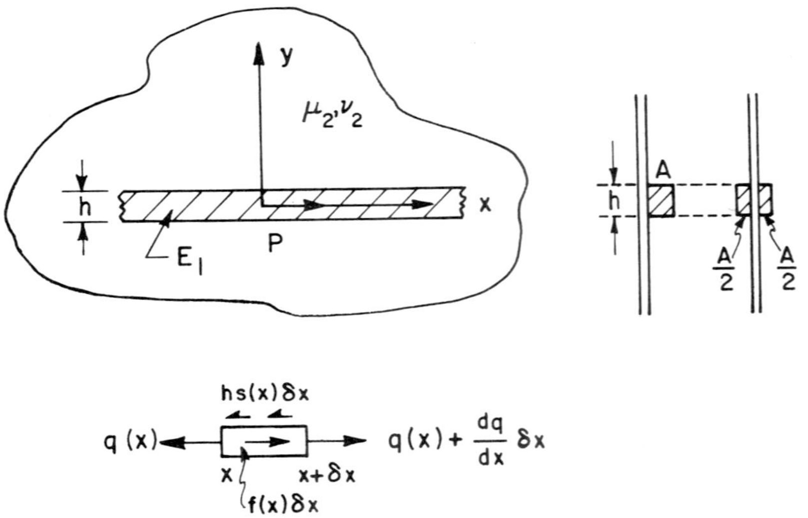

Figure 7.2.3 - Melan's second problem

Martynovich and Yurinets (1975), (1976), (1977) solved Melan's problem for nonhomogeneous and nonisotropic half-planes. Sarkisian and Avanesian (1977) consider Melan's problem for an infinite stringer attached to two elastic strips with fixed edges.

Examples 7.2

1. Show that

$$\lim_{x\to 0\pm} [\cos\bar{x}\,\mathrm{si}(\bar{x}) - \sin\bar{x}\,\mathrm{Ci}(\bar{x})] = \mp \frac{\pi}{2} .$$

2. Show that for small r the stresses in the edge-stiffened plate of Figure 7.2.1 are

$$\tau_{xx} = \frac{PE_2}{2\pi AE_1}\left\{\frac{\sin 2\theta}{2} - \pi + 2\theta\right\} + o(1) ,$$

$$\tau_{xy} = \frac{PE_2}{2\pi AE_1} \ln r + O(1) , \quad \tau_{yy} = -\frac{PE_2}{4\pi AE_1}\sin 2\theta + o(1) .$$

Use the integrals

$$\int_0^\infty \exp(-ax)\sin(bx)dx = b/(a^2+b^2) ,$$

$$\int_0^\infty x^{-1}\exp(-ax)\sin bx dx = \tan^{-1}(b/a) , \quad a > 0 .$$

3. Show that for large r the stresses in the edge-stiffened plate of Figure 7.2.1 are

$$\tau_{xx} = -\frac{2P}{\pi hr}\left\{\cos^3\theta + \frac{a\sin 2\theta(1-4\cos^2\theta)}{2\omega r} + O(r^{-2})\right\} ,$$

$$\tau_{xy} = -\frac{2P}{\pi hr}\left\{\sin\theta\cos^2\theta + \frac{a\cos 4\theta}{2\omega r} + O(r^{-2})\right\} ,$$

$$\tau_{yy} = -\frac{2P}{\pi hr}\left\{\sin^2\theta\cos\theta + \frac{a\sin\theta\cos 3\theta}{\omega r} + O(r^{-2})\right\} ,$$

where $a = A/h_1$, $\omega = h_2E_2/(2h_1E_1)$. Note that the first terms in these expressions are the stresses in the unstiffened plate (see Example 2.1.1).

7.3 *Melan's problems for stiffeners of finite length*

Consider a plate in a state of generalized plane stress (§1.11) and occupying the upper half-plane $y > 0$. Suppose that the edge $y = 0$ is stiffened over the interval $[-a,a]$ by an elastic stiffener of thickness h and cross-sectional area A, as shown in Figure 7.3.1. It is assumed that the bending stiffness of the stiffener, and hence the normal pressure exerted by the stiffener on the plate, may be neglected, so that the stiffener again acts only in tension. The stiffener is thus in a state of uniaxial stress with Young's modulus E_1, and Hooke's law gives the strain $e_{xx}^{(1)}$ in the stiffener as

$$e_{xx}^{(1)} = u^{(1)\prime}(x) = \frac{1}{AE_1}\left[h\int_{-a}^{x} s(t)dt - Q_0\right], \tag{7.3.1}$$

where the (mean) shearing stresses on the stiffener and plate are

$$\tau_{xy}^{(1)}(x) = \tau_{xy}^{(2)}(x) = -s(x), \quad -a \leq x \leq a, \tag{7.3.2}$$

and Q_0 is the force applied to the left-hand end of the stiffener, as shown in Figure 7.3.1. On the other hand the displacement derivatives on the boundary of the half-plane are given by Example 2.1.3 where now $\vartheta_2 = (1-\nu_2^*)/\mu_2 = 2/E_2$ and

$$e_{xx}^{(2)} = u^{(2)\prime}(x) = \frac{2}{\pi E_2}\int_{-a}^{a} \frac{s(t)dt}{t-x}. \tag{7.3.3}$$

The equality of the displacements u_1, u_2 on $[-a,a]$ gives

$$\int_{-a}^{a} \frac{s(t)dt}{t-x} = \frac{\lambda}{a}\left[\int_{-a}^{x} s(t)dt - \frac{Q_0}{h}\right], \quad \lambda = \frac{\pi h a E_2}{2AE_1}. \tag{7.3.4}$$

The total force P_0 applied to the stiffener is

$$h\int_{-a}^{a} s(t)dt = P_0+Q_0. \tag{7.3.5}$$

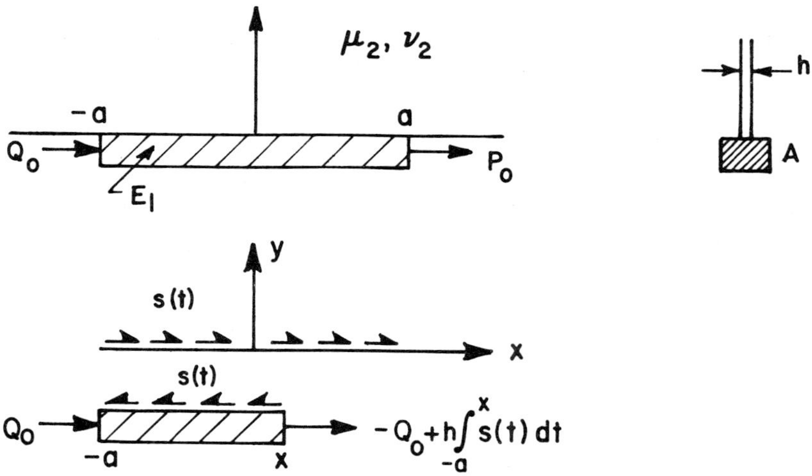

Figure 7.3.1 - A finite stiffener attached to a plate

Equation (7.3.4) may be rephrased in various ways. Thus, following Reissner (1940), we may write

$$q(x) = h \int_{-a}^{x} s(t)dt - Q_0 \ , \qquad q'(x) = hs(x) \ , \tag{7.3.6}$$

so that equation (7.3.4) becomes an integrodifferential equation

$$\int_{-a}^{a} \frac{q'(t)dt}{t-x} = \frac{\lambda}{a}\, q(x) \ , \qquad -a \leq x \leq a \ , \tag{7.3.7}$$

with end conditions

$$q(a) = P_0 \ , \qquad q(-a) = -Q_0 \ . \tag{7.3.8}$$

Equation (7.3.7) appears in Prandtl's lifting line theory in Aerodynamics and is the equation to which Benscoter (1949) applied approximate methods developed for Prandtl's equation.

The equation is studied in Chapter 17 of Muskhelishvili (1953b) where reference is made to the work of earlier researchers in Germany, and especially in the U.S.S.R., and a detailed account is given of the reduction, due to I.N. Vekua and L.G. Magnaradze, of the equation to a Fredholm integral equation of the second kind. We prefer to present a series solution of the equation which may be derived simply and which leads to equations well suited to computation.

Following Morar' and Popov (1970) we may use the identity

$$\int_{-a}^{x} s(t)dt = \frac{1}{2}\int_{-a}^{x} s(t)dt - \frac{1}{2}\int_{x}^{a} s(t)dt + \frac{1}{2}\int_{-a}^{a} s(t)dt =$$

$$= \frac{P_0+Q_0}{2h} - \frac{1}{2}\int_{-a}^{a} \text{sign}(t-x)s(t)dt \qquad (7.3.9)$$

to write the equation in the form

$$\int_{-a}^{a} \left\{\frac{1}{t-x} + \frac{\lambda}{2a}\,\text{sign}(t-x)\right\} s(t)dt = \frac{\lambda(P_0-Q_0)}{2ah}\,. \qquad (7.3.10)$$

This equation has the form (6.3.25) and the shear stress will therefore have a square root singularity at $\pm$ a.

It is convenient to consider the cases of symmetrical and antisymmetrical loading separately. In the former, $Q_0 = -P_0$, s(t) is an odd function and we seek a solution of the form

$$s(t) = (P_0/h)(a^2-t^2)^{-1/2} \sum_{n=0}^{\infty} a_n T_n(t/a)\;, \qquad (7.3.11)$$

with $a_n = 0$ for even n. The integrals in (7.3.10) may be evaluated by using (2.3.26) and the result

$$\frac{1}{2}\int_{-a}^{a} \frac{\text{sign}(t-x)T_n(t/a)dt}{(a^2-t^2)^{1/2}} = \begin{cases} -\sin^{-1}(x/a) & n = 0 \\ n^{-1}(1-x^2/a^2)^{1/2}U_{n-1}(x/a) & n = 1,2,\ldots \end{cases} \qquad (7.3.12)$$

valid for $|x| \le a$, derived by Popov (1968). We find

$$\sum_{n=1}^{\infty} \left\{\pi + \lambda n^{-1}(1-x^2/a^2)^{1/2}\right\} a_n U_{n-1}(x/a) = \lambda \ . \qquad (7.3.13)$$

Now multiply throughout by $(1-x^2/a^2)^{1/2}\, U_{m-1}(x/a)$ and use the orthogonality condition of Example 2.3.2. The resulting equations are

$$a_m + \frac{2\lambda}{\pi^2} \sum_{n=1}^{\infty} K_{m,n} a_n = \frac{4\lambda}{\pi^3} b_m \ , \qquad (7.3.14)$$

where m,n are odd and

$$K_{m,n} = n^{-1} \int_{-1}^{1} (1-t^2) U_{n-1}(t) U_{m-1}(t)\,dt \ , \qquad (7.3.15)$$

$$= \begin{cases} -\dfrac{2m[1+(-)^{n-m}]}{[n^2-(m-1)^2][n^2-(m+1)^2]} \ , & n \neq m-1 \ , \ n \neq m+1 \\ 0 \ , & \text{otherwise} \end{cases} \qquad (7.3.16)$$

$$b_m = \begin{cases} 2m[1+(-)^m]/(m^2-1)^2 \ , & m \neq 1 \\ \dfrac{\pi^2}{4} \ , & m=1 \ . \end{cases} \qquad (7.3.17)$$

In the anti-symmetrical case $Q_0 = P_0$, we seek a solution of the same form (7.3.11) where now $a_n = 0$ for odd n, equation (7.3.5) gives $\pi a_0 = 2$, and (7.3.13) is replaced by

$$\sum_{n=1}^{\infty} \left\{\pi + \lambda n^{-1}(1-x^2/a^2)^{1/2}\right\} a_n U_{n-1}(x/a) = 2\lambda\pi^{-1}\sin^{-1}(x/a) \ . \qquad (7.3.18)$$

The resulting set of equations is again (7.3.14) where m,n are now even. Notice that there are numerous errors of sign in the analysis of Arutiunian and Mkhitarian (1971) which follows the same lines; the earlier solution of Arutiunian (1968) is extremely cumbersome, while the solution due to Morar' and Popov (1970) is in error after their equation (1.7). Erdogan and Gupta (1971) give an analysis very similar to that outlined above.

The solution for the corresponding problems of a stiffener of finite length 2a and width h attached to the x-axis of an infinite plate may, as discussed in §7.2, be obtained by replacing E_2 in λ by E_2^*.

The system of equations (7.3.14) with m,n both odd or both even is *fully regular* in the sense of Kantorovich and Krylov (1958) for all values of λ. The set of equations

$$x_m = \sum_{n=1}^{\infty} c_{m,n} x_n + b_m , \qquad m = 1,2,... \tag{7.3.19}$$

is fully regular if there is a number θ such that

$$s_m = \sum_{n=1}^{\infty} |c_{m,n}| \leq 1-\theta < 1 , \qquad m = 1,2,... ; \tag{7.3.20}$$

in this case, provided that the $|b_m|$ are bounded, the limit, as N tends to infinity, of the solution $\{x_1,...,x_N\}$ obtained by solving the first N equations (7.3.19) will be the solution of the set of equations (7.3.19). To prove that the system (7.3.14) is fully regular we note that when m,n have the same parity

$$K_{m,n} \begin{cases} = 4m/(4m^2-1) & n = m , \\ < 0 & n \neq m . \end{cases} \tag{7.3.21}$$

Thus equations (7.3.14) may be written

$$\left[1 + \frac{8\lambda m}{\pi^2(4m^2-1)}\right] a_m + \frac{2\lambda}{\pi^2} \sum_{n=1}^{\infty}{}' K_{m,n} a_n = \frac{4\lambda}{\pi^3} b_m \tag{7.3.22}$$

where ' means $n \neq m$. Now since $m/(4m^2-1) < 1/3$, and $K_{m,n}$ is negative

$$s_m \equiv \frac{2\lambda}{\pi^2}\left(1 + \frac{8\lambda m}{\pi^2(4m^2-1)}\right)^{-1} \sum_{n=1}^{\infty}{}' |K_{m,n}| \leq -2\lambda\left(1 + \frac{8\lambda}{\pi^2}\cdot\frac{1}{3}\right)^{-1} \sum_{n=1}^{\infty}{}' K_{m,n} . \tag{7.3.23}$$

But it may easily be shown (Example 7.3.2) that

$$-\sum_{n=1}^{\infty}{}' K_{m,n} = \frac{4m}{4m^2-1} - \sum_{n=1}^{\infty} K_{m,n} \tag{7.3.24}$$

and

$$\sum_{n=1}^{\infty} K_{m,n} = \begin{cases} \pi^2/8 & m,n \text{ both odd, } m = 1 \\ 0 & m,n \text{ both odd, } m \neq 1 \\ 2m/(m^2-1)^2 & m,n \text{ both even .} \end{cases} \tag{7.3.25}$$

Therefore

$$s_m \leq \frac{2\lambda}{\pi^2}\left(1 + \frac{8\lambda}{3\pi^2}\right)^{-1} \frac{4m}{4m^2-1} \leq \frac{8\lambda}{3\pi^2}\left(1 + \frac{8\lambda}{3\pi^2}\right)^{-1} < 1 , \tag{7.3.25}$$

so that the system is fully regular. Arutiunian and Mkhitarian (1971) and Erdogan and Gupta (1971) prove that the system is fully regular only for a restricted range of λ.

Büfler (1959), (1961) made a detailed study of Melan's problem for a finite stiffener (both rigid (1959) and elastic (1961) in tension), and considered also an edge stiffener attached to a semi-infinite plate in unilateral tension, a central stiffener in a plate under allround tension, a stiffener connecting two semi-infinite plates, a plate deformed by various combinations of tension, shear and twisting applied to an edge stiffener, and two semi-infinite plates attached along a finite length of their common boundary with various systems of forces applied to the plates. Much of the analysis in Büfler (1961) can be simplified by making greater use of the integrals involving Chebyshev polynomials which are described in this section; he gives extensive numerical results and some experimental verifications. Keer and Freedman (1973) consider a stiffener attached to a transversely isotropic strip.

Melan's problems for stiffeners of finite length

Erdogan (1971) formulated the general problem of an elastic half-plane with an arbitrary number of stiffeners (cover plates). He employs Muskhelishvili's complex variable analysis and considers, in particular, the cases of two, three, and a periodic array of stiffeners. He finds closed form solutions for one, two, or a periodic array of *inextensible* stiffeners, and numerical results for elastic stiffeners.

Problems involving a pair of identical stiffeners, whether loaded anti-symmetrically, as shown in Figure 7.3.2, or symmetrically as in Figure 7.3.3, may be solved by using the analysis developed in §4.8 for two punches. The governing equations are

$$e_{xx}^{(1)} = u_x^{(1)'}(x) = \begin{cases} -\dfrac{h}{AE_1}\displaystyle\int_x^{-b} s(t)dt & -a \le x \le -b \ , \\ \dfrac{h}{AE_1}\displaystyle\int_b^{x} s(t)dt & b \le x \le a \ , \end{cases} \qquad (7.3.27)$$

and

$$e_{xx}^{(2)} = u_x^{(2)'}(x) = \frac{2}{\pi E_2}\int_{L'} \frac{s(t)dt}{t-x} \ , \qquad (7.3.28)$$

where L' is the union of [-a,-b] and (b,a] and

$$\tau_{xy}^{(1)}(x) = \tau_{xy}^{(2)}(x) = -s(x) \text{ on } L' \ . \qquad (7.3.29)$$

Now, depending upon whether the loading is anti-symmetric (a) or symmetric (s), we have the results

$$\text{(a)} \quad P_0 = h\int_{-a}^{-b} s(t)dt = h\int_b^a s(t)dt \ , \quad s(-t) = s(t) \ , \qquad (7.3.30)$$

$$\text{(a)} \quad u^{(1)'}(x) = \frac{-h}{2AE_1}\int_{L'} \text{sign}(t-x)s(t)dt \ , \quad x \text{ in } L' \ , \qquad (7.3.31)$$

$$\text{(s)} \quad P_0 = -h\int_{-a}^{-b} s(t)dt = h\int_{b}^{a} s(t)dt \ , \qquad s(-t) = -s(t) \ , \tag{7.3.32}$$

$$\text{(s)} \quad u^{(1)'}(x) = \frac{-h}{2AE_1}\int_{L'} \text{sign}(t-x)s(t)dt + \frac{P_0}{AE_1} \ , \quad x \text{ in } L' \ , \tag{7.3.33}$$

so that

$$\text{(a)} \quad \int_{L'} \left\{\frac{1}{t-x} + \frac{\lambda}{2a}\,\text{sign}(t-x)\right\} s(t)dt = 0 \ , \quad x \text{ in } L' \ , \tag{7.3.34}$$

$$\text{(s)} \quad \int_{L'} \left\{\frac{1}{t-x} + \frac{\lambda}{2a}\,\text{sign}(t-x)\right\} s(t)dt = \frac{\lambda P_0}{ah} \ , \quad x \text{ in } L' \ . \tag{7.3.35}$$

Now s(t) will have square root singularities at each of the ends $\pm$ a, $\pm$ b. It was shown in Example 4.8.6 that if

$$\tau = (2t^2-a^2-b^2)/(a^2-b^2) \tag{7.3.36}$$

then the polynomials $T_n(\tau)$ are orthogonal over L' with weight function $|t|\{(a^2-t^2)(t^2-b^2)\}^{-1/2}$. In the anti-symmetrical case we assume a solution of the form

$$s(t) = P_0|t|h^{-1}\{(a^2-t^2)(t^2-b^2)\}^{-1/2}\sum_{n=0}^{\infty} a_n T_n(\tau) \ , \tag{7.3.37}$$

so that (7.3.30) gives $\pi a_0 = 2$. The first integral in (7.3.34) is given by Example 4.8.5 while the second may be obtained by transforming (7.3.12) by means of the substitution (7.3.36) to give

$$\int_{L'} \frac{\text{sign}(t-x)|t|T_n(\tau)dt}{\{(a^2-t^2)(t^2-b^2)\}^{1/2}} = \begin{cases} -(\pi/2+\sin^{-1}\xi)\,\text{sign}(x) \ , & n = 0 \\ n^{-1}(1-\xi^2)^{1/2}U_{n-1}(\xi)\,\text{sign}(x) \ , & n = 1,2, \ldots, \end{cases} \tag{7.3.38}$$

where $\xi = (2x^2-a^2-b^2)/(a^2-b^2)$. Thus

$$\sum_{n=1}^{\infty} \{-4\pi a|x|(a^2-b^2)^{-1}+\lambda n^{-1}(1-\xi^2)^{1/2}\}U_{n-1}(\xi) = \lambda a_0(\pi/2+\sin^{-1}\xi) , \tag{7.3.39}$$

for x on L'. Now multiply throughout by $(1-\xi^2)^{1/2}U_{m-1}(\xi)$ and use the identity

$$(1-\xi^2)^{1/2} = 2\{(a^2-x^2)(x^2-b^2)\}^{1/2}/(a^2-b^2) . \tag{7.3.40}$$

The resulting equations are

$$a_m - \lambda \sum_{n=1}^{\infty} K_{m,n}a_n = -b_m , \tag{7.3.41}$$

where

$$K_{m,n} = \frac{a^2-b^2}{2\pi^2 an}\int_0^{\pi} \frac{\sin m\theta \sin n\theta \sin\theta d\theta}{(a^2\cos^2\theta/2+b^2\sin^2\theta/2)^{1/2}} , \tag{7.3.42}$$

$$b_m = \frac{\lambda(a^2-b^2)}{\pi^3 a}\int_0^{\pi} \frac{(\pi-\theta)\sin m\theta \sin\theta d\theta}{(a^2\cos^2\theta/2+b^2\sin^2\theta/2)^{1/2}} . \tag{7.3.43}$$

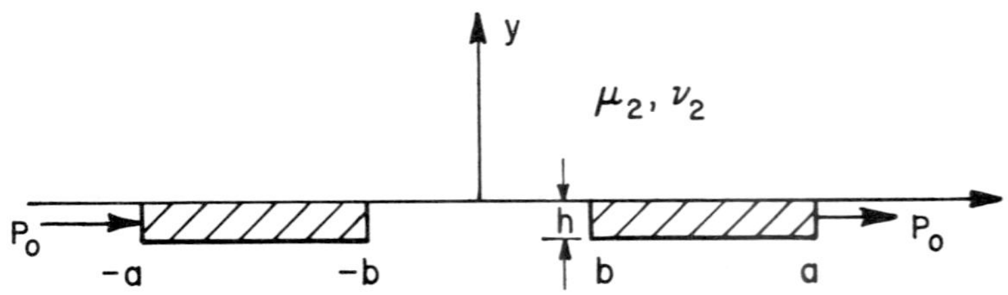

Figure 7.3.2 - Anti-symmetrical loading of two identical stiffeners

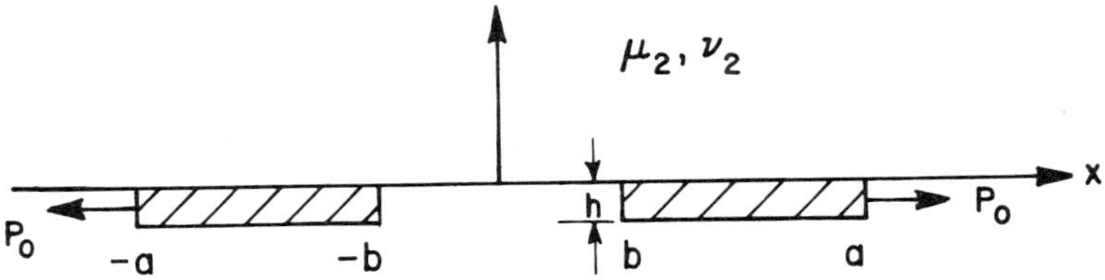

Figure 7.3.3 - Symmetrical loading of two identical stiffeners

In the symmetrical case we assume a solution of the form

$$s(t) = P_0 a h^{-1}\{(a^2-t^2)(t^2-b^2)\}^{-1/2}\operatorname{sign}(t)\sum_{n=0}^{\infty} a_n T_n(\tau) . \tag{7.3.44}$$

Working as in (7.3.9), we may write equation (7.3.35) in the form

$$\int_{L'} \frac{s(t)dt}{t-x} = \frac{\lambda}{a}\int_b^x s(t)dt , \quad b \le x \le a , \tag{7.3.45}$$

and using Example 4.8.5 with m = 0 we find

$$\sum_{n=1}^{\infty} a_n U_{n-1}(\xi) = -\frac{\lambda(a^2-b^2)}{4a^2\pi}\sum_{n=0}^{\infty} a_n c_n(\theta) , \tag{7.3.46}$$

where

$$c_n(\theta) = a\int_\theta^\pi (a^2\cos^2\theta/2+b^2\sin^2\theta/2)^{-1/2}\cos n\phi\, d\phi , \tag{7.3.47}$$

and $\cos\theta = \xi$, $0 \le \theta \le \pi$. The total load on the right-hand stiffener is P_0, giving

$$2 = \sum_{n=0}^{\infty} a_n c_n(0) . \tag{7.3.48}$$

Now multiplying throughout by $x(1-\xi^2)^{1/2}U_{m-1}(\xi)$ and integrating over [b,a], we find

$$a_m + \lambda \sum_{n=0}^{\infty} K_{m,n} a_n = 0 , \quad m = 1,2,\ldots , \tag{7.3.49}$$

where

$$K_{m,n} = \frac{a^2-b^2}{2\pi^2 a^2} \int_0^{\pi} \sin m\theta \sin\theta\, c_n(\theta)\, d\theta . \tag{7.3.50}$$

The analysis above follows closely that of Arutiunian and Mkhitarian (1971), except for various errors of sign in their work. They prove that both equations (7.3.41) and (7.3.49) are at least quasi fully regular for all values of λ. That is, the inequality (7.3.20) holds for all sufficiently large m. Also, for sufficiently small λ the system is fully regular - see Example 7.3.3. See also Arutiunian and Mkhitarian (1975).

Examples 7.3

1. Show that the axial force in the stiffener of Figure 7.3.1 is

$$q(x) = P_0\left[1-(1-x^2/a^2)^{1/2} \sum_{n=1}^{\infty} a_n U_{n-1}(x/a)\right]$$

where n is odd, in the symmetrical case, and

$$q(x) = P_0\left[(2/\pi)\sin^{-1}(x/a) - (1-x^2/a^2)^{1/2} \sum_{n=1}^{\infty} a_n U_{n-1}(x/a)\right]$$

where n is even, in the anti-symmetrical case.

2. Show that when m,n have the same parity the $K_{m,n}$ of (7.3.16) may be written

$$K_{m,n} = \frac{1}{2(m-1)}\left[\frac{1}{n-m+1} - \frac{1}{n+m-1}\right] + \frac{1}{2(m+1)}\left[\frac{1}{n+m-1} - \frac{1}{n-m-1}\right] ,$$

and show that if the sum is taken over values of n having the same parity as m then

$$\sum_{n=1}^{\infty}\left[\frac{1}{n-m+1}-\frac{1}{n+m-1}\right]=\begin{cases}0, & m,n \text{ odd}, \ m\neq 1,\\ 1/[2(m-1)], & m,n \text{ even},\end{cases}$$

and hence deduce (7.3.25).

3. Show that

$$\int_0^{\pi}\frac{\sin p\theta\, d\theta}{(a^2\cos^2\theta/2+b^2\sin^2\theta/2)^{1/2}}\leq\frac{A}{|p|}, \quad A=\frac{a+b}{b^2}\left\{\frac{b}{2}+\frac{\pi(a-b)}{4b}\right\}.$$

Hence show that, for each of the $K_{m,n}$ in (7.3.42) and (7.3.50),

$$\lim_{m\to\infty}\sum_{n=1}^{\infty}|K_{m,n}|=0$$

so that both systems are quasi fully regular. For what ranges of λ are the systems fully regular?

4. A *rigid* straight stiffener is attached to the upper half-plane along $y = 0$, $-a \leq x \leq a$. Assuming that the half-plane is in generalized plane stress, that the stiffener is twisted through a small in-plane angle ϕ, and that the normal pressure and shearing stress in the contact region are $p(x)$, $s(x)$ respectively, use Example 2.1.3 to show that

$$-(1-\nu)p(x)+\frac{2}{\pi}\int_{-a}^{a}\frac{s(t)dt}{t-x}=0, \qquad E\phi=(1-\nu)s(x)+\frac{2}{\pi}\int_{-a}^{a}\frac{p(t)dt}{t-x}.$$

Obtain the corresponding equations for two half-planes of different elastic properties, joined along $[-a,a]$. These equations, which may be solved by the methods of §3.5, are considered in detail by Büfler (1959).

5. Show that the dimensionless axial force at a point on the right-hand stiffener of Figure 7.3.2 may be written

$$q(x)/P_0 = 1-\phi/\pi-\sum_{n=1}^{\infty}(2n)^{-1}a_n\sin n\phi,$$

where $(2x^2-a^2-b^2)/(a^2-b^2) = \xi = \cos\phi$.

7.4 *Semi-infinite stiffeners*

Consider the semi-infinite stiffener occupying the region $x > 0$, $y = 0$ as shown in Figure 7.4.1 or 7.4.2. In the previous notation the longitudinal strain $e_{xx}^{(1)}$ in the stiffener is

$$e_{xx}^{(1)} = \frac{1}{AE_1}\left[h \int_0^x s(t)dt - P_0 \right] , \tag{7.4.1}$$

while that for the corresponding point of the half-plane is

$$e_{xx}^{(2)} = \frac{2}{\pi E_2} \int_0^\infty \frac{s(t)dt}{t-x} . \tag{7.4.2}$$

For the central stiffener of Figure 7.4.2, E_2 must be replaced by the E_2^* of (7.2.13). The total force on the stiffener is

$$P_0 = h \int_0^\infty s(t)dt . \tag{7.4.3}$$

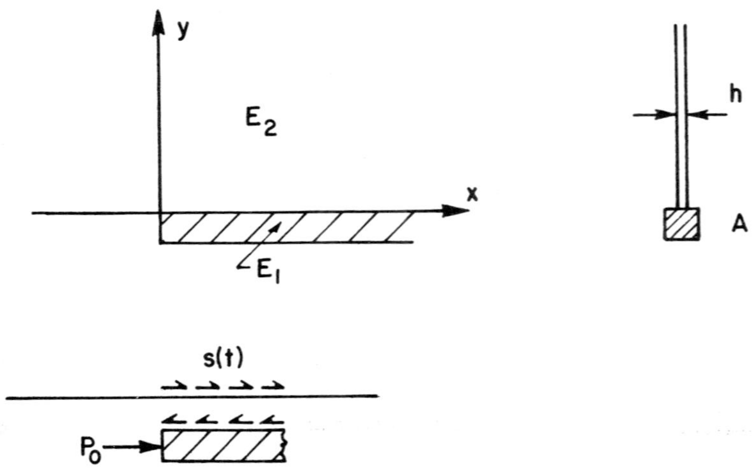

Figure 7.4.1 - Semi-infinite edge stiffener

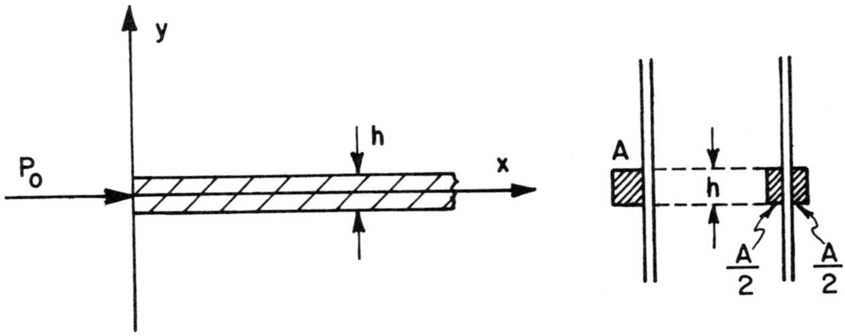

Figure 7.4.2 - Semi-infinite central stiffener

First introduce dimensionless variables

$$\xi = \left(\frac{hE_2}{AE_1}\right)x \ , \ \eta = \left(\frac{hE_2}{AE_1}\right)t \ , \ \tau(\xi) = \left(\frac{AE_1}{P_0E_2}\right)s(x), \ q(\xi) = \frac{AE_1}{P_0} e_{xx} \ , \tag{7.4.4}$$

so that $\tau(\xi)$, $q(\xi)$ are dimensionless shear and longitudinal strain (or stringer force). Then equations (7.4.1), (7.4.2) yield

$$q(\xi) = -\int_{\xi}^{\infty} \tau(\eta)d\eta \ , \qquad q(\xi) = \frac{2}{\pi}\int_{0}^{\infty} \frac{\tau(\eta)d\eta}{\eta-\xi} \ , \tag{7.4.5}$$

while equation (7.4.3) yields

$$\int_{0}^{\infty} \tau(\eta)d\eta = 1 \ . \tag{7.4.6}$$

E.L. Buell (1948) presented a purported solution for the problem by using a conformal mapping of the z-plane onto the interior of the unit circle. We shall first derive his final set of equations by using a straightforward transformation of equations (7.4.5) after noting certain defects in his solution we shall present a shortened version of a rigorous analysis derived by Koiter (1955).

Buell's method is well adapted to give numerical results and these agree with those obtained, with more difficulty, from Koiter's.

Consider the transformation

$$u = \frac{1-\eta}{1+\eta}, \quad \eta = \frac{1-u}{1+u}, \tag{7.4.7}$$

which transforms $\eta \geq 0$ into $-1 \leq u \leq 1$. Putting $\tau(t) = (1+u)S(u)$, $\xi = (1-v)/(1+v)$ we find that (7.4.5) becomes

$$(1+v)\int_{-1}^{1} \frac{S(u)du}{u-v} = \pi \int_{-1}^{v} \frac{S(u)du}{1+u}, \quad -1 \leq v \leq 1, \tag{7.4.8}$$

so that, using (7.3.9) and the fact that

$$1 = \int_0^\infty \tau(\eta)d\eta = 2\int_{-1}^{1} \frac{S(u)du}{1+u}, \tag{7.4.9}$$

we obtain

$$\int_{-1}^{1} \left\{\frac{1}{u-v} + \frac{\pi}{2}\,\frac{\mathrm{sign}(u-v)}{(1+u)(1+v)}\right\} S(u)du = \frac{1}{4(1+v)}. \tag{7.4.10}$$

Since this equation has the form (6.3.25), its solution will have a square root singularity at $u = \pm 1$; however, the presence of $(1+u)$ in the denominator of the second term of the kernel demands that $S(-1)$ must be zero. We therefore seek $S(u)$ of the form

$$S(u) = [(1+u)/(1-u)]^{1/2} \sum_{n=0}^{\infty} a_n T_n(u), \tag{7.4.11}$$

so that (7.4.9) gives $2\pi a_0 = 1$. In terms of the variables τ, η this is

$$\tau(\eta) = \frac{2\eta^{-1/2}}{1+\eta} \sum_{n=0}^{\infty} a_n T_n\left(\frac{1-\eta}{1+\eta}\right). \tag{7.4.12}$$

Now insert (7.4.12) into (7.4.10), evaluate the integrals by using (2.3.27), (7.3.12) and put $v = \cos\phi$; the result is

$$(1+\cos\phi)\left\{a_0+(1+\cos\phi)\sum_{n=1}^{\infty}\frac{a_n\sin n\phi}{\sin\phi}\right\} = a_0(\pi-\phi) - \sum_{n=1}^{\infty}\frac{a_n\sin n\phi}{n}\ , \tag{7.4.13}$$

or

$$\sum_{n=1}^{\infty}\{n^{-1}+2\cos^3\phi/2\operatorname{cosec}\phi/2\}a_n\sin n\phi = a_0\{\pi-\phi-2\cos^2\phi/2\}\ , \tag{7.4.14}$$

which agrees with the equation obtained by Buell. It should be noted that the purported solution (7.4.12) has a fundamental defect in that the longitudinal strain in the stiffener computed from the left-hand side of (7.4.5) is $O(\xi^{-1/2})$ for large $|\xi|$, while that computed from the right-hand side is $O(\xi^{-1})$; Koiter shows that the latter is correct (see (7.4.44) below).

In order to solve equation (7.4.14) Buell suggested that the equation be multiplied by $\sin\phi/2\sin m\phi$ and integrated over $[0,\pi]$ to give

$$\sum_{n=1}^{\infty} c_{m,n}a_n = b_m\ , \tag{7.4.15}$$

where $c_{m,n} = A_1(m,n)+A_2(m,n)\ ,$ (7.4.16)

$$A_1(m,n) = \frac{1}{n}\int_0^{\pi}\sin\phi/2\ \sin m\phi\sin n\phi d\phi = \frac{1}{2n}\, d(m,n,1)\ , \tag{7.4.17}$$

$$d(m,n,p) = \frac{1}{2(m-n)+p} - \frac{1}{2(m-n)-p} + \frac{1}{2(m+n)-p} - \frac{1}{2(m+n)+p}\ , \tag{7.4.18}$$

$$A_2(m,n) = 2\int_0^{\pi}\cos^3\phi/2\sin m\phi\sin n\phi d\phi = \frac{(-)^{m-n}}{4}\left\{3d(m,n,1)-d(m,n,3)\right\}\ , \tag{7.4.19}$$

$$b_m = a_0\int_0^{\pi}\{\pi-\phi-2\cos^2\phi/2\}\ \sin\phi/2\ \sin m\phi d\phi = \frac{16a_0m}{4m^2-1}\left\{\frac{1}{4m^2-1} - \frac{(-)^m}{4m^2-9}\right\}\ . \tag{7.4.20}$$

It is difficult to prove that the system (7.4.15) is regular or not regular but, on the other hand, Buell found that the computed a_n alternate in sign and steadily and rapidly decrease in magnitude so that the approximations to $q(\xi)$, $\tau(\xi)$ appear to converge rapidly.

An alternative solution of Melan's original problems for the semi-infinite stiffener may be derived by using the Mellin transform solution of §9.2. Since the positive x-axis is the line $\theta = 0$ in polar coordinates the boundary conditions for the first problem are

$$\tau_{\theta\theta}(r,\theta) = 0 \ , \ \theta = 0,\pi \ ; \ \tau_{r\theta}(r,\pi) = 0 \ , \tag{7.4.21}$$

$$\tau_{r\theta}(r,0) = -s(r) \ , \tag{7.4.22}$$

and

$$e_{rr}(\mathbf{r},0) = -\frac{h}{AE_1}\int_r^\infty s(t)dt \ . \tag{7.4.23}$$

Now introduce the dimensionless variables (7.4.4), and apply the conditions (7.4.21) to the solution (9.4.15), using $\kappa = 3-4\nu^*$, as is appropriate for generalized plane stress. We find

$$A_0+A_1 = 0 = B_0+B_1 \ , \quad A_0\sin\pi s = B_0\cos\pi s \ , \tag{7.4.24}$$

which gives the relation between the dimensionless Mellin transforms

$$T(s) = M[\tau(\eta);s] \ , \quad Q(s) = M[q(\eta);s] \tag{7.4.25}$$

of the quantities $\tau(\xi)$, $q(\xi)$ defined in (7.4.4) as

$$Q(s+1) = 2\cot\pi s T(s+1) \ . \tag{7.4.26}$$

The transform of the first of equations (7.4.5) is

$$Q(s) = -s^{-1}T(s+1) \ , \tag{7.4.27}$$

and together equations (7.4.26), (7.4.27) yield the equation

$$T(s+1) = -2s\cot\pi s\, T(s) \ . \tag{7.4.28}$$

The normalization equation satisfied by T(s) is

$$T(1) = \int_0^\infty \tau(\eta)d\eta = 1 \ . \tag{7.4.29}$$

Equation (7.4.28), which was derived by Koiter (1955) by applying Mellin transforms directly to equations (7.4.5), is a linear difference equation. Equation (7.4.28) is a particular example of the equation

$$T(s+w) = \phi(s)T(s) \ , \tag{7.4.30}$$

where $\phi(s)$ may be expressed as the ratio of two holomorphic functions. The general solution of such equations was discussed by Barnes (1904), and further references may be found in Kuczma (1968), which is concerned primarily with the existence and uniqueness of solutions of a wide class of functional equations in a single variable. Koiter solved equation (7.4.28) by using a method suggested by Alblas; in a later paper, Alblas (1965) presented a refined version of his method. We prefer to construct the solution directly.

The Gamma function $\Gamma(s)$ satisfies the equations

$$\Gamma(s+1) = s\Gamma(s) \ , \quad \Gamma(s)\Gamma(1-s) = \pi/\sin\pi s \ , \tag{7.4.31}$$

and the second relation shows that the factor $-s\cot\pi s$ may be written

$$-s\cot\pi s = \frac{\Gamma(1+s)\Gamma(1-s)}{\Gamma(3/2-s)\Gamma(s-1/2)} \ . \tag{7.4.32}$$

Alexeiewsky's G-function defined by (Whittaker and Watson (1952, p. 264))

$$G(z+1) = (2\pi)^{z/2}\exp\left\{-\frac{1}{2}[z(z+1)+\gamma z^2]\right\} \prod_{n=1}^{\infty} \left\{\left(1+\frac{z}{n}\right)^n \exp[-z+z^2/2n]\right\} , \tag{7.4.33}$$

where γ is the Euler-Mascheroni constant, is an integral function with n-tuple zeros $z = -n$, $(n = 1,2,3,...)$, and satisfies the recurrence relation

$$G(z+1) = \Gamma(z)G(z) . \tag{7.4.34}$$

Thus $G(2+s) = \Gamma(1+s)G(1+s)$, $G(5/2-s) = \Gamma(3/2-s)G(3/2-s)$, $G(2-s) = \Gamma(1-s)G(1-s)$, $G(s+1/2) = \Gamma(s-1/2)G(s-1/2)$ so that, by noting the four factors in (7.4.32) and the normalizing condition (7.4.29), we find that

$$T(s) = \frac{2^{s-1}}{\Gamma(1/2)} \frac{G(1+s)G(5/2-s)}{G(s-1/2)G(2-s)} . \tag{7.4.35}$$

This is a meromorphic function with n-tuple poles at $s = n+1$ $(n = 1,2,...)$, and k-tuple poles at $s = 3/2-k$ $(k = 1,2,3,...)$; in particular $T(s)$ is holomorphic in the strip $1/2 < \mathrm{Re}(s) < 2$, and tends to zero for $|s| \to \infty$ in the strip. The solution (7.4.35) is not the unique solution of (7.4.28) satisfying (7.4.29); in fact $T(s)$ may be multiplied by any function of period 1, holomorphic in $1/2 < \mathrm{Re}(s) < 2$ and equal to unity for $s = 1$. However, such a function is at least of order $\exp(2\pi|s|)$ as $|s| \to \infty$ in this strip, and the alternative solutions therefore violate the requirement $T(s) \to 0$ as $|s| \to \infty$ in this strip. Hence (7.4.35) is the only solution which satisfies all requirements on $T(s)$.

The (dimensionless) shear stress and stringer force may now be obtained by inverting $T(s)$ and $Q(s)$. Thus (9.2.2) shows that

$$\tau(\eta) = \frac{1}{2\pi i} \int_{c-i\infty}^{c+i\infty} T(s)\eta^{-s} ds \; , \tag{7.4.36}$$

where $1/2 < \mathrm{Re}(c) < 2$. Following Koiter (1955), we consider the contour shown in Figure 7.4.3. It can be shown that the contributions to the integral around the contour from the sides $\mathrm{Im}(s) = \pm\omega$ and $\mathrm{Re}(s) = -k + 3/4$ tend to zero as $\omega \to \infty$ and $k \to \infty$ respectively so that, since $T(s)\eta^{-s}$ has poles as $s = 3/2 - k$,

$$\tau(\eta) = \sum_{k=1}^{\infty} \{\text{residue of } [T(s)\eta^{-s}] \text{ at } s = 3/2 - k\} \; . \tag{7.4.37}$$

Since $s = 3/2 - k$ is a k-tuple pole, we have

$$\tau(\eta) = \sum_{k=1}^{\infty} \frac{1}{(k-1)!} \left\{ \frac{d^{k-1}}{ds^{k-1}} [(s+k-3/2)^k T(s)\eta^{-s}] \right\}_{s=3/2-k} \; . \tag{7.4.38}$$

In a similar way we find

$$q(\xi) = -\int_{\xi} \tau(\eta)d\eta = -1 + \int_0^{\xi} \tau(\eta)d\eta = -1 + \frac{1}{2\pi i}\int_{c-i\infty}^{c+i\infty} \frac{T(s)}{1-s}\, \xi^{1-s} ds$$

$$= -1 + \sum_{k=1}^{\infty} \frac{1}{(k-1)!} \left\{ \frac{d^{k-1}}{ds^{k-1}} \left[(s+k-3/2)^k \frac{T(s)}{1-s}\, \xi^{1-s} \right] \right\}_{s=3/2-k} \; , \tag{7.4.39}$$

where we now take $1/2 < c < 1$.

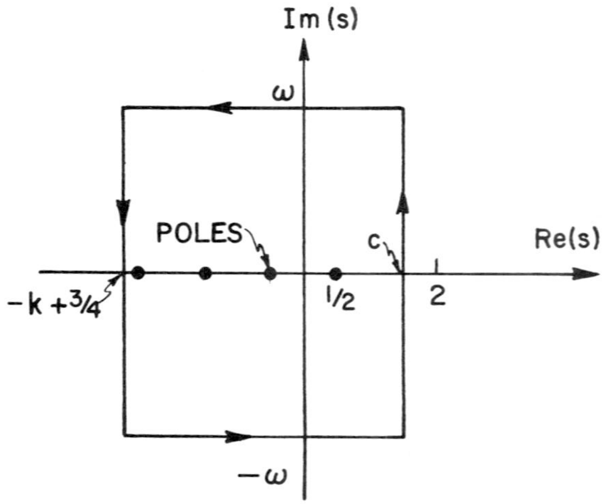

Figure 7.4.3 - Koiter used this contour to evaluate the inverse Mellin transform of T(s)

The first terms in (7.4.38) and (7.4.39) involve

$$\lim_{s\to 1/2} \{(s-1/2)T(s)\} = \lim_{s\to 1/2}\left\{\frac{2^{-1/2}}{\Gamma(1/2)}\,\frac{(s-1/2)}{G(s-1/2)}\,\frac{G(3/2)G(2)}{G(3/2)}\right\}, \quad (7.4.40)$$

but

$$\lim_{s\to 1/2}\left\{\frac{s-1/2}{G(s-1/2)}\right\} = \lim_{s\to 1/2}\left\{\frac{(s-1/2)\Gamma(s-1/2)}{G(s+1/2)}\right\} = \lim_{s\to 1/2}\left\{\frac{\Gamma(s+1/2)}{G(s+1/2)}\right\} = \frac{\Gamma(1)}{G(1)}, \quad (7.4.41)$$

so that

$$\lim_{s\to 1/2} \{(s-1/2)T(s)\} = \frac{2^{-1/2}}{\Gamma(1/2)}\,\frac{\Gamma(1)G(2)}{G(1)} = (2\pi)^{-1/2}, \quad (7.4.42)$$

and

$$\tau(\eta) = (2\pi\eta)^{-1/2}+O(\eta^{1/2}), \quad q(\xi) = -1+(2\xi/\pi)^{1/2}+O(\xi^{3/2}). \quad (7.4.43)$$

Koiter exhibits the next two terms in the q series and shows that Buell's approximate analysis yields $q(\xi) \simeq -1 + c\xi^{1/2}$ where, for n = 6, c differs from $(2/\pi)^{1/2}$ by about 2.5%. The evaluation of τ, q from equations (7.4.38), (7.4.39) involves the evaluation of complicated limits so that, for numerical purposes, Buell's analysis is superior. Koiter shows that, for large ξ, the dimensionless stringer force is

$$q(\xi) = \frac{2}{\pi\xi} - \frac{4}{\pi^2\xi^2}(\ln 2 - \gamma - \ln\xi) + \ldots . \qquad (7.4.44)$$

Benthem and van der Vooren (1959) and Alblas and Kuypers (1965/1966) studied the diffusion of load from a stiffener attached all along one side of a wedge-shaped plate, using methods similar to Koiter's. Kalandia (1969b) presented an elegant alternative solution to Koiter's problem which uses a Fourier transform of equation (7.4.5) to obtain an equation (with variable coefficients) linking the values of a sectionally holomorphic function on either side of the cut [-1,1] in the complex plane; this equation is solved by the methods described by Muskhelishvili (1953b). Bantsuri (1973) extended this analysis to the problem of a wedge with one side stiffened; (see also Bantsuri (1975), Vorob'ev and Popov (1970)). Rybakov and Cherepanov (1977) considered a semi-infinite stiffener attached to a plate at discrete points.

7.5 *Periodically stiffened plates*

Consider the plate shown in Figure 7.5.1 which is stiffened by an infinity of equal stiffeners of length 2a, with a distance 2ℓ between centres. As before it is assumed that the stiffeners are extensible and have zero bending stiffness. Following the analysis and notation of §7.3, we find that the strain $e_{xx}^{(1)}$ in any one stiffener is

$$e_{xx}^{(1)} = u^{(1)\prime}(x) = \frac{1}{AE_1}\left[h\int_{-a}^{x} s(t)dt - Q_0\right], \qquad (7.5.1)$$

and the total force applied to each stiffener is

$$P_0 + Q_0 = h\int_{-a}^{a} s(t)dt . \qquad (7.5.2)$$

The displacement derivative on the boundary of the half-plane is now given by

$$e_{xx}^{(2)} = u^{(2)\prime}(x) = \frac{2}{\pi E_2}\int_{-a}^{a} s(t) \sum_{n=-\infty}^{\infty} \frac{dt}{t+n\ell-x} . \qquad (7.5.3)$$

But

$$\sum_{n=-\infty}^{\infty} \frac{1}{t+n\ell-x} = \frac{1}{t-x} + 2(t-x)\sum_{n=1}^{\infty} \frac{1}{(t-x)^2-n^2\ell^2} = \frac{\pi}{\ell}\cot\left[\frac{\pi(t-x)}{\ell}\right], \qquad (7.5.4)$$

so that by equating the displacement derivatives we find the basic equation

$$\int_{-a}^{a} s(t)\cot\left[\frac{\pi(t-x)}{\ell}\right] dt = \lambda\left[\int_{-a}^{x} s(t)dt - \frac{Q_0}{h}\right], \quad \lambda = \frac{h\ell E_2}{2AE_1}, \qquad (7.5.5)$$

which may also be written

$$\int_{-a}^{a} \left\{\cot\left[\frac{\pi(t-x)}{\ell}\right] + \frac{\lambda}{2}\,\mathrm{sign}(t-x)\right\} s(t)dt = \lambda\frac{(P_0-Q_0)}{2h} . \qquad (7.5.6)$$

This equation was derived first by Arutiunian and Mkhitarian (1969). Their subsequent analysis was extremely cumbersome and we shall therefore follow the elegant analysis developed by Morar' and Popov (1971) which makes use of the transformation described in §4.9. An alternative solution using Muskhelishvili's complex variable analysis was given by Erdogan (1971).

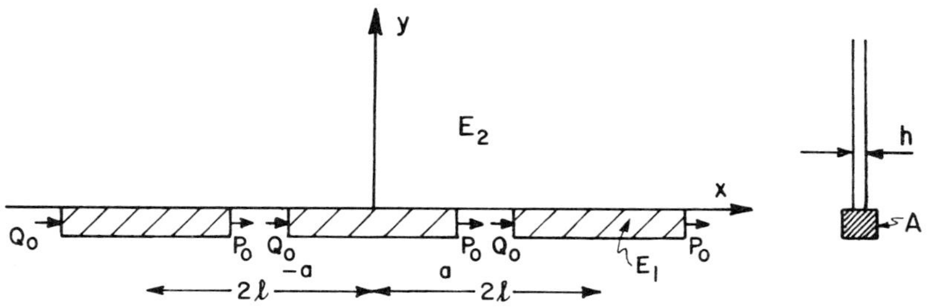

Figure 7.5.1 - A periodically stiffened plate

Start from the integral (2.3.26), namely

$$\frac{1}{\pi}\int_{-1}^{1}\frac{T_n(\tau)d\tau}{(1-\tau^2)^{1/2}(\tau-\zeta)} = U_{n-1}(\zeta) \ , \quad n = 0,1,2,\ldots \ , \ -1 \le \zeta \le 1 \ , \tag{7.5.7}$$

and make the substitutions

$$\left.\begin{array}{l} \pi t = \ell\eta \ , \ \pi x = \ell\xi \ , \ \pi a = \ell\alpha \\ \\ \tau = \tan\eta\cot\alpha \ , \ \zeta = \tan\xi\cot\alpha \end{array}\right\} \tag{7.5.8}$$

to obtain

$$A_n(\xi) = \int_{-\alpha}^{\alpha}\frac{\cot(\eta-\xi)T_n(\tan\eta\cot\alpha)d\eta}{[\cos^2\eta-\cos^2\alpha]^{1/2}\cos\eta} \ , \tag{7.5.9}$$

and

$$A_n(\xi) = \begin{cases} \pi\sec\alpha\tan\xi & , \ n = 0 \\ \\ \pi\operatorname{cosec}\alpha\sec^2\xi U_{n-1}(\tan\xi\cot\alpha) \ , & n = 1,2,\ldots \ , \end{cases} \tag{7.5.10}$$

for $|\xi| \le \alpha$. In a similar way we may start from the integral (7.3.12), namely

$$\frac{1}{2}\int_{-1}^{1}\frac{\text{sign}(\tau-\zeta)T_n(\tau)d\tau}{(1-\tau^2)^{1/2}} = \begin{cases} -\sin^{-1}\zeta & , \quad n = 0 \\ n^{-1}(1-\zeta^2)^{1/2}U_{n-1}(\zeta) & , \quad n = 1,2,\ldots, \end{cases} \tag{7.5.11}$$

valid for $-1 \leq \zeta \leq 1$, use the substitutions (7.5.8) and obtain

$$B_n(\xi) = \frac{1}{2}\int_{-\alpha}^{\alpha}\frac{\text{sign}(\eta-\xi)T_n(\tan\eta\cot\alpha)d\eta}{[\cos^2\eta-\cos^2\alpha]^{1/2}\cos\eta}, \tag{7.5.12}$$

and

$$B_n(\xi) = \begin{cases} -\sec\alpha\sin^{-1}[\tan\xi\cot\alpha] & , \quad n = 0 \\ 2[n\sin2\alpha\cos\xi]^{-1}[\cos^2\xi-\cos^2\alpha]^{1/2}U_{n-1}(\tan\xi\cot\alpha) , & n = 1,2,\ldots . \end{cases} \tag{7.5.13}$$

Now seek a solution of (7.5.6) of the form

$$s(t) = \sec\eta[\cos^2\eta-\cos^2\alpha]^{-1/2}\sum_{n=0}^{\infty} a_n T_n(\tan\eta\cot\alpha) , \tag{7.5.14}$$

substitute into (7.5.6) and use the relations (7.5.9) and (7.5.12) to obtain

$$\sum_{n=0}^{\infty} a_n A_n(\xi)+\lambda\sum_{n=0}^{\infty} a_n B_n(\xi) = \frac{\lambda(P_0-Q_0)}{2h}\cdot\frac{\ell}{\pi}, \quad -\alpha \leq \xi \leq \alpha . \tag{7.5.15}$$

Now $A_n(\xi)$, $B_n(\xi)$ may be rewritten in terms of ζ; they are

$$A_n(\xi) = \begin{cases} (\pi\sec\alpha\tan\alpha)\zeta & , \quad n = 0 \\ \pi\operatorname{cosec}\alpha(1+\zeta^2\tan^2\alpha)U_{n-1}(\zeta) & , \quad n = 1,2,\ldots, \end{cases} \tag{7.5.16}$$

$$B_n(\xi) = \begin{cases} -\sec\alpha\sin^{-1}\zeta & , \quad n = 0 \\ n^{-1}\sec\alpha(1-\zeta^2)^{1/2}U_{n-1}(\zeta) & , \quad n = 1,2,\ldots . \end{cases} \tag{7.5.17}$$

If therefore these expressions are substituted for $A_n(\xi)$, $B_n(\xi)$ in (7.5.15) and the equation is multiplied by

$\cos\alpha(1-\zeta^2)^{1/2}(1+\zeta^2\tan^2\alpha)^{-1}U_{m-1}(\zeta)$ and integrated over $[-1,1]$, the result is

$$\pi^2\cot\alpha a_m/2 + \lambda\sum_{n=1}^{\infty} c_{m,n}a_n = b_m \ , \quad m = 1,2,\ldots \ , \tag{7.5.18}$$

where

$$c_{m,n} = \frac{1}{n}\int_{-1}^{1}\frac{(1-\zeta^2)U_{n-1}(\zeta)U_{m-1}(\zeta)d\zeta}{1+\zeta^2\tan^2\alpha} \ , \tag{7.5.19}$$

$$b_m = \int_{-1}^{1}\left[\frac{\lambda(P_0-Q_0)}{2h}\frac{\ell}{\pi}\cos\alpha-\pi a_0\zeta\tan\alpha+\lambda a_0\sin^{-1}\zeta\right]\frac{(1-\zeta^2)^{1/2}U_{m-1}(\zeta)d\zeta}{(1+\zeta^2\tan^2\alpha)} \tag{7.5.20}$$

and a_0 is determined from equations (7.5.2), (7.5.14) to be

$$a_0 = (P_0+Q_0)\cos\alpha/(h\ell) \ . \tag{7.5.21}$$

The coefficients $c_{m,n}$ may be calculated from the identity

$$(1+\zeta^2\tan^2\alpha)^{-1} = 2\cot\alpha\sum_{k=0}^{\infty}(-)^k\,(\tan\alpha/2)^{2k+1}U_{2k}(\zeta) \ , \quad |\alpha| < \pi/2 \ , \quad |\zeta| \le 1 \ , \tag{7.5.22}$$

which gives

$$c_{m,n} = 2n^{-1}\cot\alpha\sum_{k=0}^{\infty}(-)^k\,(\tan\alpha/2)^{2k+1}I_{n-1,m-1,2k} \ , \tag{7.5.23}$$

where

$$I_{k,m,n} = \int_{-1}^{1}(1-x^2)U_k(x)U_m(x)U_n(x)dx$$

$$= -\frac{4(k+1)(m+1)(n+1)[\cos(k+m+n)\pi/2]^2}{[(m+1)^2-(k-n)^2][(m+1)^2-(k+n+2)^2]} \ , \quad k,m,n = 0,1,2,\ldots \ . \tag{7.5.24}$$

The coefficients b_m may be calculated by using the following integrals which may be deduced from (7.5.22), namely

$$\int_{-1}^{1} \frac{(1-\zeta^2)^{1/2}U_{m-1}(\zeta)d\zeta}{(1+\zeta^2\tan^2\alpha)} = \pi\cot\alpha\sin(m\pi/2)(\tan\alpha/2)^m \ , \qquad (7.5.25)$$

$$\int_{-1}^{1} \frac{\zeta(1-\zeta^2)^{1/2}U_{m-1}(\zeta)d\zeta}{(1+\zeta^2\tan^2\alpha)} = \pi\cot^2\alpha\sin[(m-1)\pi/2](\tan\alpha/2)^m \ , \qquad (7.5.26)$$

and the integral

$$\int_{-1}^{1} \frac{\sin^{-1}\zeta(1-\zeta^2)^{1/2}U_{m-1}(\zeta)d\zeta}{(1+\zeta^2\tan^2\alpha)} = 8m\cot\alpha\sin^2[(m-1)\pi/2]S_m \ , \qquad (7.5.27)$$

$$S_m = \sum_{k=0}^{\infty} (-)^k \ (\tan\alpha/2)^{2k+1}(2k+1)/[(2k+1)^2-m^2]^2 \ . \qquad (7.5.28)$$

Morar' and Popov show that the system is completely regular for $\lambda < \sqrt{2}\cot\alpha$, and is quasi-completely regular for all λ.

Equations (7.5.23), (7.5.24) show that $c_{m,n} = 0$ when m,n have opposite parity. In the antisymmetrical case, when each stiffener is loaded by equal forces at its ends, so that $P_0 = Q_0$, then equations (7.5.26), (7.5.27) show that $b_m = 0$, m odd, so that $a_m = 0$, m odd and

$$s(t) = (1+\tau^2\tan^2\alpha)\mathrm{cosec}\alpha(1-\tau^2)^{-1/2}\sum_{n=0}^{\infty} a_{2n}T_{2n}(\tau) \ , \qquad (7.5.29)$$

and the normal force in the stiffener is

$$AE_1e_{xx}^{(1)}(x) = h\int_{-a}^{x} s(t)dt - P_0$$

$$= \pi^{-1}\ell h\sec\alpha\int_{-1}^{\zeta} (1-\tau^2)^{-1/2}\sum_{n=0}^{\infty} a_nT_{2n}(\tau)d\tau-P_0$$

$$= \pi^{-1}\ell h\sec\alpha\left\{a_0\sin^{-1}\zeta-(1-\zeta^2)^{1/2}\sum_{n=1}^{\infty}(2n)^{-1}a_nU_{2n-1}(\zeta)\right\} . \quad (7.5.30)$$

In the symmetrical case when $Q_0 = -P_0$, $a_0 = 0$ and $b_m = 0$ for m even. Now $a_m = 0$, m even and

$$s(t) = (1+\tau^2\tan^2\alpha)\operatorname{cosec}\alpha(1-\tau^2)^{-1/2}\sum_{n=0}^{\infty} a_{2n+1}T_{2n+1}(\tau) , \quad (7.5.31)$$

$$AE_1e_{xx}^{(1)}(x) = P_0-\pi^{-1}\ell h\sec\alpha(1-\zeta^2)^{1/2}\sum_{n=0}^{\infty}(2n+1)^{-1}a_{2n+1}U_{2n}(\zeta) . \quad (7.5.32)$$

7.6 *Refinements of Melan's problems*

Muki and Sternberg (1967), in their definitive analysis of Melan's problem, considered in detail the following three problems:

(1) Melan's original problem of an infinite edge-stiffened plate

(2) an intermediate problem in which the bending stiffness of the stiffener is taken into account within the framework of elementary beam theory

(3) a problem wholly within classical thin plate theory in which the stiffener is assumed to be a plate of thickness h_1, Poisson's ratio ν_1, shear modulus μ_1 and width a, as shown in Figure 7.6.1. Problem (1) was discussed in §7.2, and problem (2), which was solved also by Büfler (1964), will now be formulated by using the solutions for the upper half-plane and strip derived in §5.4, §6.3 respectively. The boundary conditions on the line y = 0 are

$$u^{(1)}(x,0) = u^{(2)}(x,0) , \quad v^{(1)}(x,0) = v^{(2)}(x,0) , \quad (7.6.1)$$

$$h_1\tau_{xy}^{(1)}(x,0) = h_2\tau_{xy}^{(2)}(x,0) , \quad h_1\tau_{yy}^{(1)}(x,0) = h_2\tau_{yy}^{(2)}(x,0) , \quad (7.6.2)$$

and since the edge y = -a is stress-free

$$\tau_{xy}^{(1)}(x,-a) = 0 = \tau_{yy}^{(1)}(x,-a) \ . \qquad (7.6.3)$$

Following Muki and Sternberg, we assume that the applied body-force field applied to the stiffener is independent of y, parallel to the x-axis, an even function of x vanishing outside the band $-\varepsilon < x < \varepsilon$, and has resultant P, so that

$$F_x(x,y) = f(x)/h_1 \ , \quad F_y(x,y) = 0 \ , \qquad (7.6.4)$$

and

$$a\int_{-\varepsilon}^{\varepsilon} f(x)dx = P_0 \ . \qquad (7.6.5)$$

Melan's problem is obtained by taking the limit as $\varepsilon \to 0$.

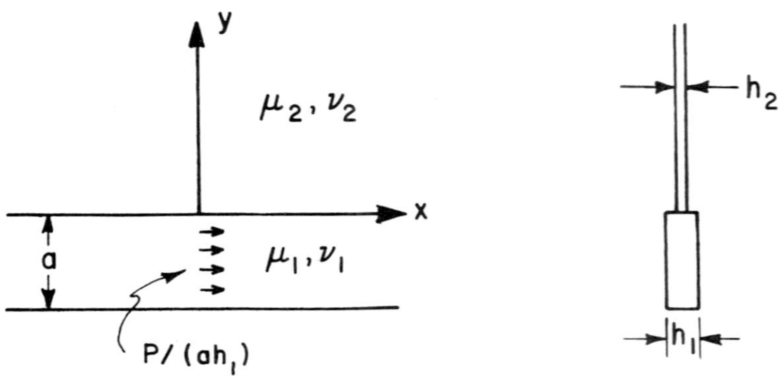

Figure 7.6.1 - An edge stiffener modelled as a thin strip

The problem is antisymmetric in the sense of Example 5.4.2 and Example 6.3.4. We therefore replace A,C of (5.4.6) and A,B,C,D of (6.3.12) by $-iA_1$, $-iC_1$, $-iA_2$, $-iB_2$, $-iC_2$, $-iD_2$ respectively; if now U, T_{12} are the F_c transforms of u, τ_{xy}, while

V, T_{11}, T_{22} are the F_s transforms of v, τ_{xx}, τ_{yy}, then the values of the transforms for the upper half-plane on $y = 0$ are given by

$$2\mu_2\xi U = C_2 , \quad -2\mu_2\xi V = (3-4\nu_2)A_2+C_2 , \tag{7.6.6}$$

$$-T_{12}^{(2)} = (1-2\nu_2)A_2+C_2 , \quad T_{22}^{(2)} = 2(1-\nu_2)A_2+C_2 . \tag{7.6.7}$$

The values of the corresponding transforms for the strip are

$$2\mu_1\xi U = C_1+D_1+(1-2\nu_1)F , \quad -2\mu_1\xi V = (3-4\nu_1)A_1+C_1+(3-4\nu_1)B_1-D_1 , \tag{7.6.8}$$

$$\left.\begin{aligned} -T_{12}^{(1)} &= (1-2\nu_1)A_1+C_1+(1-2\nu_1)B_1-D_1 , \\ T_{22}^{(1)} &= 2(1-\nu_1)A_1+C_1-2(1-\nu_1)B_1+D_1-\nu_1F . \end{aligned}\right\} \tag{7.6.9}$$

Here $F = F_1/[(1-\nu_1)\xi h_1]$, $F_1 = F_c[f(x);\xi]$ and the terms involving F are taken from Example 5.4.5. In addition it should be noted that since the plates are in generalized plane stress, ν_i $(i=1,2)$ should be interpreted as $\nu_i^* = \nu_i/(1+\nu_i)$. Finally the boundary conditions on $y = -a$ yield

$$[(1-2\nu_1)A_1-sA_1+C_1]\exp(s) + [(1-2\nu_1)B_1+sB_1-D_1]\exp(-s) = 0 , \tag{7.6.10}$$

$$[2(1-\nu_1)A_1-sA_1+C_1]\exp(s) + [-2(1-\nu_1)B_1-sB_1+D_1]\exp(-s)-\nu_1F = 0 , \tag{7.6.11}$$

where $s = a\xi$.

Equations (7.6.10), (7.6.11) yield

$$2C_1 = [2s-(3-4\nu_1)]A_1 + B_1\exp(-2s) + \nu_1F\exp(-s) , \tag{7.6.12}$$

$$2D_1 = [2s+(3-4\nu_1)]B_1 - A_1\exp(2s) + \nu_1F\exp(s) . \tag{7.6.13}$$

Now equate the two expressions for U, V to obtain

$$\frac{\mu_1}{\mu_2} C_2 = C_1+D_1+(1-2\nu_1)F \ , \quad \frac{\mu_1}{\mu_2}(3-4\nu_2)A_2 = (3-4\nu_1)(A_1+B_1)-2D_1-(1-2\nu_1)F \ . \tag{7.6.14}$$

Another expression for each of A_2, C_2 may be obtained by using the stress conditions on $y = 0$; A_2, C_2 may now be eliminated from these four equations to yield two equations for A_1, B_1, namely

$$\{4(1-\nu_1)-\gamma[1-\exp(2s)]\}A_1-2\gamma sB_1 = \{1-\nu_1-\gamma\nu_1[1-\exp(s)]\}F \ , \tag{7.6.15}$$

$$2\beta sA_1-\{4(1-\nu_1)+\beta[1-\exp(-2s)]\}B_1 = \{1-\nu_1+\beta\nu_1[1-\exp(-s)]\}F \ , \tag{7.6.16}$$

where

$$\beta = \frac{\mu_1 h_1}{\mu_2 h_2} - 1 \ , \quad \gamma = \frac{\mu_1 h_1}{\mu_2 h_2}(3-4\nu_2) + 1 \ . \tag{7.6.17}$$

The solution of equations (7.6.15), (7.6.16) may conveniently be expressed by introducing $p(s)$, $q(s)$ defined by

$$p = -1 + 4\Delta^{-1}\{\gamma\nu_1[\beta+4(1-\nu_1)]\exp(s) - \gamma\beta\nu_1(1-2s)\exp(-s) -$$

$$-\beta(1-\nu_1-\gamma\nu_1)\exp(-2s) + (1-\nu_1-\gamma\nu_1)[\beta+4(1-\nu_1)] - 2(1-\nu_1+\beta\nu_1)\gamma s\} \ , \tag{7.6.18}$$

$$q = 1-4\Delta^{-1}\{\beta\nu_1[\gamma-4(1-\nu_1)]\exp(-s) - \beta\gamma\nu_1(1+2s)\exp(s) +$$

$$+ \gamma(1-\nu_1+\beta\nu_1)\exp(2s) - (1-\nu_1+\beta\nu_1)[\gamma-4(1-\nu_1)] - 2(1-\nu_1-\gamma\nu_1)\beta s\} \ , \tag{7.6.19}$$

where Δ is the determinant of coefficients of (7.6.15), (7.6.16), namely

$$\Delta = \gamma[\beta+4(1-\nu)]\exp(2s)+\beta[\gamma-4(1-\nu)]\exp(-2s)-[\beta+4(1-\nu)][\gamma-4(1-\nu)] -$$

$$- (1+4\beta\gamma)s^2 \ . \tag{7.6.20}$$

In the limit as $\varepsilon \to 0$, $F_1 = (2\pi)^{-1/2} P_0/a$, so that

$$F = (2\pi)^{-1/2} P_0/[(1-\nu)a\xi h_1] \ , \tag{7.6.21}$$

and

$$4A_1 = (p+1)F \ , \qquad 4B_1 = (q-1)F \ . \tag{7.6.22}$$

Now equations (7.6.9) yield

$$\beta(T_{22}^{(1)} - T_{12}^{(1)}) = 4(1-\nu_1)B_1 + (1-\nu_1)F = (1-\nu_1)qF \ , \tag{7.6.23}$$

while equations (7.6.9) and (7.6.22) give

$$\gamma(T_{22}^{(1)} + T_{12}^{(1)}) = 4(1-\nu_1)A_1 - (1-\nu_1)F = (1-\nu_1)pF \ , \tag{7.6.24}$$

so that, on combining (7.6.21), (7.6.23), (7.6.24), we have

$$h_1\tau_{xy}^{(1)}(x,0) = h_2\tau_{xy}^{(2)}(x,0) = \frac{P}{2\pi a\beta\gamma}\int_0^\infty [\beta p(s) - \gamma q(s)]\, \frac{\cos(\xi x)}{\xi}\, d\xi \ , \tag{7.6.25}$$

$$h_1\tau_{yy}^{(1)}(x,0) = h_2\tau_{yy}^{(2)}(x,0) = \frac{P}{2\pi a\beta\gamma}\int_0^\infty [\beta p(s) + \gamma q(s)]\, \frac{\sin(\xi x)}{\xi}\, d\xi \ . \tag{7.6.26}$$

The remaining coefficients A_2, C_2, C_1, D_1 may now be found from equations (7.6.12)-(7.6.14).

Of particular interest is the form of the stresses in the plate for large and small r. The values of the stresses for small r may be found, as usual, by determining the form of the transforms for large ξ. It may be shown that for large ξ,s

$$A_2 \simeq -\frac{h_1}{h_2}\frac{(1-\nu_1)}{\gamma} F \ , \qquad (3-4\nu_2)A_2 + C_2 \simeq \frac{h_1}{h_2}\frac{(1-\nu_1)(1-4\nu_1)}{\beta+4(1-\nu_1)} F \ , \tag{7.6.27}$$

so that, noting the form of F, using the transforms given in Example 7.2.2, and replacing ν_1 by $\nu_1^* = \nu_1/(1+\nu_1)$, we find

$$\tau_{xx} = -\frac{P}{4\pi\gamma a h_2}\left\{(3+\alpha)(\pi-2\theta)-2\sin 2\theta\right\} + o(1) \ , \qquad (7.6.28)$$

$$\tau_{yy} = -\frac{P}{4\pi\gamma a h_2}\left\{(1-\alpha)(\pi-2\theta)+2\sin 2\theta\right\} + o(1) \ , \qquad (7.6.29)$$

$$\tau_{xy} = \frac{P}{2\pi\gamma a h_2}(1+\alpha)\ln r + O(1) \ , \qquad (7.6.30)$$

where

$$\alpha = \frac{(1-4\nu_1^*)\gamma}{\beta+4(1-\nu_1^*)} = \frac{(1-3\nu_1)\gamma}{\beta(1+\nu_1)+4} \ . \qquad (7.6.31)$$

For large r the stresses have the form

$$\tau_{xx} = -\frac{2P}{\pi h_2 r}\left\{\cos^3\theta + \frac{a}{2\omega r}\sin 2\theta[1-4\cos^2\theta+\omega\cos 2\theta]+O(r^{-2})\right\} \ , \qquad (7.6.32)$$

$$\tau_{yy} = -\frac{2P}{\pi h_2 r}\left\{\sin^2\theta\cos\theta + \frac{a}{\omega r}\sin\theta[\cos 3\theta+2\omega\sin^2\theta\cos\theta]+O(r^{-2})\right\} \ , \qquad (7.6.33)$$

$$\tau_{xy} = -\frac{2P}{\pi h_2 r}\left\{\sin\theta\cos^2\theta + \frac{a}{2\omega r}[\cos 4\theta+\omega\sin\theta\sin 3\theta]\right\} + O(r^{-2}) \ , \qquad (7.6.34)$$

where $\omega = h_2E_2/(2h_1E_1)$.

Muki and Sternberg (1967), who obtained these results, noted that the stresses in the far field obtained from Melan's original analysis (Example 7.2.3) and from the 'exact' analysis all reproduce the terms in r^{-1} arising in the unstiffened plate. Moreover the ratio of the terms in r^{-2} in the exact solution to the corresponding terms in the approximate solution is $1 + O(\omega)$ as $\omega \to 0$. Consequently, the departure of Melan's solution from the exact solution becomes less pronounced as the stringer becomes stiffer. Muki and Sternberg compute the interface stresses for different values of $\kappa = \mu_1h_1/\mu_2h_2$ and show that Melan's analysis predicts the shear stress $\tau_{xy}(x,0)$ and the axial force

$$Q(x) = h_1\int_{-a}^{0}\tau_{11}^{(1)}(x,y)dy$$

extremely closely for all the parameter values chosen. Figures 7.6.2 - 7.6.4, adapted from their paper, show the variation of $\tau_{xx}(x,0)$ and $\tau_{yy}(x,0)$ - in Melan's analysis the latter is identically zero - and the stiffener force Q(x).

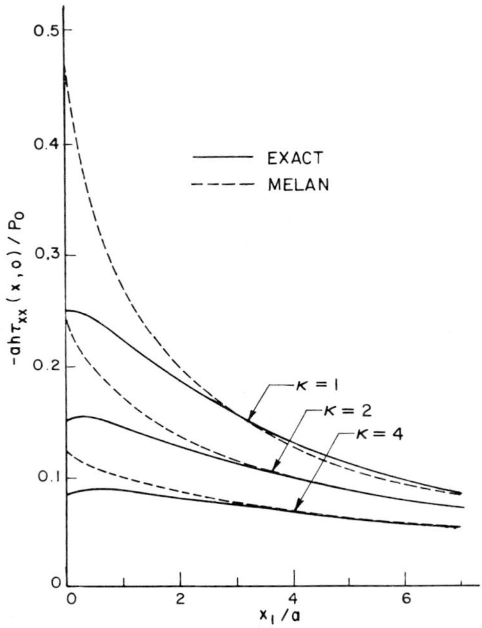

Figure 7.6.2 - Sheet stress τ_{xx} *at interface*

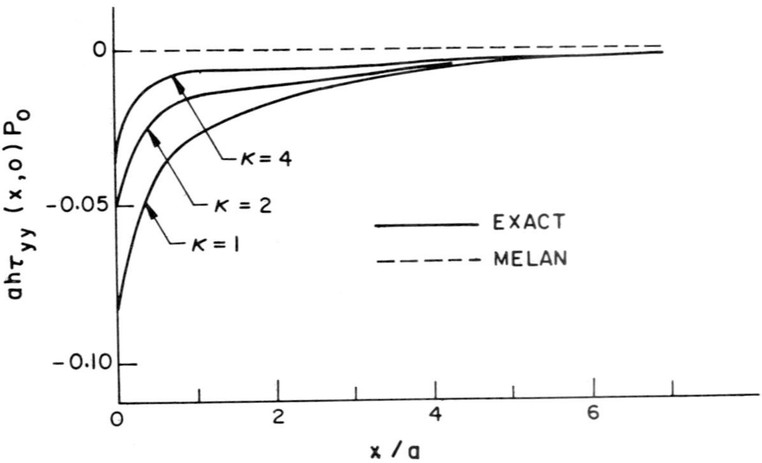

Figure 7.6.3 - Sheet stress τ_{yy} at interface

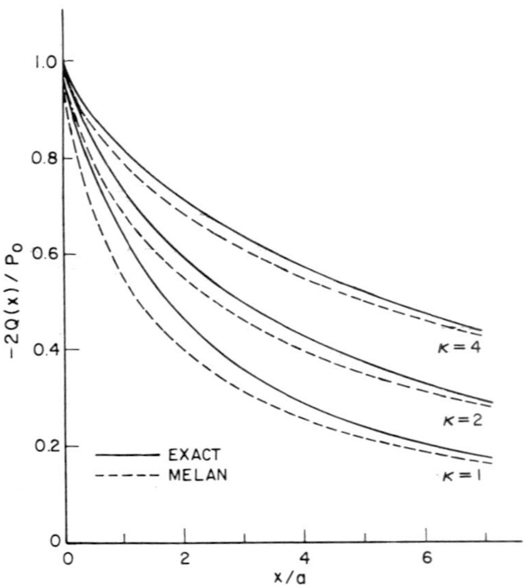

Figure 7.6.4 - Decay of axial stiffener force

Erdogan and Civelek (1974) consider an elastic strip stiffened by an orthotropic elastic stiffening strip and bonded to it either directly or by means of a third, adhesive, strip which acts only in shear. They make a careful study of the effect of the adhesive strip, give numerical results and consider limiting cases.

7.7 *Reissner's problem and its generalizations*

Reissner (1940) considered the diffusion of load from a *transverse* tension bar into a semi-infinite elastic sheet as shown in Figure 7.7.1, where the bar may equivalently be replaced by two equal bars, of half the thickness, placed one on either side of the sheet. Following Reissner's original hypothesis we assume that the stiffener width 2b is so small that the load may be assumed to act on the y-axis, and the stiffener is in a uniaxial state of stress. If $p(y)$ is the bar force at section y then the extensional strain in the bar at section y is

$$e_{yy}^{(1)}(y) = p(y)/AE_1 \ . \qquad (7.7.1)$$

One of the controversies surrounding the problem centred around the question of whether the end-load P_0 will diffuse smoothly into the plate or whether there will be concentrated forces acting on the plate at the ends $y = 0, \ell$ of the stringer. Reissner excluded such forces on physical grounds and Muki and Sternberg (1968a) demonstrated that this is in fact the case. Suppose, if possible, that there are such forces P_1, P_2 as shown, and that $q(y)$ is the shearing force, per unit length of stiffener, at section y. The equilibrium of the stiffener from 0 to y yields

$$p(y) = P_0 - P_1 - \int_0^y q(t')dt' \ , \qquad (7.7.2)$$

so that

$$p(0) = P_0 - P_1 \ , \quad p(\ell) = P_0 - P_1 - \int_0^\ell q(t')dt' = P_2 \ . \qquad (7.7.3)$$

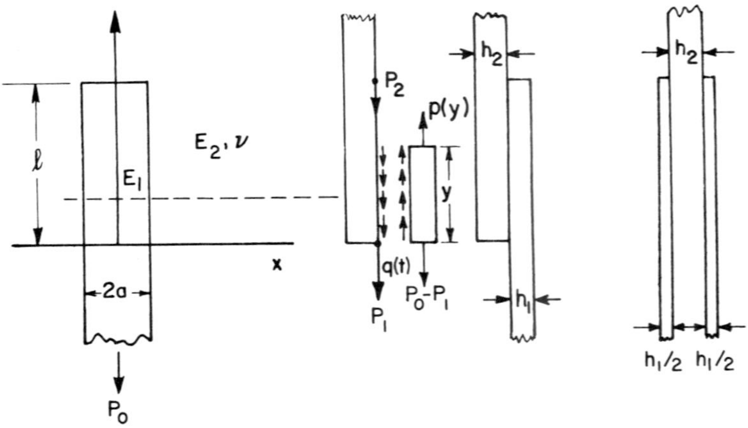

Figure 7.7.1 - A semi-infinite elastic sheet with one or two transverse tension bars attached to it

The displacement $v^{(2)}(0,y)$, and hence the strain $e_{yy}^{(2)}(0,y)$ in the plate at section y, may be determined by using the analysis of §5.8 which relates to an in-plane force applied at an interior point of a half-plane with a free edge, provided that ν is replaced by the $\nu^* = \nu/(1+\nu)$ appropriate to generalized plane stress. If the body force, i.e., applied force per unit volume, in that analysis is identified with $-q(t)/(2ah)$, $-P_1\delta(y)/(2ah)$ and $-P_2\delta(\ell-y)/(2ah)$, in turn, uniformly distributed across the stiffener width, then

$$4\pi\ell hE_2e_{yy}^{(2)}(0,y) = K(y,0)P_1+K(y,\ell)P_2 + \int_0^\ell K(y,t')q(t')dt' \ , \qquad (7.7.4)$$

where

$$K(y,t') = -4\pi\ell E_2 \frac{\partial}{\partial y} K^*_{22}(0,y,t')$$

$$= \ell\left\{-\frac{(3-\nu)(1+\nu)}{t'-y} + \frac{8-(3-\nu)(1+\nu)}{t'+y} + \frac{2(1+\nu)^2 t'(y-t')}{(t'+y)^3}\right\} \quad ,(7.7.5)$$

and K^*_{22} is defined in equation (5.8.30). Now equate the strains $e^{(1)}_{yy}$, $e^{(2)}_{yy}$, using equations (7.7.1), (7.7.4), and write

$$y = \ell\xi, \quad t' = \ell\eta, \quad q(t') = P_0 f(\eta)/\ell, \quad P_i = P_0 G_i \quad (i = 1,2) \ , \quad (7.7.6)$$

$$\lambda = AE_1/(4\pi\ell hE_2) \ , \quad (7.7.7)$$

to obtain

$$1-G_1 - \int_0^\xi f(\eta)d\eta = \lambda[G_1 K(\xi,0)+G_2 K(\xi,1) + \int_0^1 K(\xi,\eta)f(\eta)d\eta] \ . \quad (7.7.8)$$

Equation (7.7.8) is an integral equation for the determination of the shear stress distribution. Following Muki and Sternberg (1968a) we first show that the equation implies $G_1 = 0 = G_2$, i.e., that there are no concentrated forces at the points $y = 0, \ell$. This may be done by showing that the ξ^{-1} and $(1-\xi)^{-1}$ singularities in $K(\xi,0)$ and $K(\xi,1)$ respectively cannot be cancelled by corresponding singularities in the final term on the right-hand side of (7.7.8). The behaviour of this final term may be ascertained by assuming that $f(\eta)$ has the form

$$f(\eta) = \frac{f^*(\eta)}{\eta^\alpha(1-\eta)^\beta} \ , \quad 0 < \alpha < 1 \ , \quad 0 < \beta < 1 \ , \quad (7.7.9)$$

and $f^*(\eta)$ satisfies a Hölder condition on [0,1]. In this case the form of

$$I(\xi) = \int_0^1 K(\xi,\eta)f(\eta)d\eta$$

near the points 0,1 may be deduced from the analysis of §3.2. Thus we write

$$K(\xi,\eta) = -\frac{C}{\eta-\xi}+\frac{8-C}{\eta+\xi}-D\left\{\frac{1}{\eta+\xi}-\frac{3\xi}{(\eta+\xi)^2}+\frac{2\xi^2}{(\eta+\xi)^3}\right\}, \tag{7.7.10}$$

where $C = (3-\nu)(1+\nu)$, $D = 2(1+\nu)^2$. The behaviour of the integral

$$I_1(\xi) = \int_0^1 \frac{f^*(\eta)d\eta}{\eta^\alpha(1-\eta)^\beta(\eta-\xi)} \tag{7.7.11}$$

near $\xi = 0$ may be deduced directly from (3.2.24); it has the form

$$I_1(\xi) = \pi\cot(\alpha\pi)f^*(0)\xi^{-\alpha}+J_1(\xi), \tag{7.7.12}$$

where $\xi^\alpha J_1(\xi) \to 0$ as $\xi \to 0$. The integral

$$I_2(-\xi) = \int_0^1 \frac{f^*(\eta)d\eta}{\eta^\alpha(1-\eta)^\beta(\eta+\xi)} \tag{7.7.13}$$

may be considered to be the value at the point $z = -\xi$ of the integral $I_2(z)$, and note that $I_2(z)$ is continuous across the negative real axis. Now equation (3.2.32) shows that, near $\xi = 0$,

$$I_2(-\xi) = \frac{\pi\exp(i\alpha\pi)}{\sin(\alpha\pi)}f^*(0)(-\xi)^{-\alpha}+J_2(\xi), \tag{7.7.14}$$

where $(-\xi)^{-\alpha} = [\xi\exp(i\pi)]^{-\alpha} = \xi^{-\alpha}\exp(-i\alpha\pi)$ and $J_2(\xi)$ is a power series. Thus

$$I_2(-\xi) = \pi\operatorname{cosec}(\alpha\pi)f^*(0)\xi^{-\alpha}+J_2(\xi). \tag{7.7.15}$$

The remaining integrals involving $(\eta+\xi)^{-2}$ and $(\eta+\xi)^{-3}$ may be obtained by differentiating $I_2(-\xi)$. Thus

$$I_2'(-\xi) = -\int_0^1 \frac{f^*(\eta)d\eta}{\eta^\alpha(1-\eta)^\beta(\eta+\xi)^2} = -\alpha\pi\text{cosec}(\alpha\pi)f^*(0)\xi^{-\alpha-1}+J_2'(\xi), \quad (7.7.16)$$

and

$$I_2''(-\xi) = 2\int_0^1 \frac{f^*(\eta)d\eta}{\eta^\alpha(1-\eta)^\beta(\eta+\xi)^3} = \alpha(\alpha+1)\pi\text{cosec}(\alpha\pi)f^*(0)\xi^{-\alpha-2}+J_2''(\xi) \ . \quad (7.7.17)$$

Collecting the terms we find that, near $\xi = 0$,

$$I(\xi) = \pi\text{cosec}(\alpha\pi)f^*(0)\xi^{-\alpha}\{C\cos[(1-\alpha)\pi]+8-C-(1-\alpha)^2D\}+J(\xi) \ , \quad (7.7.18)$$

where $\xi^\alpha J(\xi) \to 0$ as $\xi \to 0$. Now insert this into equation (7.7.8). Since, with the assumed form (7.7.9) of $f(\eta)$, the integral

$$\int_0^\xi f(\eta)d\eta$$

will not be singular at $\xi = 0$, the coefficients of the ξ^{-1} and ξ^{-c} singularities must separately be zero, so that $G_1 = 0$ and

$$C\cos[(1-\alpha)\pi] + 8-C-(1-\alpha)^2D = 0 \ . \quad (7.7.19)$$

The behaviour near $\xi = 1$ is simpler. Equation (3.2.24) shows that, near $\xi = 1$,

$$I_1(\xi) = -\pi\cot(\beta\pi)f^*(1)(1-\xi)^{-\beta}+J_3(\xi) \ , \quad (7.7.20)$$

where $(1-\xi)^\beta J_3(\xi) \to 0$ as $\xi \to 1$. Thus $G_2 = 0$ and

$$\cot(\beta\pi) = 0 \ . \quad (7.7.21)$$

It may be readily confirmed that if $0 < \nu < 1/2$ then equation (7.7.19) has no complex root whose real part lies in (0,1) and, for this range of ν, has one and only one real root α_* in (0,1);

for $\nu = 0$, the only real root in $[0,1)$ is $\alpha = 0$. The dependence of α_* on ν is shown in Figure 7.7.2, taken from Muki and Sternberg (1968a). Equation (7.7.21) has only real roots, the only one in $(0,1)$ being $\beta = 1/2$.

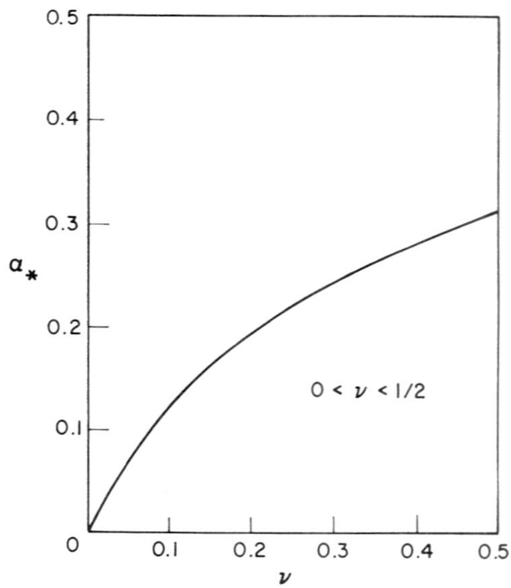

Figure 7.7.2 - Dependence of α_* *on Poisson's ratio* ν

From the above analysis we deduce that

$$\int_0^\xi f(\eta)d\eta + \lambda \int_0^1 K(\xi,\eta)f(\eta)d\eta = 1 \, , \tag{7.7.22}$$

and that $p(0) = P_0$, $p(\ell) = 0$ and

$$\int_0^1 f(\eta)d\eta = 1 \, . \tag{7.7.23}$$

Equation (7.7.22) appears to be still unsolved.

Muki and Sternberg (1968a) consider two possible ways of replacing Reissner's line-contact model of the stringer attachment by one of *area* contact. They consider the strip to be fastened to the plate throughout a strip of adhesion of width 2b and assume the bond force q(y) per unit stringer length to be uniformly distributed across the width of the strip. They then consider two possible bond conditions. In the first, originally proposed in Krahn's (1955) discussion of Goodier and Hsu (1954), the *average* plate strain over the width of the strip at each cross-section y is equated to the corresponding stringer axial strain. Thus

$$e_{yy}^{(1)}(y) = \frac{1}{2b}\int_{-b}^{b} e_{yy}^{(2)}(x,y)\,dx \ , \qquad 0 < y < \ell \ . \tag{7.7.24}$$

This formulation, which will not be carried further here, can be completed by replacing $e_{yy}^{(2)}(x,y)$ by an expression similar to (7.7.4) by using the influence function $K_{22}^{*}(x,y,t)$ of §5.8. The resulting integral equation, which is completely solved by Muki and Sternberg, is found to give results which are, except in the immediate neighbourhood of y = 0, quantitatively similar to those obtained by using the *centre-line* matching scheme, proposed by Lefevre, Mudge and Dickie (1966) and corrected by Muki and Sternberg (1968a), in which the stringer axial strain is equated to the strain on the centre line x = 0 due to the distributed bond force q(y), i.e.,

$$e_{yy}^{(1)}(y) = e_{yy}^{(2)}(0,y) \ , \qquad 0 < y < \ell \ . \tag{7.7.25}$$

In both schemes part of the load is transmitted to the plate at the ends $y = 0,\ell$, and in both cases the forces P_1, P_2 can be determined only in the course of solving the integral equation governing the contact. The centre-line matching scheme runs as follows.

The displacement $v^{(2)}(0,y)$, and hence the strain $e_{yy}^{(2)}(0,y)$ in the plate at section y, may again be determined by using the analysis of §5.4. In fact all that is needed is a modification of equation (7.7.4) to take account of the fact that the bonding forces both at the ends and along the length of the stiffener are properly distributed across the stiffener width. Thus,

$$4\pi\ell h_2 E_2 e_{yy}^{(2)}(0,y) = \overset{\circ}{K}(y,0)P_1+\overset{\circ}{K}(y,\ell)P_2 + \int_0^{\ell} \overset{\circ}{K}(y,t')q(t')dt' \quad , \qquad (7.7.26)$$

where now

$$\overset{\circ}{K}(y,t') = -\frac{4\pi\ell}{2b} E_2 \int_{-b}^{b} \frac{\partial}{\partial y} K_{22}^*(-t,y,t')dt \quad , \qquad (7.7.27)$$

and K_{22}^* is defined in (5.8.30), with ν replaced by ν^*. Now $\overset{\circ}{K}(y,t')$ involves inverse tan functions and is not a convenient kernel for numerical computation; it has, moreover, a finite jump discontinuity at $t' = y$; the discontinuity arises from the terms $-(3-4\nu^*)\ell n r_1 - x^2 r_1^{-2}$ in K_{22}^*. It may easily be shown that

$$\overset{\circ}{K}(y,y+) - \overset{\circ}{K}(y,y-) = -2\pi(1-\nu^2)\ell/b \ . \qquad (7.7.28)$$

If therefore we integrate the last term in (7.7.26) by parts, using the equation $\partial p/\partial y = -q(y)$ obtained from (7.7.2), we find

$$4\pi\ell h_2 E_2 e_{yy}^{(2)}(0,y) = \overset{\circ}{K}(y,0)P_1+\overset{\circ}{K}(y,\ell)P_2-[\overset{\circ}{K}(y,t')p(t')]_0^{y-} -$$
$$- [\overset{\circ}{K}(y,t')p(t')]_{y+}^{\ell} + \int_0^{\ell} \frac{\partial\overset{\circ}{K}(y,t')}{\partial t'} p(t')dt' \quad , \qquad (7.7.29)$$

so that, on using equations (7.7.3), (7.7.28), we obtain

$$4\pi\ell h_2 E_2 e_{yy}^{(2)}(0,y) = P_0\overset{\circ}{K}(y,0) - \frac{2\pi(1-\nu^2)\ell}{b} p(y) + \int_0^{\ell} \frac{\partial\overset{\circ}{K}(y,t')}{\partial t'} p(t')dt' \ . \qquad (7.7.30)$$

Now insert equation (7.7.1) for $e_{yy}^{(1)}(y)$ to obtain the dimensionless equation

$$\phi(\xi) - \omega\int_0^1 \hat{K}(\xi,\eta)\phi(\eta)d\eta = \omega\hat{M}(\xi) , \tag{7.7.31}$$

where ξ,η,λ are given in (7.7.6), with $h = h_2$, and

$$\phi(\xi) = \frac{p(y)}{P_0} , \quad \gamma = \frac{b}{\ell} , \quad \omega = \frac{\lambda}{4\gamma[\gamma+2(1-\nu^2)\pi\lambda]} , \tag{7.7.32}$$

$$\hat{K}(\xi,\eta) = 4\Bigg\{(1+\nu)(3-\nu)\rho_1^{-2}-2(1+\nu)^2\gamma^2\rho_1^{-4}-[4+(1-\nu)^2]\rho_2^{-2}+2(1+\nu)(3-\nu)\gamma^2\rho_2^{-4}$$

$$- 2\frac{(1+\nu)^2}{(\xi+\eta)^2}\left[\frac{\xi\eta}{\rho_2^2}+\frac{[(\xi+\eta)^2-\gamma^2][\xi^2-\xi\eta-\eta^2]}{\rho_2^4}+2\xi\eta\frac{[(\xi+\eta)^2-\gamma^2]^2}{\rho_2^6}\right]\Bigg\} , \tag{7.7.33}$$

where $\rho_1^2 = (\xi-\eta)^2+\gamma^2$, $\rho_2^2 = (\xi+\eta)^2+\gamma^2$; $\hat{K}(\xi,\eta)$ is obtained as the dimensionless form of $\partial\overset{\circ}{K}/\partial t'$ after carrying out the integration in (7.7.30). The kernel $\hat{K}(\xi,\eta)$ is analytic in the open square $0 < \xi < 1$, $0 < \eta < 1$ and continuous on the closed square except for a finite discontinuity at the origin. Near the origin the discontinuous term in square brackets is

$$- \frac{8(1+\nu)^2}{(\xi+\eta)^2}(4\xi\eta-\xi^2+\eta^2) \rightarrow -8(1+\nu)^2\frac{(4m-m^2+1)}{(m+1)^2}$$

as (ξ,η) approaches $(0,0)$ along $\xi = m\eta$. The function $\hat{M}(\xi)$ is defined by

$$\hat{M}(\xi) = \frac{16}{\gamma}(1-\eta)\tan^{-1}\left(\frac{\gamma}{\xi}\right) + \frac{16(1+\nu)\xi}{\xi^2+\eta^2} , \quad \xi \neq 0 , \quad \hat{M}(0) = \frac{8(1-\nu)\pi}{\gamma} . \tag{7.7 34}$$

Equation (7.7.31) is readily amenable to a numerical treatment and Figures 7.7.3 and 7.7.4, taken from Muki and Sternberg (1968a) show a comparison between results obtained from both methods and from experiment. Figure 7.7.3 relates to

$$\lambda = \frac{AE_1}{4\pi\ell h_2E_2} = \frac{1}{2\pi}\cdot\frac{b}{\ell}\cdot\frac{h_1E_1}{h_2E_2} = \frac{1}{2\pi}\cdot\frac{1}{12}\cdot 1 , \quad \gamma = \frac{b}{\ell} = \frac{1}{12} , \quad \nu = 0.3 , \tag{7.7.35}$$

and shows that $P_1/P_0 = 1-p(0)/P_0 = 1-\phi(0) = 0.6$, $P_2/P_0 = 0.08$, so that 60% of the load is communicated at $y = 0$, 8% at $y = \ell$. Muki and Sternberg point out that the slope of the dashed curve is finite at the two ends while that of the solid line, corresponding to the average strain matching, is infinite (with a logarithmic singularity) at both ends; the scale of the plot does not show the singularity at $y = \ell$. Figure 7.7.4 relates to the values used by Lefevre, Mudge and Dickie (1966), namely $\lambda = 1/(6\pi)$, $\gamma = 1/6$, $\nu = 0.5$; the discrepancy between theory and experiment at the ends here is not surprising since the slenderness ratio $2\ell/b = 3$ is far too small for the approximate analysis to yield realistic results.

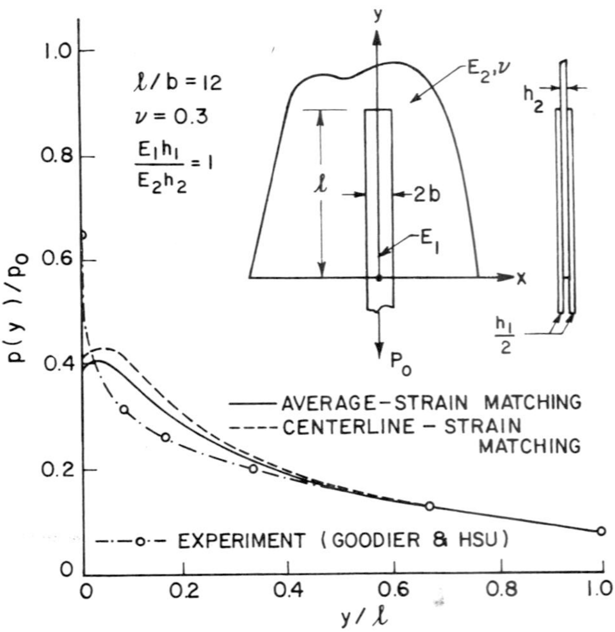

Figure 7.7.3 - Area contact. Comparison of theoretical load diffusion with experiment

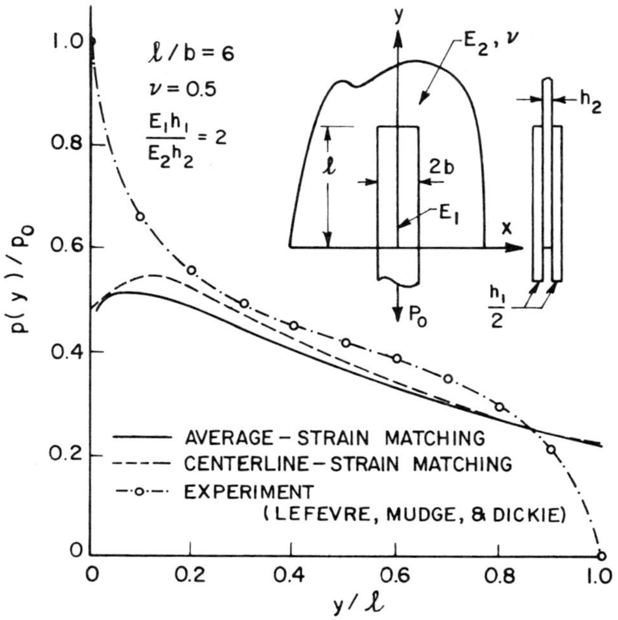

Figure 7.7.4 - Area contact. Comparison of theoretical load diffusion with experiment

Arutiunian and Mkhitarian (1972) present a broad-based view of contact problems for a half-space reinforced in various ways. They consider slender stiffeners which may be finite, semi-infinite or infinite.

CHAPTER VIII - PLANE PROBLEMS RELATING TO A CIRCLE

"Or madly squeeze a right-hand foot
Into a left-hand shoe"
Through the Looking Glass

8.1 *Introduction*

In almost all the plane contact problems which have been considered in earlier chapters it has been assumed that the contact region is *straight*, usually part of the real axis. Thus, for instance, for the indentation of a half-plane by a parabolic punch (§2.4) the boundary conditions were applied on the real axis; such problems are often loosely termed *Hertzian* contact problems. Hertz, in his study of the contact between two curved bodies in 3-dimensions, assumed that, in the region of contact, each body could be considered as a half-space with an infinitesimal displacement. The basic reason why this assumption is made is that it simplifies the analysis; the fact that results deduced on the basis of this assumption agree with experimental results for a wide range of practical problems indicates that this assumption is reasonable.

It should be noted that Hertz made *two* assumptions: first, that the bodies could be approximated by *half-spaces*, and secondly, that the boundary conditions could be applied on the *undeformed* surface, in this case $z = 0$. The first is the characteristic assumption of Hertzian contact, while the second is one of the basic assumptions of infinitesimal elasticity theory.

This chapter is concerned with plane contact problems in which one or both of the undeformed surfaces of the contacting bodies, on which the boundary conditions are applied, are *circular*. First we construct a complex variable form of the equations of plane elasticity which is appropriate to a circular boundary:

this is a severely abbreviated version of the general theory described in Muskhelishvili (1953a). Then, after constructing some particular functions which lead to an algebraic simplification of the computation of Cauchy integrals over an arc of a circle, we apply the theory to some typical indentation, inclusion and cradling problems related to a circular boundary.

8.2 *Complex variable formulation*

In the analysis of Chapter 4 it was assumed that the elastic body occupied the upper half-plane. Now instead it will be assumed that the body occupies either the inside, or the outside, of a circle, or some region which may be mapped onto the exterior of a circle by using a conformal transformation. We shall first consider in detail the case when it occupies the *interior* of the circle, which will be denoted by $\mathcal{D}^+$, and is deformed under the action of surface tractions applied to $|z| = a$. The remaining cases may then be discussed by making minor modifications to the analysis.

We first recall that, in the absence of body forces, the Cartesian displacements u, v and the polar stresses are linked to the complex potentials of §4.2 by the equations

$$2\mu(u+iv) = \kappa\phi(z)-z\overline{\phi}'(\overline{z})-\overline{\psi}(\overline{z}) \ , \qquad (8.2.1)$$

$$\Theta = \tau_{rr}+\tau_{\theta\theta} = 2[\Phi(z)+\overline{\Phi}(\overline{z})] \ , \qquad (8.2.2)$$

$$\Xi\exp(2i\theta) = \tau_{\theta\theta}-\tau_{rr}+2i\tau_{r\theta} = 2\exp(2i\theta)[\overline{z}\Phi'(z)+\Psi(z)] \ , \qquad (8.2.3)$$

where the last two equations are obtained by combining equations (4.2.15), (4.2.16) and Example 4.2.3. The functions $\Phi(z)$, $\Psi(z)$ are defined in $\mathcal{D}^+$, i.e., $|z| < a$; and by arguing as in §4.3 we may show that $\Phi(z)$, $\Psi(z)$ are holomorphic in $\mathcal{D}^+$. The function $\Phi(z)$ will be continued into $\mathcal{D}^-$, i.e., $|z| > a$ by using the equation

$$\Phi(z) = -\overline{\Phi}(z_*)+z_*\overline{\Phi}'(z_*)+(z_*/a)^2\overline{\Psi}(z_*) , \quad z \text{ in } \mathcal{D}^- , \tag{8.2.4}$$

where $z_* = a^2/z$: on C $z_* = \overline{z}$.

This equation has been chosen to make the stress combination $\tau_{rr}+i\tau_{r\theta}$ expecially simple, as shown below. In equation (8.2.4) substitute $\overline{z}_*$ for z, assuming now that z is in $\mathcal{D}^+$; then

$$\Phi(\overline{z}_*) = -\overline{\Phi}(\overline{z})+\overline{z}\overline{\Phi}'(\overline{z})+(\overline{z}/a)^2\overline{\Psi}(\overline{z}) , \quad z \text{ in } \mathcal{D}^+ . \tag{8.2.5}$$

This equation may be viewed as an equation for $\Psi(z)$, for $|z| < a$, in terms of $\Phi(z)$ and $\Phi(z_*)$, i.e.,

$$a^{-2}\Psi(z) = z^{-2}\Phi(z)+z^{-2}\overline{\Phi}(z_*)-z^{-1}\Phi'(z) , \quad z \text{ in } \mathcal{D}^+ . \tag{8.2.6}$$

Equations (8.2.2), (8.2.3) express the stresses in terms of $\Phi(z)$ and $\Psi(z)$ for z in $\mathcal{D}^+$; equation (8.2.6) expresses $\Psi(z)$ in $\mathcal{D}^+$ in terms of $\Phi(z)$ in $\mathcal{D}^+$ and $\mathcal{D}^-$, so that the stresses have been expressed entirely in terms of $\Phi(z)$. If Φ, Ψ are holomorphic in $\mathcal{D}^+$ then equation (8.2.4) gives a Φ which is holomorphic in $\mathcal{D}^-$ and, in addition, bounded at infinity. On the other hand, if $\Phi(z)$ is holomorphic in $\mathcal{D}^+$ and in $\mathcal{D}^-$ so that

$$\Phi(z) = A_0+A_1z+A_2z^2+ \cdots , \quad z \text{ in } \mathcal{D}^+ , \tag{8.2.7}$$

$$\Phi(z) = B_0+B_1z^{-1}+B_2z^{-2}+ \cdots , \quad z \text{ in } \mathcal{D}^- , \tag{8.2.8}$$

then $\Psi(z)$ defined by (8.2.6) will have a pole at $z = 0$ unless

$$A_0+\overline{B}_0 = 0 , \quad B_1 = 0 . \tag{8.2.9}$$

Equations (8.2.2), (8.2.3) and the fact that $\exp(-2i\theta) = \overline{z}z^{-1}$ yield

$$\tau_{rr}+i\tau_{r\theta} = \Phi(z)+\overline{\Phi}(\overline{z})-\overline{z}z^{-1}[z\overline{\Phi}'(\overline{z})+\overline{\Psi}(\overline{z})] , \quad z \text{ in } \mathcal{D}^+ . \tag{8.2.10}$$

But equation (8.2.5) shows that

$$\overline{\Phi}(\bar{z})-\bar{z}\overline{\Phi}'(\bar{z}) = a^{-2}\bar{z}^{2}\overline{\Psi}(\bar{z})-\Phi(\bar{z}_*) \ , \tag{8.2.11}$$

so that

$$\tau_{rr}+i\tau_{r\theta} = \Phi(z)-\Phi(\bar{z}_*)+a^{-2}\bar{z}(\bar{z}-z_*)\overline{\Psi}(\bar{z}) \ , \quad z \text{ in } \mathcal{D}^+ \ . \tag{8.2.12}$$

A similar expression for the derivatives of the displacements may be obtained by differentiating equation (8.2.1) with respect to θ; thus

$$\frac{d\phi(z)}{d\theta} = \frac{d\phi(z)}{dz}\frac{dz}{d\theta} = \Phi(z)ir\exp(i\theta) = iz\Phi(z) \ , \tag{8.2.13}$$

$$2\mu(u'+iv') = [\kappa\Phi(z)-\overline{\Phi}(\bar{z})+\bar{z}\overline{\Phi}'(\bar{z})+\bar{z}z^{-1}\overline{\Psi}(\bar{z})] \ , \tag{8.2.14}$$

where remember that $\Phi(z) = \phi'(z)$ and

$$u' = \frac{\partial u}{\partial z} \ , \quad v' = \frac{\partial v}{\partial z} \ . \tag{8.2.15}$$

Now substitute for the expression $\overline{\Phi}(\bar{z})-\bar{z}\overline{\Phi}'(\bar{z})$ by using equation (8.2.11) to obtain

$$2\mu(u'+iv') = [\kappa\Phi(z)+\Phi(\bar{z}_*)-a^{-2}\bar{z}(\bar{z}-z_*)\overline{\Psi}(\bar{z})] \ , \quad z \text{ in } \mathcal{D}^+ \ . \tag{8.2.16}$$

Note that in equations (8.2.12), (8.2.16), $\Psi(z)$ must be understood to be given by equation (8.2.6).

Now we introduce conditions to take the place of (4.4.14), (4.4.15). It will be assumed that $\Phi(z)$ is continuous on the boundary C of the circle $|z| = a$ as z approaches from $\mathcal{D}^+$ and from $\mathcal{D}^-$, except possibly at a finite number of points c_k of C near which

$$|\Phi(z)| < A|z-c_k|^{-\alpha} \ , \quad 0 \leq \alpha < 1 \ . \tag{8.2.17}$$

In addition, it will be assumed that

$$\lim_{r \to a} (a-r)\Phi'(z) = 0 \,, \quad z = r\exp(i\theta) \tag{8.2.18}$$

for all values of θ, except possibly at the points c_k. It is easily seen from equation (8.2.6) that these conditions imply that

$$\lim_{r \to a} (\bar{z}-z_*)\Psi(z) = \lim_{r \to a} \exp(-i\theta)(r-a^2r^{-1})\Psi(z) = 0 \,. \tag{8.2.19}$$

Now use the conditions to evaluate the limiting values of equations (8.2.12), (8.2.16) as z approaches $t = a\exp(i\theta)$ from $\mathcal{D}^+$; we have

$$\tau_{rr}+i\tau_{r\theta} = \Phi^+(t)-\Phi^-(t) \,, \tag{8.2.20}$$

$$2\mu(u'+iv') = \kappa\Phi^+(t)+\Phi^-(t) \,. \tag{8.2.21}$$

Equation (8.2.20) shows that $\Phi(z)$ behaves just like the corresponding $\Phi(z)$ for the half-plane; it is continuous across the unloaded part of the boundary C. It is now possible to formulate and solve a class of boundary value problems for the circle, just as was done previously for the half-plane. Let C' be the union of a finite number of arcs of the circle, and C'' the remainder of C. Suppose that the displacements are given on C', and that C'' is unloaded, i.e.,

$$u+iv = g(t) \text{ on } C' \,, \tag{8.2.22}$$

$$\tau_{rr}+i\tau_{r\theta} = 0 \text{ on } C'' \,, \tag{8.2.23}$$

then equations (8.2.20), (8.2.21) give

$$\kappa\Phi^+(t)+\Phi^-(t) = 2\mu g'(t) \text{ on } C' \,, \tag{8.2.24}$$

$$\Phi^{+}(t)-\Phi^{-}(t) = 0 \text{ on } C'' , \tag{8.2.25}$$

where

$$g'(t) = \frac{dg}{dt} . \tag{8.2.26}$$

The full solution of equations (8.2.24), (8.2.25), when C' is the union of a finite number of arcs, is given in Muskhelishvili (1953a, Chapter 18), (1953b, §104); it is strictly analogous to that appearing in the half-plane problem. When for instance C' consists of one segment of the circle from $t = t_1$ to $t = t_2$ the general solution is

$$\Phi(z) = \frac{\mu}{\pi i \kappa} X_0(z) \int_{C'} \frac{g'(t)dt}{X_0^{+}(t)(t-z)} + (Az+B)X_0(z) , \tag{8.2.27}$$

where

$$X_0(z) = (z-t_1)^{-1/2+i\alpha}(z-t_2)^{-1/2-i\alpha} , \tag{8.2.28}$$

$$\kappa = \exp(2\pi\alpha) \text{ and } \alpha = \alpha_1+i\alpha_2 , -1/2 < \alpha_2 < 1/2 .$$

By $X_0(z)$ is understood that branch that is holomorphic in the plane cut along C' and which is such that

$$\lim_{z\to\infty} zX_0(z) = 1 . \tag{8.2.29}$$

In practice the boundary condition on the stressed part of the boundary is not simply (8.2.22): in this case it is preferable to formulate the problem as a stress boundary value problem, i.e., by assuming (an initially unknown) distribution of stresses on C' so that

$$\tau_{rr}+i\tau_{r\theta} \equiv \Phi^{+}(t)-\Phi^{-}(t) = f(t) \text{ on } C' . \tag{8.2.30}$$

In this case the complex potential $\Phi(z)$ is

$$\Phi(z) = \frac{1}{2\pi i} \int_{C'} \frac{f(t)}{t-z}\, dt + A . \qquad (8.2.31)$$

The constants A_0, B_0 of (8.2.7), (8.2.8) for this function are

$$A_0 = \frac{1}{2\pi i} \int_{C'} \frac{f(t)}{t}\, dt + A , \qquad B_0 = A , \qquad (8.2.32)$$

while

$$B_1 = -\frac{1}{2\pi i} \int_{C'} f(t)dt . \qquad (8.2.33)$$

Thus

$$\Phi(z) = \frac{1}{2\pi i} \int_{C'} \frac{f(t)}{t-z}\, dt - \frac{1}{4\pi i} \int_{C'} \frac{f(t)}{t}\, dt , \qquad (8.2.34)$$

and the equation $B_1 = 0$ gives

$$\int_{C'} (\tau_{rr}+i\tau_{r\theta})e^{i\theta}d\theta = 0 , \qquad (8.2.35)$$

which states that the stresses applied to C' have zero resultant; this will usually be the case. On occasion however the stresses on C' will be equilibrated by body forces and then the whole complex variable formulation has to be slightly altered (see §4.2 and §8.6).

Now suppose that the elastic body occupies the region $\mathcal{D}^-$ outside a simple closed contour C. Serious consideration must now be given to questions of multi-valuedness, as $\mathcal{D}^-$ is not simply connected. Thus immediately the argument which led to equation (4.3.1) breaks down and a new argument must be constructed, as follows. The stresses and displacements are single-valued in $\mathcal{D}^-$. Therefore Θ, and hence $\mathrm{Re}[\Phi(z)]$, is single-valued. But $\mathrm{Im}[\Phi(z)]$ need not be, for, in one circuit around a contour L

surrounding C, $\text{Im}[\Phi(z)]$ may increase by an amount $2\pi iB$, where B is a real constant. Now assume without loss of generality that $z = 0$ lies within C, i.e., in $\mathcal{D}^+$, then ℓnz will increase by $2\pi i$ in one (anti-clockwise) circuit of L so that $\Phi(z) - B\ell nz$ will be single-valued: one may deduce that

$$\Phi(z) = B\ell nz+\Phi^*(z) \ , \tag{8.2.36}$$

where $\Phi^*(z)$ is holomorphic and single-valued in $\mathcal{D}^-$, except possibly at infinity. Integration of this equation shows that (Example 8.2.1)

$$\phi(z) = Bz\ell nz+\gamma\ell nz+\phi^*(z) \ , \tag{8.2.37}$$

where γ is a complex constant and $\phi^*(z)$ is holomorphic in $\mathcal{D}^-$ except possibly at infinity. Now the equation

$$\Xi = -2[z\overline{\Phi}'(\overline{z})+\overline{\Psi}(\overline{z})] \tag{8.2.38}$$

shows that $\Psi(z)$ is holomorphic and single-valued in $\mathcal{D}^-$, so that its integral $\psi(z)$ has the form

$$\psi(z) = \gamma'\ell nz+\psi^*(z) \ . \tag{8.2.39}$$

Now consider the single-valuedness of the displacements: in one circuit of L

$$[2\mu(u+iv)]_L = 2\pi i[(\kappa+1)Bz + (\kappa\gamma+\overline{\gamma}')] \ . \tag{8.2.40}$$

If therefore the displacements are to be single-valued then

$$B = 0 \ , \quad \kappa\gamma+\overline{\gamma}' = 0 \ . \tag{8.2.41}$$

Just as for the half-plane, the constants γ, γ' are linked to the total force applied to the boundary of the body. Thus equation

(4.2.26) shows that the total external force applied to the body over the hole C is

$$X+iY = i[\phi(z)+z\overline{\phi}'(\overline{z})+\overline{\psi}(\overline{z})]_C = -2\pi(\gamma-\overline{\gamma}') . \qquad (8.2.42)$$

Here one uses the convention of §1.4 and notes that if C is traversed anti-clockwise then $\hat{n}$ points outward so that the sign of (4.2.26) must be changed. Equations (8.2.41), (8.2.42) give

$$\gamma = -\frac{(X+iY)}{2\pi(1+\kappa)} , \quad \gamma' = \frac{\kappa(X-iY)}{2\pi(1+\kappa)} . \qquad (8.2.43)$$

Now we may return to Φ, Ψ and deduce as in §4.3 by considering the boundedness of the stresses at infinity that equations (4.3.7), (4.3.9), (4.3.10), (4.3.11) hold with the revised values of γ, γ' given by equation (8.2.43).

If $\mathcal{D}^-$ is the exterior of the unit circle then $\Phi(z)$ defined in $\mathcal{D}^+$ by means of (8.2.4) will be holomorphic in $\mathcal{D}^+$ except at $z = 0$, near which

$$\Phi(z) = \overline{\Gamma}'z^{-2}+\overline{\gamma}'z^{-1}+0(1) . \qquad (8.2.44)$$

Apart from this the above analysis holds with only obvious minor alterations (see Example 8.2.2).

Now suppose that the body occupies a region G^- which may be mapped conformally on to the exterior $\mathcal{D}^-$ of the unit circle by using the transformation $z = \omega(\zeta)$ which is such that the points $z = \infty$ and $\zeta = \infty$ correspond to each other. The most important practical case is that for which G^- is the exterior of the ellipse with axes 2a, 2b ($a > b$). Then

$$\omega(\zeta) = R(\zeta+m\zeta^{-1}) , \quad R > 0 , \quad 0 \leq m < 1 \qquad (8.2.45)$$

and

$$a = R(1+m) , \quad b = R(1-m) . \qquad (8.2.46)$$

Denote by $\phi_1(z)$, $\psi_1(z)$, $\Phi_1(z)$, $\Psi_1(z)$ the functions that were earlier written as $\phi(z)$, $\psi(z)$, $\Phi(z)$, $\Psi(z)$ respectively, and introduce the new notation

$$\phi(\zeta) = \phi_1(z) = \phi_1(\omega(\zeta)) \ , \qquad \psi(\zeta) = \psi_1(z) = \psi_1(\omega(\zeta)) \ , \tag{8.2.47}$$

$$\Phi(\zeta) = \Phi_1(z) = \Phi_1(\omega(\zeta)) = \frac{\phi'(\zeta)}{\omega'(\zeta)} \ , \qquad \Psi(\zeta) = \Psi_1(z) = \frac{\psi'(\zeta)}{\omega'(\zeta)} \ . \tag{8.2.48}$$

Now the functions $\omega'(\zeta)\Phi(\zeta)$, $\omega'(\zeta)\Psi(\zeta)$ will be holomorphic for $\zeta\epsilon\mathcal{D}^-$, i.e., $|\zeta| > 1$, and equations (8.2.1)-(8.2.3) become

$$2\mu(u+iv) = \kappa\phi(\zeta) - \frac{\omega(\zeta)}{\overline{\omega}'(\overline{\zeta})}\,\overline{\phi}'(\overline{\zeta}) - \overline{\psi}(\overline{\zeta}) \ , \tag{8.2.49}$$

$$\Theta = 2[\Phi(\zeta)+\overline{\Phi}(\overline{\zeta})] \ , \tag{8.2.50}$$

$$\Xi = 2[\overline{\omega}(\overline{\zeta})\,\frac{\Phi'(\zeta)}{\omega'(\zeta)} + \Psi(\zeta)] \ . \tag{8.2.51}$$

By putting $\zeta = \rho\exp(i\sigma)$ we may introduce a set of curvilinear coordinates in the z-plane. Thus, for the transformation (8.2.45) the curves ρ = const. are confocal ellipses (see Example 8.2.10) while the curves σ = const. are the orthogonal set of confocal hyperbolae, as shown in Figure 8.2.1. The ellipse (8.2.46) is the curve $\rho = 1$. If the directions at any point z corresponding to ρ increasing and to σ increasing are labelled $\hat{\rho}$ and $\hat{\sigma}$ respectively then the angle which $\hat{\rho}$ makes with the x-axis is δ where (Example 8.2.2)

$$\exp(i\delta) = \frac{\zeta}{\rho}\,\frac{\omega'(\zeta)}{|\omega'(\zeta)|} \ . \tag{8.2.52}$$

Thus equation (8.2.3) holds with θ replaced by δ, so that

$$\tau_{\rho\rho}-\tau_{\sigma\sigma}+2i\tau_{\rho\sigma} = 2\exp(2i\delta)[\overline{\omega}(\overline{\zeta})\,\frac{\Phi'(\zeta)}{\omega'(\zeta)} + \Psi(\zeta)] \tag{8.2.53}$$

and

$$\tau_{\rho\rho}+i\tau_{\rho\sigma} = \overline{\Phi}(\overline{\zeta})+\Phi(\zeta) - \frac{\overline{\zeta}}{\zeta}\frac{\omega(\zeta)}{\omega'(\zeta)}\overline{\Phi}'(\overline{\zeta}) - \frac{\overline{\zeta}}{\zeta}\frac{\overline{\omega}'(\overline{\zeta})}{\omega'(\zeta)}\overline{\Psi}(\overline{\zeta}) \ , \ \zeta \text{ in } G^+ \ . \tag{8.2.54}$$

We now define $\omega'(\zeta)\Phi(\zeta)$ in $\mathcal{D}^+$ by means of the equation

$$\omega'(\zeta)\Phi(\zeta) = -\omega'(\zeta)\overline{\Phi}(\zeta^{-1})+\zeta^{-2}\omega(\zeta)\overline{\Phi}'(\zeta^{-1})+\zeta^{-2}\overline{\omega}'(\zeta^{-1})\overline{\Psi}(\zeta^{-1}) \ . \tag{8.2.55}$$

In the particular case when $\omega(\zeta)$ is given by (8.2.45) then $\omega'(\zeta)\Phi(\zeta)$ will be holomorphic in $\mathcal{D}^+$ except at $\zeta = 0$. The singularity there can be found by assuming that Φ, Ψ have the form (4.3.7), (4.3.9) at infinity, hence

$$\Phi(\zeta) = \Gamma+\gamma z^{-1}+0(z^{-2}) = \Gamma+\gamma R^{-1}\zeta^{-1}+0(\zeta^{-2}) \ , \tag{8.2.56}$$

$$\Psi(\zeta) = \Gamma'+\gamma' z^{-1}+0(z^{-2}) = \Gamma'+\gamma R^{-1}\zeta^{-1}+0(\zeta^{-2}) \ , \tag{8.2.57}$$

so that, near $\zeta = 0$

$$\omega'(\zeta)\Phi(\zeta) = (m\overline{\Gamma}+\overline{\Gamma}')R\zeta^{-2}+\overline{\gamma}'\zeta^{-1}+0(1) \ . \tag{8.2.58}$$

With $\Phi(\zeta)$ given by (8.2.55), the stress combination (8.2.54) becomes

$$\tau_{\rho\rho}+i\tau_{\rho\sigma} = \Phi^-(t)-\Phi^+(t) \ , \quad t = \exp(i\sigma) \ , \tag{8.2.59}$$

which shows that Φ is continuous over the unloaded part of the boundary. It may be shown that when there is no resultant force on the ellipse, so that $\gamma = 0 = \gamma'$, $\Phi(\zeta)$ is holomorphic in $\mathcal{D}^-$, except at the zeros of $\omega'(\zeta)$ where it has simple poles, and that

$$\Phi(\zeta) = -F(\zeta) + \Gamma + \frac{\Lambda}{\zeta^2-m} \ , \tag{8.2.60}$$

where

$$F(\zeta) = \frac{1}{2\pi i}\int_C \frac{(\tau_{\rho\rho}+i\tau_{\rho\sigma})}{t-\zeta}\,dt \ , \tag{8.2.61}$$

and C is the circle $|\zeta| = 1$ traversed anti-clockwise. Now equating the coefficient of ζ^{-2} in $\omega'(\zeta)\Phi(\zeta)$ from this equation with that given by (8.2.58), we deduce that

$$\Lambda = m(\Gamma+\overline{\Gamma}) + \overline{\Gamma}' - mF(0) \ . \qquad (8.2.62)$$

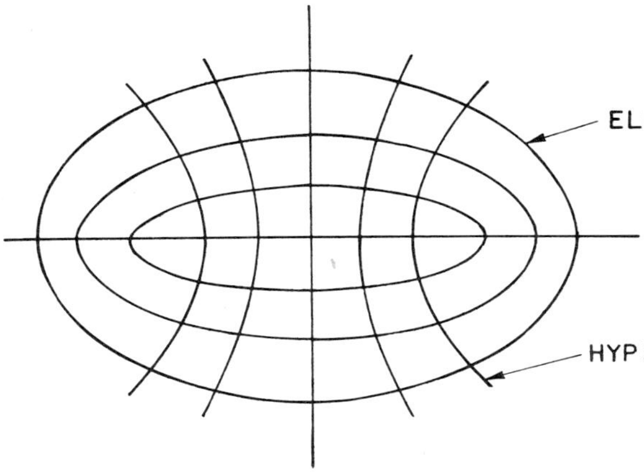

Figure 8.2.1 - The confocal ellipses and hyperbolae relating to the transformation (8.2.45)

Examples 8.2

1. $\mathcal{D}^-$ is the region $|z| > a$; $\Phi(z)$ is holomorphic and single-valued in $\mathcal{D}^-$; z, z_0 are two points in $\mathcal{D}^-$, joined by an arc lying in $\mathcal{D}^-$. Show that

$$\int_{z_0}^{z} \Phi(z)dz = c\ell nz+\phi(z)$$

where $\phi(z)$ is holomorphic and single-valued, and c is a complex constant. Hint: integrate the Laurent expansion of $\Phi(z)$.

2. Use the analysis of this section and Example 4.3.1 to show that when the region $|z| > 1$ is acted on by surface tractions $\tau_{rr}+i\tau_{r\theta} = f(t)$ on $|z| = 1$ which have zero resultant, and by stresses $\tau_{xx} = T_1$, $\tau_{yy} = T_2$ at infinity, the potential $\Phi(z)$ is

$$\Phi(z) = -\frac{1}{2\pi i}\int_{C'} \frac{f(t)}{t-z}\,dt + \frac{T_2+T_1}{4} + \frac{T_2-T_1}{2z^2} .$$

3. Show that equation (8.2.6) may be integrated to yield

$$\psi(z) = -z_*\phi(z)-\bar{\phi}(z_*) , \qquad z \text{ in } \mathcal{D}^+ .$$

4. Remembering equation (4.3.3), that $\Phi(z) = Ci$, where C is real, represents a rigid body rotation, show that the conditions (8.2.9) state that the stresses acting on a closed curve which lies within $\mathcal{D}^+$ and which circles the origin have zero moment and resultant.

5. Show that the complex potentials $\Phi(z)$, $\Psi(z)$ corresponding to a uniform compression $\tau_{rr} = -p$ on the circle $r = a$ are

$$\Phi(z) = \begin{cases} -p/2 & |z| < a \\ & \\ p/2 & |z| > a \end{cases} , \qquad \Psi(z) = 0 .$$

Hence show that the radial displacement on the circle is $-(\kappa-1)pa/(4\mu)$. Show that the corresponding radial displacement on the circle $r = a$ due to a uniform radial stress p applied to the region $|z| > a$ is $\kappa pa/(2\mu)$.

6. Two equal and opposite point loads act on the circle $|z| \leq a$, as shown in Figure 8.2.2. Show that, if $\zeta = a\exp(i\alpha)$, the complex potentials $\Phi(z)$, $\psi(z)$ may be written

$$\Phi(z) = -\frac{P}{2\pi}\left[\frac{1}{z+\bar{\zeta}} - \frac{1}{z-\zeta}\right] + \frac{P\cos\alpha}{2\pi a}$$

$$\psi(z) = \frac{P}{2\pi}\ln\left(\frac{z+\bar{\zeta}}{z-\zeta}\right) - \frac{P}{2\pi}\left[\frac{\zeta}{z+\bar{\zeta}} + \frac{\bar{\zeta}}{z-\zeta}\right] .$$

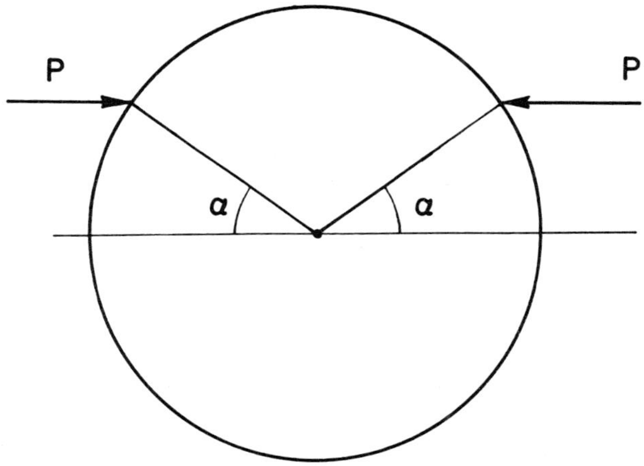

Figure 8.2.2 - Two equal and opposite forces P act on a circle of radius a

7. Show that the angle δ which $\hat{\rho}$ makes with the x-axis (see Figure 8.2.1 for the special case of (8.2.45)) is given by equation (8.2.52).

8. Use Example 4.2.4 with $\alpha = \pi/2+\theta$ to show that, on C,
$\tau_{rx}+i\tau_{ry} = (t/a)[\Phi^+(t)-\Phi^-(t)]$.

9. Show that if ' denotes differentiation with respect to $t = \exp(i\sigma)$ on C, then the derivative of equation (8.2.49) is
$2\mu(u'+iv') = \omega'(t)[\kappa\Phi^-(t)+\Phi^+(t)]$.

10. Show that the foci of the ellipses ρ = const. and the hyperbolae σ = const. are at $\pm\ 2Rm^{1/2}$.

8.3 *Basic functions for circular arcs*

Before proceeding to the solution of particular problems involving circular regions we shall first construct some basic complex potentials corresponding to loadings on one or two circular arcs.

First consider the arc C' shown in Figure 8.3.1 joining points $\exp(\pm i\theta_0)$, and let $X(z)$ denote the function

$$X(z) = [(z-\exp(i\theta_0))(z-\exp(-i\theta_0)]^{1/2} = [z^2-2z\cos\theta_0+1]^{1/2} . \qquad (8.3.1)$$

The function $1/X(z)$ is the generating function of the Legendre polynomials (Abramowitz and Stegun (1972, p. 783)), is holomorphic in the plane cut along C', satisfies the relation

$$X(z^{-1}) = -z^{-1}X(z) , \qquad (8.3.2)$$

and has the expansions

$$\frac{1}{X(z)} = -\sum_{k=0}^{\infty} p_k z^k , \quad |z| < 1 ; \quad \frac{1}{X(z)} = z^{-1}\sum_{k=0}^{\infty} p_k z^{-k} , \quad |z| > 1 , \qquad (8.3.3)$$

where

$$p_k = P_k(\cos\theta_0) . \qquad (8.3.4)$$

The boundary values of $X(z)$ as z approaches a point $t = \exp(i\theta)$ on C' from $\mathcal{D}^+$ ($|z| < 1$) or $\mathcal{D}^-$ ($|z| > 1$) satisfy

$$X^+(t) = -X^-(t) . \qquad (8.3.5)$$

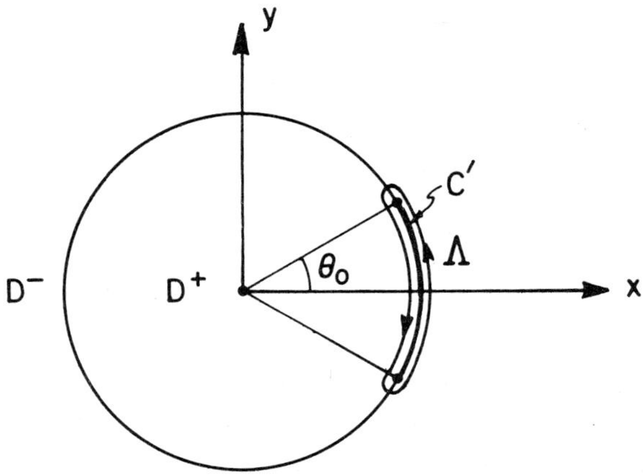

Figure 8.3.1 - Geometry of a single arc on the unit circle

The function $z^n/X(z)$ is holomorphic outside the contour Λ of Figure 8.3.1 except at infinity, where it has the form

$$\frac{z^n}{X(z)} = \sum_{k=0}^{n-1} p_{n-1-k} z^k + O(z^{-1}) \ . \tag{8.3.6}$$

But then equation (3.2.10) shows that

$$\frac{1}{2\pi i}\int_{\Lambda} \frac{\zeta^n d\zeta}{X(\zeta)(\zeta-z)} = \sum_{k=0}^{n-1} p_{n-1-k} z^k - \frac{z^n}{X(z)} \ , \tag{8.3.7}$$

so that, on contracting Λ so that it proceeds along C'^- and back along C'^+, and noting (8.3.5) we find

$$\frac{1}{\pi i}\int_{C'} \frac{t^n dt}{X^+(t)(t-z)} = \frac{z^n}{X(z)} - \sum_{k=0}^{n-1} p_{n-1-k} z^k \ . \tag{8.3.8}$$

Now put

$$W_n(z) = \frac{1}{\pi i}\int_{C'} \frac{t^n dt}{X^+(t)(t-z)} + C_{n-1} , \tag{8.3.9}$$

and choose C_{n-1} so that if $W_n(z)$ has expansions of the form (8.2.7), (8.2.8) then $A_0+\overline{B}_0 = 0$. This means that $W_n(z)$ will be a suitable candidate for $\Phi(z)$. For $W_0(z)$, $B_0 = C_{-1}$, $A_0 = -1+C_{-1}$, since p_{-1} in (8.3.8) must be taken to be zero; thus $C_{-1} = 1/2$. For $n > 0$, $B_0 = C_{n-1}$, $A_0 = -p_{n-1}+C_{n-1}$ so that $C_{n-1} = (1/2)p_{n-1}$. Thus

$$C_n = 1/2 , \quad n = -1 ; \quad C_n = (1/2)p_n , \quad n \geq 0 , \tag{8.3.10}$$

and

$$W_n(z) = z^n/X(z) - N_{n-1}(z) , \tag{8.3.11}$$

where

$$N_n(z) = \sum_{k=0}^{n}{}' p_{n-k} z^k , \quad n \geq 0 , \quad N_{-1}(z) = -1/2 , \tag{8.3.12}$$

and ' denotes that the constant term (k = 0) is halved. Proceeding in the same way we may show that

$$W_{-n+1}(z) = \frac{1}{\pi i}\int_{C'} \frac{t^{-n+1}dt}{X^+(t)(t-z)} + C_{n-1} = \frac{z^{-n+1}}{X(z)} + N_{n-1}(z^{-1}) = -W_n(z^{-1}) . \tag{8.3.13}$$

For certain indentation and inclusion problems, the loading occurs on the two symmetrically situated arcs of Figure 8.3.2. Now we consider the function

$$Y(z) = [(z-\exp(i\theta_0))(z-\exp(-i\theta_0))(z+\exp(i\theta_0))(z+\exp(-i\theta_0))]^{1/2}$$

$$= [z^4-2\cos 2\theta_0 z^2+1]^{1/2} , \tag{8.3.14}$$

which is clearly $X(z^2)$ for the angle $2\theta_0$. If C' denotes the pair of arcs on $|z| = 1$, traversed anti-clockwise, then

$$V_n(z) = \frac{1}{\pi i}\int_{C'} \frac{t^n dt}{Y^+(t)(t-z)} + A_{n-2} = \frac{z^n}{Y(z)} - M_{n-2}(z) \;, \tag{8.3.15}$$

$$A_{-2} = 1/2 \;, \quad M_{-2}(z) = -1/2 \;, \quad A_{-1} = 0 \;, \quad M_{-1}(z) = 0 \;, \tag{8.3.16}$$

and, for $n \geq 0$ $A_n = (1/2)p_m$, n even; $A_n = 0$, n odd, and

$$M_n(z) = \sum_{k=0}^{m}{}' p_k z^{n-2k} \;, \quad p_k = P_k(\cos 2\theta_0) \;. \tag{8.3.17}$$

Here $m = [n/2]$ denotes the greatest integer less than or equal to $n/2$, and $'$ denotes that the constant term (if n is even) in (8.3.17) is halved. Similarly

$$V_{-n+2}(z) = \frac{1}{\pi i}\int_{C'} \frac{t^{-n+2} dt}{Y^+(t)(t-z)} + A_{n-2} = \frac{z^{-n+2}}{Y(z)} + M_{n-2}(z^{-1}) = -V_n(z^{-1}) \;. \tag{8.3.18}$$

These functions will be used in §§8.4 - 8.6.

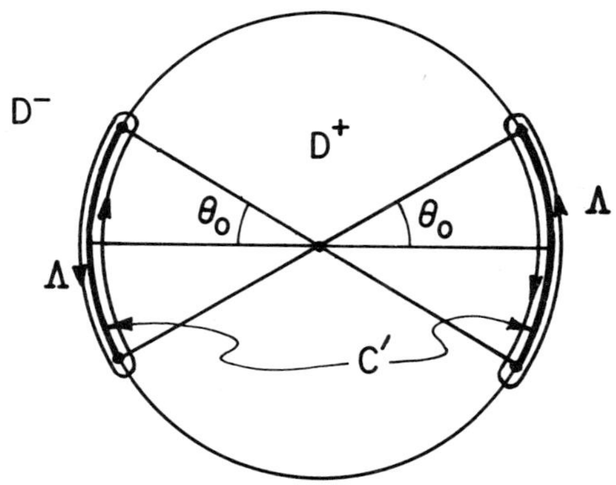

Figure 8.3.2 - Geometry of a pair of arcs on the unit circle

Examples 8.3

1. The function $(z-z_0)^{1/2}$ must be defined in a plane cut along some line from z_0 to ∞. Show that, if the two factors in $X(z)$ are defined in planes cut along the arcs shown in Figure 8.3.3, then one branch of $X(z)$ has boundary values on $0 \leq \theta \leq \theta_0$ given by

    $$X^+(t) = -X^-(t) = -\exp(i\theta/2)[2\cos\theta-2\cos\theta_0]^{1/2} ,$$

    and $X(z)$ is continuous across the cut from 1 to ∞. Use equation (8.3.2) to calculate the values at the point corresponding to $2\pi-\theta$ on the lower half of C'; remember that if $|z| < 1$ then $|z^{-1}| > 1$. Show that the value of this branch of $X(z)$ on $\theta_0 \leq \theta \leq \frac{\pi}{2}$ is

    $$X(t) = i\exp(i\theta/2)[2\cos\theta_0-2\cos\theta]^{1/2} .$$

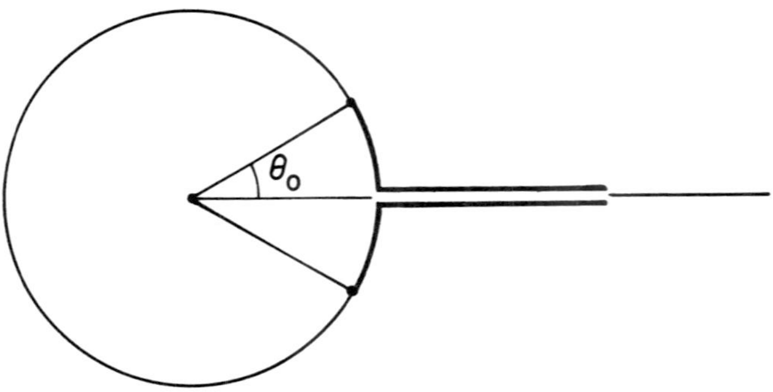

Figure 8.3.3 - Cut in the complex plane used for the definition of X(z)

2. Show that if $X(z)$ is chosen as in Example 8.3.1 then the boundary values of $W_n(z)$ as z approaches $t = \exp(i\theta)$ $(0 \leq \theta \leq \theta_0)$ satisfy

$$W_n^+(t)-W_n^-(t) = -2\exp[i(n-1/2)\theta][2\cos\theta-2\cos\theta_0]^{-1/2} ,$$

while those of $V_n(z)$ satisfy

$$V_n^+(t)-V_n^-(t) = -\exp[i(n-1)\theta][\cos^2\theta-\cos^2\theta_0]^{-1/2} .$$

3. Show that

$$\sum_{k=0}^{\infty} \cos(k+1/2)\theta P_k(\cos\theta_0) = \begin{cases} [2(\cos\theta-\cos\theta_0)]^{-1/2} , & 0 < \theta < \theta_0 \\ 0 , & \theta_0 < \theta < \pi . \end{cases}$$

4. Use the result of Example 8.3.3 to show that

$$\int_0^{\pi/2} \sum_{k=0}^{\infty} (-)^k \cos(2k+1)\theta(1+\cos2\theta)P_k(\cos2\theta_0)d\theta$$

$$= \int_0^{\pi/2-\theta_0} \frac{\cos^2\theta d\theta}{(\cos^2\theta-\sin^2\theta_0)^{1/2}} = E(\cos\theta_0) .$$

8.4 *Compression of an elastic cylinder*

It is well known that Hertz's theory of contact (§2.9) for two 3-dimensional elastic bodies gives remarkably accurate results in practice, even though it is based on the assumption that, for the purpose of computing displacements from given surface tractions, each body may be approximated by a half-space. Numerous attempts have been made to find an equally satisfactory 2-dimensional theory. The simple theory described in §2.4, which may be viewed as a limiting case of Hertz's theory (Examples 2.9.2, 2.9.3), suffers from the drawback that the two half-planes representing the cylinder in plane-strain are subject to unbalanced forces on their plane boundaries and in two-dimensional elasticity such forces produce displacements which, through meaningful locally, are unrealistic away from the region in which the forces are applied and, in particular, become infinite at infinity; the displacements are also arbitrary to within a constant translation. This has two immediate consequences: it is not possible to formulate the condition, corresponding to equation (2.9.3) in Hertz's theory, that the 'cylinders' do not interpenetrate outside the contact region; nor is it possible to find the *approach* of the two bodies; note that in Example 2.9.2 the expression for the approach δ becomes meaningless when the bodies are curved in only one direction.

To eliminate this indeterminacy, particularly in δ, it is essential to take into account the shape and size of at least one of the bodies in contact. Of particular practical interest is the compression of an elastic cylindrical roller of finite radius between two cylindrical bodies, which need not necessarily be identical, and which may have elastic properties distinct from those of the roller.

Loo (1958) considered the compression of a roller between two identical cylinders of arbitrary radius and not necessarily of the same material as the roller. As his basic tool Loo used the displacement field, given first by Hertz (1883), in a cylinder due to two equal and opposite non-diametrical collinear forces (Example 8.2.6 and Muskhelishvili (1953a, §80a)). He correctly formulated the integral equation for the contact, assuming that the outside cylinders also were subject to systems of equal and opposite forces in the direction of the line of centres of the cylinders. He then neglected higher-order terms in this equation, reducing it to one with the simple $\ell n|x-\xi|$ kernel characteristic of the half-plane problems (see for instance equation (6.2.16)), but with a modified right-hand side that depends on the curvatures, elastic constants and total load. In this way he obtained a first approximation to the approach of the outside cylinders to each other.

The first precise formulation and closed-form solution of a problem in the class under consideration appears in the masterly paper of Sternberg and Turteltaub (1972). They note that the terms neglected by Loo are of the same order as those neglected in

Hertz's analysis so that his results are suspect. They therefore consider the closed-form solution of a special case of Loo's problem, that in which the roller, of radius R, is pressed between flat rigid plates. We shall use their formulation, but show that the problem, which, as they prove, has a unique solution, may be solved by using a simple combination of the basic complex potentials $V_n(z/R)$ derived in §8.3.

The geometry of the problem in shown in Figure 8.4.1. The boundary conditions are taken in the form

$$\tau_{rx} = 0 \text{ on } C'' , \quad \tau_{ry} = 0 \text{ on } C, \tag{8.4.1}$$

$$u = R(\operatorname{sign}x\cos\theta_* - \cos\theta) \text{ on } C' . \tag{8.4.2}$$

Here C is the circle $r = R$, C' is the contact region and C'' is the remainder of C. Equations (8.4.1) assert that no loads are applied to the surface of the cylinder outside the contact region, and that the contact is frictionless. Equation (8.4.2) states that the horizontal component of the displacement of the cylinder is known in the contact region; here θ_* is related to the relative approach δ of the plates, given by

$$\delta = 2R(1-\cos\theta_*) . \tag{8.4.3}$$

Symmetry considerations show that $u(x,y)$ is an odd function in x, even in y, while $v(x,y)$ is even in x, odd in y; thus with these provisos equation (8.4.2) may be replaced by

$$u = R(\cos\theta_* - \cos\theta) , \quad 0 \le \theta \le \theta_0 . \tag{8.4.4}$$

The remaining conditions on the solution, again subject to the symmetry relations, are the separation condition

$$u \le R(\cos\theta_* - \cos\theta) , \quad \theta_0 \le \theta \le \frac{\pi}{2} , \tag{8.4.5}$$

which states that the deformed roller does not penetrate the plates, the condition precluding tensile tractions between roller and plates in the contact region, viz.

$$\tau_{rx} \leq 0 \ , \quad 0 \leq \theta \leq \theta_0 \ , \tag{8.4.6}$$

the equilibrium condition

$$-R \int_{-\theta_0}^{\theta_0} \tau_{rx} d\theta = P \ , \tag{8.4.7}$$

and the condition which states that the contact stress is zero at the edge of the contact region, namely

$$\tau_{rx} = 0 \ , \quad \theta = \theta_0 \ . \tag{8.4.8}$$

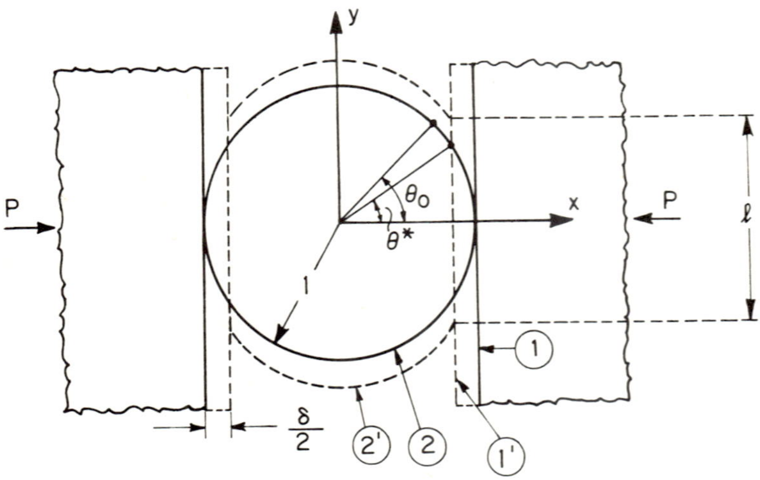

Figure 8.4.1 - Circular roller compressed between rigid plates:
1 initial plate position 1' final plate position
2 undeformed roller 2' deformed roller

It should be noted that the above formulation differs from the standard formulation of contact problems in infinitesimal elasticity. In the conventional formulation the boundary conditions are referred wholly to the *undeformed* boundary, whereas the second of (8.4.1) in conjunction with (8.4.2) amounts to a specification of the tractions tangential, and of the displacements normal, to a portion of the *deformed* surface of the roller. This departure from the usual practice is necessary because the displacements normal to the original cylindrical boundary are not known beforehand, while those normal to the plate faces are determinate to within an initially unknown constant $\cos\theta_*$. Note however that equations (8.4.1), (8.4.2) are applied on the *undeformed* surface, i.e., on the cylinder, so that the solution holds only in the limit of infinitesimal elasticity.

The first step is to determine the condition on the complex potential $\Phi(z)$ of §8.2 which must be satisfied if $\tau_{ry} = 0$ on *C*. Example 8.2.8 shows that on *C*

$$\tau_{rx}+i\tau_{ry} = (t/R)[\Phi^+(t)-\Phi^-(t)] \ , \tag{8.4 9}$$

so that the condition is that $t[\Phi^+(t)-\Phi^-(t)]$ must be *real* on *C*. Since equations (8.4.1) imply that $\Phi(z)$ is continuous across *C''*, the complex potential will be sought in the form

$$\Phi(z) = \sum_{n=0}^{N} [a_n V_n(z/R)+b_n V_{-n}(z/R)] \ . \tag{8.4.10}$$

Example 8.3.2 shows that, for this Φ, at $t = R\exp(i\theta)$, where $0 \leq \theta \leq \theta_0$,

$$\Phi^+(t)-\Phi^-(t) = -S(\theta)\sum_{n=0}^{N}\{a_n\exp[i(n-1)\theta]+b_n\exp[-i(n+1)\theta]\} \tag{8.4.11}$$

where $S(\theta) = (\cos^2\theta-\cos^2\theta_0)^{-1/2}$. Symmetry considerations show that $a_n = b_n$ and that $a_n = 0$ when n is odd, while the condition

$\tau_{ry} = 0$ shows that a_n must be real. Therefore

$$\Phi(z) = \sum_{n=0}^{N} a_n[V_n(z/R)+V_{-n}(z/R)] \ , \quad n \text{ is even} \ , \tag{8.4.12}$$

and it may be verified that this $\Phi(z)$ satisfies both of equations (8.2.9).

Now consider the displacement condition. Equations (8.3.15) and (8.3.18) show that, for $0 \leq \theta < \theta_0$,

$$it[V_n^{\pm}(t)+V_{-n}^{\pm}(t)] = \mp Ri\cos n\theta S(\theta)+it[M_n(R/t)-M_{n-2}(t/R)] \ , \tag{8.4.13}$$

so that Example 8.2.9 gives, for $0 \leq \theta < \theta_0$,

$$2\mu\left(\frac{du}{d\theta} + i\frac{dv}{d\theta}\right) = (1-\kappa)RiS(\theta) \sum_{n=0}^{N} a_n\cos n\theta+(1+\kappa)it \sum_{n=0}^{N} a_n[M_n(R/t)-M_{n-2}(t/R)] \ . \tag{8.4.14}$$

Since the first term is imaginary, $du/d\theta$ is given by the real part of the second; equating $du/d\theta = R\sin\theta$ with this real part, and taking the values of M_{-2}, M_0, M_2 from (8.3.16), (8.3.17), we find

$$2\mu R\sin\theta = (1+\kappa)R\text{Re}\{i\exp(i\theta)[a_0+a_2(-1/2+\exp(-2i\theta)+p_1/2)+\cdots]\} \ . \tag{8.4.15}$$

The terms not shown explicitly will involve exponentials of multiple angles so that the corresponding a_n must be zero. Equating the coefficients of $\sin\theta$ on both sides we find

$$4\mu = (1+\kappa)[-2a_0+(3-p_1)a_2] \ . \tag{8.4.16}$$

Return to equation (8.4.9) and evaluate τ_{rx}; it is

$$\tau_{rx} = -2S(\theta)[a_0+a_2\cos 2\theta] \ , \quad 0 \leq \theta < \theta_0 \ . \tag{8.4.17}$$

Equation (8.4.8) states that $\tau_{rx}(\theta_0) = 0$, so that

$$a_0+a_2\cos 2\theta_0 = 0 . \qquad (8.4.18)$$

Since $p_1 = \cos 2\theta_0$, equations (8.4.16), (8.4.18) give

$$a_2 = 1/[2\vartheta(1+\cos^2\theta_0)] , \qquad (8.4.19)$$

where $\vartheta = (1+\kappa)/4\mu = (1-\nu)/\mu$. Thus

$$\tau_{rx} = -\frac{2R(\cos^2\theta-\cos^2\theta_0)^{1/2}}{\vartheta(1+\cos^2\theta_0)} , \quad 0 \le \theta \le \theta_0 , \qquad (8.4.20)$$

and the maximum pressure is

$$p_0 = 2R\sin\theta_0/[\vartheta(1+\cos^2\theta_0)] . \qquad (8.4.21)$$

The equilibrium condition (8.4.7) yields the relation between P and θ_0; the integral is of elliptic type, and yields

$$P = \frac{4R}{\vartheta(1+\cos^2\theta_0)} [E(\sin\theta_0)-\cos^2\theta_0 K(\sin\theta_0)] , \qquad (8.4.22)$$

where $K(\sin\theta_0) = F(\sin\theta_0, \pi/2)$ and $E(\sin\theta_0) = E(\sin\theta_0, \pi/2)$ are complete elliptic integrals of the first and second kinds respectively.

Equations (8.4.14), (8.4.19) show that

$$\frac{dv}{d\theta} = \frac{R\sin^2\theta_0\cos\theta}{1+\cos^2\theta_0} - R\varphi\frac{(\cos^2\theta-\cos^2\theta_0)^{1/2}}{1+\cos^2\theta_0} , \quad 0 \le \theta \le \theta_0 . \qquad (8.4.23)$$

This may be integrated with the condition $v(0) = 0$ to yield

$$v(\theta) = \frac{R\sin^2\theta_0\sin\theta}{1+\cos^2\theta_0} - \frac{R\varphi}{1+\cos^2\theta_0}[E(\sin\theta_0,\phi)-\cos^2\theta_0 F(\sin\theta_0,\phi)] ,$$
$$0 \le \theta \le \theta_0 , \qquad (8.4.24)$$

where F,E are the incomplete elliptic integrals and ϕ is related to θ by $\sin\theta = \sin\theta_0\sin\phi$. In order to find $u(\theta)$ on $\theta_0 \leq \theta \leq \pi/2$ it is necessary to find $u'(\theta)$ on $\theta_0 \leq \theta \leq \pi/2$ and integrate it in conjunction with the condition $u(\pi/2) = 0$. Equations (8.3.15), (8.3.18) together with the boundary value

$$Y(t/R) = 2i\exp(i\theta)[\cos^2\theta_0-\cos^2\theta]^{1/2}\ , \quad \theta_0 \leq \theta \leq \pi/2\ , \tag{8.4.25}$$

derived as in Example 8.3.1, give

$$i(t/R)[V_n(t/R)+V_{-n}(t/R)] = T(\theta)\cos n\theta+i(t/R)[M_n(R/t)-M_{n-2}(t/R)]\ , \tag{8.4.26}$$

where $T(\theta) = (\cos^2\theta_0-\cos^2\theta)^{-1/2}$, and $\theta_0 \leq \theta < \pi/2$. Thus

$$2\mu(u'+iv') = (\kappa+1)R\{T(\theta)(a_0+a_2\cos 2\theta)+a_0 i\exp(i\theta) +$$

$$+ a_2 i[(p_1-1)/2)\exp(i\theta)+\exp(-i\theta)]\}\ , \qquad \theta_0 \leq \theta < \pi/2 \tag{8.4.27}$$

When simplified with the aid of (8.4.18), (8.4.19), this gives

$$u'(\theta) = R\sin\theta-2R(\cos^2\theta_0-\cos^2\theta)^{1/2}/(1+\cos^2\theta_0)\ , \quad \theta_0 \leq \theta \leq \pi/2\ , \tag{8.4.28}$$

$$v'(\theta) = R\sin^2\theta_0\sin\theta/(1+\cos^2\theta_0)\ , \quad \theta_0 \leq \theta \leq \pi/2\ . \tag{8.4.29}$$

Thus by combining the expression for $u(\theta)$ in $0 \leq \theta \leq \theta_0$ with that for $u'(\theta)$ in $\theta_0 \leq \theta \leq \pi/2$ we find

$$u(\pi/2)-u(0) = R(1-\cos\theta_*) = R - \frac{2R}{1+\cos^2\theta_0}[E(\cos\theta_0)-\sin^2\theta_0 K(\cos\theta_0)]\ , \tag{8.4.30}$$

so that the approach of the two plates, given by (8.4.3), is

$$\delta = 2R\{1 - 2[E(\cos\theta_0)-\sin^2\theta_0 K(\cos\theta_0)]/(1+\cos^2\theta_0)\}\ . \tag{8.4.31}$$

The displacement on $0 \leq \theta \leq \pi/2$ is therefore

$$u(\theta) = \begin{cases} 2R[E(\cos\theta_0) - \sin^2\theta_0 K(\cos\theta_0)]/(1+\cos^2\theta_0) - R\cos\theta \;, \\ \qquad\qquad 0 \leq \theta \leq \theta_0, \\ 2R[E(\cos\theta_0,\chi) - \sin^2\theta_0 F(\cos\theta_0,\chi)]/(1+\cos^2\theta_0) - R\cos\theta \;, \\ \qquad\qquad \theta_0 \leq \theta \leq \pi/2 \;, \end{cases} \tag{8.4.32}$$

where χ is related to θ by $\cos\theta = \cos\theta_0 \sin\chi$.

Equation (8.4.24) yields the *contact length* ℓ, shown in Figure 8.4.1, as

$$\ell = 2R\sin\theta_0 + 2v(\theta_0) = \frac{2R\sin\theta_0}{1+\cos^2\theta_0} - \frac{R\varphi}{1+\cos^2\theta_0}[E(\sin\theta_0) - \cos^2\theta_0 K(\sin\theta_0)] \;. \tag{8.4.33}$$

This effectively completes the solution. The complex potential $\Phi(z)$ is known, so that the stresses in the cylinder may be computed. The displacements inside the cylinder may be exhibited explicitly as elliptic integrals. The most important equations are (8.4.22), (8.4.31), (8.4.33) yielding the total load, the approach and the contact length as functions of R, μ, ν and θ_0. For small angles θ_0 it is convenient to use the dimensionless parameter

$$\gamma = \left(\frac{2\vartheta P}{\pi R}\right)^{1/2} \;. \tag{8.4.34}$$

By using the expansions of the elliptic integrals $K(\cos\theta_0)$, $E(\cos\theta_0)$, $K(\sin\theta_0)$, $E(\sin\theta_0)$, we find the following dimensionless equations for the maximum pressure, approach and contact pressure:

$$\vartheta p_0 = \gamma + \frac{3\gamma^3}{16} + O(\gamma^5) \;, \tag{8.4.35}$$

$$\frac{\delta}{R} = \gamma^2\left[\ln\left(\frac{4}{\gamma}\right) - \frac{1}{2}\right] + \frac{9\gamma^4}{32} + O(\gamma^6 \ln\gamma) \;, \tag{8.4.36}$$

$$\frac{\ell}{R} = 2\gamma - \frac{1}{2}\varphi\gamma^2 + O(\gamma^3) \;. \tag{8.4.37}$$

The leading terms in (8.4.35), (8.4.37) agree with the results of Example 2.9.3 for Hertzian contact. The leading term in δ/R agrees with Loo's first approximation for the special case. (Note that $(4\beta/P)^{1/2}$ in his equation (24) should read $(4B/P)^{1/2}$).

The rigorous analysis described above for the problem of Figure 8.4.1 may be applied to give approximate results for the relative approach of two rigid cylinders pressing on either side of an elastic roller, or of two elastic cylinders. These problems were considered by Schwartz and Harper (1971) by using matched asymptotic expansions.

Thus, in the first problem, shown in Figure 8.4.2, if the angle of contact is small, the frictionless condition may still be taken as $\tau_{ry} = 0$, and the displacement condition in the contact region is approximately

$$u(\theta) = R(1+R/R')(1-\cos\theta)-\delta/2 . \qquad (8.4.38)$$

Thus, starting with equation (8.4.4), there is a factor $(1+R/R')$ appearing throughout, leading eventually to an approximation to δ/R given by the first term of (8.4.36) with a new

$$\gamma = \left[\frac{2\vartheta P}{\pi R(1+R/R')}\right]^{1/2} . \qquad (8.4.39)$$

Belokon' (1969) considers the problem of an elastic disc pressed by two closely-fitting circular rigid stamps.

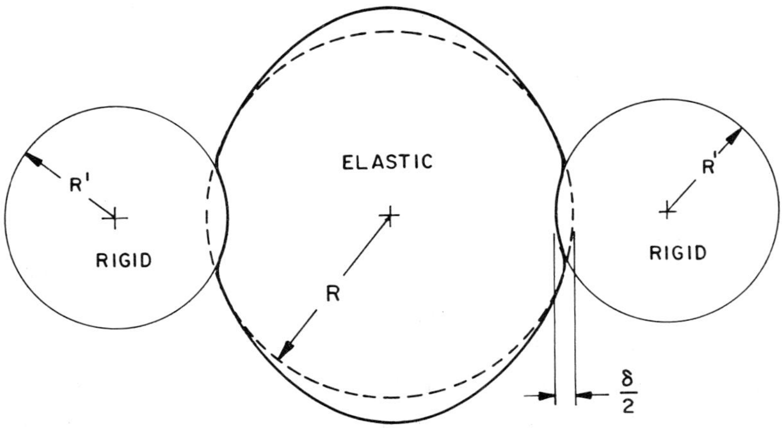

Figure 8.4.2 - Elastic roller compressed by two rigid cylinders

Examples 8.4

1. Apply the analysis of this section to the problem, also considered by Sternberg and Turteltaub (1972), of a roller symmetrically *indented* by two flat rigid punches of lengths $\ell = 2R\sin\theta_0$. Show that the punch will be in complete contact with the roller if and only if

$$P \geq 4R \frac{[4R^2E(\sin\theta_0)-(4R^2-\ell^2)K(\sin\theta_0)]}{\vartheta(8R^2-\ell^2)} .$$

8.5 *Inclusion problems*

The second application of the theory of §8.2, §8.3 is to the class of inclusion problems in which a rigid or elastic disc is inserted in a hole in a plate. The disc is smooth, and the disc fits exactly into the hole in the unstressed state. The plate is subjected to stresses at infinity so that the insert separates from the hole along part of the boundary. It is required to find the arcs of contact and the pressure distribution along these arcs.

The inclusion problem for a circular insert was formulated by Sheremet'ev (1952) and solved by him in an unwieldy paper (1961). Stippes, Wilson and Krull (1962), who noted that Szelagowski's (1956) formulation was incorrect, independently formulated it and derived the closed form solution for the case in which insert and plate are composed of identical materials; their analysis can be greatly simplified by a systematic use of the functions derived in §8.3, as will be shown below. Wilson (1964a) and Noble and Hussain (1967), (1969a) gave numerical solutions for the case when insert and plate have different elastic properties. Keer, Dundurs and Kiattikomol (1973) developed an alternative analysis in which the problem is reduced to a Fredholm integral equation of the second kind with a weakly singular kernel. They present a comprehensive set of numerical results. Wilson (1964b) formulated the problem for a rigid elliptical insert and Goree (1965) solved this problem numerically.

The goemetry of the inclusion problem for a frictionless circular insert is shown in Figure 8.5.1. Stresses T_1, T_2 at infinity cause partial separation between the insert, body 1, and the plate, body 2, so that the contact region C' consists of two symmetrically placed arcs. First of all it is necessary to ascertain when this partial contact situation will occur. For this we may consider the problem of Example 8.5.1 in which the insert and plate are in contact over the *whole of the boundary*. The plate and insert will separate if the solution of Example 8.5.1 gives a normal contact stress which is somewhere compressive and elsewhere tensile. When the plate and insert do separate then the analysis of Example 8.5.1 no longer holds and the following analysis, which is particularized to the configuration of Figure 8.5.1, must be followed. For both bodies the analysis of §8.2 applies; the potential $\Phi_1(z)$ will be continuous across C'' and satisfy

$$\Phi_1^+(t)-\Phi_1^-(t) = f(t) \text{ on } C' , \tag{8.5.1}$$

$$\kappa_1\Phi_1^+(t)+\Phi_1^-(t) = 2\mu_1(u_1'+iv_1') \text{ on } C , \tag{8.5.2}$$

where $f(t) = \tau_{rr}(t)$ is real. The displacements in body 1 will be given by (8.2.1), i.e.,

$$2\mu_1(u_1+iv_1) = \kappa_1\phi_1(z)-z\overline{\phi_1'}(\overline{z})-\overline{\psi}_1(\overline{z}) \text{ in } \mathcal{D}^+ . \tag{8.5.3}$$

The appropriate solution of equation (8.5.1) is that given by equation (8.2.34), namely

$$\Phi_1(z) = \frac{1}{2\pi i}\int_{C'} \frac{f(t)}{t-z}\,dt + B , \quad B = \frac{-1}{4\pi i}\int_{C'} \frac{f(t)}{t}\,dt . \tag{8.5.4}$$

The potential $\Psi_1(z)$ is given by equation (8.2.6) and will be holomorphic in $\mathcal{D}^+$ ($|z| < 1$) since equation (8.2.9) will be satisfied. For body 2, the plate, the corresponding equations are

$$\Phi_2^-(t)-\Phi_2^+(t) = f(t) \text{ on } C' , \tag{8.5.5}$$

$$\kappa_2\Phi_2^-(t)+\Phi_2^+(t) = 2\mu_2(u_2'+iv_2') \text{ on } C, \tag{8.5.6}$$

$$2\mu_2(u_2+iv_2) = \kappa_2\phi_2(z) - z\overline{\phi_2'}(\overline{z}) - \overline{\psi}_2(\overline{z}) \text{ in } \mathcal{D}^- , \tag{8.5.7}$$

where now $\Phi_2(z)$ is given by Example 8.2.2 as

$$\Phi_2(z) = -\frac{1}{2\pi i}\int_{C'} \frac{f(t)}{t-z}\,dt + \frac{T_2+T_1}{4} + \frac{T_2-T_1}{2z^2} = -\Phi_1(z)+A+2Gz^{-2}/3 , \tag{8.5.8}$$

where

$$A = (T_2+T_1)/4+B , \quad G = 3(T_2-T_1)/4 , \tag{8.5.9}$$

and $\Psi_2(z)$ is given in terms of $\Phi_2(z)$ in $\mathcal{D}^-$ by equation (8.2.5).

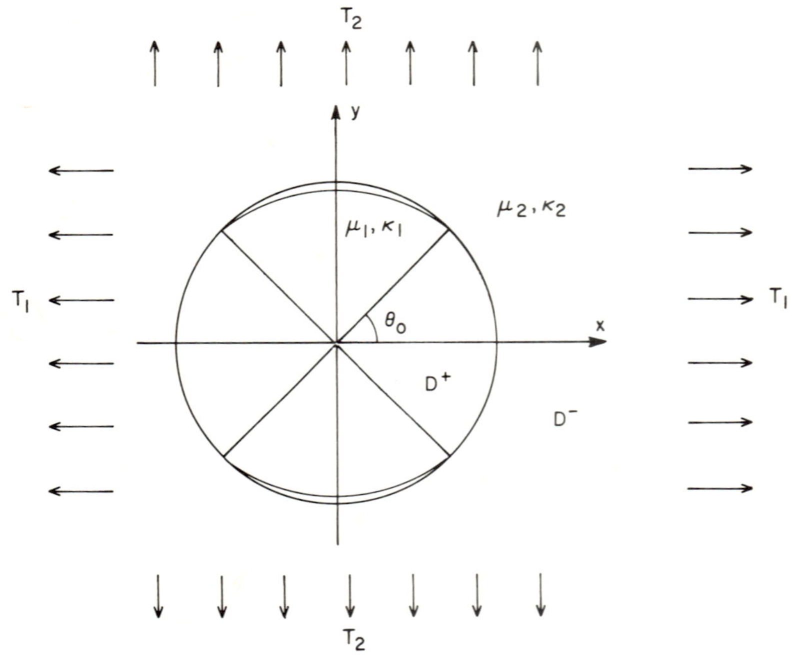

Figure 8.5.1 - Geometry of the circular inclusion problem

The contact is smooth, f(t) is real, and (8.5.1) gives

$$\Phi_1^+(t)-\Phi_1^-(t) = \overline{\Phi_1^+(t)}-\overline{\Phi_1^-(t)} = \overline{\Phi}_1^-(t^{-1})-\overline{\Phi}_1^+(t^{-1}) \text{ on } C' , \tag{8.5.10}$$

so that

$$\Phi_1^+(t)+\overline{\Phi}_1^+(t^{-1}) = \Phi_1^-(t)+\overline{\Phi}_1^-(t^{-1}) \text{ on } C' , \tag{8.5.11}$$

which states that $\Phi_1(z)+\overline{\Phi}_1(z^{-1})$ is continuous across C'; this function, being holomorphic in the whole plane, is a constant. But since $\Phi_1(z)$ satisfies (8.2.9a), this constant is zero and

$$\bar{\Phi}_1(z^{-1}) = -\Phi_1(z) , \quad \overline{\Phi_1^{\pm}(t)} = -\Phi_1^{\mp}(t) . \tag{8.5.12}$$

The displacement condition on C' is

$$(u_r)_1 = (u_r)_2 . \tag{8.5.13}$$

Since

$$2u_r = t^{-1}(u+iv)+t(u-iv) , \tag{8.5.14}$$

it follows readily that

$$L(u_r) \equiv 2\frac{d}{dt}\left[t^3\frac{d}{dt}(t^{-1}u_r)\right] = -(u'+iv')+t(u''+iv'') +$$

$$+ 3t^2(u'-iv')+t^3(u''-iv'') . \tag{8.5.15}$$

Now u,v are real functions of t so that $u(t) = \bar{u}(\bar{t})$ and

$$u'(t) = \frac{d}{dt}u(t) = \frac{d}{dt}\bar{u}(\bar{t}) = \frac{d}{dt}\bar{u}(t^{-1}) = -t^{-2}\bar{u}'(\bar{t}) . \tag{8.5.16}$$

The complex conjugate of equation (8.2.21) is therefore

$$2\mu_1(u_1'-iv_1') = -t^{-2}[\kappa_1\overline{\Phi_1^{+}(t)}+\overline{\Phi_1^{-}(t)}] = t^{-2}[\kappa_1\Phi_1^{-}(t)+\Phi_1^{+}(t)] , \tag{8.5.17}$$

and differentiation of equation (8.5.17) now gives

$$2\mu_1 L_1(u_r) = (1-\kappa_1)(\Phi_1^{+}(t)-\Phi_1^{-}(t))+(1+\kappa_1)t\frac{d}{dt}(\Phi_1^{+}(t)+\Phi_1^{-}(t)) . \tag{8.5.18}$$

A similar manipulation for the plate gives

$$2\mu_2 L_2(u_r) = (1-\kappa_2)(\Phi_1^{+}(t)-\Phi_1^{-}(t))-(1+\kappa_2)t\frac{d}{dt}(\Phi_1^{+}(t)+\Phi_1^{-}(t)) -$$

$$- 2(1+\kappa_2)[A+G(t^{-2}+t^2)] , \tag{8.5.19}$$

so that the condition

$$L_1(u_r) = L_2(u_r) \text{ on } C' , \qquad (8.5.20)$$

when expressed in terms of the mismatch parameters α, β of (4.10.13), is

$$\beta(\Phi_1^+(t)-\Phi_1^-(t))-t\frac{d}{dt}(\Phi_1^+(t)+\Phi_1^-(t)) = (1-\alpha)\{A+G(t^2+t^{-2})\} . \qquad (8.5.21)$$

The fact that this equation depends only on the elastic constant combinations α, β agrees with the general theory of plane receding contacts discussed in §4.10. This theory states moreover that the extent of the contact region, i.e., the angle θ_0, will be independent of the level of loading, and will thus depend only on the quantities α, β and T_1/T_2.

Equation (8.5.21), i.e., (8.5.20) depends on the derivatives of the displacements, and in order to ensure that equation (8.5.13) is satisfied it is necessary to satisfy further conditions. Because of the symmetry of the problem it is necessary and sufficient to impose the condition

$$u_1(1) + iv_1(1) = u_2(1) + iv_2(1) . \qquad (8.5.22)$$

It may be shown that this condition may be written

$$(1-\beta)\int_1^\infty \Phi_1^*(x)dx-(1+\beta)\int_0^1 \Phi_1^*(x)dx = (1+\alpha)B+(1-\alpha)(T_2-3T_1)/4 , \qquad (8.5.23)$$

where

$$\Phi_1^*(x) = \Phi_1(x) - B . \qquad (8.5.24)$$

First consider the case in which the materials are identical, so that $\alpha = 0 = \beta$. Note however (Noble and Hussain (1969b)) that the following analysis has to be modified in only a trivial manner to cover the case $\alpha = 0$, $\beta \neq 0$; see Example 8.5.2. Writing $F(z) = d\Phi_1/dz$ in (8.5.21) we find

$$F^+(t)+F^-(t) = -At^{-1}-G(t+t^{-3}) \text{ on } C' . \qquad (8.5.25)$$

This equation is a linear relation of the type considered in §3.4 but there only for the case in which C' was an interval of the real axis. However, the analysis used to change (3.4.1) into (3.4.20) may be adapted here, replacing $X_0(z)$ by the fundamental solution of the homogeneous equation

$$F^+(t) + F^-(t) = 0 \text{ on } C' , \qquad (8.5.26)$$

which is just the function $[Y(z)]^{-1}$ given by equation (8.3.14). Thus the general solution of (8.5.25) which has integrable singularities at the ends of C' is

$$F(z) = -\frac{1}{2\pi iY(z)}\int_{C'}\frac{Y^+(t)[At^{-1}+G(t+t^{-3})]}{t-z}dt + \frac{P(z)}{Y(z)} , \qquad (8.5.27)$$

where $P(z)$ is a polynomial. Since $\Phi_1(z)$ has the form (8.2.8) with $B_1 = 0$ (see (8.2.35)), $F(z) = O(z^{-3})$ at infinity so that $P(z) = 0$. The integral in (8.5.27) may be evaluated by writing

$$Y^+(t) = (t^4-2t^2\cos2\theta_0+1)/Y^+(t) , \qquad (8.5.28)$$

expanding the numerator in powers of t and t^{-1} and using the integrals $V_n(z)$ of §8.3. Thus

$$F(z) = -\frac{1}{2Y(z)}[A\{V_3(z)-2\cos 2\theta_0 V_1(z)+V_{-1}(z)\} +$$

$$+ G\{V_5(z)-2\cos 2\theta_0 V_3(z)+V_1(z)\}+G\{V_1(z)-2\cos 2\theta_0 V_{-1}(z) +$$

$$+ V_{-3}(z)\}] = -\frac{A}{2z} - \frac{G}{2}(z+z^{-3}) - \frac{1}{2Y(z)}[A\{-M_1(z)+M_1(z^{-1})\} +$$

$$+ G\{-M_3(z)+2\cos 2\theta_0 M_1(z)-2\cos 2\theta_0 M_1(z^{-1})+M_3(z^{-1})\}]$$

$$= -\frac{A}{2z} - \frac{G}{2}(z+z^{-3}) - \frac{1}{2Y(z)}\{A(-z+z^{-1})+G[-z^3+z^{-3}+\cos 2\theta_0(z-z^{-1})]\} \tag{8.5.29}$$

and the expansions of $[Y(z)]^{-1}$ for large and small $|z|$ show that $F(z) = O(z)$ for small $|z|$, and $O(z^{-3})$ for large $|z|$. The potential $\Phi_1(z)$ is obtained by integrating $F(z)$, and Gradshteyn and Ryzhik (1965, 2.261, 2.266) give

$$\int (z^{-1}-z)\frac{dz}{Y(z)} = -\ln\left[\frac{1+z^2+Y(z)}{z}\right] \tag{8.5.30}$$

$$\int \{z^{-3}-z^3+\cos 2\theta_0(z-z^{-1})\}\frac{dz}{Y(z)} = -\frac{1}{2}(1+z^{-2})Y(z) \tag{8.5.31}$$

so that

$$\Phi_1(z) = \frac{A}{2}\ln\left[\frac{1+z^2+Y(z)}{z^2}\right] + \frac{G}{4}(z^{-2}-z^2) + \frac{G}{4}(1+z^{-2})Y(z)+D , \tag{8.5.32}$$

where D is a constant that must be determined so that $\Phi_1(z)$ satisfies the conditions

$$\Phi_1(\infty) = -\Phi(0) = B . \tag{8.5.33}$$

Use of the expansions of $Y(z)$ for large and small $|z|$ shows that

$$\Phi_1(0) = \frac{A}{2}\ln(1+\cos 2\theta_0) - \frac{G}{4}(1-\cos 2\theta_0)+D , \tag{8.5.34}$$

$$\Phi_1(\infty) = \frac{A}{2}\,\ell n2 + \frac{G}{4}\,(1-\cos 2\theta_0)+D\ , \tag{8.5.35}$$

giving

$$A(2+\ell n\cos\theta_0) = \frac{T_1+T_2}{2} + \frac{G}{4}\sin^2\theta_0\ , \qquad D = -\frac{A}{2}\,\ell n(2\cos\theta_0)\ . \tag{8.5.36}$$

The normal stress on the arc $0 \leq \theta \leq \theta_0$ is now given by equation (8.5.1) as

$$f(t) = -A\ell n\left[\frac{\cos\theta+(\cos^2\theta-\cos^2\theta_0)^{1/2}}{\cos\theta_0}\right] - 2G\cos\theta(\cos^2\theta-\cos^2\theta_0)^{1/2}\ . \tag{8.5.37}$$

It remains to determine the contact angle θ_0, and for this it is somewhat simpler to fix θ_0 and find the ratio T_1:T_2 for which this is the contact angle. When $\alpha = 0 = \beta$, equation (8.5.23) may be integrated by parts to give

$$\int_1^\infty (x-x^{-1})F(x)dx = B - \frac{(T_2-3T_1)}{4}\ . \tag{8.5.38}$$

$F(x)$ is given by equation (8.5.29), and the integral may be calculated by expanding $[Y(x)]^{-1}$ in powers of x^{-1} and using Example 8.3.4 and a similar result for

$$\int_0^{\pi/2-\theta_0} \frac{\sin^2\theta\cos^2\theta d\theta}{(\cos^2\theta-\sin^2\theta_0)^{1/2}} = E(\cos\theta_0) + \frac{1}{3}\,[\sin^2\theta_0 K(\cos\theta_0)-2(1+\sin^2\theta_0)E(\cos\theta_0 \tag{8.5.39}$$

The resulting equation may be reduced, with the aid of (8.5.33)-(8.5.36), to

$$\frac{T_2+T_1}{T_2-T_1} = (1+\sin^2\theta_0)\ell n\cos\theta_0 + \frac{4+\sin^2\theta_0}{2} - 2\sin^2\theta_0(2+\ell n\cos\theta_0)\,\frac{K(\cos\theta_0)}{E(\cos\theta_0)}\ . \tag{8.5.40}$$

When $T_1 = 0$ this equation reduces to that found by Stippes et al (1962). Figure 8.5.2 shows the variation of θ_0 with $(T_2+T_1)/(T_2-T_1)$

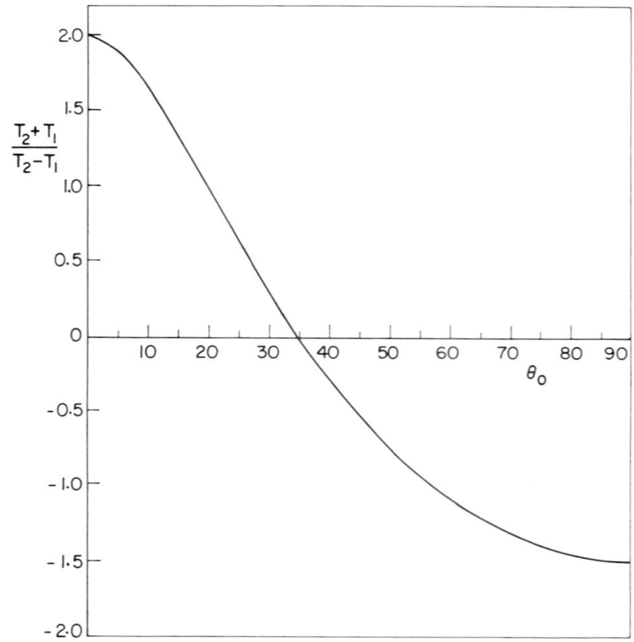

Figure 8.5.2 - The variation of θ_0 *with* $(T_2+T_1)/(T_2-T_1)$

When the insert and plate have different elastic properties we note that the Plemelj formulae (3.2.17), (3.2.18) applied to equation (8.5.4) show that

$$\Phi_1^+(t)+\Phi_1^-(t) = \frac{1}{\pi i}\int_{C'} \frac{f(s)}{s-t}\,ds - \frac{1}{2\pi i}\int_{C'} \frac{f(s)}{s}\,ds\ , \qquad (8.5.41)$$

and Muskhelishvili (1953b, p. 376) shows that when f(s) = 0 at the ends of C' then

$$\frac{d}{dt}\int_{C'} \frac{f(s)}{s-t}\,ds = \int_{C'} \frac{f'(s)}{s-t}\,ds\ , \qquad (8.5.42)$$

so that equation (8.5.21) may be rewritten

$$\beta f(t) - \frac{t}{\pi i}\int_{C'} \frac{f'(s)}{s-t}\,ds = (1-\alpha)\{A+G(t^2+t^{-2})\} \text{ on } C' . \tag{8.5.43}$$

This is the type of integro-differential equation considered in Chapter 17 of Muskhelishvili (1953b), primarily for the cases in which C' is $(-1,1)$, for which see further references: the equation is often referred to as the *airfoil* equation.

Wilson (1964a) solved equation (8.5.43) numerically, and we now present a modified version of his analysis. The starting point is a suitable expansion for the contact stress $f(t)$. When $\alpha = 0 = \beta$ the contact stress is given by equation (8.5.37). Now a consideration of the boundary values of the equation

$$(1-2zt+t^2)^{-1/2}\,\ell n\left[\frac{z-t+(1-2zt+t^2)^{1/2}}{(z^2-1)^{1/2}}\right] = \sum_{k=0}^{\infty} t^k Q_k(z) \tag{8.5.44}$$

valid for $|t| < 1$ and $\mathrm{Re}(z) > 1$, given in Gradshteyn and Ryzhik (1965, 8.7912), shows that

$$\ell n\left[\frac{\cos\theta+(\cos^2\theta-\cos^2\theta_0)^{1/2}}{\cos\theta_0}\right] = -2(\cos^2\theta-\cos^2\theta_0)^{1/2}\sum_{k=0}^{\infty} Q_k(\cos 2\theta_0)\cos(2k \tag{8.5.45}$$

when $0 < \theta < \theta_0$. Here $Q_k(z)$ is the associated Legendre function. This suggests that $f(t)$ be sought in the form

$$f(t) = (\cos^2\theta-\cos^2\theta_0)^{1/2}\sum_{n=0}^{N} a_n\cos(2n+1)\theta , \quad 0 \le \theta \le \theta_0 . \tag{8.5.46}$$

It is somewhat easier to take $f(t)$ in the form

$$f(t) = (\cos^2\theta-\cos^2\theta_0)^{-1/2}\sum_{n=0}^{N} a_n\cos(2n+1)\theta$$

$$= -\sum_{n=0}^{N} a_n(t^{2n+2}+t^{-2n})/Y^+(t) , \tag{8.5.47}$$

and then to impose the condition $f(\exp(i\theta_0)) = 0$. With this $f(t)$, equations (8.3.15), (8.3.18) give the $\Phi_1(z)$ of (8.5.4) as

$$2\Phi_1(z) = -\sum_{n=0}^{N} a_n V_{2n+2}(z) - \sum_{n=0}^{N} a_n V_{-2n}(z) , \tag{8.5.48}$$

so that

$$[\Phi_1^+(t)+\Phi_1^-(t)] = \sum_{n=0}^{N} a_n[M_{2n}(t)-M_{2n}(t^{-1})] ,$$

$$= \sum_{n=0}^{N} a_n \sum_{k=0}^{n} p_k 2i\sin(2n-2k)\theta , \tag{8.5.49}$$

and

$$t\frac{d}{dt}\left[\Phi_1^+(t)+\Phi_1^-(t)\right] = 4\sum_{n=0}^{N} a_n \sum_{k=0}^{n} (n-k)p_k\cos(2n-2k)\theta . \tag{8.5.50}$$

The approximate form of equation (8.5.21) is therefore

$$\beta(\cos^2\theta-\cos^2\theta_0)^{-1/2}\sum_{n=0}^{N} a_n\cos(2n+1)\theta - 4\sum_{n=0}^{N} a_n\sum_{k=0}^{n}(n-k)p_k\cos(2n-2k)\theta$$

$$= (1-\alpha)\{A+2G\cos 2\theta\}+\varepsilon_N(\theta) . \tag{8.5.51}$$

This equation must hold for $0 \leq \theta < \theta_0$; observing the symmetry of the problem we note that for an approximate solution it is sufficient to demand that the error $\varepsilon_N(\theta)$ be orthogonal to $\cos 2m\theta$ over $0 \leq \theta < \theta_0$ where $m = 0,1,\ldots,(N-1)$. Now Gradshteyn and Ryzhik (1965, 8.823) gives

$$\int_0^{\theta_0} (\cos^2\theta-\cos^2\theta_0)^{-1/2}\cos(2n+1)\theta d\theta = \pi P_n(\cos 2\theta_0) = \pi p_n , \tag{8.5.52}$$

so that on multiplying (8.5.51) by $\cos 2m\theta$ and integrating over $(0,\theta_0)$, we find

$$\beta \frac{\pi}{2} \sum_{n=0}^{N} a_n [p_{2n+2m+1} + p_{|2n-2m+1|}] - 4 \sum_{n=0}^{N} \sum_{k=0}^{n} (n-k) p_k c_{m,n-k} a_n$$

$$= (1-\alpha)\{Ac_{m,0} + 2Gc_{m,1}\}\ , \quad m = 0,1,\ldots,N-1\ , \tag{8.5.53}$$

where

$$c_{m,n} = \int_0^{\theta_0} \cos 2m\theta \cos 2n\theta d\theta\ . \tag{8.5.54}$$

Equations (8.5.53) together with the equation

$$\sum_{n=0}^{N} a_n \cos(2n+1)\theta_0 = 0 \tag{8.5.55}$$

yield N+1 equations which may be written

$$\sum_{n=0}^{N} d_{m,n} a_n = Be_m^{(0)} + (T_2+T_1)e_m^{(1)} + (T_2-T_1)e_m^{(2)}\ , \tag{8.5.56}$$

when A is expressed as $B+(T_1+T_2)/4$ by using (8.5.9). The solution of equation (8.5.56) may be written

$$a_n = Ba_n^{(0)} + (T_2+T_1)a_n^{(1)} + (T_2-T_1)a_n^{(2)}\ . \tag{8.5.57}$$

Now $\Phi_1(z)$, although automatically satisfying $\Phi_1(0) = -\Phi_1(\infty)$, must still be made to satisfy $\Phi_1(\infty) = B$. Equation (8.5.48) shows that

$$\Phi_1(\infty) = -\frac{1}{2} \sum_{n=0}^{N} a_n p_n\ . \tag{8.5.58}$$

This equation and the displacement condition (8.5.23), which may be rewritten

$$\beta \int_0^{\theta_0} f(\theta)\sin\theta d\theta + \frac{2}{\pi} \int_0^{\theta_0} f(\theta)\cos\theta \ln(\cot(\theta/2))d\theta = -(1+\alpha)B + (1-\alpha)\frac{(T_2-3T_1)}{\ell} \tag{8.5.59}$$

provide two homogeneous equations linking B, T_2+T_1, T_2-T_1 from which B may be eliminated, giving an equation for $(T_2+T_1)/(T_2-T_1)$ as a function of θ_0. When $\alpha = 0 = \beta$, equation (8.5.59) reduces to an approximation to equation (8.5.40).

Wilson (1964a) uses the representation (8.5.46) for f(t) and finds the a_n so that the integral

$$\int_0^{\theta_0} \varepsilon_N^2(\theta)d\theta$$

should be minimized. He specifies $T_1:T_2$ and finds the angle θ_0 corresponding to this ratio by iteration. He examines the numerical accuracy of the approximation by using different values of N and gives extensive graphical and tabular results for θ_0 and the normal and circumferential stresses on the edge of the hole and insert.

The inclusion problem for a *rigid* insert in an *elliptical* hole was formulated by Wilson (1964b); the analysis follows from that given in §8.2. The complex potential for the plate is

$$\Phi(\zeta) = -\frac{1}{2\pi i}\int_{C'} \frac{f(t)}{t-\zeta}dt + \Gamma + \frac{\Lambda}{\zeta^2-m}, \tag{8.5.60}$$

where

$$\Lambda = m(\Gamma+\overline{\Gamma}) + \overline{\Gamma}' + 2mB, \quad B = -\frac{1}{4\pi i}\int_{C'} \frac{f(t)}{t-\zeta}dt, \tag{8.5.61}$$

and C' is the pair of arcs on the unit circle $|\zeta| = 1$ which are mapped by the function (8.2.45) onto the contact arcs of the ellipse. The constants Γ, Γ' are linked to the stresses at infinity (see (8.2.56), (8.2.57)) and are

$$\Gamma = (T_2+T_1)/4, \quad \Gamma' = (T_2-T_1)/2. \tag{8.5.62}$$

The displacement derivatives on C are given by

$$2\mu(u'+iv') = \omega'(t)[\kappa\Phi^{-}(t)+\Phi^{+}(t)] \;, \tag{8.5.63}$$

where ' denotes differentiation with respect to $t = \exp(i\theta)$. The displacement component u_ρ normal to the contact arc must vanish since the insert is rigid. This condition is expressible as

$$2|\omega'(t)|u_\rho = (u+iv)(t^{-1}-mt) + (u-iv)(t-mt^{-1}) = 0 \text{ on } C' \;. \tag{8.5.64}$$

Repeated differentiation of this equation gives

$$(t-mt^3)(u''+iv'')+(t^3-mt)(u''-iv'')-(3mt^2+1)(u'+iv') +$$

$$+ (3t^2+m)(u'-iv') = 0 \;, \tag{8.5.65}$$

and on substituting the expression (8.5.63) for $u'+iv'$ we find

$$\beta(1-m^2)f(t)+3m(t^2-t^{-2}) \frac{1}{\pi i}\int_{C'} \frac{f(s)ds}{s-t} - t[1+m^2-m(t^2+t^{-2})] \frac{1}{\pi i}\int_{C'} \frac{f'(s)ds}{s-t}$$

$$= [2m+3(t^2+t^{-2})]\Lambda+2(1-m^2)\Gamma+2[1-m^2-3m(t^2-t^{-2})]B \;. \tag{8.5.66}$$

This equation with $m = 0$ is identical to equation (8.5.43) with $\alpha = -1$, as is the case if the insert is rigid. Goree (1965) shows that equation (8.5.66) may be effectively solved by using the same representation (8.5.46) for $f(\theta)$.

There is a whole group of inclusion problems, which we classify as *force-fit* rather than contact problems, in which the tractions across the *whole of the boundary between matrix and inclusion* are equal, and there is a discontinuity of displacement across the boundary. Reference to early work on these problems may be found in Muskhelishvili (1953a, §109), and to later work in

Theocaris and Ioakimidis (1977a). Margetson and Morland (1970a) consider the case of an initially compressed circular disc, of initial radius $1+\varepsilon$, in a plate with a hole of radius 1. The tension at infinity is in the y-direction. As the tension T is increased monotonically from zero, separation occurs first at $\theta_0 = \pm\,\pi/2$, and the separation arcs gradually expand. Margetson and Morland plot θ_0 as a function of $T/\{\varepsilon[(\kappa_1+1)/\mu_1+(\kappa_2+1)/\mu_2]\}$, obtaining the angle $\theta_0 = 19.62°$ shown in Figure 8.5.2 for $T_1 = 0$ when $\varepsilon = 0$. The authors discuss the numerical solution of the integral equation governing the contact in this paper and in Margetson and Morland (1970b), Morland (1970). See also Morar' and Popov (1976).

Examples 8.5

1. In order to examine partial contact between a circular inclusion and an infinite plate we consider the following boundary value problem

$$(u_r)_1 = (u_r)_2 \,, \quad (\tau_{r\theta})_1 = 0 = (\tau_{r\theta})_2 \,, \quad (\tau_{rr})_1 = (\tau_{rr})_2 \text{ on } r =$$

with the stresses in the plate (medium 2) satisfying

$$\tau_{xx} = T_1 \,, \quad \tau_{yy} = T_2 \quad \text{at } \infty \,.$$

Equation (8.5.12) shows that $\Phi_1(z) = -\overline{\Phi}_1(z^{-1})$ so that symmetry considerations give $\Phi_1(z) = \sum_{n=0}^{\infty} a_n z^{2n}$. Show that

$$a_0 = \frac{1-\alpha}{1-\alpha+2\beta}\left(\frac{T_1+T_2}{4}\right) \,, \qquad a_1 = \frac{3(1-\alpha)}{2-\beta}\left(\frac{T_1-T_2}{4}\right)$$

and $a_n = 0$, $n > 1$. Hence show that the

$$\tau_{rr} = \frac{3(1-\alpha)}{2-\beta}\left\{(T_1-T_2)\cos^2\theta + \frac{T_2-kT_1}{1+k}\right\} \text{ on } r = 1$$

where $k = (1-3\alpha+7\beta)/(5-3\alpha+5\beta)$ is a parameter satisfying $-0.5 \le k \le 1$. Hence show that

$\tau_{rr} \geq 0$ for all θ if $T_2 - kT_1 > 0$ and $kT_2 - T_1 < 0$

$\tau_{rr} \leq 0$ for all θ if $kT_2 - T_1 > 0$ and $T_2 - kT_1 < 0$

and τ_{rr} is sometimes positive and sometimes negative otherwise. The four cases are shown in Figure 8.5.3 for a case in which $k > 0$. Notice that this solution does not give the correct solution for the partial contact cases as in these situations the boundary conditions no longer apply on the whole of the circle $r = 1$. (Note that this diagram is (incorrectly) given in Keer, Dundurs and Kiattikomol (1973).)

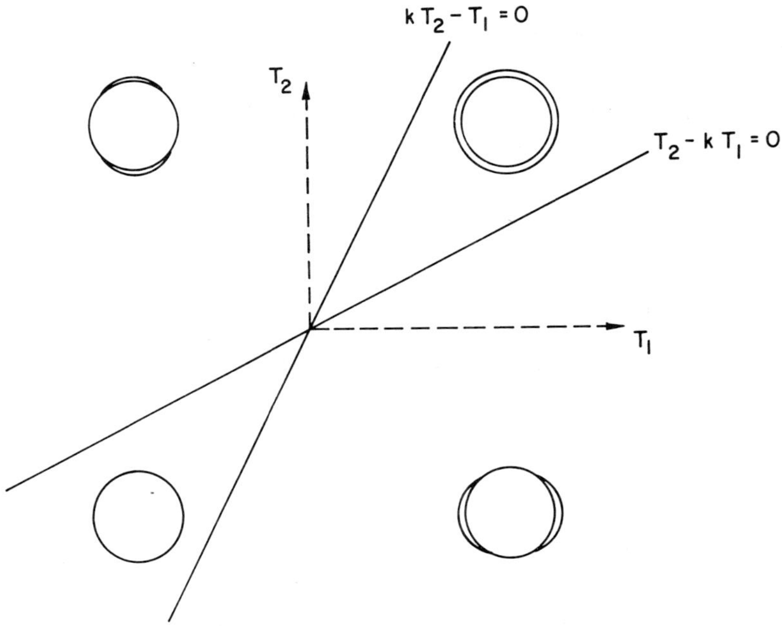

Figure 8.5.3 - The possible contact configurations: full contact, no contact and partial contact

2. Use the result of Example 8.2.5 to analyse a smooth elastic disc of radius R and elastic constants μ_1, κ_1 compressed and inserted into a hole of radius a in an infinite plate with elastic constants μ_2, κ_2. Assuming that R > a and R-a << a show that the pressure across the interface is

$$p = \frac{4\mu_1\mu_2}{2\mu_1\kappa_2+\mu_2(\kappa_1-1)}\left(\frac{R}{a} - 1\right)$$

and find the radius of the interface.

8.6 *Cradling contact problems*

Cradling refers to the situation in which a body of circular section, in plane strain or plane stress, rests in or is pressed against another, the cradle. Kalandia (1957) formulated two such problems for a rigid or elastic cylinder pressed by a force applied at its centre into a hole of the same radius in an infinite plate. He obtained an approximate solution to the problems by reducing the integro-differential equation governing the contact to one for the interval [-1,1]; both equations are studied in depth in Kalandia (1957). Wilson and Hill (1972) considered a heavy elastic cylinder pressed by its own weight into a rigid cradle which need not necessarily be circular. We shall formulate these problems and briefly discuss how the governing equations may be solved.

All three problems have essentially the same geometry, which is shown in Figure 8.6.1. The cylinder occupies the region $\mathcal{D}^+$ ($|z| < 1$); the cradle, if it is elastic, occupies $|z| > 1$. First consider a rigid cylinder in frictionless contact with an elastic cradle. The theory of §8.2 shows that if P is the total force applied to the cradle then the complex potentials $\Phi_2(z)$, $\Psi_2(z)$ have the form

$$\Phi_2(z) = \gamma z^{-1}+O(z^{-2}) , \quad \Psi_2(z) = -\kappa\gamma z^{-1}+O(z^{-2}) , \tag{8.6.1}$$

for large $|z|$, where

$$\gamma = -P/[2\pi(1+\kappa_2)] . \tag{8.6.2}$$

Now $\Phi_2(z)$ defined in $\mathcal{D}^+$ by means of (8.2.4) will be holomorphic in $\mathcal{D}_+$ except at 0, near which

$$\Phi_2(z) = -\kappa_2\gamma z^{-1}+O(1) . \tag{8.6.3}$$

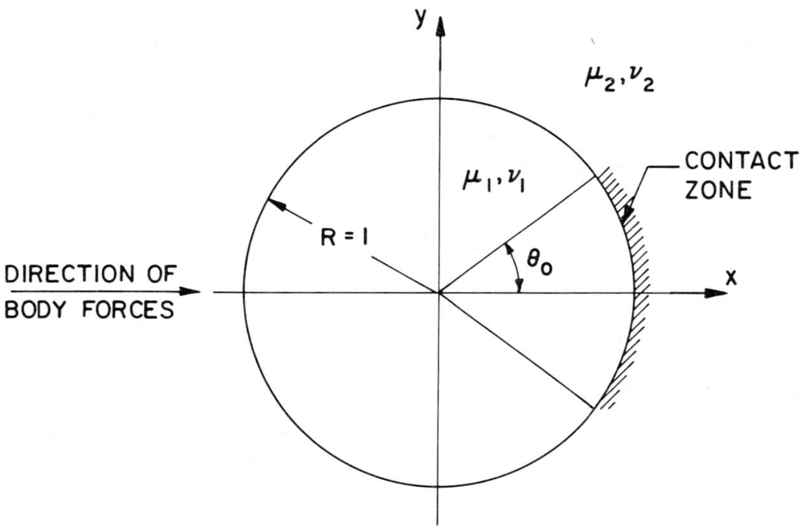

Figure 8.6.1 - Geometry of cradling contact

The boundary conditions on C ($|z| = 1$) are

$$\tau_{r\theta} = 0 \text{ on } C, \quad \tau_{rr} = 0 \text{ on } C'' , \tag{8.6.4}$$

$$u_r = 0 \text{ on } C' , \tag{8.6.5}$$

where again C' is the contact region and C'' is the remainder of C. Proceeding as in §8.2 we find that

$$\Phi_2^-(t)-\Phi_2^+(t) = \tau_{rr} = f(t) \text{ on } C' , \tag{8.6.6}$$

$$\kappa_2\Phi_2^-(t)+\Phi_2^+(t) = 2\mu_2(u_2'+iv_2') \text{ on } C , \tag{8.6.7}$$

$$2\mu_2(u_2+iv_2) = \kappa_2\phi_2(z)-z\overline{\phi}_2'(\overline{z})-\overline{\psi}_2(\overline{z}) \text{ in } \mathcal{D}^- , \tag{8.6.8}$$

and combining equations (8.6.3), (8.6.6) we deduce that $\Phi_2(z)+\overline{\Phi}_2(z^{-1})$ is holomorphic in the whole plane except at 0 and at infinity. But

$$\Phi_2(z) = \gamma z^{-1}+0(z^{-2}) \text{ at } \infty , \tag{8.6.9}$$

$$\Phi_2(z) = -\kappa_2\gamma z^{-1}+A+Bz+0(z^2) \text{ near } 0 , \tag{8.6.10}$$

so that

$$\Phi_2(z)+\overline{\Phi}_2(z^{-1}) = \gamma z^{-1}-\kappa_2\gamma z+A+Bz^{-1}+0(z^{-2}) \text{ at } \infty , \tag{8.6.11}$$

$$\Phi_2(z)+\overline{\Phi}_2(z^{-1}) = -\kappa_2\gamma z^{-1}+A+Bz+\gamma z+0(z^2) \text{ near } 0 , \tag{8.6.12}$$

and therefore $B+\gamma = -\kappa_2\gamma$, $B = -(1+\kappa_2)\gamma$ and

$$\Phi_2(z)+\overline{\Phi}_2(z^{-1}) = -\kappa_2\gamma(z+z^{-1})+A , \tag{8.6.13}$$

so that, on combining this with equation (8.2.11) we deduce that

$$z\Psi_2(z)+\Phi_2'(z) = -\kappa_2\gamma(z+z^{-1})+A . \tag{8.6.14}$$

From equation (8.6.6) we find

$$\Phi_2(z) = -\frac{1}{2\pi i}\int_{C'} \frac{f(t)}{t-z}\,dt + A_0 + B_0 z^{-1} , \qquad (8.6.15)$$

and on expanding this for large and small $|z|$ we have

$$A_0 = 0, \quad B_0 = -\kappa_2\gamma, \quad A = -\frac{1}{2\pi i}\int_{C'} \frac{f(t)}{t}\,dt , \quad i\int_{C'} f(t)dt = P \qquad (8.6.16)$$

the last of which is the equilibrium equation. Since contact occurs over one arc, rather than two as in inclusion problems, the contact displacement equation may be replaced by the derivative form $L_2(u_r) = 0$, where L is defined in (8.5.15). Proceeding as in §8.5, and noting that equation (8.6.13) now implies that

$$\Phi_2^{\pm}(t)+\overline{\Phi_2^{\mp}}(\bar{t}) = -\kappa_2\gamma(t+t^{-1}) + A \text{ on } C , \qquad (8.6.17)$$

we find

$$2\mu_2 L_2(u_r) \equiv (1-\kappa_2)f(t)-(1+\kappa_2)\frac{t}{\pi i}\int_{C'} \frac{f'(s)}{s-t}\,ds +$$

$$+ \frac{(1+\kappa_2)}{2\pi i}\int_{C'} \frac{f(s)}{s}\,ds + 2(1+\kappa_2)\kappa_2\gamma_2(t+t^{-1}) = 0 \text{ on } C' . \qquad (8.6.18)$$

The problem is thus reduced to solving equation (8.6.18) subject to the conditions

$$f(\exp(\pm i\theta_0)) = 0 , \qquad i\int_{C'} f(t)dt = P . \qquad (8.6.19)$$

Now assume that the cylinder is elastic and is acted upon by a concentrated force P at its centre, as before. Equations (8.5.1)-(8.5.2) hold, i.e.,

$$\Phi_1^{+}(t)-\Phi_1^{-}(t) \equiv \tau_{rr} = f(t) \text{ on } C' , \qquad (8.6.20)$$

$$\kappa_1\Phi_1^+(t)-\Phi_1^-(t) = 2\mu_1(u_1'+iv_1') \text{ on } C . \tag{8.6.21}$$

The complex potentials $\Phi_1(z)$, $\Psi_1(z)$ will have a singularity at $z = 0$ (see Example 4.2.5);

$$\Phi_1(z) = \gamma_1 z^{-1}+O(1) , \quad \Psi_1(z) = \kappa_1\gamma_1 z^{-1} + O(1) \text{ near } 0 , \tag{8.6.22}$$

where

$$\gamma_1 = -P/[2\pi(1+\kappa_1)] . \tag{8.6.23}$$

An analysis similar to (8.6.9)-(8.6.13) shows that $\Phi_1(z)+\bar{\Phi}_1(z^{-1}) = -\gamma_1(z+z^{-1})$ and

$$\Phi_1(z) = \frac{1}{2\pi i}\int_{C'} \frac{f(t)}{t-z}\,dt + \gamma_1(z^{-1}-2z) + \frac{A}{2} , \tag{8.6.24}$$

where A is given in (8.6.16). Now

$$2\mu_1 L_1(u_r) = (1-\kappa_1)f(t) + \frac{(1+\kappa_1)t}{\pi i}\int_{C'} \frac{f'(s)}{s-t}\,ds - 2(1+\kappa_1)\gamma_1(t+t^{-1}) , \tag{8.6.25}$$

so that the equation

$$L_1(u_r) = L_2(u_r) \text{ on } C' \tag{8.6.26}$$

is

$$\beta f(t) - \frac{t}{\pi i}\int_{C'} \frac{f'(s)}{s-t}\,ds = \frac{(1-\alpha)}{2} A + (1-\beta)\frac{P}{2\pi}(t+t^{-1}) . \tag{8.6.27}$$

This equation depends only on α, β even though the contact between the two bodies is not receding; on the other hand the extent of the contact region depends on P, as is to be expected in advancing contact.

If the force P is applied to the cylinder as a uniform body force $p = P/\pi$ per unit area, then the complex potentials for the cylinder may be constructed as follows. First take a field corresponding to the body force F_x. Since there are many such, choose one so that $\Xi = 0$. Equations (4.2.15)-(4.2.18) show that

$$V = -p(z+\bar{z})/2, \quad W = -p(2z^2+z\bar{z})/4, \quad \Phi_0(z) = -\varphi pz/2, \quad \Psi_0(z) = 0 , \tag{8.6.28}$$

and

$$\Theta_0 = -2px, \quad \Xi_0 = 0, \quad 2\mu D_0 = -(1-2\nu)pz^2/2 . \tag{8.6.29}$$

Now construct a field produced by surface tractions which have zero resultant and which exactly counterbalances the surface tractions due to the field (8.6.29) on C''. According to the theory of §8.2, one must find $\Phi(z)$ so that

$$\Phi_1^+(t)-\Phi_1^-(t) = p(t+t^{-1})/2+f(t) \text{ on } C , \tag{8.6.30}$$

(note that $\Xi = 0$ implies $\tau_{\theta\theta} = \tau_{rr}$, $\tau_{r\theta} = 0$, $\Theta = 2\tau_{rr}$) where $f(t) = 0$ on C'' (and since $\tau_{r\theta} = 0$, $f(t)$ will be real). Since the normal stress due to $\Phi_0(z)$ is $-px = -p(t+t^{-1})/2$, $f(t)$ is the normal stress on the cylinder. The field $\Phi_1(z)$ is given by

$$\Phi_1(z) = \frac{1}{2\pi i}\int_{C'} \frac{f(t)}{t-z}dt - \frac{1}{4\pi i}\int_{C'} \frac{f(t)}{t}dt + \frac{pz}{2} , \quad |z| < 1 , \tag{8.6.31}$$

$$\Phi_1(z) = \frac{1}{2\pi i}\int_{C'} \frac{f(t)}{t-z}dt - \frac{1}{4\pi i}\int_{C'} \frac{f(t)}{t}dt - \frac{p}{2z} , \quad |z| > 1 . \tag{8.6.32}$$

The condition (8.2.9) gives the equilibrium equation

$$\frac{1}{\pi i}\int_{C'} f(t)dt+p = 0 , \tag{8.6.33}$$

which is the same as the last of (8.6.16), and

$$2\mu_1 L_1(u_r) = (1-\kappa_1)(\Phi_1^+(t)-\Phi_1^-(t)) + (1+\kappa_1)t\,\frac{d}{dt}\,(\Phi_1^+(t)+\Phi_1^-(t))$$

$$= (1-\kappa_1)f(t) + \frac{(1+\kappa_1)}{\pi i}\,t\int_{C'}\frac{f'(s)}{s-t}\,ds + \frac{p}{2}\,(t+t^{-1})\ , \qquad (8.6.34)$$

so that the equation obtained from (8.5.20) has exactly the same form as (8.6.27) except that the last term does not depend on α,β alone; in fact $(1-\beta)$ is replaced by $(\mu_2+2\kappa_2\mu_1)/[\mu_2(1+\kappa_1)+\mu_1(1+\kappa_2)]$.

Equation (8.6.27) has essentially the same form as equation (8.5.43). Now C' consists of a single arc instead of two. When $\beta = 0$, so that cylinder and cradle have identical properties, then there is a closed form solution similar to that found by Stippes et al (1962). When $\beta \neq 0$ the solution may be solved by taking

$$f(t) = 2(2\cos\theta-2\cos\theta_0)^{-1/2}\sum_{n=0}^{N} a_n\cos(n+1/2)\theta$$

$$= -\sum_{n=0}^{N} a_n(t^{n+1}+t^{-n})/X^+(t) \qquad (8.6.35)$$

and proceeding as in (8.5.48)-(8.5.58). Kalandia (1957) first mapped C' onto the line segment $[-1,1]$ in the ζ-plane by using the transformation

$$z = \frac{i\sigma-\zeta}{i\sigma+\zeta}\ , \qquad \sigma = \cot(\theta_0/2)\ . \qquad (8.6.36)$$

Equation (8.6.27) becomes

$$\frac{\beta\sigma}{\chi^2+\sigma^2}\,f(\chi) + \frac{1}{2\pi}\int_{-1}^{1}\frac{f'(\xi)}{\xi-\chi}\,d\xi = -\frac{(1-\alpha)}{2}\,\frac{\sigma^2}{(\chi^2+\sigma^2)}\int_{-1}^{1}\frac{f(\xi)}{\xi^2+\sigma^2}\,d\xi$$

$$-\ (1-\beta)\,\frac{P}{\pi}\,\sigma\,\frac{(\chi^2-\sigma^2)}{(\chi^2+\sigma^2)^2}\ , \qquad -1 \leq \chi \leq 1\ , \qquad (8.6.37)$$

while the auxiliary conditions (8.6.19) become

$$f(-1) = 0 = f(1) , \qquad 2\sigma \int_{-1}^{1} \frac{(\xi^2-\sigma^2)f(\xi)}{(\xi^2+\sigma^2)^2} d\xi = P . \tag{8.6.38}$$

Kalandia considers the equation

$$\frac{f(x)}{B(x)} - \frac{1}{2\pi} \int_{-1}^{1} \frac{f'(t)}{t-x} dt = g(x) , \qquad -1 \le x \le 1 , \tag{8.6.39}$$

and supposes that, for computational purposes, f(x) is replaced by a Lagrangian approximation

$$f_N(x) = \frac{1}{N+1} \sum_{n=1}^{N} (-)^{n+1} f(x_n) \frac{\sin(N+1)\theta \sin\theta_n}{\cos\theta - \cos\theta_n} \tag{8.6.40}$$

where

$$x = \cos\theta , \quad x_n = \cos\theta_n , \quad \theta_n = n\pi/(N+1) , \quad 0 \le \theta \le \pi , \quad n = 1,2,\ldots,N . \tag{8.6.41}$$

This has the property

$$f_N(x_n) = f(x_n) , \qquad f_N(\pm 1) = 0 . \tag{8.6.42}$$

The expression $f_N(x)$ may be written in the alternative form

$$f_N(x) = \frac{2}{N+1} \sum_{n=1}^{N} f(x_n) \sum_{k=1}^{N} \sin k\theta \sin k\theta_n \tag{8.6.43}$$

so that

$$\frac{1}{\pi} \int_{-1}^{1} \frac{f_N'(t)dt}{t-x} = \frac{1}{\pi} \int_{0}^{\pi} \frac{df}{d\theta} \frac{d\theta}{\cos\theta - \cos\phi} = \frac{2}{N+1} \sum_{n=1}^{N} f(x_n) \sum_{k=1}^{N} k \sin k\theta_n I_k , \tag{8.6.44}$$

where

$$I_k = \frac{1}{\pi}\int_0^{\pi} \frac{\cos k\theta d\theta}{\cos\theta - \cos\phi} = \frac{\sin k\phi}{\sin\phi} \qquad (8.6.45)$$

on account of (2.3.27). Thus the equation obtained by equating both sides of (8.6.39) at x_m is

$$\frac{f(x_m)}{B(x_m)} - \sum_{n=1}^{N} f(x_n)\left\{\frac{1}{N+1}\sum_{k=1}^{N} \frac{k\sin k\theta_n \sin k\theta_m}{\sin\theta_m}\right\} = g(x_m) , \qquad (8.6.46)$$

which, for m = 1,2,...,N, provides N equations for the $f(x_m)$.

Wilson and Hill (1972) consider an elastic cylinder resting in a rigid cradle of arbitrary shape, use effectively the representation (8.6.35) for the contact stress, and construct a set of equations for the a_n from the condition that the difference between the displacement of the cylinder and the shape of the cradle should be minimal. One of the important conclusions of their numerical study is that the variation of the contact stress $f(\theta)$ may be made nearly to fit the equation

$$\left[\frac{f(\theta)}{f(0)}\right]^m + \left(\frac{\theta}{\theta_0}\right)^2 = 1 \qquad (8.6.47)$$

where m is about equal to 1.5; the Hertz theory valid for contact length very small compared to the radii of cradle and cylinder would give m = 2.

CHAPTER IX - CONTACT PROBLEMS INVOLVING WEDGES AND STRAIGHT EDGES

"You don't know how to manage Looking-glass cakes," the Unicorn remarked. "Hand it round first, and cut it afterwards."
-*Through the Looking Glass*

9.1 *Introduction*

In this chapter we shall present a brief review of research in the area covered by the title. The chapter will follow the general pattern followed by the others: presentation of basic theory, solution of sample problems, reference to recent advances; but by the nature of the topic the first two phases will be abbreviated while the last will be extended. The reasons for the change are threefold: the analysis involved is essentially lengthy; it appears in generally accessible journals; and for a thorough treatment it requires a higher level of rigour than is the norm for this book.

Section 9.2 will present a formal account of the Mellin transform; a review of the theory of elasticity relating to a wedge appears in §9.3; and the application of the Mellin transform is described in §9.4.

9.2 *Mellin transforms*

The Mellin transform (Tranter (1956), Sneddon (1972)) is defined by

$$f^*(s) = M[f(x);s] = \int_0^\infty f(x)x^{s-1}dx \ . \tag{9.2.1}$$

The parameter s in (9.2.1) is a complex number $s = \sigma+i\tau$. The existence of an infinite integral such as (9.2.1) will depend on three factors: the existence of the integral

$$\int_a^b f(x)x^{s-1}dx$$

over any finite range $0 < a \leq x \leq b < \infty$, the behaviour of $|f(x)x^{s-1}| = |f(x)|x^{\sigma-1}$ for large x - here σ must not be too large, i.e., $\sigma < \sigma_2$, and the behaviour of $|f(x)x^{s-1}|$ for small x - and here σ must not be too small, i.e., $\sigma > \sigma_1$. These two facts imply that the integral will converge when σ lies in a strip $\sigma_1 < \sigma < \sigma_2$. The Mellin transform is said to *converge absolutely* if

$$\int_0^\infty |f(x)|x^{\sigma-1}dx$$

exists. Although much stronger results may be proved, the following are sufficient for our purposes.

1. If $M[f(x);s]$ is absolutely convergent in a strip $\alpha_1 \leq \sigma \leq \alpha_2$ then $f^*(s)$ is holomorphic in the strip $\alpha_1 < \sigma < \alpha_2$. This may be proved by modifying the corresponding theorem for the two-sided Laplace transform (Sneddon (1972), p. 137 ff).
2. If, in addition, $f(x)$ is continuous at x then

$$f(x) = \frac{1}{2\pi i}\int_{c-i\infty}^{c+i\infty} f^*(s)x^{-s}ds \qquad (9.2.2)$$

for all c in $\alpha_1 \leq c \leq \alpha_2$ (Doetsch (1950, Vol. I, p. 212)). The path of integration is shown in Figure 9.2.1.

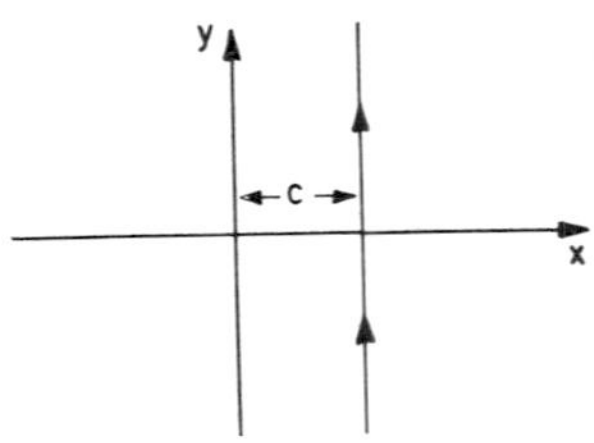

Figure 9.2.1 - The path of integration for the inverse Mellin transform

3. Suppose that $f^*(s)$ is holomorphic in the strip $\alpha_1 < \sigma < \alpha_2$, that, for arbitrary $\eta > 0$, $f^*(s)$ tends uniformly to zero as $|\tau| \to \infty$ in the strip $\alpha_1 + \eta \leq \sigma \leq \alpha_2 - \eta$ and that

$$\int_{-\infty}^{\infty} f^*(\sigma+i\tau)d\tau$$

is absolutely convergent for each σ in $\alpha_1 < \sigma < \alpha_2$. If $f(x)$ is defined by (9.2.2) then, for all s such that $\alpha_1 < \mathrm{Re}(s) < \alpha_2$, $f^*(s)$ is given by (9.2.1) (Sneddon (1972), p. 273).

Perhaps the simplest example of a Mellin transform may be obtained by starting from the integral

$$\int_0^{\infty} x^{n-1}\exp(-x)dx = (n-1)! \ , \qquad n = 1,2,\dots \ . \tag{9.2.3}$$

The integral

$$\int_0^{\infty} x^{s-1}\exp(-x)dx$$

exists for all complex s with $\mathrm{Re}(s) = \sigma > 0$, and this is used to define the Gamma function

$$\int_0^{\infty} x^{s-1}\exp(-x)dx = \Gamma(s) \ , \qquad \sigma > 0 \ , \tag{9.2.4}$$

where, for $s = n$, $\Gamma(n) = (n-1)!$ Note that since the Mellin transform (9.2.4) exists in the half-plane $\sigma > 0$, $\alpha_1 = 0$ and $\alpha_2 = \infty$.

The Mellin transform of the function x^{α} does not exist for *any* s since the integrals over the ranges $(0,1)$ and $(1,\infty)$ will exist only if s lies in the mutually exclusive ranges $\mathrm{Re}(s) > \mathrm{Re}(-\alpha)$ and $\mathrm{Re}(s) < \mathrm{Re}(-\alpha)$ respectively. If however

$$f(x) = \begin{cases} x^{\alpha} & 0 < x < 1 \\ 0 & x > 1 \end{cases} \tag{9.2.5}$$

then equation (9.2.1) immediately gives

$$f^*(s) = (s+\alpha)^{-1}, \quad \mathrm{Re}(s) > \mathrm{Re}(-\alpha), \tag{9.2.6}$$

so that $\alpha_1 = -\alpha$, $\alpha_2 = \infty$. In order to verify the inverse (9.2.2) we apply Cauchy's theorem to the paths shown in Figure 9.2.2. Write $x = \exp(u)$, $s = R\exp(i\theta)$ then $|x^{-s}| = |\exp[-uR\exp(i\theta)]| = \exp(-uR\cos\theta)$. Thus if $0 < x < 1$ so that $u < 0$, the integral over the large semi-circle to the left, where $\cos\theta < 0$, will tend to zero as $R \to \infty$ and the integral (9.2.2) will be the residue x^{α} of $x^{-s}(s+\alpha)^{-1}$ at $s = -\alpha$. When $x > 1$ so that $u > 0$, the integral over the large semi-circle to the right, where $\cos\theta > 0$, will tend to zero as $R \to \infty$ and the integral (9.2.2) will be zero, so that $M^{-1}[s+\alpha)^{-1};x] = x^{\alpha}$ if $0 < x < 1$ and will be zero if $x > 1$, in accordance with equation (9.2.5). See also Example 9.2.3.

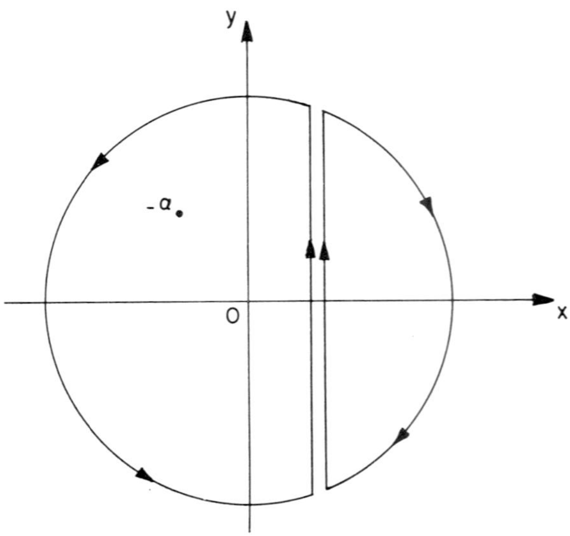

Figure 9.2.2 - The paths used for the evaluation of the inverse of f(s) given in (9.2.6)*

The rules

$$M[f(ax);s] = a^{-s}f^*(s) , \qquad (9.2.7)$$

$$M[x^a f(x);s] = f^*(s+a) , \qquad (9.2.8)$$

$$M[f(x^a);s] = a^{-1}f^*(s/a) , \quad a > 0 , \qquad (9.2.9)$$

where s and a must be such that the transforms exist, may easily be proved. The rule for the Mellin transform of a derivative may be derived as follows. By definition

$$M[f'(x);s] = \int_0^\infty f'(x)x^{s-1}dx \qquad (9.2.10)$$

so that, on integrating by parts, we obtain

$$M[f'(x);s] = [f(x)x^{s-1}]_0^\infty - (s-1)\int_0^\infty f(x)x^{s-2}dx . \qquad (9.2.11)$$

If s is such that

$$\lim_{x\to 0} f(x)x^{s-1} = 0 = \lim_{x\to\infty} f(x)x^{s-1} , \qquad (9.2.12)$$

then

$$M[f'(x);s] = -(s-1)M[f(x);s-1] . \qquad (9.2.13)$$

Examples 9.2

1. Use the transformations $u = ax$ and $u = x^a$ respectively to demonstrate equations (9.2.7), (9.2.9).
2. Show that $M[xf'(x);s] = -sf^*(s)$.
3. If

$$I(c) = \frac{1}{2\pi i}\int_{c-i\infty}^{c+i\infty} (s+\alpha)^{-1}x^{-s}ds ,$$

show that if $\mathrm{Re}(c) < \mathrm{Re}(-\alpha)$, then $I(c) = 0$ if $0 < x < 1$, $I(c) = -x^{\alpha}$ if $x > 1$.

9.3 *The theory of elasticity for a plane wedge*

In the absence of body forces the equations governing the plane strain of an elastic body in plane polar coordinates are the equilibrium equations (1.5.2) - (1.5.3), namely

$$r\,\frac{\partial\tau_{rr}}{\partial r} + \frac{\partial\tau_{r\theta}}{\partial\theta} + \tau_{rr} - \tau_{\theta\theta} = 0\ , \tag{9.3.1}$$

$$r\,\frac{\partial\tau_{r\theta}}{\partial r} + \frac{\partial\tau_{\theta\theta}}{\partial\theta} + 2\tau_{r\theta} = 0\ , \tag{9.3.2}$$

and the stress-strain equations

$$(1-2\nu)\tau_{rr} = 2\mu[(1-\nu)e_{rr}+\nu e_{\theta\theta}]\ , \qquad \tau_{r\theta} = 2\mu e_{r\theta}\ , \tag{9.3.3}$$

$$(1-2\nu)\tau_{\theta\theta} = 2\mu[(1-\nu)e_{\theta\theta}+\nu e_{rr}]\ . \tag{9.3.4}$$

These equations, together with the definitions of the strains imply, amongst other equations, the compatibility equation Example 1.11.6, namely

$$\nabla^2(\tau_{rr}+\tau_{\theta\theta}) = 0\ . \tag{9.3.5}$$

There are basically three ways in which these equations have been solved. Michell (1899) expressed the stresses in the form (Example 1.11.6)

$$\tau_{rr} = \frac{1}{r}\frac{\partial U}{\partial r} + \frac{1}{r^2}\frac{\partial^2 U}{\partial\theta^2}\ , \qquad \tau_{\theta\theta} = \frac{\partial^2 U}{\partial r^2}\ , \qquad \tau_{r\theta} = -\frac{\partial}{\partial r}\left(\frac{1}{r}\frac{\partial U}{\partial\theta}\right)\ . \tag{9.3.6}$$

These automatically satisfy equations (9.3.1), (9.3.2), and equation (9.3.5) shows that U must be biharmonic, i.e.,

$$\left\{\frac{1}{r}\frac{\partial}{\partial r}\left(r\frac{\partial}{\partial r}\right)+\frac{1}{r^2}\frac{\partial^2}{\partial\theta^2}\right\}^2 U = 0 . \qquad (9.3.7)$$

Carothers (1913) considered the wedge of Figure 9.3.1, bounded by angles $\pm\gamma$, and used separable solutions (Example 9.3.1)

$$U = r^n[A\sin n\theta + B\cos n\theta + C\sin(n-2)\theta + D\cos(n-2)\theta] , \qquad (9.3.8)$$

to obtain solutions corresponding to certain simple distributions of stress. In particular he presents

$$U = \frac{Q(2\theta\cos 2\gamma - \sin 2\theta)}{2(\sin 2\gamma - 2\gamma\cos 2\gamma)} \qquad (9.3.9)$$

as the stress function appropriate to a concentrated moment Q applied at the origin. Equations (9.3.6) show that U yields the stresses

$$\tau_{rr} = \frac{2Q\sin 2\theta}{r^2(\sin 2\gamma - 2\gamma\cos 2\gamma)} , \quad \tau_{\theta\theta} = 0 , \quad \tau_{r\theta} = \frac{Q(\cos 2\theta - \cos 2\gamma)}{r^2(\sin 2\gamma - 2\gamma\cos 2\gamma)} , \qquad (9.3.10)$$

so that the faces of the wedge are stress-free in the sense

$$\tau_{r\theta}(r,\pm\gamma) = 0 = \tau_{\theta\theta}(r,\pm\gamma) , \quad 0 < r < \infty . \qquad (9.3.11)$$

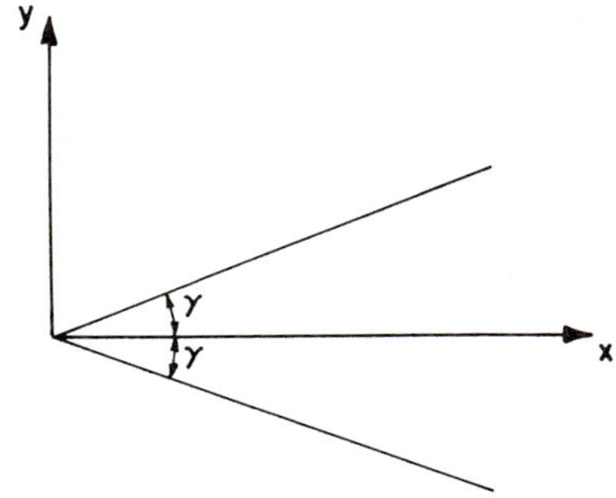

Figure 9.3.1 - Wedge bounded by angles $\theta = \pm\gamma$

Let X(r), Y(r), M(r) be respectively the Cartesian components of the resultant force, and the resultant moment applied to the finite region bounded by the lines $\theta = \pm\gamma$ and the circular arc of radius r. Then

$$X(r) = -r\int_{-\gamma}^{\gamma}(\tau_{rr}\cos\theta-\tau_{r\theta}\sin\theta)d\theta\ , \qquad Y(r) = -r\int_{-\gamma}^{\gamma}(\tau_{rr}\sin\theta+\tau_{r\theta}\cos\theta)d\theta\ , \tag{9.3.12}$$

$$M(r) = -r^2\int_{-\gamma}^{\gamma}\tau_{r\theta}d\theta\ , \tag{9.3.13}$$

and the stresses (9.3.10) yield

$$X(r) = 0 = Y(r)\ , \quad M(r) = Q\ , \quad 0 < r < \infty\ . \tag{9.3.14}$$

The stress-function (9.3.9) is clearly meaningless when γ is equal to the unique non-zero root $\gamma^* \simeq 0.715\pi$ of the equation $\tan 2\gamma = 2\gamma$. In an important paper, Sternberg and Koiter (1958) show that Carothers' solution is but one of an infinity of solutions satisfying the three conditions: (a) the stresses conform to the two-dimensional equilibrium and compatibility equations throughout the wedge region ($0 < r < \infty$, $-\gamma < \theta < \gamma$), are regular in the closed region ($0 \leq r < \infty$, $-\gamma \leq \theta \leq \gamma$) except for a point singularity at the origin, and vanish at infinity

$$\tau_{rr}, \tau_{r\theta}, \tau_{\theta\theta} \to 0 \quad \text{as} \quad r \to \infty\ .$$

(b) The boundaries are free from traction, i.e., equation (9.3.11) holds. (c) The singularity at the origin has as stress resultant the prescribed moment Q.

Sternberg and Koiter show moreover that if one attempts to obtain the concentrated moment Q solution as the limit as $a \to 0$ of a anti-symmetrical normal load distribution p(r) on $\theta = \pm\gamma$, $0 < r < a$, as shown in Figure 9.3.1, then this limit can be interpreted in a way that is independent of the particular distribution $p(r)$ only if $0 < \gamma < \gamma^*$. When $\gamma^* < \gamma < \pi$ the limit will not exist unless another condition, which does not admit a useful physical interpretation, is met, and in general all stress components become infinite throughout the field. The importance of Sternberg and Koiter's paper rests on the following conclusions that they draw. (1) Extreme caution must be used in applying St. Venant's Principle: in the wedge problem for $\gamma^* < \gamma < \pi$ there is a *self-equilibrated stress field* which decays at infinity *less rapidly* than the field corresponding to (9.3.9) which has a *couple* as its stress resultant. (2) For some wedge problems, namely for $\gamma < \gamma^* < \pi$, *no meaning* may be attached to the idea of a concentrated couple. Note that if $p(r)$ has the required resultant couple Q, but is not restricted to be antisymmetric, then γ must be restricted to the range $0 < \gamma < \pi$. (3) Great care must sometimes be exercised in applying and inverting Mellin transforms. The serious reader should study this paper.

Brahtz (1933) used solutions of the form

$$U = \int_0^\infty A(m) \begin{Bmatrix} x \\ y \end{Bmatrix} \cdot \begin{Bmatrix} \cos m\xi \\ \sin m\xi \end{Bmatrix} \cdot \begin{Bmatrix} \cosh m\theta \\ \sinh m\theta \end{Bmatrix} dm , \qquad (9.3.15)$$

where $\xi = \ell nr$, and applied them to find stresses due to a concentrated load on the boundary of a plane with a right-angled re-entrant corner. Shepherd (1935) carried out more detailed calculations with such solutions for an arbitrary wedge angle.

The third solution of equations (9.3.1) - (9.3.4) is provided by the Papkovich-Neuber solution of (1.10.11), which for plane problems may be written

$$2\mu u = \kappa(\psi_1\cos\theta+\psi_2\sin\theta) - r\cos\theta \frac{\partial\psi_1}{\partial r} - r\sin\theta \frac{\partial\psi_2}{\partial r} - \frac{\partial\phi}{\partial r} , \qquad (9.3.16)$$

$$2\mu v = \kappa(-\psi_1\sin\theta+\psi_2\cos\theta) - \cos\theta \frac{\partial\psi_1}{\partial\theta} - \sin\theta \frac{\partial\psi_2}{\partial\theta} - \frac{1}{r}\frac{\partial\phi}{\partial\theta} , \qquad (9.3.17)$$

where $u \equiv u_\theta$, $v \equiv u_r$ and the functions ϕ, ψ_1, ψ_2 are plane harmonic. This representation will be used in the following section in conjunction with the Mellin transform.

Finally, M.L. Williams (1956) and England (1971b) have applied the complex variable analysis of Chapter 4 to wedge problems. This analysis will be briefly noted in §9.5.

Examples 9.3

1. Show that (9.3.8) is a solution of equation (9.3.7) for arbitrary n. What are the solutions corresponding to the special cases n = 0,1,2? Use the solution for n = 2 to obtain a stress distribution satisfying $\tau_{r\theta}(r,\pm\gamma) = 0$, and *either* $\tau_{\theta\theta}(r,\pm\gamma) = \pm P$, *or* $\tau_{\theta\theta}(r,\pm\gamma) = -P$.

9.4 *The application of the Mellin transform to wedge problems*

Tranter (1948) used the stress function solution (9.3.6), in conjunction with the Mellin transform, to find the formal solution for the wedge of Figure 9.4.1 subject to arbitrary pressures and shear stresses on the faces $\theta = \pm\gamma$,

$$p_1(r) = -\tau_{\theta\theta}(r,\gamma)\ , \qquad q_1(r) = \tau_{r\theta}(r,\gamma)\ , \tag{9.4.1}$$

$$p_2(r) = -\tau_{\theta\theta}(r,-\gamma)\ , \qquad q_2(r) = \tau_{r\theta}(r,-\gamma)\ . \tag{9.4.2}$$

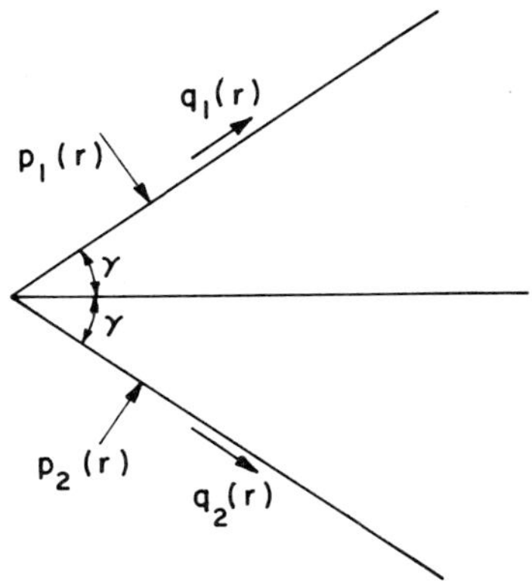

Figure 9.4.1 - Normal and shearing stresses acting on the faces of a wedge

He notes that $\mathcal{U}(r,\theta)$ will be biharmonic if $\mathcal{U}^*(s,\theta) = M[\chi(r,\theta);r \to s]$ satisfies

$$\frac{\partial^4 \mathcal{U}^*}{\partial\theta^4} + [(s+2)^2+s^2]\frac{\partial^2 \mathcal{U}^*}{\partial\theta^2} + s^2(s+2)^2\mathcal{U}^* = 0 \qquad (9.4.3)$$

the solution of which is

$$\mathcal{U}^*(s,\theta) = A\sin s\theta+B\cos s\theta+C\sin(s+2)\theta+D\cos(s+2)\theta\ , \qquad (9.4.4)$$

where A,B,C,D, which are functions of s, may be determined from the boundary conditions (9.4.1) - (9.4.2). Thus by using (9.2.13) twice, and putting $\partial/\partial r \equiv {}'$, we find

$$P_n(s) \equiv M[p_n(r);s+2] = -M[\mathcal{U}''(r,\pm\gamma);r \to s+2]$$

$$= -(s+1)sM[\mathcal{U}(r,\pm\gamma);r \to s] = -(s+1)s\mathcal{U}^*(r,\pm\gamma) \qquad (9.4.5)$$

and by using (9.2.8), (9.2.13) we find

$$Q_n(s) \equiv M[q_n(r);s+2] = -M[(r^{-1}\partial U/\partial\theta)';r \to s+2]$$

$$= (s+1)\,\frac{\partial U^*}{\partial\theta}\,(r,\pm\gamma) \tag{9.4.6}$$

where the sign is + or - for n = 1,2 respectively. Thus

$$2(s+1)\begin{bmatrix} GA \\ GB \\ HC \\ HD \end{bmatrix} = \begin{bmatrix} (s+2)\cos(s+2)\gamma, & -(s+2)\cos(s+2)\gamma, & s\sin(s+2)\gamma, & s\sin(s+2)\gamma \\ -\cos s\gamma, & \cos s\gamma, & -\sin s\gamma, & -\sin s\gamma \\ -(s+2)\sin(s+2)\gamma, & -(s+2)\sin(s+2)\gamma, & \cos(s+2)\gamma, & -\cos(s+2)\gamma \\ \sin s\gamma, & \sin s\gamma, & -\cos s\gamma, & \cos s\gamma \end{bmatrix} \begin{bmatrix} P_1(s) \\ P_2(s) \\ Q_1(s) \\ Q_2(s) \end{bmatrix}. \tag{9.4.7}$$

where

$$\left.\begin{aligned} G(s,\gamma) &= (s+1)\sin 2\gamma - \sin 2(s+1)\gamma, \\ H(s,\gamma) &= (s+1)\sin 2\gamma + \sin 2(s+1)\gamma. \end{aligned}\right\} \tag{9.4.8}$$

Tranter considers in detail the case in which $q_1 = 0 = q_2$ and

$$p_1(r) = p_2(r) = \begin{cases} 1, & r < a \\ 0, & r > a \end{cases} \tag{9.4.9}$$

and shows that when $\gamma = \pi/2$ the known results for the half-plane are recovered. Sternberg and Koiter (1958) show how the equations (9.4.7) may be used in practice. Harrington and Ting (1971), Tsamasphyros and Theocaris (1976) make a critical examination of the Mellin transform analysis.

Ufliand (1965) used the Papkovich-Neuber solution (1.10.11) for mixed boundary value problems - of which category contact problems form a part. It is permissible to introduce one relation linking ψ_1, ψ_2, ϕ; following Ufliand we assume that ψ_1, ψ_2 are linked by the Cauchy-Riemann equations

$$\frac{\partial\psi_1}{\partial r} = \frac{1}{r}\frac{\partial\psi_2}{\partial\theta}, \qquad \frac{\partial\psi_2}{\partial r} = -\frac{1}{r}\frac{\partial\psi_1}{\partial\theta}, \tag{9.4.10}$$

so that the stresses $\tau_{\theta\theta}$, $\tau_{r\theta}$ become

$$\tau_{\theta\theta} = \frac{\partial^2\phi}{\partial r^2} + \cos\theta\frac{\partial\psi_1}{\partial r} + \sin\theta\,\frac{\partial\psi_2}{\partial r} - \frac{1}{r}\left(\cos\theta\,\frac{\partial^2\psi_1}{\partial\theta^2} + \sin\theta\,\frac{\partial^2\psi_2}{\partial\theta^2}\right) , \quad (9.4.11)$$

$$\tau_{r\theta} = -\frac{\partial}{\partial r}\left(\frac{1}{r}\frac{\partial\phi}{\partial\theta} + \cos\theta\,\frac{\partial\psi_1}{\partial\theta} + \sin\theta\,\frac{\partial\psi_2}{\partial\theta}\right) + \frac{1}{r}\left(\cos\theta\,\frac{\partial\psi_1}{\partial\theta} + \sin\theta\,\frac{\partial\psi_2}{\partial\theta}\right). \quad (9.4.12)$$

Now write

$$\overline{\phi} \equiv \phi^*(s,\theta) = M[\phi(r,\theta);r \to s] , \quad (9.4.13)$$

$$\overline{\psi}_n \equiv \psi_n^*(s+1,\theta) = M[\psi_n(r,\theta);r \to s+1] , \quad n = 1,2 ; \quad (9.4.14)$$

then since ϕ, ψ_1, ψ_2 are harmonic and linked by (9.4.10), we find

$$\left.\begin{aligned} \overline{\phi}(s,\theta) &= A_0\cos s\theta + B_0\sin s\theta , \\ \overline{\psi}_1(s,\theta) &= A_1\cos(s+1)\theta + B_1\sin(s+1)\theta , \\ \overline{\psi}_2(s,\theta) &= -A_1\sin(s+1)\theta + B_1\cos(s+1)\theta . \end{aligned}\right\} \quad (9.4.15)$$

Ufliand considers the boundary conditions

$$u(r,-\gamma) = u_0(r) , \quad v(r,-\gamma) = v_0(r) , \quad (9.4.16)$$

$$\tau_{\theta\theta}(r,\gamma) = -p(r) , \quad \tau_{r\theta}(r,\gamma) = q(r) . \quad (9.4.17)$$

Write

$$\overline{u},\overline{u}_0,\overline{v},\overline{v}_0 = M[u(r,\theta),u_0(r),v(r,\theta),v_0(r);r \to s+1] , \quad (9.4.18)$$

$$\overline{\tau}_{\theta\theta},\overline{p},\overline{\tau}_{r\theta},\overline{q} = M[\tau_{\theta\theta}(r,\theta),p(r),\tau_{r\theta}(r,\theta),q(r);r \to s+2] ; \quad (9.4.19)$$

then the boundary conditions (9.4.16), (9.4.17) may be written

$$\overline{u}(-\gamma) = \overline{u}_0(s) , \quad \overline{v}(-\gamma) = \overline{v}_0(s) , \quad (9.4.20)$$

$$\overline{\tau}_{\theta\theta}(\gamma) = -\overline{p}\ , \quad \overline{\tau}_{r\theta}(\gamma) = \overline{q} \tag{9.4.21}$$

while the transforms of equations (1.10.11), (9.4.11), (9.4.12) are

$$2\mu\overline{u} = (\kappa+s+1)[\overline{\psi}_1\cos\theta+\overline{\psi}_2\sin\theta] + s\overline{\phi}\ , \tag{9.4.22}$$

$$2\mu\overline{v} = (\kappa-s-1)[-\overline{\psi}_1\sin\theta+\overline{\psi}_2\cos\theta] - \partial\overline{\phi}/\partial\theta\ , \tag{9.4.23}$$

$$\overline{\tau}_{\theta\theta} = s(s+1)[\overline{\phi}+\overline{\psi}_1\cos\theta+\overline{\psi}_2\sin\theta]\ , \tag{9.4.24}$$

$$\overline{\tau}_{r\theta} = (s+1)\frac{\partial\overline{\phi}}{\partial\theta} + (s+2)\left[\frac{\partial\overline{\psi}_1}{\partial\theta}\cos\theta + \frac{\partial\overline{\psi}_2}{\partial\theta}\sin\theta\right]. \tag{9.4.25}$$

Note that the arguments in (9.4.13), (9.4.14) and (9.4.19) are s, s+1, s+2 respectively.

The four boundary conditions (9.4.16)-(9.4.17) now yield four equations for A_0, B_0, A_1, B_1 from which the displacements and stresses may be found by inverting the Mellin transforms; but it is necessary to find the strip in which c must be taken, i.e., the common strip in which the Mellin transforms are holomorphic. If we assume that u_0, v_0, $p(r)$, $q(r)$ are continuous functions and that

$$u_0, v_0 = O(r^{-\delta})\ , \quad p(r), q(r) = O(r^{-1-\delta}) \tag{9.4.26}$$

for large r then the integrals occuring in the Mellin transforms $\overline{u}_0$, $\overline{v}_0$, $\overline{p}$, $\overline{q}$ will converge absolutely at the upper limit if $\mathrm{Re}(s) < \delta-1$. Further, if the displacements are bounded as $r \to 0$ while the stresses have integrable singularities, i.e.,

$$p(r), q(r) = O(r^{-\varepsilon})\ , \quad \varepsilon < 1\ , \tag{9.4.27}$$

then the integrals will converge absolutely at the lower limit when $\mathrm{Re}(s) > -1$. Thus the Mellin transforms $\overline{u}_0$, $\overline{v}_0$, $\overline{p}$, $\overline{q}$ will be holomorphic in the strip $-1 < \sigma < -1 + \delta$. The transforms of the stresses are obtained by solving equations (9.4.20) - (9.4.25) for A_0, B_0, A_1, B_1; each of these quantities will be expressed as a ratio. The numerator will be a combination of $\bar{u}_0$, $\bar{v}_0$, $\bar{p}$, $\bar{q}$ and sines and cosines: the denominator will be the determinant of the equations, and may be found by seeking those values of s for which the stress-functions satisfy the clamped-free conditions

$$u(r,-\gamma) = 0 = v(r,-\gamma) = \tau_{\theta\theta}(r,\gamma) = \tau_{r\theta}(r,\gamma) \ . \qquad (9.4.28)$$

The determinant is

$$\mathcal{D}(s+1) = 4\kappa\sin^2 2(s+1)\gamma + 4(s+1)^2\sin^2 2\gamma - (\kappa+1)^2 \ . \qquad (9.4.29)$$

Let s_1 be that zero of $\mathcal{D}(s+1)$ which has least real part, but such that $\mathrm{Re}(s_1) < -1$; then each of the transforms $\bar{u}$, $\bar{v}$, $\bar{\tau}_{r\theta}$, $\bar{\tau}_{\theta\theta}$ will be holomorphic in the strip $-1 < \sigma < \sigma_0$, where $\sigma_0 = \min(\mathrm{Re}(s_1), -1+\delta)$:c must be taken in this strip.

In order to investigate the singularities, if any, in the stresses, we consider how the integrals

$$\left\{\tau_{r\theta}(r,\theta),\tau_{\theta\theta}(r,\theta)\right\} = \frac{1}{2\pi i}\int_{c-i\infty}^{c+i\infty}\left\{\bar{\tau}_{r\theta}(\theta),\bar{\tau}_{\theta\theta}(\theta)\right\}r^{-s-2}ds \qquad (9.4.30)$$

are evaluated. For small r they will be found by closing $\mathrm{Re}(s) = c$ by a large semi-circle to the left, as in Figure 9.2.2. If one of the transforms $\bar{\tau}_{r\theta}$, $\bar{\tau}_{\theta\theta}$ has a simple pole, i.e., if $\mathcal{D}(s+1)$ has a simple zero, s_0, satisfying $\mathrm{Re}(s_0) < -1$, then the corresponding stress will contain a term $C(s_0)r^{-s_0-2}$ arising from the residue of the integral at s_0, and this stress will be unbounded at the origin if $\mathrm{Re}(s_0) > -2$. The investigation of the zeroes of $\mathcal{D}(s+1)$ is part of singularity analysis, and will be deferred to §9.5.

The formal solution of wedge problems given by equations (9.4.13) - (9.4.25) does not lend itself to computation; for this

Lutchenko (1966) and Lutchenko and Popov (1970) have derived more suitable equations. They considered the particular case of a wedge fixed along one edge, $\theta = -\gamma$, and in contact with a frictionless stamp over part of the other edge, $\theta = \gamma$; thus

$$u(r,-\gamma) = 0 = v(r,-\gamma) \quad ; \quad \tau_{r\theta}(r,\gamma) = 0 \; ; \quad r > 0 \, , \tag{9.4.31}$$

$$\left.\begin{aligned} v(r,\gamma) &= f^*(r) \, , \quad 0 \le r \le a \; ; \\ \tau_{\theta\theta}(r,\gamma) &= \begin{cases} -p(r) \, , & 0 < r < a \, , \\ 0 \, , & r > a \, , \end{cases} \end{aligned}\right\} \tag{9.4.32}$$

where $f^*(r)$ defines the shape of the punch. After satisfying the boundary conditions (9.4.31) we find

$$\bar{v}(s,\gamma) = \frac{\vartheta(\kappa\sin 4s\gamma - s\sin 4\gamma)}{s[8(1-\nu)^2 - 2\kappa\sin^2 2s\gamma - 2s^2\sin^2 2\gamma]} \cdot \bar{\tau}_{\theta\theta}(s+1,\gamma) \, , \tag{9.4.33}$$

so that after transforming the integrals and assuming that the Mellin inverse may be taken along the imaginary axis $\mathrm{Re}(s) = 0$, we obtain

$$-\frac{\vartheta}{\pi}\int_0^a v_\gamma^*\left[\ell n\left(\frac{x}{r}\right)\right]p(x)dx = f^*(r) \, , \quad 0 \le r \le a \, , \tag{9.4.34}$$

where

$$v_\gamma^*(z) = \int_0^\infty \frac{\phi_\gamma(t)}{t} \cos zt dt \, , \tag{9.4.35}$$

$$\phi_\gamma(t) = \frac{\kappa\sinh 4t\gamma - t\sin 4\gamma}{8(1-\nu)^2 + 2\kappa\sinh^2 2t\gamma + 2t^2\sin^2 2\gamma} \, . \tag{9.4.36}$$

Note that the *indentation* of the wedge is $-v(r,\gamma) \equiv -f^*(r)$.

Lutchenko and Popov note that v_γ^* has a singularity at $x = r$, i.e., $z = 0$. They write

$$v_\gamma^*(z) = \int_0^\infty \phi_\gamma(t/\pi)\,\frac{\cos zt\,dt}{t} = k_0(z/\pi) + k(z/\pi) , \qquad (9.4.37)$$

where

$$k_0(z) = \int_0^\infty \frac{\tanh t}{t}\cos zt\,dt , \qquad k(z) = \int_0^\infty \frac{\phi_\gamma(t/\pi)-\tanh t}{t}\cos zt\,dt , \qquad (9.4.38)$$

and note that $k(z)$ is regular on the real axis. Now introduce dimensionless variables ξ,ρ such that

$$x = a\xi, \quad r = a\rho, \quad \chi(\xi) = ap(x) , \quad f(\rho) = \pi f^*(r)/\vartheta , \qquad (9.4.39)$$

$$-\int_0^1 \left\{k_0\left[\frac{1}{\pi}\ln\left(\frac{\xi}{\rho}\right)\right] + k\left[\frac{1}{\pi}\ln\left(\frac{\xi}{\rho}\right)\right]\right\}\chi(\xi)d\xi = f(\rho) . \qquad (9.4.40)$$

The kernel k_0 is a polynomial kernel in the sense of §6.12, since

$$k_0\left[\frac{1}{\pi}\ln\left(\frac{\xi}{\rho}\right)\right] = \ln\left[\frac{\xi^{1/2}+\rho^{1/2}}{|\xi^{1/2}-\rho^{1/2}|}\right] = \int_0^{\min(\xi,\rho)} \frac{d\sigma}{[(\xi-\sigma)(\rho-\sigma)]^{1/2}} , \qquad (9.4.41)$$

and it possesses the spectral relation

$$\int_0^1 k_0\left[\frac{1}{\pi}\ln\left(\frac{\xi}{\rho}\right)\right]\frac{U_{2m}[(1-\xi)^{1/2}]}{(1-\xi)^{1/2}} = \pi\rho^{1/2}\,\frac{U_{2m}[(1-\rho)^{1/2}]}{m+1/2} , \qquad (9.4.42)$$

where the $U_m(x)$ are the Chebyshev polynomials of the second kind. Equation (9.4.34) may thus be solved approximately by assuming an expansion

$$\chi(\xi) = (1-\xi)^{-1/2}\sum_{n=0}^{N} a_n U_{2n}[(1-\xi)^{1/2}] , \qquad (9.4.43)$$

multiplying equation (9.4.34) throughout by $(1-\rho)^{-1/2}U_{2m}[(1-\rho)^{1/2}]$, and integrating over $(0,1)$, using either the continuous or the discrete orthogonality property of the U_{2m}. Lutchenko and Popov use the former method; the latter follows the lines laid down in §6.4. Lutchenko and Popov give detailed calculations for a stamp with a flat base indenting the wedge normally, and for various cases of a beam in flexure lying on the face $\theta = \alpha$. See also Lutchenko (1972).

Brief mention will now be made of a number of other papers. Matczynski (1963) obtains an approximate solution for a symmetrically loaded wedge with a specified displacement condition $v = \delta r^{\lambda}$, $0 < r < 1$; he uses Koiter's (1954) approximate solution of Wiener-Hopf integral equations. Benthem (1963) considers an anisotropic wedge. Srivastav and Prem Narain (1965) show how the equations such as (9.4.33), (9.4.34) obtained from the Mellin transform may be manipulated to form a Fredholm integral equation of the second kind for a transform of the contact pressure. Kurshin (1959) and Hetenyi (1960) consider mixed boundary value problems for a quadrant (i.e., a quarter plane). Such problems have been reviewed and analyzed by Sneddon (1971). V.M. Aleksandrov (1967a), (1976b) finds an approximate solution for a rigid punc indenting a wedge over an interval $a < r < b$ of one face. See also V.M. Aleksandrov and Kopasenko (1968). Nuller (1972) considers problems associated with the application of forces to a wedge through a rigid yoke attached to the apex. Popov and Tikhonenko (1974), (1975) consider a wedge with a beam lying on one face; Nuller (1974) discusses the case when a beam is attached to the mid-line $\theta = 0$ of the wedge. Erdogan and Arin (1976a), (1976b) make a careful study, with numerical results, of a number of contact problems for the wedge. Babeshko and Berkovich (1972), Ufliand (1972) and Berkovich (1974) consider the three-dimensional wedge.

9.5 *Stress singularities at a sharp corner*

The question of stress singularities at a sharp corner has attracted the attention of researchers for more than fifty years; this section will now give a brief review of this work.

Almost every one of the researchers has commented on the fact that *stress singularities*, where the stresses become infinite, occur in infinitesimal linear elasticity theory, which is applicable only if displacements and stresses are *small*, in some sense. Dempsey and Sinclair (1979), for instance, make the following statement: 'In solid mechanics stress concentrations are of paramount interest. Typically the theory of linear elasticity models such concentrations with *stress singularities*. These stress singularities reflect the limited physical application of the theory since they cannot be sustained by a real material; further they are at variance with the mathematical assumptions of the theory. Nonetheless, singular elasticity solutions can still furnish useful information on stress concentrations. Of course, if the singularities are ignored then the limited mathematical model that is elasticity may be rendered quite useless. The point is that, while stress singularities are not of the real world, they are a real fact of the theory of linear elasticity and must be taken into account in any elastic analysis.'

The basic problem of singularity analysis for a wedge is the search for stress fields which satisfy the equations of plane-strain and a set of *homogeneous* boundary conditions on the two faces $\theta = \pm\gamma$. Knein (1926) carried out such an analysis, using the Airy stress function of (9.3.6), for a rectangular wedge ($\gamma = \pi/4$). M.L. Williams (1952) considered a wedge of general angle and supposed that each edge was either *free*, i.e., $\tau_{\theta\theta} = 0 = \tau_{r\theta}$ or *clamped*, i.e., $u = 0 = v$. Following Huth (1953) M.L. Williams (1956) rephrased this analysis in terms of complex variables; see also Kalandia (1969a). England (1971b) provided a definitive study of the problem.

The condition on each side $\theta = \pm\gamma$ of the wedge can be one of the following:

(a) stress-free: $\tau_{r\theta} = 0 = \tau_{\theta\theta}$.

Using Example 4.2.6 we may write this in the form

$$\phi(z) + z\overline{\phi}'(\overline{z}) + \overline{\psi}(\overline{z}) = 0 . \tag{9.5.1}$$

(b) zero displacement: $u = 0 = v$.

Equation (4.2.5) gives

$$\kappa\phi(z) - z\overline{\phi}'(\overline{z}) - \overline{\psi}(\overline{z}) = 0 . \tag{9.5.2}$$

(c) frictionless: $u_\theta = 0 = \tau_{r\theta}$.

Williams and England take the complex potentials in the form

$$\phi(z) = A_1 z^{\lambda} + A_2 z^{\overline{\lambda}} , \quad \psi(z) = B_1 z^{\lambda} + B_2 z^{\overline{\lambda}} . \tag{9.5.3}$$

A lower bound for λ may be obtained by requiring that the displacement at the wedge tip be continuous, or equivalently that the strain energy in any region of the wedge be finite; both yield $\mathrm{Re}(\lambda) > 0$. If $\mathrm{Re}(\lambda) > 1$ the stresses at the tip will be finite. Thus the stress singularities at the wedge tip are governed by the eigenvalue λ for which $\mathrm{Re}(\lambda)$ is a minimum and which satisfies

$$0 < \mathrm{Re}(\lambda) < 1 . \tag{9.5.4}$$

England examines four cases, in which the conditions on the faces $\theta = +\gamma$, $\theta = -\gamma$ are respectively

$$(1)\ a - a \quad (2)\ b - b \quad (3)\ c - a \quad (4)\ a - b . \tag{9.5.5}$$

In case (4), for example, equation (9.5.1) on $\theta = \gamma$ gives

$$A_1 r^{\lambda}\exp(i\lambda\gamma) + \bar{\lambda}\bar{A}_1 r^{\bar{\lambda}}\exp[-(\bar{\lambda}-2)i\gamma] + B_1 r^{\bar{\lambda}}\exp(-i\bar{\lambda}\gamma) +$$

$$+ A_2 r^{\bar{\lambda}}\exp(i\bar{\lambda}\gamma) + \lambda\bar{A}_2 r^{\lambda}\exp[-(\lambda-2)i\gamma] + \bar{B}_2 r^{\lambda}\exp(-i\lambda\gamma) = 0 , \quad (9.5.6)$$

and on equating to zero the powers of r^{λ} and $r^{\bar{\lambda}}$ we obtain

$$A_1\exp(i\lambda\gamma) + \lambda\bar{A}_2\exp[-(\lambda-2)i\gamma] + \bar{B}_2\exp(-i\lambda\gamma) = 0 , \quad (9.5.7)$$

$$\lambda A_1\exp[(\lambda-2)i\gamma] + \bar{A}_2\exp(-i\lambda\gamma) + B_1\exp(i\lambda\gamma) = 0 . \quad (9.5.8)$$

In the same way, condition b, i.e., equation (9.5.2), on $\theta = -\gamma$, gives

$$\kappa A_1\exp(-i\lambda\gamma) - \lambda\bar{A}_2\exp[(\lambda-2)i\gamma] - \bar{B}_2\exp(i\lambda\gamma) = 0 , \quad (9.5.9)$$

$$-\lambda A_1\exp[-(\lambda-2)i\gamma] + \kappa\bar{A}_2\exp(i\lambda\gamma) - B_1\exp(-i\lambda\gamma) = 0 , \quad (9.5.10)$$

and the determinant of equations (9.5.7) - (9.5.10) is

$$\mathcal{D}(\lambda) = 4\kappa\sin^2 2\lambda\gamma + 4\lambda^2\sin^2 2\gamma - (\kappa+1)^2 , \quad (9.5.11)$$

which yields (9.4.29) under the substitution $\lambda = -s-1$. We first examine certain particular cases. When $2\gamma = \pi$, so that the two edges of the wedge are the two halves of an infinite straight line, the equation $\mathcal{D}(\lambda) = 0$ becomes

$$\sin\lambda\pi = \pm (\kappa+1)\kappa^{-1/2}/2 , \quad (9.5.12)$$

so that the dominant eigenvalue is

$$\frac{1}{2} + \frac{i\ln\kappa}{2\pi} . \quad (9.5.13)$$

This is the exponent encountered in the adhesive contact problem for the half-plane - see equations (4.6.13). Figure 9.5.1, taken from England (1971b), shows contours of the dominant root of equation (9.5.11). In the region *C* the eigenvalue is complex and the real part is almost entirely independent of the elastic constant κ, and decreases as the wedge angle γ increases. In the regions *R* the eigenvalue is real and depends on both κ and γ. It will be seen that for a wedge of angle $2\gamma < \pi/4$ no singularity is possible. For the right-angled wedge $2\gamma = \pi/2$ the eigenvalue is real, while for $2\gamma = \pi$ the eigenvalues are complex. The information about the roots λ such that $0 < \text{Re}(\lambda) < 1$ may be carried over into the Mellin transform analysis of §9.4; a root λ gives a corresponding pole $s = -\lambda - 1$ satisfying $-2 < \text{Re}(s) < -1$ in the stress transforms $\bar{\tau}_{r\theta}$, $\bar{\tau}_{\theta\theta}$, $\bar{\tau}_{rr}$.

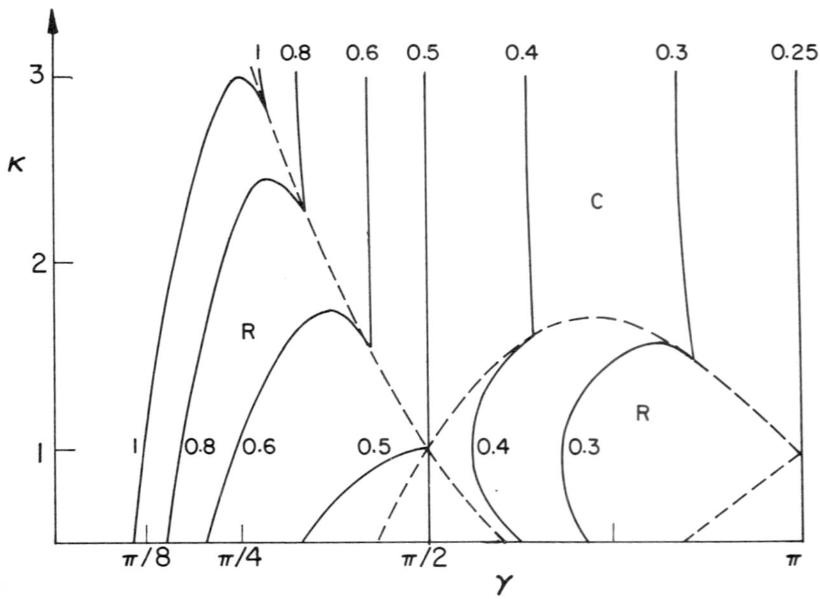

Figure 9.5.1 - Contours of the real part of the dominant root of $\mathcal{D}(\lambda) = 0$

The following conclusion can be drawn concerning the order of the tip stresses for different roots λ of the determinantal equation $\mathcal{D}(\lambda) = 0$.

$$\tau_{rr}, \tau_{r\theta}, \tau_{\theta\theta} = 0(r^{\lambda-1}) \text{ for } \lambda \text{ real and } 0 < \lambda < 1 , \quad (9.5.14)$$

$$= 0\,[r^{\xi-1}\cos(\eta \ln r)] \text{ or } 0[r^{\xi-1}\sin(\eta \ln r)]$$

$$\text{for } \lambda = \xi + i\eta \text{ complex and } 0 < \xi < 1 , \quad (9.5.15)$$

$$= 0(\ln r) \text{ for } \lambda = 1 \text{ and } \partial\mathcal{D}/\partial\lambda = 0 \text{ at } \lambda = 1 , \quad (9.5.16)$$

$$= 0(1) \text{ for no zeroes of } \mathcal{D} \text{ in } 0 < \mathrm{Re}(\lambda) < 1 \text{ and}$$

$$\partial\mathcal{D}/\partial\lambda \neq 0 \text{ at } \lambda = 1 . \quad (9.5.17)$$

Dempsey and Sinclair (1979) have made an extensive classification of these and other singular solutions.

Many other Mellin transform and/or singularity analyses have been carried out, amongst which are the detailed and informative paper of Koiter and Alblas (1954-7), M.L. Williams (1959), Kurshin (1959), Karp and Karal (1962) and Zak and Williams (1963). Bogy and Sternberg (1968), in considering the effect of shear loading on a right-angled wedge, encountered case 1 above (see Example 9.5.1). Bogy (1968) considered two dissimilar orthogonal elastic wedges bonded together and subject to normal and shear loading. It was in a published discussion of this paper that Dundurs (1969) showed that the elastic constants of the two media enter only through two mismatch parameters α, β (see §4.10). Bogy (1970) reanalyzed and extended his 1968 paper to take account of Dundurs' comment. Bogy (1971), Bogy and Wang (1971) and Hein and Erdogan (1971) made similar analyses for two edge-bonded wedges of different materials and arbitrary angles. Rao (1971) classified a large group of problems governed by harmonic or biharmonic equations and concerned with different materials bonded together and meeting at a corner. Among the problems he considered were wedges bonded together and having flexible inextensible stiffeners attached to their edges (Figure 9.5.2), and an elastic punch indenting an elastic half-plane (Figure 9.5.3).

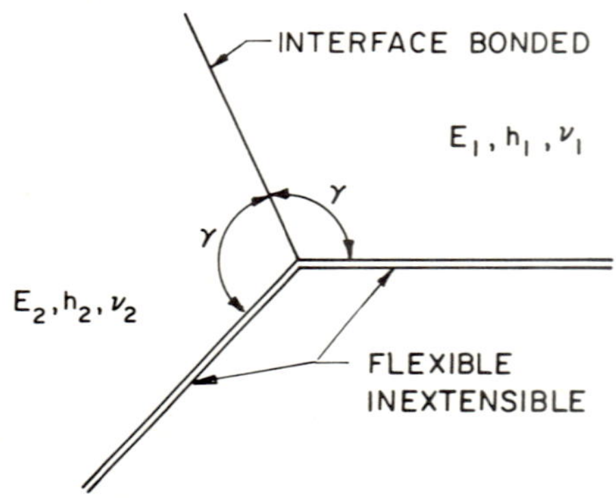

Figure 9.5.2 - Two wedges bonded together and having inextensible edge supports

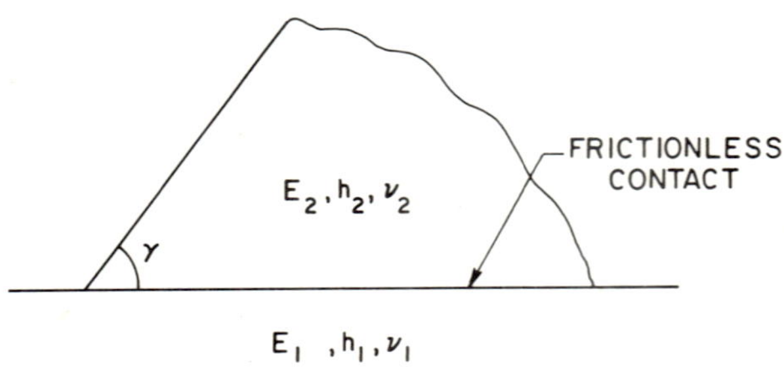

Figure 9.5.3 - An elastic punch indenting a half-plane

Note that Rao states that the Poisson's ratios ν_1, ν_2 of the contacting bodies have 'no influence' on the eigenvalues, which he therefore computes and displays for different values of

$e = E_2 h_2 / E_1 h_1$ only. However, Dundurs and Lee (1972) noted that since the contact is frictionless and receding it falls into the category discussed in §4.10 (equation (4.10.22)) for which the elastic constants appear only through the parameter α of equation (4.10.13) where now (Example 4.10.1) $\Gamma = \mu_2 h_2 / \mu_1 h_1$. Rao s observation is therefore empirical and should be rephrased with 'no appreciable influence' replacing 'no influence'. The determinant $\mathcal{D}(\lambda;\alpha,\gamma)$ is

$$\mathcal{D}(\lambda;\alpha,\gamma) = (1+\alpha)\cos\lambda\pi(\lambda^2\sin^2\gamma - \sin^2\lambda\gamma) - (1-\alpha)\sin\lambda\pi(\lambda\sin 2\gamma + \sin 2\lambda\gamma)/2. \tag{9.5.18}$$

Although Dundurs and Lee were unable to prove analytically that $\mathcal{D}$ possesses only real roots λ they found no evidence of complex roots in an extensive numerical search. The equation $\mathcal{D}(\lambda;\alpha,\gamma) = 0$ always has the root $\lambda = 1$ but in general this is not the *dominant* root, i.e., there is another root λ satisfying $0 < \lambda < 1$. The boundary between values of α, γ that, on the one hand, lead to bounded stresses and, on the other hand, are associated with singular fields at $r = 0$, may be found by demanding that this smaller root also be 1. There will then be a double root at $\lambda = 1$ and $\partial\mathcal{D}/\partial\lambda = 0$ at $\lambda = 1$. This occurs when

$$\alpha = \frac{(\pi+\gamma)\cos\gamma - \sin\gamma}{(\pi-\gamma)\cos\gamma + \sin\gamma}. \tag{9.5.18}$$

Figure 9.5.4 shows the contours of λ for the physical ranges of α, γ values. Note that the smallest value of λ, namely $\lambda = 1/2$, occurs for a rigid wedge, $\alpha = 1$, irrespective of wedge angle γ. Dundurs and Lee conclude that if sharp edges between contacting bodies cannot be avoided, locally high stresses may be significantly reduced by undercutting the edge, so that the combination of α, γ falls in the empty region on the left-hand side of Figure 9.5.4.

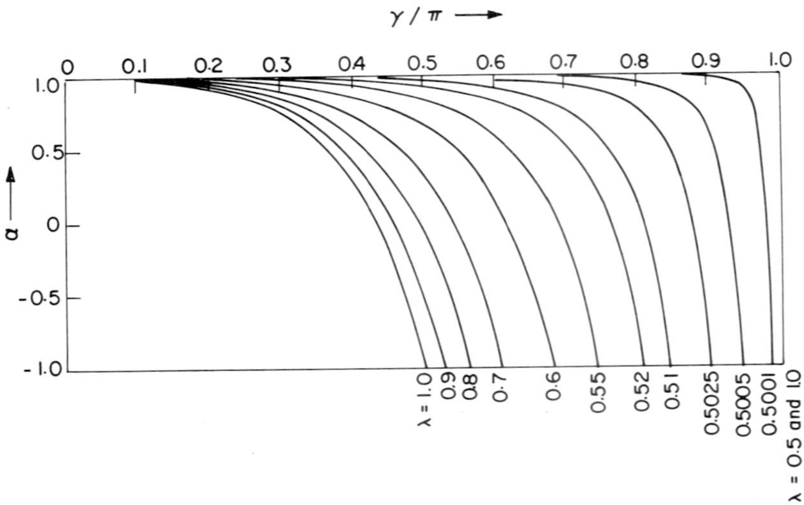

Figure 9.5.4 - The dominant real root of equation (9.5.18)

Gdoutos and Theocaris (1975) investigated the action of friction on the singularity occurring when an elastic wedge is in contact with a half-plane. Comninou (1976) explored the nature of this singularity in more detail; see also Theocaris and Gdoutos (1976). The study of singularities occurring when more than two wedges meet at a point was considered by Theocaris (1974) and developed further by Dempsey and Sinclair (1979).

Problems involving contact between elastic strips, or between a strip and a half-plane, have been considered by Bogy (1975a) (1975b), Adams and Bogy (1976a), (1976b), (1977).

The nature of the singularity at the tip of a wedge-shaped rigid punch indenting a half-space, or the corresponding singularity in closely related or analogous problems, has been studied by numerous researchers including Galin (1947a), Rvachev (1959a), Galin (1961), S.N. Brown and Stewartson (1969), V.M. Aleksandrov and Babeshko (1972), Bueckner (1973), Bažant (1974), Bažant and

Keer (1974), Morrison and Lewis (1976), Keer and Parihar (1978), Parihar and Keer (1978), (1979).

Examples 9.5

1. Use the complex potentials (9.5.3) to show that the determinant $\mathcal{D}(\lambda)$ for case 1 of equation (9.5.6), in which both edges $\theta = \pm\gamma$ of the wedge are stress free, is

   $$\mathcal{D}(\lambda) = \lambda^2\sin^2 2\gamma - \sin^2 2\lambda\gamma .$$

   Show that when $2\alpha < \pi$, the root with least real part in $0 < \mathrm{Re}(\lambda) \leq 1$ is $\lambda = 1$, and that when $\pi < 2\alpha < 2\pi$, the root with least real part is real and satisfies $1/2 < \lambda < \pi/(2\alpha)$ (England (1971b)).
2. Show that for the wedge of Figure 9.5.3, which falls into case 3 of equation (9.5.5), the determinant $\mathcal{D}$ is given by (9.5.18).

CHAPTER X - AXISYMMETRIC CONTACT PROBLEMS FOR THE ELASTIC HALF-SPACE

"Therefore, to test his patience -
How much he can endure -
Mention no places, names or dates,
And evermore be sure
Throughout the poem to be found
Consistently obscure".

-*Poeta fit, non nascitur*

10.1 *Introduction*

This chapter is concerned almost entirely with the quasi-static contact between a rigid axisymmetric punch and an elastic half-space occupying the region $z \geq 0$.

Such contact problems may be classified according to the *type of punch*, i.e., whether it is circular or annular and whether its face is flat or curved; the *type of indentation*, i.e., whether it is a rotation or a translation about one or other of the axes; the *type of contact*, i.e., whether it is frictionless, adhesive, or in limiting friction, and if frictionless, whether the contact is complete or incomplete.

Following Gladwell (1968) it may be assumed that the type of indentation, specified by the displacement of the punch, is either

(1) a rotation about the z-axis
(2) a translation in the z-direction
(3) a rotation about the y-axis
(4) a translation in the x-direction.

Each of cases (2)-(4) gives rise to two distinct extreme problems in which the contact is assumed to be either completely adhesive (a) or completely frictionless (f). The frictionless version of (1) is trivial in the sense that the half-space is not deformed at all.

The classification may also be made on purely mathematical grounds. All the contact problems may be solved by using Hankel transforms and they then reduce to the solution of a number of sets (or pairs) of dual integral equations. Consider the problems for the important case of an indenter with a *plane* face. Those leading to one pair include 1(a), first solved by Reissner and Sagoci (1944) using oblate spheroidal coordinates, and then by Sneddon (1947) using Hankel transforms; 2(f), of (2.6.25), the solution of which was known to Boussinesq and to Chong (1952); 3(f), solved by Abramov (1939) and 4(f) solved by Mindlin (1949). Problem 1(a) is solved in §10.2, 3(f) and 4(f) in §10.3. Those leading to two pairs include 2(a), which was solved by Mossakovskii (1954a), (1963), Abramian, Arutiunian and Babloian (1966), and Spence (1968a). The last two papers use a reduction to an equation which may be solved by the Wiener-Hopf method. In §10.3 it is shown that the equation may be reduced to a singular integral equation of the type considered in §3.5. Problems 3(a) and 4(a) involve three pairs of dual integral equations; 4(a) was solved by Ufliand (1956) by using the Mehler-Fok transform; both problems were solved by Gladwell (1969) by using Hankel transforms and subsequently using Khadem's (1967) reduction to the type of equation considered by Spence (1968b): a simpler solution is given in §10.4.

Contact problems for the half-space have led to the development of a number of important mathematical tools. Thus oblate spheroidal coordinates are introduced in §10.5, a number of diverse representations in §10.6 and ellipsoidal coordinates in §11.3.

10.2 *Axisymmetric Hankel-transform solutions for the half-space*

The general form of the Papkovich-Neuber solution appropriate to axisymmetric problems may be obtained from (1.10.11) by using polar coordinates (r, θ, z) and writing

$$\vec{\psi} = \psi_1 \hat{r} + \psi_2 \hat{\theta} + \psi_3 \hat{k} , \quad \vec{r} = r\hat{r} + z\hat{k} . \tag{10.2.1}$$

One component of $\vec{\psi}$ may be taken as zero; let it be ψ_1; the problem is axisymmetric, so that $\psi_2 \equiv 0$. Put $\psi_3 = \psi$ and obtain

$$2\mu u = -\frac{\partial}{\partial r}(z\psi + \phi) , \quad 2\mu w = 4(1-\nu)\psi - \frac{\partial}{\partial z}(z\psi + \phi) , \tag{10.2.2}$$

where, in this chapter, u, v, w denote u_r, u_θ and u_z respectively. It will be assumed that there are no body forces; equation (1.10.13) shows that ϕ and ψ are harmonic. Equation (5.6.20) gives

$$H_0[\nabla^2 f(r,z); r \to \xi] = (D_z^2 - \xi^2)\overline{f}_0(\xi,z) , \tag{10.2.3}$$

where

$$\overline{f}_0(\xi,z) = H_0[f(r,z); r \to \xi] . \tag{10.2.4}$$

This shows that if $f(r,z)$ is harmonic in the upper half-space $z \geq 0$, then $(D_z^2 - \xi^2)\overline{f}_0(\xi,z) = 0$ so that, by inversion,

$$f(r,z) = H_0[A(\xi)\exp(-\xi z); \xi \to r] . \tag{10.2.5}$$

Taking

$$\psi(r,z) = H_0[\xi^{-1}A(\xi)\exp(-\xi z); \xi \to r] , \tag{10.2.6}$$

$$\phi(r,z) = H_0[\xi^{-2}C(\xi)\exp(-\xi z); \xi \to r] , \tag{10.2.7}$$

we find the displacements and stresses in the form

$$u(r,z) = H_1[U(\xi,z); \xi \to r] , \quad w(r,z) = H_0[W(\xi,z); \xi \to r] , \tag{10.2.8}$$

$$\tau_{rz} = H_1[T_{13}(\xi,z);\xi \to r] \ , \quad \tau_{zz}(r,z) = H_0[T_{33}(\xi,z);\xi \to r] \ , \quad (10.2.9)$$

where

$$\begin{bmatrix} 2\mu U \\ 2\mu W \\ -T_{13} \\ -T_{33} \end{bmatrix} = \begin{bmatrix} \{\xi z A + C\}\xi^{-1} \\ \{(3-4\nu)A + \xi z A + C\}\xi^{-1} \\ (1-2\nu)A + \xi z A + C \\ (2-2\nu)A + \xi z A + C \end{bmatrix} \exp(-\xi z) \ . \quad (10.2.10)$$

The only other non-zero stresses are τ_{rr} and $\tau_{\theta\theta}$, which may be expressed as an H_0 transform plus an H_1 transform (Example 10.2.2). (Note that Harding and Sneddon (1945), who were perhaps the first to use Hankel transform solutions for the problem, use the solution

$$2\mu u = -\frac{\partial^2\phi}{\partial r\partial z} \ , \quad 2\mu v = 2(1-\nu)\nabla^2\phi - \frac{\partial^2\phi}{\partial z^2} \ , \quad \nabla^4\phi = 0 \ , \quad (10.2.11)$$

which leads to the stress representation (Example 10.2.5) used in many standard texts. See e.g., Timoshenko and Goodier (1951, Chapter 13), Sneddon (1951, Chapter 10).) When only a normal stress acts on z = 0, so that

$$\tau_{rz}(r,0) = 0 \ , \quad \tau_{zz}(r,0) = -p(r) \ , \quad 0 \le r < \infty \ , \quad (10.2.12)$$

then equation (10.2.10) shows that if $P(\xi) = H_0[p(r);\xi]$, then

$$(1-2\nu)A + C = 0 \ , \quad (2-2\nu)A + C = P(\xi) \ , \quad (10.2.13)$$

so that on the plane z = 0

$$w(r,0) = \vartheta H_0[\xi^{-1}P(\xi);r] \quad (10.2.14)$$

$$\tau_{zz}(r,0) = -H_0[P(\xi);r] \ . \quad (10.2.15)$$

Equation (10.2.14), and the corresponding results for $w(r,z)$, $u(r,z)$ may be expressed in terms of $p(r)$ to yield the axisymmetric versions of (2.5.28), (2.5.30) and the parts of (2.6.4), (2.6.5) referring to $p(r)$.

Equations (10.2.14), (10.2.15) show that the boundary conditions for problem 2(f) of §10.1, often called the axisymmetric Boussinesq problem, namely

$$w(r,0) = d + f(r) \ , \quad 0 \leq r < a \ , \tag{10.2.16}$$

$$\tau_{rz}(r,0) = 0 \ , \quad 0 \leq r < \infty \ , \quad \tau_{zz}(r,0) \equiv -p(r) = 0 \ , \quad r > a \ , \tag{10.2.17}$$

(and without loss of generality $f(0)$ may be taken as zero) lead to the pair of dual integral equations

$$w(r,0) \equiv \vartheta H_0[\xi^{-1}P(\xi);r] = d + f(r) \ , \quad 0 \leq r < a \ , \tag{10.2.18}$$

$$p(r) \equiv H_0[P(\xi);r] = 0 \ , \quad r > a \ . \tag{10.2.19}$$

This pair, like the pair (6.2.20), (6.2.21) encountered in plane problems, is a particular example of the pair of dual integral equations

$$H_\nu[\xi^\alpha A(\xi);x] = f(x) \quad , \quad 0 \leq x < 1 \ , \tag{10.2.20}$$

$$H_\nu[A(\xi);x] = g(x) \quad , \quad x > 1 \ . \tag{10.2.21}$$

It is appropriate at this stage to outline the history of these equations which occur in numerous problems in electrostatics and elastostatics; ν, α typically take the values $\nu = 0, \pm 1/2, 1, 2$, $\alpha = \pm 1$ and α must lie in the range $-2 < \alpha < 2$.

The equations may be reduced to a form in which $g(x) = 0$. For, following Gordon (1954), we may write

$$A(\xi) = A_1(\xi) + A_2(\xi) , \tag{10.2.22}$$

where $A_2(\xi)$ is chosen so that

$$H_\nu[A_2(\xi);x] = g(x)H(x-1) , \quad x \geq 0 , \tag{10.2.23}$$

and $A_2(\xi)$ may be found by H_ν inversion. Then $A_1(\xi)$ must be found so that

$$H_\nu[\xi^\alpha A_1(\xi);x] = f_1(x) , \quad 0 \leq x < 1 , \tag{10.2.24}$$

$$H_\nu[A_1(\xi);x] = 0 , \quad x > 1 , \tag{10.2.25}$$

where $f_1(x) = f(x) - H_\nu[\xi^\alpha A_2(\xi);x]$. Equations (10.2.24), (10.2.25) have the original form with $g(x) = 0$; this will be taken as the standard form of the equations.

The history of the problem has two strands: derivation of methods which may be applied uniformly to as wide a range of ν, α values as possible and attempts to produce methods which are simple in application and may be understood with little background mathematics. Various particular solutions for $\nu = 0$, $\alpha = \pm 1$ and $f(x) =$ constant were given by H. Weber (1873). Beltrami gave a solution for $\nu = 0$, $\alpha = -1$ and arbitrary $f(x)$. MacDonald (1895) solved the case $\nu = n$, $\alpha = -1$, $f(x) = J_n(cx)$; King (1935) solved the case $\nu = n$, $\alpha = -1$, $f(x)$ arbitrary. Titchmarsh (1937) gave a general solution of the equations by using Mellin transforms which Busbridge (1938) showed was valid for $0 < \alpha < 2$, provided that ν was suitably restricted; Busbridge gave another solution valid for $-2 < \alpha < 0$. Copson (1947) gave a simple solution involving only Abel transforms for the case $\nu = n$, $\alpha = -1$ and $f(x)$ arbitrary. Tranter (1951) presented a formal solution for $\alpha = \pm 1$ without reducing the equation to the case in which $g(x) = 0$. Gordon (1954) derived solutions by using the discontinuous Weber-

Schafheitlin integral (Abramowitz and Stegun (1972, p. 487)) which are valid for $\nu > -1$ and $0 < \alpha < 2$ or $-2 < \alpha < 0$. Noble (1958) derived another form of solution valid for $-2 < \alpha < 0$, $2\nu + \alpha + 2 > 0$ by using Busbridge's method, but his attempts at extending the solution to $\alpha > 0$ were complicated. Noble (1955) extended Copson's method to give solutions for restricted ranges of ν; for $-2 < \alpha < 0$ he required $\nu > 0$; for $0 < \alpha < 2$ he required $2\nu - \alpha + 2 > 0$. Sneddon (1960), using a procedure suggested by Lebedev and Ufliand (1958), gave simple solutions for the cases $\nu = 0$, $\alpha = \pm 1$ by taking $A(\xi)$ as a Fourier cosine or sine transform, as appropriate, to ensure that equation (10.2.25) is satisifed automatically, and then inverting the Abel integral equation which results from (10.2.24); this is basically the method we shall describe below. Sneddon considered the trigonometric cases $\nu = \pm 1/2$, $\alpha = -1$ separately in (1961); Copson (1961) and Lowengrub and Sneddon (1963) extended Sneddon's method to the general problem. W.E. Williams (1962) simplified the Mellin transform solution of the equations. Burlak (1963a), (1963b) examined the ranges of validity of the various methods which had been proposed, and refined Copson's method. Noble (1963), generalizing results due to Cooke (1956), derived an operator approach for solving the equations which generalizes the use of Abel transforms A_1, A_2 described below. For the definitive treatment of the problem and the use of fractional derivatives and Kober-Erdelyi operators the reader may consult Sneddon (1966a).

Now return to the particular pair (10.2.18), (10.2.19). First note that the kernel $k_{33}(r,s)$ of equation (2.6.8) may be obtained by writing

$$P(\xi) = \mathcal{H}_0[p(s);s \to \xi] = \int_0^a sp(s)J_0(s\xi)ds \ , \qquad (10.2.26)$$

and substituting this into (10.2.18) to obtain

$$w(r,0) = \int_0^a sk_{33}(r,s)p(s)ds \ , \qquad (10.2.27)$$

where

$$k_{33}(r,s) = \vartheta \int_0^\infty J_0(r\xi)J_0(s\xi)d\xi \qquad (10.2.28)$$

may be expressed in terms of elliptic integrals; see equations (2.6.2) and (2.6.8). A family of Legendre polynomial solutions for w, p(r) is given in §2.6.

A second form for the solution of (10.2.18), (10.2.19) may be obtained by using the operators in Table 5.7.1; this is only superficially different from the method employed by Sneddon (1960), (1965). First make the equations dimensionless by writing

$$\rho = r/a \ , \quad x = a\xi \ , \quad d + f(r) = \vartheta P_0 a^{-1}(\delta+\chi(\rho)) \ , \qquad (10.2.29)$$

$$\Phi(x) = P(\xi)/P_0 \ , \quad \phi(\rho) = a^2 p(r)/P_0 \ , \qquad (10.2.30)$$

so that

$$(\vartheta P_0 a^{-1})^{-1} w(r,0) \equiv H_0[x^{-1}\Phi(x);\rho] = \delta + \chi(\rho) \ , \quad 0 \le \rho < 1 \ , \qquad (10.2.31)$$

$$a^2\tau_{zz}(r,0)/P_0 \equiv -H_0[\Phi(x);\rho] = -\phi(\rho)H(1-\rho) \ , \quad \rho \ge 0 \ . \qquad (10.2.32)$$

Now operate on (10.2.32) by entry 10 to obtain

$$F_c[\Phi(x);\tau] = A_2[\rho\phi(\rho)H(1-\rho);\tau] = \psi(\tau) \qquad (10.2.33)$$

and since $\psi(\tau) = 0$ when $\tau > 1$ this yields

$$\Phi(x) = (2/\pi)^{1/2} \int_0^1 \psi(\tau)\cos\tau x d\tau \ . \qquad (10.2.34)$$

Now operate on (10.2.31) with entry 1 to obtain

$$F_c[\Phi(x);\tau] = \psi(\tau) = A_1^{-1}[\delta + \chi(\rho);\tau] \ , \tag{10.2.35}$$

which yields $\psi(\tau)$. Now equation (10.2.33) gives the dimensionless contact pressure in the form

$$\rho\phi(\rho) = -(2/\pi)^{1/2} D_\rho \int_\rho^1 \tau(\tau^2-\rho^2)^{-1/2}\psi(\tau)d\tau \ . \tag{10.2.36}$$

Carrying out an integration by parts before differentiation we find

$$\phi(\rho) = (2/\pi)^{1/2} \left\{ \psi(1)(1-\rho^2)^{-1/2} - \int_\rho^1 (\tau^2-\rho^2)^{-1/2}\psi'(\tau)d\tau \right\} \ . \tag{10.2.37}$$

Equations (10.2.30), (10.2.34) yield the total load

$$P = 2\pi \int_0^a rp(r)dr = 2\pi P_0 \int_0^1 \rho\phi(\rho)d\rho = 2\pi(2/\pi)^{1/2} P_0 \int_0^1 \psi(\tau)d\tau \ , \tag{10.2.38}$$

and equation (10.2.35) shows that

$$\int_0^1 \psi(\tau)d\tau = \int_0^1 D_\tau A_1\{\rho(\delta+\chi(\rho));\tau\}d\tau =$$

$$= A_1\{\rho(\delta+\chi(\rho));1\} = (2/\pi)^{1/2} \int_0^1 \rho(\delta+\chi(\rho))(1-\rho^2)^{-1/2}d\rho =$$

$$= (2/\pi)^{1/2} \left\{ \delta + \int_0^1 (1-\rho^2)^{1/2}\chi'(\rho)d\rho \right\} \ . \tag{10.2.39}$$

When the contact between punch and half-space is incomplete, the condition $\phi(1) \to 0$ as $\rho \to 1-$ provides an expression for the contact depth or penetration d. It may be shown that if $\psi(\tau)$ is differentiable in the neighbourhood of 1, then the integral in (10.2.37) will be finite and will tend to zero as $\rho \to 1-$ so that the condition for incomplete contact is

$$\psi(1) = 0 . \tag{10.2.40}$$

But equation (10.2.35) shows that

$$\psi(\tau) = (2/\pi)^{1/2} D_\tau \int_0^\tau \rho[\delta+\chi(\rho)](\tau^2-\rho^2)^{-1/2} d\rho =$$

$$= (2/\pi)^{1/2} \left\{ \delta + \tau \int_0^\tau (\tau^2-\rho^2)^{-1/2} \chi'(\rho) d\rho \right\} , \tag{10.2.41}$$

which, with (10.2.40), yields

$$\delta = - \int_0^1 (1-\rho^2)^{-1/2} \chi'(\rho) d\rho , \tag{10.2.42}$$

and when this is substituted into (10.2.39) we obtain

$$P = -4P_0 \int_0^1 \rho^2 (1-\rho^2)^{-1/2} \chi'(\rho) d\rho = -4\vartheta^{-1} \int_0^a r^2 (a^2-r^2)^{-1/2} f'(r) dr \tag{10.2.43}$$

and this agrees with (2.8.23), (2.8.24).

The investigation of the relationship between load and penetration for axisymmetric punches of various profiles has been the subject of numerous investigations. Love (1929) considers the cases of constant pressure, constant displacement (plane face) and Hertzian (parabolic) pressure, using essentially the method of §2.5, and Love (1939) considers Boussinesq's problem for a rigid cone, using oblate spheroidal coordinates (§10.5) and other special solutions. The systematic use of spheroidal (§10.5) and of ellipsoidal (§11.3) coordinates for contact problems is due largely to Lur'e (1941), (1964). Harding and Sneddon (1945) use Hankel transforms to reduce the problem to a pair of dual integral equations and then use Titchmarsh's somewhat cumbersome solution; this is essentially the analysis described in Sneddon (1951).

Sneddon (1946) considers the flat-ended cylindrical punch in detail. Galin (1946) considers many punch configurations, with and without friction. Mossakovskii (1954a) develops a comprehensive treatment of a large class of contact problems; his analysis can be greatly simplified by using modern operational methods. Segedin (1957) starts from a Papkovich-Neuber solution, finds the solution for a flat-ended punch, and then uses this to build up solutions for parabolic, conical and other punches with general polynomial profiles; a similar analysis was given by Mossakovskii (1963). Deich (1962) considers the case when the punch profile is given by $f(r) = Ar^{2\lambda}$, and λ is not necessarily an integer. Sneddon (1965) uses essentially the analysis described above and gives detailed results for punches with flat, parabolic, spherical (Example 10.2.8) and conical (Example 10.2.9) profiles, and rederives Segedin's results. An alternative analysis, mostly phrased in terms of the Mehler-Fok transform, may be found in Ufliand (1965). A punch with a *rounded* tip will have a profile $z = f(r)$, where $f(r)$ may be approximated by a polynomial *even* in r. For such punches the analysis of §2.6, in particular the expressions for p, w in (2.6.12), (2.6.13), gives perhaps the simplest form of solution; the corresponding displacements outside the contact region are given in §10.5. The only important case not covered by that analysis is the conical punch; if the semi-vertical angle is γ, as shown in Figure 10.2.1, then (Sneddon (1948))

$$f(r) = -\varepsilon r/a \ , \quad d = b+\varepsilon \ , \quad \varepsilon = a\cot\gamma \ , \qquad (10.2.44)$$

and equation (10.2.42) yields $d = \varepsilon\pi/2$; the details of the solution are given in Example 10.2.9.

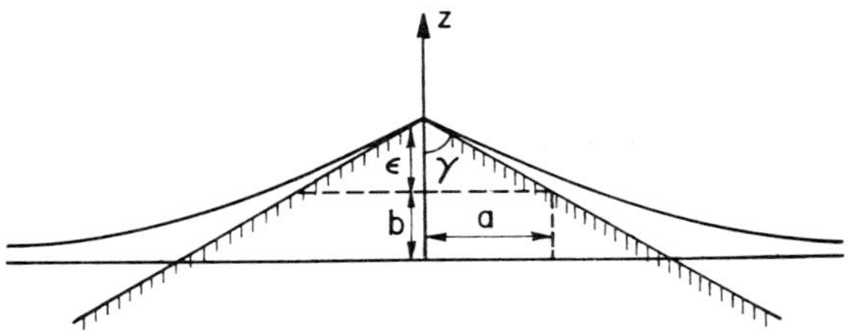

Figure 10.2.1 - Conical punch with semi-vertical angle γ

Selvadurai (1978) finds the interaction between a rigid circular punch on an elastic half-space and a force applied at a point inside the half-space.

The Reissner-Sagoci Problem

The analysis of the torsional contact problem 1(a), the so-called *Reissner-Sagoci* problem, follows much the same lines as that for 2(f). The Hankel-transform analysis was given by Sneddon (1947), (1951), but was later simplified (1966b). The original analysis of Reissner and Sagoci (1944) and Sagoci (1944) was phrased in terms of oblate spheroidal coordinates for which see §10.5 and Example 10.5.6.

The Papkovich-Neuber solution

$$\vec{\psi} = \psi(r,z)\hat{\theta} , \qquad \phi = 0 \qquad (10.2.45)$$

yields the displacements and stresses

$$2\mu v = 4(1-\nu)\psi , \qquad \tau_{\theta z} = \mu \frac{\partial v}{\partial z} , \qquad \tau_{r\theta} = \mu\left(\frac{\partial v}{\partial r} - \frac{v}{r}\right) . \qquad (10.2.46)$$

The vector $\vec{\psi}$ is harmonic (thus $\psi\cos\theta$, $\psi\sin\theta$ are harmonic, but not ψ); remembering that $\partial^2\hat{\theta}/\partial\theta^2 = -\hat{\theta}$ we find that

$$\nabla^2\vec{\psi} = \hat{\theta}(D_r^2+r^{-1}D_r-r^{-2}+D_z^2)\psi = \hat{\theta}(\mathcal{B}_1+D_z^2)\psi = 0 , \qquad (10.2.47)$$

where the notation is that of (5.6.18). Now using (5.6.18) we obtain

$$\mathcal{H}_1[(\mathcal{B}_1+D_z^2)\psi(r,z);r \to \xi] = (D_z^2-\xi^2)\mathcal{H}_1[\psi(r,z);r \to \xi] = 0 , \qquad (10.2.48)$$

so that the appropriate form for $\psi(r,z)$ in the half-space $z \geq 0$ is

$$\psi(r,z) = \mathcal{H}_1[\xi^{-1}A(\xi)\exp(-\xi z);\xi \to r] . \qquad (10.2.49)$$

Now suppose that a circular punch with a plane face adheres to the surface $r \leq a$, $z = 0$ of the half-space and is given a twist γ about its axis. The boundary conditions are

$$v(r,0) = \gamma r , \quad 0 \leq r < a ; \quad \tau_{r\theta}(r,0) = 0 , \quad r > a . \qquad (10.2.50)$$

Writing $\tau_{\theta z}(r,0) = -q(r)H(a-r)$ we find

$$\mu v(r,0) = \mathcal{H}_1[\xi^{-1}Q(\xi);\xi \to r] = \mu f(r) , \quad 0 \leq r < a , \qquad (10.2.51)$$

$$-\tau_{\theta z}(r,0) = \mathcal{H}_1[Q(\xi);\xi \to r] = q(r)H(a-r) , \qquad (10.2.52)$$

where in this particular problem $f(r) = \gamma r$.

Again there are a number of ways of solving these equations, which form another particular case of (10.2.24), (10.2.25). Using the operator approach, with entries 2 and 11 of Table 5.7.1 we find

$$\mu s^{-1}A_1^{-1}[rv(r,0);s] = \mathcal{F}_s[Q(\xi);s] = g(s) , \quad 0 \leq s < a , \quad (10.2.53)$$

$$-sA_2[\tau_{\theta z}(r,0);s] = \mathcal{F}_s[Q(\xi);s] = 0 , \quad s > a , \qquad (10.2.54)$$

where $g(s) = \mu s^{-1} A_1^{-1}[rf(r);s]$ has the value $2\mu\gamma(2/\pi)^{1/2}s$ in the problem. These equations have the solution

$$Q(\xi) = (2/\pi)^{1/2} \int_0^a g(s)\sin(s\xi)ds = \frac{4\mu\gamma a}{\pi\xi}\left[\frac{\sin(a\xi)}{a\xi} - \cos(a\xi)\right] . \quad (10.2.55)$$

Actually it is unnecessary to calculate $Q(\xi)$ in order to find the stress distribution since equations (10.2.53), (10.2.54) give

$$sA_2[q(r);s] = g(s)H(a-s) , \quad (10.2.56)$$

so that

$$q(r) = -\frac{4\mu\gamma}{\pi} D_r \int_r^a s(s^2-r^2)^{-1/2}ds = \frac{4\mu\gamma r}{\pi}(a^2-r^2)^{-1/2} , \quad (10.2.57)$$

and the total torque is

$$T^0 = 2\pi \int_0^a r^2 q(r)dr = \frac{16\mu\gamma a^3}{3} . \quad (10.2.58)$$

Equation (10.2.57) shows that the shear-stress is zero for $r = 0$ and has a square root singularity as $r \to a-$. The displacement u_θ outside the contact region may also be obtained from equations (10.2.53), (10.2.54); thus

$$\mu s^{-1} A_1^{-1}[rv(r,0);s] = g(s)H(a-s) , \quad (10.2.59)$$

so that

$$\mu r v(r,0) = (2/\pi)^{1/2} \int_0^a (r^2-s^2)^{-1/2} s g(s)ds , \quad (10.2.60)$$

giving

$$v(r,0) = \frac{2\gamma}{\pi}\left[r\sin^{-1}\left(\frac{a}{r}\right) - a\left(1 - \frac{r^2}{a^2}\right)^{1/2}\right] , \quad (10.2.61)$$

which has the expected limiting values γ and 0 as r tends to a+ and ∞ respectively.

Examples 10.2

1. Show that the state vector transform corresponding to the shear stress distribution $\tau_{rz}(r,0) = -t(r)$, $0 \leq r < \infty$ is given by

$$\begin{bmatrix} 2\mu U \\ 2\mu W \\ -T_{13} \\ -T_{33} \end{bmatrix} = \begin{bmatrix} (-\xi z+2-2\nu)\xi^{-1} \\ (-\xi z-1+2\nu)\xi^{-1} \\ -\xi z+1 \\ -\xi z \end{bmatrix} T(\xi)\exp(-\xi z) \ ,$$

where $T(\xi) = \mathcal{H}_1[t(r);\xi]$.

2. Show that the stresses τ_{rr}, $\tau_{\theta\theta}$ corresponding to (10.2.10) are

$$\tau_{rr} = \mathcal{H}_0\{(-2\nu A+\xi zA+C)\exp(-\xi z);\xi\to r\}-r^{-1}\mathcal{H}_1\{\xi^{-1}(\xi zA+C)\exp(-\xi z);\xi\to r\}$$

$$\tau_{\theta\theta} = \mathcal{H}_0\{-2\nu A\exp(-\xi z);\xi\to r\}+r^{-1}\mathcal{H}_1\{\xi^{-1}(\xi zA+C)\exp(-\xi z);\xi\to r\} \ .$$

3. Show that if $U(\xi)$, $P(\xi)$ etc. denote $U(\xi,0)$, $P(\xi,0)$ etc. then

$$\begin{bmatrix} U(\xi) \\ W(\xi) \end{bmatrix} = \vartheta\xi^{-1} \begin{bmatrix} 1 & -\varphi \\ -\varphi & 1 \end{bmatrix} \begin{bmatrix} T(\xi) \\ P(\xi) \end{bmatrix}$$

where $\varphi = (1-2\nu)/(2-2\nu)$.

4. Use the results of equations (10.2.13) and Example 10.2.1, and rewrite the transforms in (10.2.10) in terms of p(r), t(r) to recover equations (2.6.4), (2.6.5).

5. Show that the stresses corresponding to the solution (10.2.11) are

$$\tau_{rr} = \frac{\partial}{\partial z}\left(\nu\nabla^2\phi - \frac{\partial^2\phi}{\partial r^2}\right) \ , \quad \tau_{\theta\theta} = \frac{\partial}{\partial z}\left(\nu\nabla^2\phi - \frac{1}{r}\frac{\partial\phi}{\partial r}\right)$$

$$\tau_{zz} = \frac{\partial}{\partial z}\left[(2-\nu)\nabla^2\phi - \frac{\partial^2\phi}{\partial z^2}\right] \ , \ \tau_{rz} = \frac{\partial}{\partial r}\left[(1-\nu)\nabla^2\phi - \frac{\partial^2\phi}{\partial z^2}\right] \ .$$

If $\phi(r,z) = H_0[\Phi(\xi,z);\xi \to r]$, show that $(D_z^2 - \xi^2)^2\Phi(\xi,z) = 0$, and hence show that $\Phi(\xi,z) = (A+B\xi z)\exp(\xi z) + (C+D\xi z)\exp(-\xi z)$, where A, B, C, D are functions of ξ.

6. Complete the solution for a punch with plane face. Show that

$$w(r,0) = \begin{cases} d, & 0 \le r < a, \\ \dfrac{2d}{\pi}\sin^{-1}\left(\dfrac{a}{r}\right), & r > a, \end{cases}$$

and that the functions $\Phi(x)$, $\psi(\tau)$ and $\phi(\rho)$ are

$$\Phi(x) = \frac{2\delta}{\pi}\frac{\sin x}{x}, \quad \psi(\tau) = \left(\frac{2}{\pi}\right)^{1/2}\delta, \quad \phi(\rho) = \frac{2\delta}{\pi}(1-\rho^2)^{-1/2}$$

where d is related to δ by equation (10.2.29). Show that the surface radial displacement is

$$u(r,0) = -\frac{2\varphi d}{\pi}\begin{cases} [a-(a^2-r^2)^{1/2}]/r, & r \le a, \\ a/r, & r > a. \end{cases}$$

7. Establish equation (10.2.38) for the total load on the half-space, and show that $P_0 = 4ad/\vartheta$ for a punch with a plane face.

8. For a spherical indenter of radius R the function $f(r) = (R^2-r^2)^{1/2}-R$. Show that the total indentation d and force P_0 are linked to R and the radius of contact a by the equations

$$d = \frac{a}{2}\ln\left(\frac{R+a}{R-a}\right), \quad P_0 = \vartheta^{-1}\left[(a^2+R^2)\ln\left(\frac{R+a}{R-a}\right) - aR\right].$$

9. Show that for the conical punch of Figure 10.2.1

$$\psi(\tau) = \left(\frac{\pi}{2}\right)^{1/2}\frac{a}{\vartheta P_0}(1-\tau), \quad \tau < 1; \quad \phi(\rho) = \frac{\varepsilon a}{\vartheta P_0}\cosh^{-1}\left(\frac{1}{\rho}\right), \quad \rho < 1$$

Use entry 1 of Table 5.7.2 to show that the displacement $w(r,0)$ of (10.2.31) may be written

$$w(r,0) = \frac{\vartheta P_0}{a} A_1[\psi(\tau);r],$$

and hence show that the displacement outside the contact area for the conical punch is

$$w(r,0) = \frac{2d}{\pi a}\left[a\sin^{-1}\frac{a}{r} - r + (r^2-a^2)^{1/2}\right] .$$

10. Show that for large r the displacement v of (10.2.61) satisfies

$$v = \frac{4\gamma a^3}{3\pi r^2} + 0\left(\frac{1}{r^4}\right) .$$

11. Show that if $\tau_{rz}(r,0) = 0$ for all r then ϕ of equation (10.2.2) must be chosen so that $\partial\phi/\partial z = (1-2\nu)\psi$ and that then

$$2\mu w = 2(1-\nu)\psi - z\frac{\partial\psi}{\partial z}, \qquad \tau_{zz} = \frac{\partial\psi}{\partial z} - z\left(\frac{\partial^2\psi}{\partial z^2}\right).$$

12. Show that the displacement (10.2.27) has the properties

$$w(0,0) = \vartheta\int_0^a p(s)ds \ ; \qquad w(r,0) = \frac{\vartheta}{r}\int_0^a sp(s)ds + 0\left(\frac{1}{r^2}\right) ,$$

as $r \to \infty$. The second relation shows that, to leading order, the displacement is the same as that for the total load concentrated at the origin.

13. Use equation (10.2.55) to find the stress distribution within the half-space $z > 0$ for the Reissner-Sagoci problem. Use the integral (Gradshteyn and Ryzhik (1965, 6.7521)).

$$\int_0^\infty x^{-1}\exp(-ax)J_1(cx)\sin bx dx = b(1-r)/c ,$$

where $b^2 = c^2/(1-r^2)-a^2/r^2$.

10.3 *Some asymmetric problems for the half-space*

In order to solve problems of types 3 and 4 of §10.1 it is necessary to derive appropriate Papkovich-Neuber solutions. An inspection of the general equations (1.10.2) shows that there are solutions in cylindrical polar coordinates for which the θ-dependence is

$$u \sim \cos m\theta\ , \quad v \sim \sin m\theta\ , \quad w \sim \cos m\theta\ . \tag{10.3.1}$$

The explicit form of the Papkovich-Neuber solution in cylindrical polar coordinates is

$$2\mu u = 4(1-\nu)\psi_1 - \frac{\partial}{\partial r}(r\psi_1+z\psi_3+\phi)\ , \tag{10.3.2}$$

$$2\mu v = 4(1-\nu)\psi_2 - \frac{\partial}{r\partial\theta}(r\psi_1+z\psi_3+\phi)\ , \tag{10.3.3}$$

$$2\mu w = 4(1-\nu)\psi_3 - \frac{\partial}{\partial z}(r\psi_1+z\psi_3+\phi)\ , \tag{10.3.4}$$

where $\vec{\psi} = \psi_1\hat{r} + \psi_2\hat{\theta} + \psi_3\hat{k}$ is harmonic. For the θ-dependence (10.3.1) the potentials will have the form

$$\psi_1 = \Psi_1\cos m\theta\ , \quad \psi_2 = \Psi_2\sin m\theta\ , \quad \psi_3 = \Psi_3\cos m\theta\ , \quad \phi = \Phi\cos m\theta\ , \tag{10.3.5}$$

where Ψ_i, Φ are functions of r, z. Now in Cartesian coordinates

$$\vec{\psi} = \left[\frac{\Psi_1-\Psi_2}{2}\right]\left[\hat{i}\cos(m-1)\theta - \hat{j}\sin(m-1)\theta\right] +$$
$$+ \left[\frac{\Psi_1+\Psi_2}{2}\right]\left[\hat{i}\cos(m+1)\theta + \hat{j}\sin(m+1)\theta\right] + \hat{k}\Psi_3\cos m\theta\ . \tag{10.3.6}$$

Each of the Cartesian components of $\vec{\psi}$ must be harmonic in $z \geq 0$ so that, noting (5.6.20), we must seek solutions of the form

$$\Psi_1-\Psi_2 = 2H_{m-1}[\xi^{-1}A_1(\xi)\exp(-\xi z);\xi \to r]\ , \tag{10.3.7}$$

$$\Psi_1+\Psi_2 = 2H_{m+1}[\xi^{-1}A_2(\xi)\exp(-\xi z);\xi \to r]\ , \tag{10.3.8}$$

$$\Psi_3 = 2H_m[\xi^{-1}B(\xi)\exp(-\xi z);\xi \to r]\ , \tag{10.3.9}$$

$$\Phi = 2H_m[\xi^{-2}C(\xi)\exp(-\xi z);\xi \to r]\ . \tag{10.3.10}$$

There is one redundancy amongst the four components of the Papkovich-Neuber solution and this may be removed by taking

$A_1(\xi) = A_2(\xi) = A(\xi)$. After some algebraic manipulations it is found that

$$u(r,\theta,z) = (u^*+v^*)\cos m\theta \ , \quad v(r,\theta,z) = (-u^*+v^*)\sin m\theta \ , \tag{10.3.11}$$

$$w(r,\theta,z) = w^*\cos m\theta \ , \tag{10.3.12}$$

$$\tau_{rz}(r,\theta,z) = -(s+t)\cos m\theta \ , \quad \tau_{\theta z}(r,\theta,z) = (s-t)\sin m\theta \ , \tag{10.3.13}$$

$$\tau_{zz}(r,\theta,z) = -p\cos m\theta \ , \tag{10.3.14}$$

where

$$u^* = H_{m-1}[U(\xi,z);\xi \to r] \ , \quad v^* = H_{m+1}[V(\xi,z);\xi \to r] \ , \tag{10.3.15}$$

$$w^* = H_m[W(\xi,z);\xi \to r] \ , \tag{10.3.16}$$

$$s = H_{m-1}[S(\xi,z);\xi \to r] \ , \quad t = H_{m+1}[T(\xi,z);\xi \to r] \ , \tag{10.3.17}$$

$$p = H_m[P(\xi,z);\xi \to r] \ , \tag{10.3.18}$$

and

$$\begin{bmatrix} 2\mu U \\ 2\mu V \\ 2\mu W \\ S \\ T \\ P \end{bmatrix} = \begin{bmatrix} \xi^{-1}\{(4-4\nu-m)A - \xi zB - C\} \\ \xi^{-1}\{(4-4\nu+m)A + \xi zB + C\} \\ \xi^{-1}\{2mA + 2(3-4\nu)B + 2\xi zB + 2C\} \\ (2-2\nu-m)A - (1-2\nu)B - \xi zB - C \\ (2-2\nu+m)A + (1-2\nu)B + \xi zB + C \\ 2mA + 4(1-\nu)B + 2\xi zB + 2C \end{bmatrix} \exp(-\xi z) \ . \tag{10.3.19}$$

The Cartesian components and stresses relating to the solution for m = 1, appropriate to the tilting of a rigid indenter, are

$$u_x = u^* + v^*\cos 2\theta \ , \quad u_y = v^*\sin 2\theta \ , \quad u_z = w^*\cos\theta \ , \tag{10.3.20}$$

$$\tau_{xz} = -s - t\cos 2\theta \ , \quad \tau_{yz} = -t\sin 2\theta, \quad \tau_{zz} = -p\cos\theta \ . \tag{10.3.21}$$

This analysis may be applied to the contact problems for the rigid circular indenter with a plane face. In problem 3(f), first solved by Abramov (1939), the boundary conditions are

$$\tau_{xz} = 0 = \tau_{yz} \quad \text{on} \quad z = 0 \text{ ,} \tag{10.3.22}$$

$$w(r,\theta,0) = \gamma r\cos\theta \quad \text{on} \quad 0 \leq r < a \text{ ;} \quad \tau_{zz}(r,\theta,0) = 0 \text{ ,} \quad r > a \text{ ,} \tag{10.3.23}$$

where γ is the angle of rotation about the y-axis. In practice this displacement will be produced by an indenter with circular section whose axis is slightly inclined to the z-axis, as shown in Figure 10.3.1a or, equivalently, by the one with inclined face, as shown in Figure 10.3.1b. In either case the normal displacement for $0 \leq r < a$ will be $w = d + \gamma r\cos\theta$; the elastic displacement field in the half-space caused by the indenter may be found as the superposition of the axisymmetric field of Example 10.2.6, and that given below. With $m = 1$, these translate into

$$S(\xi,0) = 0 = T(\xi,0) \text{ ,} \tag{10.3.24}$$

$$w^*(r,0) = \gamma r \text{ ,} \quad 0 \leq r < a \text{ ;} \quad p(r,0) = 0 \text{ ,} \quad r > a \text{ ,} \tag{10.3.25}$$

or, after choosing A,B,C, so that equations (10.3.24) are satisfied, we find

$$w^*(r,0) \equiv \vartheta H_1[\xi^{-1}P(\xi);r] = \gamma r \text{ ,} \quad 0 \leq r < a \text{ ,} \tag{10.3.26}$$

$$p(r,0) \equiv H_1[P(\xi);r] = 0 \text{ ,} \qquad r > a \text{ .} \tag{10.3.27}$$

These equations are formally the same as equations (10.2.51), (10.2.52); the solution may be found in §10.2 and is discussed in Example 10.3.2. This problem is used as an example by Muki (1960) in a general formulation of asymmetric problems for a circular contact area; see also Keer (1964c). Muki starts from a different general solution of the equations of elastic equilibrium (see Example 10.3.3). He gives detailed analytical solutions for a number of asymmetric contact problems and gives graphical results for stresses at points beneath the contact surface; Westman (1965) used Muki's analysis to solve problem 4(f) and others.

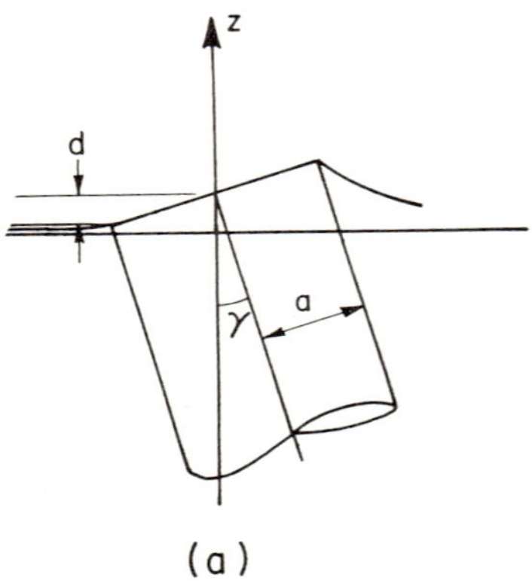

Figure 10.3.1 - Punch with circular section indenting the half-space inclined at an angle γ

Problem 4(f), the displacement due to a translation of a plane frictionless rigid punch in the direction of the x-axis, was solved by Mindlin (1949) using Boussinesq's theory (§2.5). The boundary conditions are

$$\tau_{yz} = 0 = \tau_{zz} \quad \text{on} \quad z = 0 \,, \tag{10.3.28}$$

$$u_x(r,\theta,0) = d \,, \quad 0 \le r < a \,; \quad \tau_{xz}(r,\theta,0) = 0 \,, \quad r > a \,, \tag{10.3.29}$$

which translate, with m = 1, into

$$T(\xi,0) = 0 = P(\xi,0) \,, \tag{10.3.30}$$

$$u^*(r,0) = d\ ,\quad v^*(r,0) = 0\ ,\quad 0 \le r < a\ ;\quad s(r,0) = 0\ ,\quad r > a\ . \tag{10.3.31}$$

These equations have an unusual feature in that the solution of the equations

$$u^*(r,0) = d\ ,\quad 0 \le r < a\ ;\quad s(r,0) = 0\ ,\quad r > a\ , \tag{10.3.32}$$

which are equivalent to the pair

$$2\mu u^*(r,0) \equiv (2-\nu)\mathcal{H}_0[\xi^{-1}S(\xi,0);r] = 2\mu d\ ,\quad 0 \le r < a\ , \tag{10.3.33}$$

$$s(r,0) \equiv \mathcal{H}_0[S(\xi,0);r] = 0\ ,\qquad r > a\ , \tag{10.3.34}$$

is as given in §10.2, namely

$$S(\xi,0) = \frac{2}{\pi}\cdot\frac{2\mu d}{2-\nu}\cdot\frac{\sin a\xi}{\xi}\ , \tag{10.3.35}$$

and this makes

$$2\mu v^*(r,0) \equiv \nu\mathcal{H}_2\{\xi^{-1}S(\xi,0);r\} \equiv 0\ ,\quad 0 \le r < a\ . \tag{10.3.36}$$

Notice also that $v^*(r,0)$ is not zero for $r > a$, and nor is $w^*(r,0)$ zero for $r < a$ or $r > a$ (Example 10.3.4).

Examples 10.3

1. Use equations (10.3.19) to show that if $P(\xi)$, $U(\xi)$ etc. denote $P(\xi,0)$, $U(\xi,0)$ etc. respectively then, whatever the value of m,

$$2\mu\begin{bmatrix} U(\xi) \\ W(\xi) \\ V(\xi) \end{bmatrix} = \xi^{-1}\begin{bmatrix} 2-\nu & (1-2\nu)/2 & \nu \\ 1-2\nu & 2-2\nu & -(1-2\nu) \\ \nu & -(1-2\nu)/2 & 2-\nu \end{bmatrix}\begin{bmatrix} S(\xi) \\ P(\xi) \\ T(\xi) \end{bmatrix}$$

2. Show that the pressure distribution for the inclined rigid punch for which $w = d+\gamma r\cos\theta$, $r \leq a$, is given by

$$p = \frac{2}{\pi\vartheta(a^2-r^2)^{1/2}}\{d+2\gamma r\cos\theta\} .$$

This is a particular case of the result stated earlier in (2.7.10). Show that the ratio of torque to rotation is

$$\frac{M}{\gamma} = \frac{8a^3}{3\vartheta} .$$

3. Show that the static equations of elastic equilibrium in the absence of body forces are satisfied by the expressions

$$2\mu u = -\frac{\partial^2\phi}{\partial r\partial z} + \frac{2}{r}\frac{\partial\psi}{\partial\theta}, \quad 2\mu v = -\frac{\partial^2\phi}{r\partial\theta\partial z} - \frac{2\partial\psi}{\partial r},$$

$$2\mu w = 2(1-\nu)\nabla^2\phi - \frac{\partial^2\phi}{\partial z^2},$$

where ϕ, ψ satisfy $\nabla^4\phi = 0 = \nabla^2\psi$. If $\phi = \Phi(r,z)\cos m\theta$, $\psi = \Psi(r,z)\sin m\theta$, show that

$$(B_m^2+D_z^2)^2\Phi(r,z) = 0 = (B_m^2+D_z^2)\Psi(r,z) = 0 ,$$

and that the general solution of these equations may be written

$$\Phi(r,z) = \mathcal{H}_m[G(\xi,z);\xi \to r] , \quad \Psi(r,z) = \mathcal{H}_m[H(\xi,z);\xi \to r] ,$$

where

$$G(\xi,z) = (A+B\xi z)\exp(\xi z) + (C+D\xi z)\exp(-\xi z) ,$$

$$H(\xi,z) = E\exp(\xi z) + F\exp(-\xi z) .$$

4. Use Example 10.3.1 and equation (10.3.35) to show that the displacements on the surface of the half-space in problem 4(f)

for r < a and r > a are respectively

$$u^*(r,0) = d\ ,\quad v^*(r,0) = 0\ ,\quad w^*(r,0) = \frac{2d}{\pi}\cdot\frac{1-2\nu}{2-\nu}\cdot$$
$$\cdot\left[1+\left\{1-\frac{r^2}{a^2}\right\}^{1/2}\right]^{-1}\ ,$$

$$u^*(r,0) = \frac{2d}{\pi}\sin^{-1}\left(\frac{a}{r}\right),\quad v^*(r,0) = \frac{2d}{\pi}\cdot\frac{\nu}{2-\nu}\cdot\frac{a}{r}\left(1-\frac{a^2}{r^2}\right)^{1/2}\ ,$$

$$w^*(r,0) = \frac{2d}{\pi}\cdot\frac{1-2\nu}{2-\nu}\cdot\frac{a}{r}\ ,$$

and show that the stress distribution τ_{xz} is

$$\tau_{xz} = -s = -\frac{4\mu d}{\pi(2-\nu)}(a^2-r^2)^{-1/2}\ ,$$

so that the ratio of total force to displacement is

$$\frac{T}{d} = \frac{8\mu a}{2-\nu}\ .$$

5. Use Example 10.3.1 to show that the surface normal displacement due to a pressure $\tau_{zz}(r,0) = -p(r)\cos m\theta$ acting over $|r| \leq a$ is given by

$$w^*(r) = \vartheta\int_0^a sW_m^0(r,s)p(s)ds$$

where W_m^0 is defined in (6.11.8).

6. Show that if $\tau_{yz} = 0$ and τ_{xz} is independent of θ, then $\tau_{rz} = \tau_{xz}\cos\theta$, $\tau_{\theta z} = -\tau_{xz}\sin\theta$ so that the displacements are given by (10.3.11)-(10.3.19) with m = 1 and t = 0 = p, $s = -\tau_{xz}$. Show moreover that the Cartesian displacements are $u_x = u^* + v^*\cos2\theta$, $u_y = v^*\sin2\theta$.

7. Use Example 10.3.6 to show that the displacements for $r \leq a$ due to $\tau_{xz} = -A\left(1-\frac{r^2}{a^2}\right)^{1/2}$ over the circle $r \leq a$ are given by

$$32\mu a u^* = 2(2-\nu)A\pi(2a^2-r^2)\ ,\quad 32\mu a v^* = \nu A\pi r^2\ .$$

10.4 *Adhesive problems for the half-space*

In this section we consider problems 2(a), 3(a), 4(a) of §10.1, that is, adhesive contact of a punch with a half-space, when the displacement of the punch is either (2) a translation in the z-direction, (3) a rotation about the y-axis or (4) a translation in the x-direction. We shall consider these problems for a *rigid* indenter with a plane face, the general problem 2(a) for a rigid indenter with (axisymmetric) arbitrary or polynomial profile, and certain conclusions which can be drawn regarding the adhesive contact of two elastic bodies.

Mossakovskii's solution

Problem 2(a) for a rigid indenter with a plane face was solved by Mossakovskii (1954a). The boundary conditions for this axisymmetric problem are

$$u(r,0) = 0 \ , \quad w(r,0) = w_0 \ , \quad 0 \leq r < a \ , \tag{10.4.1}$$

$$\tau_{rz}(r,0) = 0 = \tau_{zz}(r,0) \quad , \qquad r > a \ . \tag{10.4.2}$$

Mossakovskii based his paper on a potential solution of the elasticity equations due to Trefftz (1928) (Example 1.10.3). The displacement $w(r,0)$ has the form

$$w(r,0) = w_0 \Phi(r/a) \ .$$

The function $\Phi(r/a)$ will be derived below and is given in equations (10.4.50) - (10.4.51). (Ufliand (1956), (1965) gave an entirely independent solution in terms of toroidal coordinates and the Mehler-Fok transform.)

Problem 2(a) for a general rigid indenter is not immediately formulated, because the radial displacement $u(r,0)$ is initially unknown. Goodman (1962) developed an approximate incremental formulation, while Mossakovskii (1963) devised the following precise incremental method.

Mossakovskii's method

Suppose that the contact force P_0, or equivalently the contact radius a, is gradually increased; the surface displacements u(r,0) and w(r,0) will be functions of both r and a: denote them by u(r;a), w(r;a) respectively. Once a point of the surface has been brought into contact with the indenter, its radial displacement u(r;a) cannot change further with a, i.e.,

$$\frac{\partial u}{\partial a}(r;a) = 0 , \quad r < a , \tag{10.4.3}$$

as indicated in Figure 10.4.1, taken from Spence (1968a).

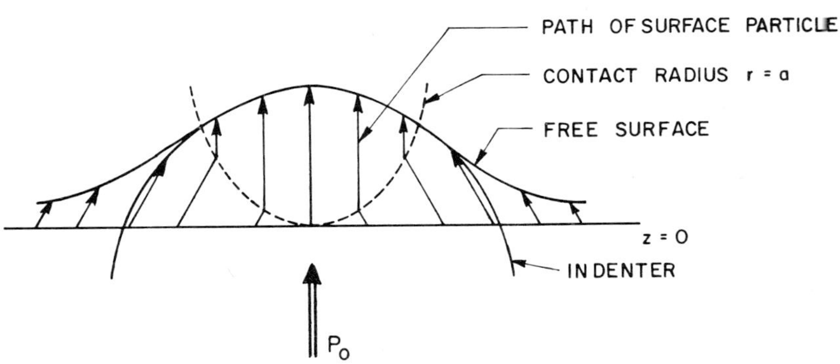

Figure 10.4.1 - Incremental deformation of a half-space

The displacement w(r;a) is determined by the shape of the body and the penetration depth $w_0(a)$ by means of

$$w(r;a) = w_0(a) + f(r) , \quad r < a , \tag{10.4.4}$$

where $z = f(r)$ is the equation of the indenter relative to the vertex, so that $f(0) = 0$. Now consider a small increase a in the contact radius; then

$$w(r;a + \delta a) = w_0(a + \delta a) + f(r) , \quad r < a + \delta a ,$$

so that

$$\frac{\partial w(r;a)}{\partial a} = \lim_{\partial a \to 0} \left[\frac{w(r;a+\delta a) - w(r,a)}{\delta a}\right] = w_0'(a) , \quad r < a . \tag{10.4.5}$$

But this means that $\partial u/\partial a$, $\partial w/\partial a$ satisfy conditions (10.4.1) for the plane indenter with indentation $w_0'(a)$ so that, in particular,

$$\frac{\partial w(r;a)}{\partial a} = w_0'(a)\Phi(r/a) , \tag{10.4.6}$$

and

$$w(r;a) = \int_0^a w_0'(t)\Phi(r/t)dt . \tag{10.4.7}$$

Equations (10.4.4), (10.4.7) now lead to Mossakovskii's integral equation

$$w_0(a) + f(r) = \int_0^a w_0'(t)\Phi(r/t)dt , \quad r < a \tag{10.4.8}$$

for the unknown function $w_0(a)$. Noting that $\Phi(r/t) = 1$ if $r < t$ (i.e., if $t > r$), we may write the equation

$$f(r) = -w_0(r) + r\int_0^1 w_0'(rx)\Phi(1/x)dx . \tag{10.4.9}$$

Mossakovskii noted (Example 10.4.1) that if $f(r) = -Br^n$ then $w_0(r) = Ar^n$; he also found an equation similar to (10.4.8) for one of the potentials for the *frictionless* contact of a rigid punch. We shall not pursue Mossakovskii's solution further, preferring instead to describe that due to Spence, and then to obtain the solution by reduction to a Fredholm integral equation.

Spence's solution

Spence (1968a) developed a similarity argument, independent of Mossakovskii's for the problem 2(a) for the class of indenters having a profile $z = -Br^n$. This class includes spheres ($n = 2$, $B = 1/(2R)$) and cones ($n = 1$, $B = \gamma$, where $\pi - \gamma$ is the semi-angle). (For small displacement theory to hold in these two cases, we must have $a << R$ and $\gamma << 1$ respectively.) For indenters of this type the progressive loading problem possesses a self-similar solution, in the sense that each of the dimensionless quantities

$$\frac{u}{a}, \quad \frac{w}{a}, \quad \frac{\tau_{rz}}{2\mu}, \quad \frac{\tau_{zz}}{2\mu}$$

on the plane $z = 0$ can be expressed as a function of r/a only, multiplied by a scale factor $S(a)$ say, the same for all four, in a way that ensures that the boundary conditions (10.4.3), (10.4.4) are satisfied. If the function $f(r) = -Br^n$, then equation (10.4.4) shows that

$$\frac{w}{a} = \frac{w_0(a)}{a} - \frac{Br^n}{a} = \frac{w_0(a)}{a} - a^{n-1}B\left(\frac{r}{a}\right)^n, \quad r < a, \tag{10.4.10}$$

and this will have the required similarity form only if $w_0(a) = Aa^n$ where A is independent of a, so that

$$\frac{w}{a} = a^{n-1}(A - B\rho^n), \quad \rho < 1, \tag{10.4.11}$$

where $\rho = r/a$; the displacement u/a will now have the form

$$\frac{u}{a} = a^{n-1}C\rho^n , \qquad \rho < 1 , \tag{10.4.12}$$

where C is independent of a. This gives

$$u = Cr^n , \qquad r < a , \tag{10.4.13}$$

which satisfies equation (10.4.3). The boundary conditions (10.4.11), (10.4.12) are now supplemented by the conditions

$$\tau_{rz}(\rho,0) = 0 = \tau_{zz}(\rho,0) , \qquad \rho > 1 . \tag{10.4.14}$$

The solution of the problem specified by equation (10.4.11), (10.4.12), (10.4.14) is arbitrary to the extent of the constants A, C and additional physical conditions must be imposed to render the solution unique. These are supplied by the conditions that the strains should be continuous, and the stresses be bounded, actually zero, at the circle $r = a$. This yields a pair of linear equations for A and C, which can therefore be related uniquely to B, with constants depending only on the exponent n and the Poisson's ratio of the half-space. The case $n = 2$ is considered in Example 10.4.4.

When the indenter has a general polynomial form

$$z = f(r) = - \sum_{n=1}^{N} B_n r^n \tag{10.4.15}$$

we look for a solution in which

$$u(r) = \sum_{n=1}^{N} C_n r^n , \qquad r < a \tag{10.4.16}$$

$$w_0(a) = \sum_{n=1}^{N} A_n a^n \tag{10.4.17}$$

and the boundary conditions (10.4.2) - (10.4.4) are satisfied. In this case the stresses τ_{rz}, τ_{zz} at r = a will depend linearly on all the constants A_n, B_n, C_n and the condition that these stresses should be zero at r = a will lead to two equations

$$\sum_{n=1}^{N} a^{n-1}(a_{n1}A_n+b_{n1}B_n+c_{n1}C_n) = 0 = \sum_{n=1}^{N} a^{n-1}(a_{n2}A_n+b_{n2}B_n+c_{n2}C_n) \qquad (10.4.18)$$

where the coefficients a_{ni}, b_{ni}, c_{ni} will depend only on n and the Poisson's ratio ν. These equations must hold for all a, and therefore the coefficients of a^{n-1} in each equation must vanish, leading to two equations relating A_n, C_n to B_n and a unique solution as before.

The argument above, by deriving the form of u(r,0) for r < a, reduces the adhesive contact problem to the standard mixed boundary-value problem in which u, w are specified inside r = a, while $\tau_{rz} = 0 = \tau_{zz}$ outside. A.Ia. Aleksandrov and Solov'ev (1964a) showed that this problem could be reduced to a two-dimensional problem which could be solved by using the methods of Muskhelishvili (1953a). Their reduction technique, described in §10.6, is equivalent to that described below. Abramian, Arutiunian and Babloian (1966) and Spence (1968a) reduced the problem to the solution of the single integral equation

$$\phi(x) + \lambda \int_0^\infty k(x-t)\phi(t)dt = \chi(x) , \qquad (10.4.19)$$

where $k(x) = (\sin x)/x$. The former authors used a solution of this equation derived by Rapoport (1948); the latter, in Spence (1968b), solved the equation by using the Wiener-Hopf technique. We shall show that the problem may be reduced to the solution of the Fredholm integral equation (3.5.1); this shows, incidentally, that equation (10.4.19) may be solved by reduction to (3.5.1).

Reduction to the Fredholm integral equation (3.5.1)

The boundary conditions for problem 2(a) are

$$u(r,0) = e(r) \ , \quad w(r,0) = w_0(a)+f(r) \ , \quad 0 \le r < a \ , \qquad (10.4.20)$$

$$\tau_{rz}(r,0) = 0 \ , \quad \tau_{zz}(r,0) = 0 \ , \quad r < a \ . \qquad (10.4.21)$$

The general theory of §10.2, in particular Example 10.2.3, shows that

$$u(r,0) \equiv \vartheta H_1[\xi^{-1}\{T(\xi) - \varphi P(\xi)\};r] = e(r) \ , \quad 0 \le r < a \ , \qquad (10.4.22)$$

$$w(r,0) \equiv \vartheta H_0[\xi^{-1}\{P(\xi) - \varphi T(\xi)\};r] = w_0(a)+f(r) \ , \quad 0 \le r < a \ , \qquad (10.4.23)$$

$$\tau_{rz}(r,0) \equiv -H_1[T(\xi);r] = 0 \ , \quad r > a \ , \qquad (10.4.24)$$

$$\tau_{zz}(r,0) \equiv -H_0[P(\xi);r] = 0 \ , \quad r > a \ . \qquad (10.4.25)$$

Applying operators 2, 1, 11, 10, of Table 5.7.1, respectively we obtain

$$s^{-1}A_1^{-1}\{ru(r,0);s\} \equiv \vartheta F_s[T(\xi)-\varphi P(\xi);s] = e^*(s) \ , \quad 0 \le s < a, \qquad (10.4.26)$$

$$A_1^{-1}\{w(r,0);s\} \equiv \vartheta F_c[P(\xi)-\varphi T(\xi);s] = f^*(s) \ , \quad 0 \le s < a \ , \qquad (10.4.27)$$

$$-sA_2\{\tau_{rz}(r,0);s\} \equiv F_s[T(\xi);s] = 0 \ , \quad s > a \ , \qquad (10.4.28)$$

$$-A_2\{r\tau_{zz}(r,0);s\} \equiv F_c[P(\xi);s] = 0 \ , \quad s > a \ , \qquad (10.4.29)$$

where $e^*(s) = s^{-1}A_1^{-1}\{re(r);s\}$, $f^*(s) = A_1^{-1}\{w_0(a)+f(r);s\}$. Writing

$$F_s[T(\xi);s] = \psi(s) \ , \quad F_c[P(\xi);s] = \phi(s) \ , \quad 0 \le s < a \ , \qquad (10.4.30)$$

extending $\psi(s)$, $\phi(s)$ to negative s by the equations

$$\psi(-s) = -\psi(s) , \quad \phi(-s) = \phi(s) , \tag{10.4.31}$$

and using equation (5.5.12) and Example 5.5.1 we find

$$\vartheta\{\psi(s) + \varphi h[\phi(\sigma);s]\} = e^*(s) , \quad |s| < a , \tag{10.4.32}$$

$$\vartheta\{\phi(s) - \varphi h[\psi(\sigma);s]\} = f^*(s) , \quad |s| < a , \tag{10.4.33}$$

where $e^*(-s) = -e^*(s)$, $f^*(-s) = f^*(s)$. Equations (10.4.32), (10.4.33) may be combined into one complex equation for $\chi(s) = \psi(s)+i\phi(s)$, namely

$$I(s) \equiv \vartheta\left\{\chi(s) + \frac{\varphi}{\pi i}\int_{-a}^{a} \frac{\chi(\sigma)d\sigma}{\sigma-s}\right\} = e^*(s)+if^*(s) , \quad |s| < a . \tag{10.4.34}$$

Equation (10.4.34) is very similar to the corresponding equation (6.2.43) encountered in the adhesive contact problem for a punch indenting a half-plane - with two differences. The displacement, rather than its derivative, appears on the right; this is linked to the fact that in space problems, unlike plane problems, it is possible to find the absolute penetration of the punch. The second difference, that φ appears as the multiplier of the second term rather than the first, also has important consequences. It means that the equation falls into the category of Example 3.5.4, and in particular, that the fundamental solution $X_0(z)$ is non-singular at the points $\pm$ 1.

If e(r), f(r) are polynomials then $e^*(s)$, $f^*(s)$ will also be polynomials (see Example 5.7.10). In that case the solution of equation (10.4.34) follows directly from Example 3.5.4. In equation (10.4.34) put

$$x = s/a , \quad t = \sigma/a , \quad \omega(x) = \chi(s) , \tag{10.4.35}$$

$$F(x) = \frac{g(x)}{1-\varphi} = \frac{e^*(s)+if^*(s)}{\vartheta(1-\varphi)} = 2\mu\{e^*(s)+if^*(s)\} , \tag{10.4.36}$$

then

$$\Omega(x) = \frac{1}{2\pi i}\int_{-1}^{1}\frac{\omega(t)dt}{t-x} = \frac{1}{2\pi i}\int_{-a}^{a}\frac{\chi(\sigma)d\sigma}{\sigma-s} =$$

$$= \frac{1}{2\vartheta\varphi} A_1^{-1}\{rs^{-1}u(r,0)+iw(r,0);s\} , \quad s > a . \tag{10.4.37}$$

This equation shows that $\Omega(x)$ must be zero for large $|x|$, so that the A of Example 3.5.4 must be zero. Thus

$$\omega(x) = X_0^+(x)S(x) \tag{10.4.38}$$

where $S(x)$ is the polynomial conjugate to $F(x)$. The stresses in the contact region are given by

$$\tau_{rz}(r,0) = -A_2^{-1}[s^{-1}\psi(s);r] = \left(\frac{2}{\pi}\right)^{1/2} D_r\int_r^a \frac{\psi(s)ds}{(s^2-r^2)^{1/2}} , \tag{10.4.39}$$

$$r\tau_{zz}(r,0) = -A_2^{-1}[\phi(s);r] = \left(\frac{2}{\pi}\right)^{1/2} D_r\int_r^a \frac{s\phi(s)ds}{(s^2-r^2)^{1/2}} , \tag{10.4.40}$$

where $\psi(s)$, $\phi(s)$ are the real and imaginary parts of $\chi(s) \equiv \omega(x)$. This essentially completes the solution; it remains to find the unknown parameters in $e(r)$ and $w_0(a)$ so that $\tau_{rz}(a,0)$, $\tau_{zz}(a,0)$ are zero (except when the indenter is flat) as described below.

For a punch with a plane face $e(r) = 0 = f(r)$ so that $f^*(s) = A_1^{-1}\{w_0(a);s\} = (2/\pi)^{1/2}w_0(a)$ and

$$F(x) = if_0 = S(x) , \quad \text{where } f_0 = (2/\pi)^{1/2}2\mu w_0 , \tag{10.4.41}$$

$$\omega(x) = if_0X_0^+(x) = if_0[(1-x)\exp(i\pi)]^{i\alpha}(1+x)^{-i\alpha} , \tag{10.4.42}$$

$$\Omega(x) = (\kappa-1)^{-1} i f_0[1-X_0(x)] , \qquad |x| > 1 , \tag{10.4.43}$$

where $\exp(2\pi\alpha) = \kappa$.

In order to evaluate the contact stresses, put

$$\frac{1+x}{1-x} = \frac{a+s}{a-s} = \exp(2\theta) , \qquad |x| < 1 , \tag{10.4.44}$$

so that

$$\omega(x) = i f_0 \exp(-\pi\alpha)\exp(-i2\alpha\theta) , \tag{10.4.45}$$

and

$$\psi(s) = f_0\exp(-\pi\alpha)\sin 2\alpha\theta , \qquad \phi(s) = f_0\exp(-\pi\alpha)\cos 2\alpha\theta , \tag{10.4.46}$$

so that, remembering that $\exp(\pi\alpha) = \kappa^{1/2}$, we find

$$\left.\begin{aligned} \tau_{rz}(r,0) &= \frac{4\mu w_0}{\pi\kappa^{1/2}} D_r \int_r^a \frac{\sin 2\alpha\theta ds}{(s^2-r^2)^{1/2}} , \\ r\tau_{zz}(r,0) &= \frac{4\mu w_0}{\pi\kappa^{1/2}} D_r \int_r^a \frac{s\cos 2\alpha\theta ds}{(s^2-r^2)^{1/2}} , \end{aligned}\right\} \tag{10.4.47}$$

which agrees with the results found by Spence (1968a). The surface displacements outside the contact area may be found by noting that the I(s) of equation (10.4.34) is

$$I(s) = s^{-1}A_1^{-1}[ru(r,0);s] + iA_1^{-1}[w(r,0);s] = \begin{cases} F(x)/(2\mu) , & s < a , \\ 2\vartheta\varphi\Omega(x) , & s > a . \end{cases} \tag{10.4.48}$$

For the plane punch therefore

$$s^{-1}A_1^{-1}[ru(r,0);s] = \begin{cases} 0 & s < a , \\ -(2/\pi)^{1/2} w_0 \sin 2\alpha\theta , & s > a , \end{cases} \tag{10.4.49}$$

where

$$\frac{x+1}{x-1} = \frac{s+a}{s-a} = \exp(2\theta) , \qquad s > a , \qquad (10.4.50)$$

and

$$A_1^{-1}[w(r,0);s] = \begin{cases} (2/\pi)^{1/2} w_0 , & s < a , \\ (2/\pi)^{1/2} w_0(1-\cos 2\alpha\theta) , & s > a , \end{cases} \qquad (10.4.51)$$

from which $u(r,0)$ and $w(r,0)$ may be computed.

Details of the surface displacements and stress fields for indenters with plane, parabolic and conical profiles may be found in Spence (1968a); see also Gladwell (1969), Linz and Noble (1971) and Abramian (1975). A few words are in order here. For the *frictionless* contact of an indenter with plane face there is only a normal contact stress $p = -\tau_{zz}(r,0)$ and this stress has a square root singularity $(a^2-r^2)^{-1/2}$ at the boundary of the indenter. In this case there is a surface radial displacement (Example 10.2.6) which however has a term behaving like $(a^2-r^2)^{1/2}$ for $r \to a^-$. For the *adhesive* contact of an indenter with a plane face the surface radial displacement under the indenter is zero, but now there is a tangential contact stress in addition to the normal stress. Both stresses $\tau_{rz}(r,0)$, $\tau_{zz}(r,0)$ have a square root singularity $(a^2-r^2)^{-1/2}$ but now they are multiplied by terms, $\cos\gamma$, $\sin\gamma$ respectively. Here γ is a function of r which tends to infinity as $r \to a$. This means that the stresses oscillate more and more rapidly as $r \to a$. The behaviour is similar, but not identical, to that found by Muskhelishvili in the corresponding plane problem (4.6.11). For indenters in *incomplete* adhesive contact with the half-space the surface radial displacement is *not* zero, and is found by satisfying the conditions that the contact stresses should be zero at the boundary of the contact region. Spence (1968a) showed that a formal solution $w_0(r)$ for a non-polynomial shape can be found by using the Mellin transform (Example 10.4.2). Spence solved the general problem of Hertzian contact with finite friction in (1975b); see also Gubenko, Grabko, and Nakashidze (1971) and Nakashidze (1975). Nayak and Srinivasan (1974) solved the problem of a 'perfectly flexible' circular plate in adhesive contact with a half-space, and subjected to a uniform load. Here the boundary conditions are $u(r,0) = 0$, $\tau_{zz}(r,0) = -p$, $0 \leq r < a$; $\tau_{rz}(r,0) = 0 = = \tau_{zz}(r,0)$, $r > a$.

Both Mossakovskii (1963) and Spence (1968a) considered the normal adhesive contact of *two* axisymmetric elastic bodies which, in the region of contact, are so flat that each may be considered as a half-space. If axes Ox, Oy, Oz_1 and Ox, Oy, Oz_2 are taken with Oz_1, Oz_2 pointing into the two bodies, as in §2.9, then one may argue as before that, within the circle of contact $r = a$,

$$\frac{\partial}{\partial a}[u_1(r;a)-u_2(r;a)] = 0 \ , \quad w_1(r;a)+w_2(r;a) = w_0(a)+f_1(r)+f_2(r) \ , \tag{10.4.52}$$

$$\tau_{zz1}(r,0) = \tau_{zz2}(r,0) = -p(r) \ , \quad \tau_{rz1}(r,0) = -\tau_{rz2}(r,0) = -t(r) \ , \tag{10.4.53}$$

and the condition on u_1, u_2 implies that

$$u_1(r;a) - u_2(r;a) = e(r) \ , \quad 0 \leq r < a \ . \tag{10.4.54}$$

For each half-space there will be a set of four equations like (10.4.22) - (10.4.25) linking u_i, w_i, τ_{rzi}, τ_{zzi} to $T_i(\xi)$, $P_i(\xi)$. Equations (10.4.53) show that

$$P_1(\xi) = P_2(\xi) = P(\xi) \ , \quad T_1(\xi) = -T_2(\xi) = T(\xi) \ , \tag{10.4.55}$$

and then equations (10.4.52) yield

$$u_1(r,0)-u_2(r,0) \equiv \vartheta H_1[\xi^{-1}\{T(\xi)-\varphi P(\xi)\};r] = e(r) \ , \quad 0 \leq r < a \ , \tag{10.4.56}$$

$$w_1(r,0)+w_2(r,0) = \vartheta H_0[\xi^{-1}\{P(\xi)-\varphi T(\xi)\};r] = w_0(a)+f(r) \ , \quad 0 \leq r < a \ , \tag{10.4.57}$$

where now

$$\vartheta = \vartheta_1+\vartheta_2 \ , \quad \varphi = \frac{\vartheta_1\varphi_1-\vartheta_2\varphi_2}{\vartheta_1+\vartheta_2} \ , \quad f(r) = f_1(r)+f_2(r) \ . \tag{10.4.58}$$

These equations are now identical to equations (10.4.22), (10.4.23). Note that φ is the β of equation (4.10.13), so that, when the materials of the two half-spaces are identical, $\beta = 0$ and equations (10.4.56), (10.4.57) uncouple.

The formulation of problems 3(a) and 4(a) may be carried out by using the analysis of §10.3. For an indenter with a plane face the boundary conditions for the combined problem in which the entire contact area is shifted by an amount d in the x-direction and rotated through an angle γ about the y-axis, while the indenter completely adheres to the half-space, are as follows:

$$u_x(r,0) = d \ , \ u_y(r,0) = 0 \ , \quad u_z(r,0) = \gamma r\cos\theta \ , \quad 0 \leq r < a \ , \tag{10.4.59}$$

$$\tau_{rz}(r,0) = 0 = \tau_{\theta z}(r,0) = \tau_{zz}(r,0) \ , \quad r > a \ , \tag{10.4.60}$$

which may be written in the notation of (10.3.20), (10.3.21) as

$$u^*(r,0) = d \ , \quad v^*(r,0) = 0 \ , \quad w^* = \gamma r \ , \quad 0 \leq r < a \ , \tag{10.4.61}$$

$$s(r,0) = 0 = t(r,0) = p(r,0) \ , \qquad r > a \ , \tag{10.4.62}$$

and then by using Example 10.3.1 as

$$2\mu u^*(r,0) \equiv H_0[\xi^{-1}\{(2-\nu)S(\xi)+(1-2\nu)P(\xi)+\nu T(\xi)\};r] = 2\mu d \ , \quad 0 \leq r < a \ , \tag{10.4.63}$$

$$2\mu w^*(r,0) \equiv H_1[\xi^{-1}\{(1-2\nu)S(\xi)+4(1-\nu)P(\xi)-(1-2\nu)T(\xi)\};r] = 2\mu\gamma r \ , \quad 0 \leq r < a \ , \tag{10.4.64}$$

$$2\mu v^*(r,0) \equiv H_2[\xi^{-1}\{\nu S(\xi)-(1-2\nu)P(\xi)+(2-\nu)T(\xi)\};r] = 0 , \quad 0 \le r < a , \tag{10.4.65}$$

$$s(r,0) \equiv H_0[S(\xi);r] = 0 , \quad r > a , \tag{10.4.66}$$

$$p(r,0) \equiv H_1[P(\xi);r] = 0 , \quad r > a , \tag{10.4.67}$$

$$t(r,0) \equiv H_2[T(\xi);r] = 0 , \quad r > a . \tag{10.4.68}$$

Now by using entries 1, 2, 3, 10, 11, 12 respectively of Table 5.7.1, we obtain

$$2\mu A_1^{-1}[u^*(r);\rho] \equiv F_c[\{(2-\nu)S(\xi)+(1-2\nu)P(\xi)+\nu T(\xi)\};\rho] = (2/\pi)^{1/2}2\mu d , \quad 0 \le \rho < a , \tag{10.4.69}$$

$$2\mu\rho^{-1}A_1^{-1}[rw^*(r);\rho] \equiv F_s[\{(1-2\nu)S(\xi)+4(1-\nu)P(\xi)-(1-2\nu)T(\xi)\};\rho] =$$

$$= 2(2/\pi)^{1/2}2\mu\gamma\rho , \quad 0 \le \rho < a , \tag{10.4.70}$$

$$2\mu\rho A_1[r^{-2}D_r r^2 v^*(r);\rho] \equiv -F_c[\{\nu S(\xi)-(1-2\nu)P(\xi)+(2-\nu)T(\xi)\};\rho] =$$

$$= -(2/\pi)^{1/2}2\mu c , \quad 0 \le \rho < a , \tag{10.4.71}$$

$$A_2[rs(r);\rho] \equiv F_c[S(\xi);r] = 0 , \quad \rho > a , \tag{10.4.72}$$

$$\rho A_2[p(r);\rho] \equiv F_s[P(\xi);r] = 0 , \quad \rho > a , \tag{10.4.73}$$

$$\int_\rho^\infty uA_2\{r^{-2}D_r r^2 t(r);u\}du \equiv F_c[T(\xi);\rho] = 0 , \quad \rho > a , \tag{10.4.74}$$

where c is a constant to be determined. Now put

$$F_c[S(\xi);\rho] = \psi_1(\rho) , \quad F_c[T(\xi);\rho] = \psi_2(\rho) , \quad 0 \le \rho < a , \tag{10.4.75}$$

$$F_s[P(\xi);\rho] = \phi(\rho) , \quad 0 \le \rho < a , \tag{10.4.76}$$

where $\psi_1(\rho)$, $\psi_2(\rho)$ are even and $\phi(\rho)$ is odd in $-a < \rho < a$, and use equation (5.5.12) and Example 5.5.1 to obtain

$$(2-\nu)\psi_1(\rho)+(1-2\nu)h[\phi(\sigma);\rho]+\nu\psi_2(\rho) = (2/\pi)^{1/2}2\mu d \ , \tag{10.4.77}$$

$$-(1-2\nu)h[\psi_1(\sigma);\rho]+4(1-\nu)\phi(\rho)+(1-2\nu)h[\psi_2(\sigma);\rho] = 2(2/\pi)^{1/2}2\mu\gamma\rho \ , \tag{10.4.78}$$

$$-\nu\psi_1(\rho)+(1-2\nu)h[\phi(\sigma);\rho]-(2-\nu)\psi_2(\rho) = -(2/\pi)^{1/2}2\mu c \ , \tag{10.4.79}$$

where h is the Hilbert transform defined in equation (5.5.6). Subtracting equation (10.4.79) from equation (10.4.77) we obtain

$$2[\psi_1(\rho)+\psi_2(\rho)] = (2/\pi)^{1/2}2\mu(d+c) \ , \tag{10.4.80}$$

and putting

$$2\psi(\rho) = \psi_1(\rho)-\psi_2(\rho) \tag{10.4.81}$$

we may write (10.4.77) + (10.4.79) and (10.4.78) in the form

$$\vartheta\{\psi(\rho)+\varphi h[\phi(\sigma);\rho]\} = (2/\pi)^{1/2}(d-c)/2 \ , \quad |\rho| < a \ , \tag{10.4.82}$$

$$\vartheta\{\phi(\rho)-\varphi h[\psi(\sigma);\rho]\} = (2/\pi)^{1/2}\gamma\rho, \quad |\rho| < a \ , \tag{10.4.83}$$

which have exactly the same form as equations (10.4.32), (10.4.33). The evaluation of the constant c and the solution of these equations is the subject of Example 10.4.5.

If the total force on the indenter in the x-direction is T and the moment about the y-axis is M then it is found that

$$16\mu a^2 d = (2+\beta+3\omega\alpha)aT+3\omega M \ , \tag{10.4.84}$$

$$16\mu a^2\gamma = 3\omega T+3\omega M/(\alpha a) \ , \tag{10.4.85}$$

where

$$\exp(2\pi\alpha) = \kappa \ , \quad \beta = (1-2\nu)/(\pi\alpha) \ , \quad \omega = \alpha\beta/(1+\alpha^2) \ . \qquad (10.4.86)$$

This agrees with the result obtained by Ufliand (1956) except that he has the equivalent (in the present notation) of 2β, instead of β, in the middle term of (10.4.84). The error can be traced to an arithmetical mistake in his equation (4.19).

A partial check of equations (10.4.84), (10.4.85) can be provided by taking $\nu = 1/2$; then

$$\alpha \to 0 \ , \quad \beta \to 1 \ , \quad \omega \to 0 \ , \quad \omega/\alpha \to 1 \qquad (10.4.87)$$

so that

$$16\mu a^2 d = 3aT \ , \quad 16\mu a^2\gamma = 3M/a \ , \qquad (10.4.88)$$

which agrees, as it should, with the results for frictionless contact, given in Example 10.3.2 and Example 10.3.4. A convenient compendium of these and other contact impedances is given in Gladwell (1968).

Examples 10.4

1. Show that equation (10.4.9) has the solution

$$f(r) = -Br^n \ , \quad w_0(r) = Ar^n \ , \quad \text{where}$$
$$B = A\left[1-n\int_0^1 x^{n-1}\phi(1/x)dx\right].$$

2. If $f^*(s) = M[f(r);s]$, $w_0^*(s) = M[w_0(r);s]$ show that equation (10.4.9) yields the relation $f^*(s) = -[1+s\phi^*(s)]w_0^*(s)$.
3. Show that in problem 2(a) the ratio of total load P_0 to displacement w_0 for the adhesive indenter with plane face is $P_0/w_0 = 16\pi\mu a\alpha/(\kappa-1)$. The integral involved is

$$\int_0^\infty \cos 2\alpha\theta \operatorname{sech}^2\theta d\theta = \pi\alpha \operatorname{cosech}(\pi\alpha) \ ;$$

see Gradshteyn and Ryzhik (1965, 3.9821). The quantity $q = aw_0/(\vartheta P_0)$ was defined in Example 2.6.2. Compute this quantity for the adhesive plane punch in problem 2(a) for $\nu = 0, 0.1, 0.5$, and show that the value for $\nu = 0.5$ is equal to that for the frictionless punch of Example 2.6.2.

4. For a punch with parabolic face the normal and radial displacements are given by equations (10.4.11), (10.4.13) as

$$w(r,0) = Aa^2 - Br^2 \ , \quad u(r,0) = Cr^2 \ , \quad 0 \le r < a$$

where $B = 1/(2R)$ and R is the radius of curvature of the punch. Show that the conditions $\tau_{rz}(a,0) = 0 = \tau_{zz}(a,0)$ lead to two simultaneous equations for A, C, the solution of which is $A = 1/[I(\alpha)R]$, $C = -16\alpha^2\kappa^{1/2}/[3(\kappa-1)I(\alpha)R]$, where

$$I(\alpha) = 1 - 4\alpha^2 \operatorname{cosec}(\pi\alpha) \int_0^{\pi/2} \cot(t)\sinh(2\alpha t)dt \ .$$

Show that the total normal load P_0 is

$$P_0 = \frac{32\pi\mu a^3\alpha}{3R(\kappa-1)}$$

and that this equals the value for frictionless contact when $\nu = 0.5$.

5. Show that the solution of equations (10.4.82)-(10.4.83) is

$$\psi(\rho) + i\phi(\rho) = X_0^+(x)G(x) \ ,$$

where $x = \rho/a$, $X_0^+(x)$ is given by (10.4.42), $G(x)$ is the polynomial conjugate to $F(x)$ and

$$F(x) = \left(\frac{2}{\pi}\right)^{1/2} \frac{1}{\vartheta} \left\{\frac{d-c}{2} + i\gamma a x\right\} \ .$$

Show that the shear stress $t(r)$ for $r > a$ is given by

$$r^2 t(r) = 2\left(\frac{2}{\pi}\right)^{1/2} \int_0^a \psi_2(\rho)d\rho = 0 .$$

Now by using the integral in Example 10.4.3 show that this leads to

$$c(2+\beta) = d(2-\beta) - 4\gamma a\alpha ,$$

and hence deduce equations (10.4.84), (10.4.85). [Hint: use

$$r \frac{d}{dr} (r^{-1} J_1(\xi r)) = -\xi J_2(\xi r) \text{ and}$$

$$\int_0^\infty J_1(\xi r)\cos(\xi\rho)d\xi = r^{-1} \text{ if } r > \rho .]$$

10.5 *The use of oblate spheroidal coordinates*

It was shown in §2.6 that a normal pressure distributed over a circle $r \leq a$ on the surface of the half-space $z \geq 0$ according to the form $(1-r^2/a^2)^{-1/2} P_{2n}[(1-r^2/a^2)^{1/2}]$ produced within the region $r \leq a$, $z = 0$ a displacement proportional to $P_{2n}[(1-r^2/a^2)^{1/2}]$. In this section we show how this result may be derived by the use of separated solutions in oblate spheroidal coordinates, and consider some generalizations of it.

The definitive treatment of the background material of this section may be found in Hobson (1931); reference may also be made to Chapter 10 of Morse and Feshbach (1953) or Whittaker and Watson (1952). The use of separated solutions has its roots in the work of Inglis (1913) and Griffith (1920) in crack problems. Their explicit use in elasticity problems is to be found in Love (1939), Lur'e (1941), Reissner and Sagoci (1944) and Sack (1946). The axisymmetric (i.e., m = 0) cases of the interior punch and crack problems below are given by Payne (1953). Hobson (1931) and Payne (1953) also consider solutions in toroidal coordinates, in which the problem of the spherical punch is easily solved; these solutions may be found in many papers of Ufliand, and are discussed in full in Ufliand (1965, part IX). An illuminating discussion of the relation between the separated solutions, and those used by Green and Zerna (1954) and Collins (1959a) is to be found in England and Shail (1977).

The oblate spheroidal coordinates ξ, η, θ are related to cylindrical coordinates r, z, θ by the equations (Example 10.5.1)

$$r = a[(1+\xi^2)(1-\eta^2)]^{1/2} , \quad z = a\xi\eta , \tag{10.5.1}$$

where $-1 \leq \eta \leq 1$, $\xi \geq 0$. The surfaces ξ = const and η = const are the ellipsoids and hyperboloids (of one sheet)

$$\frac{r^2}{a^2(1+\xi^2)} + \frac{z^2}{a^2\xi^2} = 1 , \quad \frac{r^2}{a^2(1-\eta^2)} - \frac{z^2}{a^2\eta^2} = 1 , \tag{10.5.2}$$

respectively; the surfaces $\xi = 0$ and $\eta = 0$ are the interior and exterior of the circle $r = a$, $z = 0$ respectively.

Laplace's equation in oblate spheroidal coordinates is

$$\frac{\partial}{\partial\xi}\left[(1+\xi^2)\frac{\partial\psi}{\partial\xi}\right] + \frac{\partial}{\partial\eta}\left[(1-\eta^2)\frac{\partial\psi}{\partial\eta}\right] + \frac{(\xi^2+\eta^2)}{(1+\xi^2)(1-\eta^2)}\frac{\partial^2\psi}{\partial\theta^2} = 0 , \tag{10.5.3}$$

which has separable solutions

$$\psi(\xi,\eta,\theta) = \Xi(\xi)H(\eta)\Theta(\theta) , \tag{10.5.4}$$

where Ξ, H, Θ satisfy

$$\frac{d}{d\xi}\left[(1+\xi^2)\frac{d\Xi}{d\xi}\right] - \left[n(n+1) - \frac{m^2}{1+\xi^2}\right]\Xi = 0 , \tag{10.5.5}$$

$$\frac{d}{d\eta}\left[(1-\eta^2)\frac{dH}{d\eta}\right] + \left[n(n+1) - \frac{m^2}{1-\eta^2}\right]H = 0 , \tag{10.5.6}$$

$$\frac{d^2\Theta}{d\theta^2} = -m^2\Theta . \tag{10.5.7}$$

The required periodicity of $\Theta(\theta)$ and the convergence of the solution $\Xi(\xi)$, $H(\eta)$ require that m, n be integral so that

$$\psi(\xi,\eta,\theta) = \begin{Bmatrix} P_n^m(i\xi) \\ \\ Q_n^m(i\xi) \end{Bmatrix} \cdot \begin{Bmatrix} P_n^m(\eta) \\ \\ Q_n^m(\eta) \end{Bmatrix} \cdot \begin{Bmatrix} \cos m\theta \\ \\ \sin m\theta \end{Bmatrix} \tag{10.5.8}$$

where P_n^m, Q_n^m are the Legendre functions of the first and second kinds respectively, and $m \le n$ (Hobson (1931)).

If the tractions τ_{xz}, τ_{yz} are zero on $z = 0$, then the Papkovich-Neuber solution of the static elasticity equations may be expressed in terms of a single potential $\psi_3 \equiv \psi$ related to ϕ by equation (2.5.10), namely

$$(1-2\nu)\psi = \frac{\partial \phi}{\partial z}\ . \tag{10.5.9}$$

The normal displacement and normal stress may be written

$$2\mu w = 2(1-\nu)\psi - z\,\frac{\partial \psi}{\partial z}\ , \qquad \tau_{zz} = \frac{\partial \psi}{\partial z} - z\,\frac{\partial^2 \psi}{\partial z^2}\ , \tag{10.5.10}$$

so that, in particular, on $z = 0$,

$$w(x,y,0) = \vartheta\psi\ , \qquad p(x,y,0) \equiv -\tau_{zz}(x,y,0) = -\frac{\partial \psi}{\partial z}\ . \tag{10.5.11}$$

There are four basic mixed boundary value problems relating to w,p and the circle $r = a$; they are

interior punch (i.p.) w given $0 \le r < a$; $p = 0$, $r > a$, (10.5.12)

exterior punch (e.p.) $p = 0$, $0 \le r < a$; w given $r > a$, (10.5.13)

interior crack (i.c.) p given $0 \le r < a$; $w = 0$, $r > a$, (10.5.14)

exterior crack (e.c.) $w = 0$, $0 \le r < a$; p given $r > a$. (10.5.15)

The latter two relate to a crack produced between two half-spaces by the action of pressure (Sneddon and Lowengrub (1969)); solutions to all four problems will be needed for the approximate analysis of annular problems in §10.7.

Now consider which of the combinations shown in equation (10.5.8) is eligible to represent ψ. $P_n^m(i\xi)$ is unbounded at $\xi = \infty$ ($r = \infty$, $z = 0$), while $Q_n^m(\eta)$ has logarithmic singularities at $\eta = \pm 1$ ($r = 0$, $z = 0$); both of these are therefore excluded. This leaves the combination

$$\psi(\xi,\eta,\theta) = b_{n,m}Q_n^m(i\xi)P_n^m(\eta)\cos m(\theta+\gamma) \ . \qquad (10.5.16)$$

Equations (10.5.1) show that

$$\frac{\partial}{\partial z} = \frac{1}{a(\xi^2+\eta^2)}\left\{\eta(1+\xi^2)\frac{\partial}{\partial\xi} + \xi(1-\eta^2)\frac{\partial}{\partial\eta}\right\} , \qquad (10.5.17)$$

so that on $z = 0$

$$\frac{\partial}{\partial z} = \frac{1}{a\eta}\frac{\partial}{\partial\xi} , \text{ for } \xi = 0 \ (r < a) \ ; \ \frac{\partial}{\partial z} = \frac{1}{a\xi}\frac{\partial}{\partial\eta} \text{ for } \eta = 0 \ (r > a) , \qquad (10.5.18)$$

and the w, p on $z = 0$ appropriate to problem i.p. are

$$w = \begin{cases} \vartheta b_{n,m}Q_n^m(i0)P_n^m(\eta)\cos m(\theta+\gamma) , & 0 \le r \le a , \\ \vartheta b_{n,m}Q_n^m(i\xi)P_n^m(0)\cos m(\theta+\gamma) , & r > a , \end{cases} \qquad (10.5.19)$$

$$p = \begin{cases} -\dfrac{b_{n,m}}{a\eta}\dfrac{d}{d\xi}Q_n^m(i0)P_n^m(\eta)\cos m(\theta+\gamma) , & 0 \le r < a , \\ -\dfrac{b_{n,m}}{a\xi}Q_n^m(i\xi)\dfrac{d}{d\eta}P_n^m(0)\cos m(\theta+\gamma) , & r > a . \end{cases} \qquad (10.5.20)$$

The pressure will be zero for $r > a$ if $d\,P_n^m(\eta)/d\eta|_0 = 0$, i.e., if m, n have the same parity. Noting that $\eta = (1-r^2/a^2)^{1/2}$ if $\xi = 0$, and $\xi = (r^2/a^2-1)^{1/2}$ if $\eta = 0$ we find that

$$w = \begin{cases} \vartheta b_{n,m} Q_n^m(i0) P_n^m[(1-r^2/a^2)^{1/2}] \cos m(\theta+\gamma) \ , & 0 \le r \le a \ , \\ \vartheta b_{n,m} Q_n^m(i(r^2/a^2-1)^{1/2}] P_n^m(0) \cos m(\theta+\gamma) \ , & r > a \ , \end{cases} \tag{10.5.21}$$

$$p = -b_{n,m}(a^2-r^2)^{-1/2} \frac{d}{d\xi} Q_n^m(i0) P_n^m[(1-r^2/a^2)^{1/2}] \cos m(\theta+\gamma) \ , \quad 0 \le r < a \ . \tag{10.5.22}$$

When m, n have the same parity, the formulae of Gradshteyn and Ryzhik (1965, 8·73-8·75) show that

$$Q_n^m(i0) = -\frac{\pi}{2} i \exp(-mi\pi/2) P_n^m(0) \ , \quad \frac{d}{d\xi} Q_n^m(i0) = i \exp(-mi\pi/2) \frac{(n+m)!}{(n-m)!} \cdot \frac{1}{P_n^m(} \tag{10.5.23}$$

so that, on writing

$$-i \exp(-mi\pi/2) \frac{(n+m)!}{(n-m)!} \frac{b_{n,m}}{P_n^m(0)} = a a_{n,m} \ , \tag{10.5.24}$$

we find

$$w = \vartheta \frac{\pi}{2} a a_{n,m} [P_n^m(0)]^2 \frac{(n-m)!}{(n+m)!} \cos m(\theta+\gamma) \mathcal{W}(r) \tag{10.5.25}$$

where

$$\mathcal{W} = \begin{cases} P_n^m[(1-r^2/a^2)^{1/2}] & 0 \le r \le a \ , \\ \frac{2i}{\pi} \exp(mi\pi/2) Q_n^m[i(r^2/a^2-1)^{1/2}] \ , & r > a \ , \end{cases} \tag{10.5.26}$$

and

$$p(r) = a_{n,m}(1-r^2/a^2)^{-1/2} P_n^m[(1-r^2/a^2)^{1/2}] \cos m(\theta+\gamma) \ . \tag{10.5.27}$$

When m = 0, and n, even, is replaced by 2n, these results agree with equations (2.6.15), (2.6.16).

The corresponding results for the interior crack, i.c., are given in Example 10.5.2.

The use of oblate spheroidal coordinates

In order to find solutions corresponding to the external punch and crack problems we use Kelvin's theorem (Hobson 1931, p. 120) which states that if $V(x,y,z)$ is a solution of Laplace's equation in Cartesian coordinates, and $R^2 = x^2+y^2+z^2$, then $R^{-1}V[a^2x/R^2,\ a^2y/R^2,\ a^2z/R^2]$ is another solution. Now $R^2 = r^2+z^2 = a^2(1+\xi^2-\eta^2)$, so that, starting from the solution $V(x,y,z) = Q_n^m(i\xi)P_n^m(\eta)\cos m(\theta+\gamma)$, we find the solution (Example 10.5.3)

$$\psi(\xi,\eta,\theta) = b_{n,m}(1+\xi^2-\eta^2)^{-1/2}P_n^m[\xi/(1+\xi^2-\eta^2)^{1/2}]\,\cdot$$
$$Q_n^m[i\eta/(1+\xi^2-\eta^2)^{1/2}]\cos m(\theta+\gamma)\ . \qquad (10.5.28)$$

Write

$$\sigma = \frac{\xi}{(1+\xi^2-\eta^2)^{1/2}}\ , \qquad \tau = \frac{\eta}{(1+\xi^2-\eta^2)^{1/2}}\ , \qquad (10.5.29)$$

so that $(1-\sigma^2+\tau^2) = 1/(1+\xi^2-\eta^2)$; then

$$\psi(\xi,\eta,\theta) = b_{n,m}(1-\sigma^2+\tau^2)^{1/2}P_n^m(\sigma)Q_n^m(i\tau)\cos m(\theta+\gamma)\ . \qquad (10.5.30)$$

Consider the external crack problem (Lowengrub and Sneddon (1965)). The z-derivative is (Example 10.5.4)

$$\frac{\partial}{\partial z} = \frac{1+\tau^2}{a\tau}\frac{\partial}{\partial\sigma}\ ,\ \text{for } \sigma = 0\ (r < a)\ ;\ \frac{\partial}{\partial z} = \frac{1-\sigma^2}{a\sigma}\frac{\partial}{\partial\tau}\ ,\ \text{for } \tau = 0\ (r > a)\ , \qquad (10.5.31)$$

so that the potential (10.5.28) leads to

$$w = \begin{cases} \vartheta b_{n,m}(1+\tau^2)^{1/2}P_n^m(0)Q_n^m(i\tau)\cos m(\theta+\gamma)\ , & 0 \le r < a\ , \\ \vartheta b_{n,m}(1-\sigma^2)^{1/2}P_n^m(\sigma)Q_n^m(i0)\cos m(\theta+\gamma)\ , & r > a\ , \end{cases} \qquad (10.5.32)$$

$$p = \begin{cases} -b_{n,m} \dfrac{(1+\tau^2)^{3/2}}{a\tau} \dfrac{d}{d\sigma} P_n^m(0) Q_n^m(i\tau)\cos m(\theta+\gamma) , & 0 \le r < a , \\ -b_{n,m} \dfrac{(1-\sigma^2)^{3/2}}{a\sigma} P_n^m(\sigma) \dfrac{d}{d\tau} Q_n^m(i0)\cos m(\theta+\gamma) , & r > a . \end{cases} \tag{10.5.33}$$

The displacement will be zero for $r < a$ if $P_n^m(0) = 0$, i.e., if m,n have the opposite parity. Noting that, in this case,

$$\frac{d}{d\sigma} P_n^m(\sigma)\Big|_0 = -P_n^{m+1}(0) , \quad \frac{d}{d\tau} Q_n^m(i\tau)\Big|_0 = -\frac{\pi}{2}\exp(-im\pi/2)P_n^{m+1}(0) , \tag{10.5.34}$$

$$Q_n^m(i0) = \exp(-im\pi/2)\frac{(n+m)!}{(n-m)!}\frac{1}{P_n^{m+1}(0)} , \tag{10.5.35}$$

and writing

$$\exp[-im\pi/2]\frac{(n+m)!}{(n-m)!}\frac{b_{n,m}}{P_n^{m+1}(0)} = aa_{n,m} , \tag{10.5.36}$$

we find

$$w = \vartheta \frac{a^2}{r} a_{n,m} P_n^m[(1-a^2/r^2)^{1/2}]\cos m(\theta+\gamma) , \quad r > a , \tag{10.5.37}$$

$$p = \frac{\pi}{2} a_{n,m}[P_n^{m+1}(0)]^2 \frac{(n-m)!}{(n+m)!} \cos m(\theta+\gamma)P(r) , \tag{10.5.38}$$

$$P(r) = \begin{cases} \dfrac{2a^3}{\pi r^2(a^2-r^2)^{1/2}} \exp(im\pi/2)Q_n^m[i(a^2/r^2-1)^{1/2}] , & 0 < r < a , \\ \dfrac{a^3}{r^2(r^2-a^2)^{1/2}} P_n^m[(1-a^2/r^2)^{1/2}] , & r > a . \end{cases} \tag{10.5.39}$$

The corresponding results for the exterior punch problem are given in Example 10.5.5. It should be noted that, for each of the four problems, the quantities $w(x,y,z)$ and $p(x,y,z)$ off the plane $z = 0$ are given explicitly by equation (10.5.16), while the remaining displacements and stresses involve the potential ϕ of equation (10.5.9) which is an integral of ψ.

The use of oblate spheroidal coordinates

The results presented in this section do not provide complete solutions to the four punch and crack problems listed. Instead, the results provide a family of solutions of comparatively simple form which may be used to build up solutions to the problems in some cases. The restrictions on the form of the solutions are exemplified by equations (10.5.26) - (10.5.27) for the interior punch. The restriction is basically that m,n have the same parity. In practice however this restriction is not important since if the punch, in contact over $r \le a$, has a shape $z = f(x,y)$, where f is a polynomial, then the prescribed normal displacement over $r \le a$ is

$$w(x,y,0) = d+f(r\cos\theta, r\sin\theta) , \qquad 0 \le r \le a \tag{10.5.40}$$

which may be expressed in the form

$$w(x,y,0) = \sum c_{n,m} P_n^m[(1-r^2/a^2)^{1/2}]\cos m(\theta+\gamma) \tag{10.5.41}$$

where m,n have the same parity. Note that, when m,n have the same parity, $P_n^m[(1-r^2/a^2)^{1/2}]$ is a polynomial in r/a with the same parity as m,n. This provides a simple proof of Galin's theorem (Galin, (1961)) that the pressure distributed over $r \le a$ required to produce a polynomial displacement $w(x,y,0)$ on $r \le a$ has the form

$$(1-r^2/a^2)^{-1/2} \times (\text{polynomial in } x,y) . \tag{10.5.42}$$

Galin did not show how to find the coefficients of the pressure polynomial from those of the displacement polynomial; such an explicit relation is provided by equations (10.5.26), (10.5.27). Some extensions of Galin's theorem are given in §12.1, §12.2.

Examples 10.5

1. Show that an alternative form of equations (10.5.1) is provided by

$x = a\cosh\alpha\sin\beta\cos\theta$, $y = a\cosh\alpha\sin\beta\sin\theta$, $z = a\sinh\alpha\cos\beta$, or $z+ir = a\sinh(\alpha+i\beta)$.

2. Show that when m, n have the opposite parity, equation (10.5.16) provides a solution of the interior crack problem. Show that if $b_{n,m}$ is chosen according to equation (10.5.36) then

$$w = \vartheta a a_{n,m} P_n^m[(1-r^2/a^2)^{1/2}]\cos m(\theta+\gamma) ,$$

and p is given by (10.5.38), where now

$$P(r) = \begin{cases} (1-r^2/a^2)^{-1/2} P_n^m[(1-r^2/a^2)^{1/2}] , & 0 \le r < a , \\ (2/\pi)\exp(in\pi/2)(r^2/a^2-1)^{-1/2} Q_n^m[i(r^2/a^2-1)^{1/2}] , & r > a . \end{cases}$$

3. Show that τ, σ of equation (10.5.29) are the oblate spheroidal coordinates ξ, η of the point a^2x/R^2, a^2y/R^2, a^2z/R^2 inverse to (x,y,z).

4. Show that

$$\frac{\partial}{\partial z} = \frac{1}{a(\sigma^2+\tau^2)} \left\{ \tau(1-\sigma^2)(1+\sigma^2+\tau^2)\frac{\partial}{\partial\sigma} + \sigma(1+\tau^2)(1-\sigma^2-\tau^2)\frac{\partial}{\partial\tau} \right\} .$$

5. Show than, when m, n have the same parity, equation (10.5.28) provides a solution for the exterior punch problem (10.5.13). If $b_{n,m}$ is chosen according to (10.5.24) then w is given by (10.5.25) where

$$W(r) = \begin{cases} (2i/\pi)(a/r)\exp(im\pi/2) Q_n^m[i(a^2/r^2-1)^{1/2}] , & 0 < r \le a , \\ (a/r) P_n^m[(1-a^2/r^2)^{1/2}] , & r > a , \end{cases}$$

$$p(r) = -a_{n,m}(a/r)^3(1-a^2/r^2)^{-1/2} P_n^m[(1-a^2/r^2)^{1/2}]\cos m(\theta+\gamma) , \quad r > a .$$

6. Show that the appropriate potential ψ for the Reissner-Sagoci problem of equations (10.2.45) et seq. is

$\psi(r,z) = Q_n^1(i\xi)P_n^1(\eta)\cos(\theta+\gamma)$

where n is *odd*. Hence show that the surface displacements for the boundary conditions (10.2.50) are as given in equation (10.2.61).

7. Show that the normal displacement w(r) due to the Hertzian contact pressure

$$p(r) = \frac{3P}{2\pi a^2}\left(1 - \frac{r^2}{a^2}\right)^{1/2}$$

is $w_i(r) = 3\vartheta P(2a^2-r^2)/(16a^3)$ for $r < a$ and

$$w_e(r) = \frac{3\vartheta P}{8\pi a}\left[\left(2 - \frac{r^2}{a^2}\right)\sin^{-1}\left(\frac{a}{r}\right) + \frac{r}{a}\left(1 - \frac{a^2}{r^2}\right)^{1/2}\right], \qquad r > a .$$

Show that if $w_i(r)$ is defined, as above, for $r > a$, then $w_e(r) > w_i(r)$ for $r > a$, with equality at $r = a$. This proves the 'separation' condition for Hertzian contact in this case.

10.6 *Some other integral representations for axisymmetric problems*

The Papkovich-Neuber solution of the equations of elasticity appropriate for axisymmetric problems is (equation (10.2.2))

$$2\mu u = -\frac{\partial}{\partial r}(z\psi+\phi) , \qquad 2\mu w = 4(1-\nu)\psi - \frac{\partial}{\partial z}(z\psi+\phi) . \qquad (10.6.1)$$

Here $u \equiv u_\theta$, $w = u_z$ in cylindrical polar coordinates r, θ, z and ψ, φ are harmonic. The analysis of §10.2 was based on the representation

$$\psi(r,z) = H_0[\xi^{-1}A(\xi)\exp(-\xi z);\xi \to r] , \qquad (10.6.2)$$

$$\phi(r,z) = H_0[\xi^{-2}C(\xi)\exp(-\xi z);\xi \to r] . \qquad (10.6.3)$$

From these equations one may derive a number of integral representations which appear in the literature: the analysis is formal. Some of the results appear in Sneddon (1966a, p. 77f).

Consider first the simplest problem, the frictionless interior punch problem (i.p.) for a half-space. It was shown in Example 10.2.11 that this is equivalent to the problem of finding $\psi(r,z)$, harmonic for $z \geq 0$, such that

$$\psi(r,0) = f(r) , \quad r \leq a , \tag{10.6.4}$$

$$\frac{\partial\psi}{\partial z}(r,0) = 0 , \quad r > a , \tag{10.6.5}$$

and with the representation (10.6.2) this gives

$$H_0[\xi^{-1}A(\xi);r] = f(r) , \quad 0 \leq r \leq a ; \quad H_0[A(\xi);r] = 0 , \quad r > a . \tag{10.6.6}$$

Now apply operators 1, 10 respectively of Table 5.7.1 to the equations to obtain

$$F_c[A(\xi);x] = F(x) , \quad 0 \leq x \leq a ; \quad F_c[A(\xi);x] = 0 , \quad x > a , \tag{10.6.7}$$

where $F(x) = A_1^{-1}[f(r);x]$. Thus

$$A(\xi) = F_c[F(x);\xi] = \left(\frac{2}{\pi}\right)^{1/2} \int_0^a F(x)\cos(x\xi)dx , \tag{10.6.8}$$

so that

$$\psi(r,z) = \left(\frac{2}{\pi}\right)^{1/2} \int_0^a F(x)\left\{\int_0^\infty J_0(r\xi)\exp(-\xi z)\cos x\xi d\xi\right\} dx . \tag{10.6.9}$$

Now

$$\int_0^\infty \exp(-p\xi)J_0(r\xi)d\xi = (p^2+r^2)^{-1/2} , \quad \mathrm{Re}(p) > 0 , \tag{10.6.10}$$

so that, on defining g(x) as an *even* function in [-a,a] we find

$$\psi(r,z) = \frac{1}{2}\int_{-a}^{a} \frac{g(x)dx}{[r^2+(z+ix)^2]^{1/2}} , \quad g(x) = \left(\frac{2}{\pi}\right)^{1/2} F(x) . \tag{10.6.11}$$

On z = 0 the root is interpreted as

$$[r^2+(z+ix)^2]^{1/2} = \begin{cases} (r^2-x^2)^{1/2} , & r > |x| , \\ i(x^2-r^2)^{1/2}, & r < |x| . \end{cases} \tag{10.6.12}$$

This representation and others like it have their origins in papers of A.E.H. Love (1939), Copson (1947), Green (1949) and Rostovtsev (1953), (1955), (1957). An in-depth study of the convergence of the integrals appears in E.R. Love (1949); the representation is discussed in Green and Zerna (1954, p. 173 ff) and has been extensively used by Collins (1959a, 1959b), (1961), (1963), Keer (1964c), (1967). See also Example 10.6.1, 10.6.2.

The second class of representations is obtained by operating on (10.6.2) by one of the entries in Table 5.7.1; choosing entry 10 we find

$$A_1^{-1}[\psi(r,z);r \to x] = F_c[A(\xi)\exp(-\xi z);\xi \to x] = \Psi(x,z) . \tag{10.6.13}$$

The function $\Psi(x,z)$ is harmonic in the upper half-plane $z \geq 0$ and, being a cosine transform, is symmetric in x. This equation allows one to construct a plane harmonic function from an axisymmetric harmonic function, and vice-versa. Now suppose that the potential $\psi(r,z)$ corresponds to an axisymmetric half-space $z \geq 0$ subject only to normal tractions on the boundary z = 0; the displacement w and stress τ_{zz} are given by Example 10.2.11 as

$$\begin{aligned} 2\mu w(r,z) &= 2(1-\nu)\psi(r,z) - z\frac{\partial\psi}{\partial z}(r,z) , \\ \tau_{zz}(r,z) &= \frac{\partial\psi}{\partial z}(r,z) - z\frac{\partial^2\psi}{\partial z^2}(r,z) . \end{aligned} \tag{10.6.14}$$

The potential $\Psi(x,z)$ may be related to a plane state of stress in the half-plane $z \geq 0$ of the x, z plane again subject to normal tractions only on the boundary. Equations (2.1.1), (2.1.6), (2.1.8) show that the displacement w^p and stress τ^p_{zz} corresponding to this potential are

$$2\mu w^p(x,z) = 2(1-\nu)\Psi(x,z) - z\frac{\partial\Psi}{\partial z}(x,z) ,$$

$$\tau^p_{zz}(x,z) = \frac{\partial\Psi}{\partial z}(x,z) - z\frac{\partial^2\Psi}{\partial z^2}(x,z) . \qquad (10.6.15)$$

Now equation (10.6.13) may be used to relate the displacement $w(r,z)$ to $w^p(x,z)$ and the stress $\tau_{zz}(r,z)$ to $\tau^p_{zz}(x,z)$. Thus

$$A_1^{-1}[w(r,z);r \to x] = w^p(x,z) ,$$

$$A_1^{-1}[\tau_{zz}(r,z);r \to x] = \tau^p_{zz}(x,z) . \qquad (10.6.16)$$

If, for example, the half-plane is acted upon by two equal concentrated normal loads P at ± a (the loading must be symmetric about x = 0), then the corresponding loading in the half-space is

$$\tau_{zz}(r,0) = A_1[\tau_{yy}(x,0);r] = -\left(\frac{2}{\pi}\right)^{1/2}\int_0^r \frac{P\delta(a)dx}{(r^2-x^2)^{1/2}} , \qquad (10.6.17)$$

so that

$$\tau_{zz}(r,0) = 0 , \quad 0 \leq r < a , \quad \tau_{zz} = -\left(\frac{2}{\pi}\right)^{1/2}\frac{P}{(r^2-a^2)^{1/2}} , \quad r > a . \qquad (10.6.18)$$

The study of the relationship between plane and axisymmetric problems, which is contained in (10.6.13), has a long history. C. Weber (1925) used the relation

$$\psi(r,z) = \frac{1}{2\pi}\int_0^2 \Psi(r\cos\theta,z)d\theta \qquad (10.6.19)$$

to obtain a space harmonic ψ from a plane harmonic Ψ, (see Example 10.6.6) and explored this relation and its inverse in C. Weber (1940): see also Lebedev (1937). The fact that plane and axisymmetric problems are closely connected was evident to Marguerre (1937). Mossakovskii (1954a), (1954b) derived (10.6.13) and similar relations for problems with $\cos m\theta$ dependence, and used them to obtain effective solutions to the problems by reducing them to problems of a complex variable. A. Ia. Aleksandrov (1959a) considered the axisymmetric state of stress obtained by physically rotating a plane strain state of stress which is symmetrical in x in the x, z plane about the z-axis through an angle π. In such a state of stress the displacements and stresses will be the same in each plane y = const., so that, for instance, a concentrated normal stress $\tau^p_{zz}(x,0) = -P\delta(a)$ is in fact a line of concentrated stress $\tau_{zz}(x,y,0) = -P\delta(a)$ on the line x = a, z = 0 in space, as shown in Figure 10.6.1. Aleksandrov showed that the stress $\tau^p_{zz}(x,0) = -P\delta(a)-P\delta(-a)$ when rotated gives an axisymmetric stress

$$\tau_{zz}(r,0) = \begin{cases} 0 & r < a , \\ -2P(r^2-a^2)^{-1/2} , & r > a , \end{cases} \tag{10.6.20}$$

which differs only by a factor from the state (10.6.18).

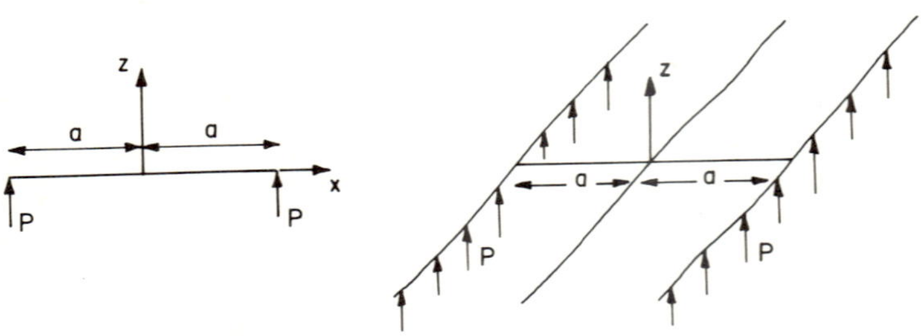

Figure 10.6.1 - The two concentrated loads P actually represent two line loads on the plane

Since the act of obtaining an axisymmetric state of stress by rotating a plane state does not facilitate the solution of the axisymmetric problem, it seems desirable to proceed purely on the basis of the Fourier and Hankel transforms involved. Equation (5.4.6) and Example 5.4.2 together show that if $A(\xi)$, $C(\xi)$ are replaced by $\xi A(\xi)$, $\xi C(\xi)$ respectively then, in a state *symmetrical about $x = 0$*,

$$\left.\begin{aligned}
2\mu u^p(x,z) &= F_s\{[\xi zA+C]\exp(-\xi z);\xi \to x\} , \\
2\mu w^p(x,z) &= F_c\{[\kappa A+\xi zA+C]\exp(-\xi z);\xi \to x\} , \\
-\tau^p_{xx}(x,z) &= F_c\{\xi[2\nu A-\xi zA-C]\exp(-\xi z);\xi \to x\} , \\
-\tau^p_{xz}(x,z) &= F_s\{\xi[(1-2\nu)A+\xi zA+C]\exp(-\xi z);\xi \to x\} , \\
-\tau^p_{zz}(x,z) &= F_c\{\xi[2(1-\nu)A+\xi zA+C]\exp(-\xi z);\xi \to x\} .
\end{aligned}\right\} \quad (10.6.21)$$

On the other hand, the axisymmetric quantities are given by equation (10.2.10), i.e.,

$$\left.\begin{aligned}
2\mu u(r,z) &= H_1\{\xi^{-1}[\xi zA+C]\exp(-\xi z);\xi \to r\} , \\
2\mu w(r,z) &= H_0\{\xi^{-1}[\kappa A+\xi zA+C]\exp(-\xi z);\xi \to r\} , \\
-\tau_{rz}(r,z) &= H_1\{[(1-2\nu)A+\xi zA+C]\exp(-\xi z);\xi \to r\} , \\
-\tau_{zz}(r,z) &= H_0\{[2(1-\nu)A+\xi zA+C]\exp(-\xi z);\xi \to r\} ,
\end{aligned}\right\} \quad (10.6.22)$$

and the operators of Table 5.7.1 show that

$$x^{-1}A_1^{-1}\{ru(r,z);x\} = u^p(x,z) \ , \ A_1^{-1}\{w(r,z);x\} = w^p(x,z) \ , \quad (10.6.23)$$

$$x^{-1}A_1^{-1}\{r\tau_{rz}(r,z);x\} = \tau^p_{xz}(x,z) \ , \ A_1^{-1}\{\tau_{zz}(r,z);x\} = \tau^p_{zz}(x,z) \ . \quad (10.6.24)$$

Some other integral representations for axisymmetric problems

Other similar relations may be found by taking a plane stress state anti-symmetrical about $x = 0$, or by replacing $A(\xi)$, $C(\xi)$ in the plane state by $\xi^{-1}A(\xi)$, $\xi^{-1}C(\xi)$ respectively; then the two states are linked by the operator A_2. Variants of these results may be found in A. Ia. Aleksandrov (1959b), (1961a), (1961b), where the analysis is mostly based on an actual physical rotation of the stress state; Belenkii (1960) uses an operator approach; Mossakovskii and Rybka (1964) solve an axisymmetric crack problem using the method. Many relations between surface displacements and stresses in plane and axisymmetric problems may be found in Noble and Spence (1971).

Another relation similar to (10.6.19) is based on the integral

$$J_0(\xi r) = \frac{1}{\pi}\int_0^{\pi} \exp(-\xi r\cos\theta)d\theta \qquad (10.6.25)$$

which leads to $\psi(r,z)$ of (10.6.2) being expressed in the form

$$\psi(r,z) = \frac{1}{\pi}\int_0^{\pi}\int_0^{\infty} A(\xi)\exp[-\xi(z+ir\cos\theta)]d\xi d\theta \ . \qquad (10.6.26)$$

This may be written in an alternative form by noting that

$$\psi(0,z) = \int_0^{\infty} A(\xi)\exp(-\xi z)d\xi \qquad (10.6.27)$$

so that

$$\psi(r,z) = \frac{1}{\pi}\int_0^{\pi} \psi(0,z+ir\cos\theta)d\theta \ . \qquad (10.6.28)$$

This representation is due to Poisson, but is usually called Whittaker's solution. An account of the development of this and related representations has been given by Temple (1958). Other forms of (10.6.28) have been given by Heins and MacCamy (1960a), (1960b), among which is

$$\psi(r,z) = \frac{1}{\pi}\int_{-r}^{r} \frac{\psi(0,z+ip)\,dp}{(r^2-p^2)^{1/2}} . \tag{10.6.29}$$

Another form of (10.6.26) appears in much of the Russian literature and stems from A. Ia. Aleksandrov (1959b). Introduce the complex variable $t = z+ir$ and let ζ be an arbitrary point in the z,r plane. Then

$$I \equiv \exp(-\xi z)J_0(\xi r) = \frac{1}{\pi}\int_0^{\pi} \exp[-\xi(z+ir\cos\theta)]\,d\theta . \tag{10.6.30}$$

The integral may be transformed into an integral along the line *L* joining $\bar{t}$ to t as shown in Figure 10.6.2; on that line $\zeta = z+ir\cos\theta$ so that

$$I = \frac{1}{\pi}\int_{\bar{t}}^{t} \exp(-\xi\zeta)\,\frac{d\zeta}{ir\sin\theta} . \tag{10.6.31}$$

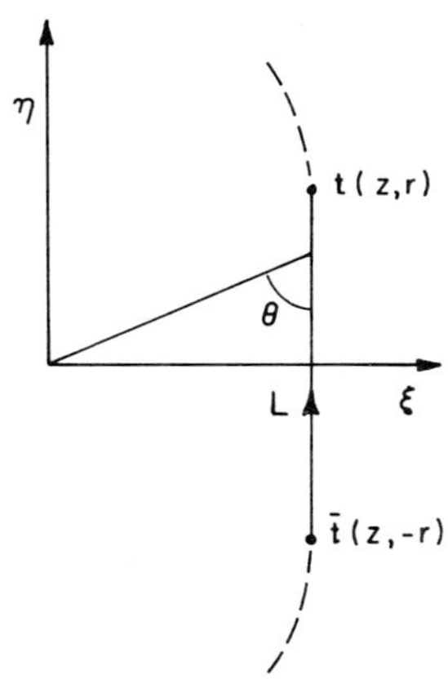

Figure 10.6.2 - The function $[(\zeta-t)(\zeta-\bar{t})]^{1/2}$ is defined in the plane cut along lines joining t, $\bar{t}$ to ∞

Now in order to express $\sin\theta$ in terms of ζ it is necessary to choose a branch of the function $[(\zeta-t)(\zeta-\bar{t})]^{1/2}$ which is holomorphic in a region around L and real on L. The root may be defined in the plane cut along lines joining each of t, $\bar{t}$ to ∞. Then, on L,

$$[(\zeta-t)(\zeta-\bar{t})]^{1/2} = [ir(\cos\theta-1)(ir)(\cos\theta+1)]^{1/2} = r\sin\theta \ , \qquad (10.6.32)$$

so that

$$\psi(r,z) = \frac{1}{\pi i}\int_{\bar{t}}^{t}\int_{0}^{\infty} A(\xi)\exp(-\xi\zeta)d\xi\frac{d\zeta}{[(\zeta-t)(\zeta-\bar{t})]^{1/2}} =$$

$$= \frac{1}{\pi i}\int_{\bar{t}}^{t}\frac{\Psi(\zeta)d\zeta}{[(\zeta-t)(\zeta-\bar{t})]^{1/2}} \ , \qquad (10.6.33)$$

where

$$\Psi(\zeta) = \int_{0}^{\infty} A(\xi)\exp(-\xi\zeta)d\xi \ , \qquad (10.6.34)$$

and by writing $\zeta = z+ir\cos\theta$ we see that

$$\mathrm{Re}[\Psi(\zeta)] = \mathrm{Re}[\Psi(\bar{\zeta})] \ , \quad \mathrm{Im}[\Psi(\zeta)] = -\mathrm{Im}[\Psi(\bar{\zeta})] \ . \qquad (10.6.35)$$

The integral (10.6.33) may now be calculated along any path in the region in which $\Psi(\zeta)$ is holomorphic and which does not cut the branch cuts of the root.

If the displacements are given in terms of the two potentials ψ, ϕ we may write each in the form (10.6.3), i.e.,

$$\phi(r,z) = \frac{1}{\pi i}\int_{t}^{t}\frac{\Phi(\zeta)d\zeta}{[(\zeta-t)(\zeta-\bar{t})]^{1/2}} \ . \qquad (10.6.36)$$

Then

$$\frac{\partial\phi}{\partial r} = \frac{1}{\pi i r}\int_{t}^{t} \frac{(\zeta-z)\Phi'(\zeta)d\zeta}{[(\zeta-t)(\zeta-\bar{t})]^{1/2}} \tag{10.6.37}$$

so that the displacements u, w are

$$2\mu u(r,z) = -\frac{1}{\pi i r}\int_{\bar{t}}^{t} \frac{[z\Psi'(\zeta)+\Phi'(\zeta)](\zeta-z)d\zeta}{[(\zeta-t)(\zeta-\bar{t})]^{1/2}}, \tag{10.6.38}$$

$$2\mu w(r,z) = \frac{1}{\pi i}\int_{\bar{t}}^{t} \frac{[\kappa\Psi(\zeta)-z\Psi'(\zeta)-\Phi'(\zeta)]d\zeta}{[(\zeta-t)(\zeta-\bar{t})]^{1/2}}. \tag{10.6.39}$$

These results were derived by A. Ia. Aleksandrov and Solov'ev in a number of papers in (1959) - (1968); see A. Ia. Aleksandrov (1959b), (1961b), (1968), A. Ia. Aleksandrov and Solov'ev (1962, (1964a). The agreement with the formulae in the latter papers may be obtained by putting

$$\Psi(\zeta) = 2\phi(\zeta), \quad \Phi'(\zeta) = \kappa\phi(\zeta)-\zeta\phi'(\zeta)+\psi(\zeta). \tag{10.6.40}$$

In the 1964a paper the functions $\phi(\zeta)$, $\psi(\zeta)$ are expressed as Cauchy integrals along the curve in the r, z plane that, when rotated about the z-axis, forms the boundary of the elastic body. In (1964b), A. Ia. Aleksandrov and Solov'ev consider an elastic state with $\cos m(\theta+\alpha)$ dependence on θ and rewrite the Hankel transform solution (10.3.19) as an integral from $\bar{t}$ to t in the ζ-plane; the basic result required is

$$J_n(z) = \frac{i^n}{\pi}\int_0^{\pi} \exp(-iz\cos\theta)\cos n\theta\, d\theta \tag{10.6.41}$$

which leads to

$$\exp(-\xi z)J_n(\xi r) = \frac{i^n}{\pi i}\int_{\bar{t}}^{t} \exp(-\xi\zeta)T_n\left[\frac{\zeta-z}{ir}\right]\frac{d\zeta}{[(\zeta-t)(\zeta-\bar{t})]^{1/2}}. \tag{10.6.42}$$

This representation may be compared with those used by Keer (1965).

Further research on these representations involves the concept of generalized analytic functions to which references may be found in Chemeris (1965), Iu. I. Solov'ev (1966), Kapshivyi and Maslyuk (1967), Maslyuk (1967) and especially Polozhii (1965), (1967).

Examples 10.6

1. Show that a solution of the exterior punch problem (see 10.5)

$$\frac{\partial\psi(r,0)}{\partial z} = 0 \,, \quad 0 \le r < a \,; \qquad \psi(r,0) = f(r) \,, \quad r > a \,,$$

is given by

$$\psi(r,z) = \frac{1}{2i}\int_{-a}^{a} \frac{h(x)dx}{[r^2+(z-ix)^2]^{1/2}} \,, \quad h(x) = \left(\frac{2}{\pi}\right)^{1/2} F(x) \,,$$

$$F(x) = A_2^{-1}[f(r);x]$$

and $h(x)$ is defined as an *odd* function in $[-a,a]$.

2. Show that a solution of the interior crack problem

$$\frac{\partial\psi}{\partial z}(r,0) = f(r) \,, \quad 0 \le r < a \,; \quad \psi(r,0) = 0 \,, \quad r > a \,,$$

is given by

$$\psi(r,z) = \frac{1}{2i}\left\{\int_{-\infty}^{-a} + \int_{a}^{\infty}\right\} \frac{h(x)dx}{[r^2+(z-ix)^2]^{1/2}} \,,$$

$$h(x) = \left(\frac{2}{\pi}\right)^{1/2} F(x) \,, \quad F(x) = A_1[rf(r);x]$$

where $h(x)$ is defined in $(-\infty,-a]$ as $h(x) = -h(-x)$.

3. Use the last result of Example 5.7.10 to show that if $f(r)$ of (10.6.4) is $P_{2n}[(1-r^2/a^2)^{1/2}]$ then $g(x)$ of (10.6.11) is $(2/\pi)P_{2n}(x/a)/P_{2n}(0)$. Thus, by comparing the solutions

for $\psi(r,z)$ given by (10.5.16) with m = 0 and by (10.6.11) deduce the result, due to E. R. Love (1949),

$$P_{2n}(\eta)Q_{2n}(i\xi) = \frac{1}{2i}\int_{-a}^{a} \frac{P_{2n}(t/a)dt}{\{r^2+(z+it)^2\}^{1/2}} .$$

See Gladwell and England (1977), England and Shail (1977) and England (1979) for further results.

4. Show that if $A(\xi)$, $C(\xi)$ in (10.6.21) are replaced by $\xi^{-1}A(\xi)$, $\xi^{-1}C(\xi)$ then the plane and axisymmetric states are related by

$$xA_2\{u(r,z);x\} = u^p(x,z) , \qquad A_2\{rw(r,z);x\} = w^p(x,z) ,$$

$$xA_2\{\tau_{rz}(r,z);x\} = \tau^p_{xz}(x,z) , \qquad A_2\{r\tau_{zz}(r,z);x\} = \tau^p_{zz}(x,z) .$$

5. Show that if $\Psi(x,z)$ is a plane harmonic function then $\psi(r,z)$ given by (10.6.19) is a space harmonic function.

10.7 *Contact problems for the circular annulus*

The Boussinesq (i.e., frictionless) contact or punch problem for the annular region $b \leq r \leq a$, $z = 0$ is the determination of the solution of the equations of elasticity for the half-space $z \geq 0$ subject to the boundary conditions

$$\tau_{rz}(r,0) = 0 , \quad r \geq 0 ; \quad \tau_{zz}(r,0) = 0 , \quad 0 \leq r < b , \quad r > a ; \tag{10.7.1}$$

$$w(r,0) = d+f(r) , \quad b \leq r \leq a . \tag{10.7.2}$$

The corresponding Reissner-Sagoci or *torsion* problem has boundary conditions

$$v(r,0) = \gamma r , \quad b \leq r \leq a ; \quad \tau_{z\theta}(r,0) = 0 , \quad 0 \leq r < b , \quad r > a . \tag{10.7.3}$$

Research on these problems may be divided into three main categories: precise formulation as one or more integral equations and solution for some values of $\varepsilon \equiv b/a$, other approximate solutions, some based on the reduction of the problems to a series of simpler problems, and asymptotic results for small and large ε.

Gubenko and Mossakovskii (1960) solved the punch problem by using the representation (10.6.13) of axisymmetric harmonic functions in terms of plane harmonic functions. (See also Gubenko (1960) for the corresponding problem of an elastic layer.) They thus obtained equations (10.7.25), (10.7.26) below which they combined into one equation and solved for small ε. Collins (1963) obtained effectively the same equations by using the representation (10.6.11). Borodachev and Borodacheva (1966) formulated the torsion problem by using the Hankel transforms. Both the punch and the torsion problems are particular examples of the triple integral equations considered by Tranter (1960), Cooke (1963a), (1963b), (1965) and W.E. Williams (1963). Further references may be found in these papers and in the discussion of the electrostatic potential problem for the annulus (formally equivalent to the punch problem) to be found in Chapter VIII of Sneddon (1966a); see also V.M. Aleksandrov (1967d), Valov (1968) and Jain and Kanwal (1971), (1972).

Olesiak (1965) attempted to solve the punch problem by reducing it to a series of dual integral equations; his method, while appearing to converge satisfactorily, is extremely cumbersome even for the flat punch. Moss and Kobayashi (1971) employ a similar iterative technique, but base their solution on the representation (10.6.13). Parlas and Michalopoulos (1972) consider the annular problem of a rigid bolt of stem-radius a and outer radius b pressed into an elastic half-space by an axial load P. Shibuya, Koizumi and Nakahara (1974), in treating the punch problem, assume a series expansion with parameters a_1, a_2, ... for the contact pressure under the punch and obtain an infinite system of simultaneous equations for the a_n. However, since the coefficients of this system are integrals of products of four Bessel functions, it can hardly be reckoned that they have achieved their stated aim of a simple solution. They do however present extensive graphical results for displacements and stresses both on and off the plane $z = 0$. Shibuya (1976), Hara, Shibuya, Koizumi and Nakahara (1975), (1977), (1978a), (1978b) and Shibuya, Koizumi, Nakahara and Hara (1979) used a similar method in other problems.

Collins (1963) and Cooke (1963b) obtained an expansion of the total load for the punch problem in powers of ε. Borodachev and Borodacheva (1966) obtained series in powers of ε for the contact stress and the total torque in the torsion problem. Borodachev (1976a), (1976c) obtained the stress concentration factors for the punch problem, expressing the results in terms of solutions of integral equations which had to be solved numerically.

Barber (1976) pays particular attention to concave punches, where the magnitude of the inner contact radius b has to be determined so that the contact pressure is zero at $r = b$. He uses a previously proved result (Barber (1974)) which states that if a solution has been found for the Boussinesq problem with given normal displacement over $z = 0$, $b \leq r \leq a$, and if $P(b)$ is the total normal load on the annulus $b \leq r \leq a$, then the correct internal radius b may be found by making P stationary, i.e., $\partial P/\partial b = 0$. He employs this result in conjunction with Collins' (1963) integral equation (effectively (10.7.27) below) when $\varepsilon = b/a$ is small, or with Grinberg and Kuritsyn's (1962) expansion when ε is nearly unity, to relate ε to the total load P.

When the shear stress τ_{rz} is zero on $z = 0$ the Papkovich-Neuber solution (10.2.2) gives (Example 10.2.11)

$$2\mu w = 2(1-\nu)\psi - z\frac{\partial\psi}{\partial z}, \qquad \tau_{zz} = \frac{\partial\psi}{\partial z} - z\frac{\partial^2\psi}{\partial z^2}, \tag{10.7.4}$$

so that the boundary conditions (10.7.1), (10.7.2) lead to the following Dirichlet problem for the harmonic function ψ

$$\psi(r,0) = F(r) \equiv \vartheta^{-1}\{d+f(r)\}, \qquad b \leq r \leq a, \tag{10.7.5}$$

$$\frac{\partial \psi}{\partial z}(r,0) = 0 , \qquad 0 \le r < b , \quad r > a . \tag{10.7.6}$$

Now assume that $F(r)$ may be represented in the form

$$F(r) = \sum_{n=0}^{\infty} a_n r^n + \sum_{n=1}^{\infty} a_{-n} r^{-n} = F_i(r) + F_e(r) . \tag{10.7.7}$$

Such a representation will be unique and $F_i(r)$ may be extended to zero, $F_e(r)$ to infinity. Now introduce the harmonic functions $\psi_{iaD}(r,z)$, $\psi_{ebD}(r,z)$ which are the solutions of the interior and exterior mixed boundary-value problems specified by

$$\psi_{iaD}(r,0) = F_i(r) , \quad 0 \le r \le a ; \quad \frac{\partial \psi_{iaD}}{\partial z}(r,0) = 0 , \quad r > a , \tag{10.7.8}$$

$$\psi_{ebD}(r,0) = F_e(r) , \quad r \ge b ; \quad \frac{\partial \psi_{ebD}}{\partial z}(r,0) = 0 , \quad r < b . \tag{10.7.9}$$

These solutions are labelled D for Dirichlet. The analysis of (10.2.31) - (10.2.36) shows that

$$\psi_{iaD}(r,z) = \mathcal{H}_0[\xi^{-1}\Phi_{iaD}(\xi)\exp(-\xi z);r] \tag{10.7.10}$$

where

$$\Phi_{iaD}(\xi) = \mathcal{F}_c[\Psi_{iaD}(t);\xi] , \tag{10.7.11}$$

$$\Psi_{iaD}(t) = 0 , \quad t > a , \quad \Psi_{iaD}(t) = A_1^{-1}[F_i(r);t] , \quad t < a . \tag{10.7.12}$$

A similar analysis shows that if

$$\psi_{ebD}(r,z) = \mathcal{H}_0[\xi^{-1}\Phi_{ebD}(\xi)\exp(-\xi z);r] \tag{10.7.13}$$

then the use of entries 4,7 of Table 5.7.1 gives

$$\Phi_{ebD}(\xi) = \mathcal{F}_s[\Psi_{ebD}(t);\xi] , \quad \Psi_{ebD}(t) = 0 , \quad t < b , \tag{10.7.14}$$

$$\Psi_{ebD}(t) = A_2^{-1}[F_e(r);t] , \qquad t > b . \tag{10.7.15}$$

The solution ψ may now be written in the form

$$\psi = \psi_{iaD}+\psi_{ebD}+\psi_{eaN}+\psi_{ibN} \tag{10.7.16}$$

where ψ_{eaN}, ψ_{ibN} are solutions of Neumann-type problems for the exterior of $r = a$ and interior of $r = b$ respectively. Thus

$$\psi_{eaN}(r,0) = 0 , \quad r \leq a ; \qquad \psi_{ibN}(r,0) = 0 , \quad r \geq b , \tag{10.7.17}$$

so that equation (10.7.5) is fulfilled. The remaining condition (10.7.6) will be fulfilled provided that

$$\begin{aligned} I_i(r) &\equiv \frac{\partial}{\partial z} (\psi_{iaD}+\psi_{eaN}+\psi_{ibN}) = 0 , \quad r < b ; \\ I_e(r) &\equiv \frac{\partial}{\partial z} (\psi_{ebD}+\psi_{eaN}+\psi_{ibN}) = 0 , \quad r > a . \end{aligned} \tag{10.7.18}$$

The potentials ψ_{eaN}, ψ_{ibN} may be written

$$\begin{aligned} \psi_{eaN}(r,z) &= \mathcal{H}_0[\xi^{-1}\Phi_{eaN}(\xi)\exp(-\xi z);r] , \\ \psi_{ibN} &= \mathcal{H}_0[\xi^{-1}\Phi_{ibN}(\xi)\exp(-\xi z);r] , \end{aligned} \tag{10.7.19}$$

where entries 1,7 of Table 5.7.1 show that

$$\Phi_{eaN}(\xi) = \mathcal{F}_c[\Psi_{eaN}(t);\xi] , \quad \text{where } \Psi_{eaN}(t) = 0 , \quad t < a , \tag{10.7.20}$$

$$\Phi_{ibN}(\xi) = \mathcal{F}_s[\Psi_{ibN}(t);\xi] , \quad \text{where } \Psi_{ibN}(t) = 0 , \quad t > b . \tag{10.7.21}$$

Equations (10.7.18) are equivalent to

$$\begin{aligned} \mathcal{H}_0[(\Phi_{ia}+\Phi_{ea}+\Phi_{ib});r] &= 0 , \quad r < b ; \\ \mathcal{H}_0[(\Phi_{eb}+\Phi_{ea}+\Phi_{ib});r] &= 0 , \quad r > a . \end{aligned} \tag{10.7.22}$$

Now use operators 4, 10 of Table 5.7.1 to obtain

$$\begin{aligned} F_s[(\Phi_{ia}+\Phi_{ea}+\Phi_{ib});t] &= 0\ , \qquad t < b\ ; \\ F_c[(\Phi_{eb}+\Phi_{ea}+\Phi_{ib});t] &= 0\ , \qquad t > a\ . \end{aligned} \tag{10.7.23}$$

It was shown in §5.5 that the $F_s(F_c)$ transform of an $F_c(F_s)$ transform may be written as a Hilbert transform. Thus Example 5.5.1 and equation (5.5.12) show that

$$F_s[\Phi_{iaD}(\xi);t] = -\frac{1}{\pi}\int_{-a}^{a}\frac{\Psi_{iad}(\tau)d\tau}{\tau-t} = -\frac{2t}{\pi}\int_{0}^{a}\frac{\Psi_{iaD}(\tau)d\tau}{\tau^2-t^2}\ , \tag{10.7.24}$$

where $\Psi_{iaD}(\tau) = \Psi_{iaD}(-\tau)$ for $-a \le \tau \le 0$. The other terms in equations (10.7.23) may be treated likewise to yield

$$\Psi_{ibN}(t) = \frac{2t}{\pi}\int_0^a \frac{\Psi_{iaD}(\tau)d\tau}{\tau^2-t^2} + \frac{2t}{\pi}\int_a^\infty \frac{\Psi_{eaN}(\tau)d\tau}{\tau^2-t^2}\ , \qquad t < b\ , \tag{10.7.25}$$

$$\Psi_{eaN}(t) = -\frac{2}{\pi}\int_0^b \frac{\tau\Psi_{ibN}(\tau)d\tau}{\tau^2-t^2} - \frac{2}{\pi}\int_b^\infty \frac{\tau\Psi_{ebD}(\tau)d\tau}{\tau^2-t^2}\ , \qquad t > a\ , \tag{10.7.26}$$

which may be combined into the single equation

$$\Psi_{eaN}(t) = \frac{2}{\pi^2}\int_a^\infty \frac{1}{t^2-\tau^2}\left\{\tau \ln\left[\frac{\tau+b}{\tau-b}\right] - t\ln\left[\frac{t+b}{t-b}\right]\right\}\Psi_{eaN}(\tau)d\tau + \Phi(t)\ , \tag{10.7.27}$$

where

$$\Phi(t) = \frac{2}{\pi^2}\int_0^a \frac{1}{t^2-\tau^2}\left\{\tau \ln\left[\frac{\tau+b}{\tau-b}\right] - t\ln\left[\frac{t+b}{t-b}\right]\right\}\Psi_{iaD}(\tau)d\tau - \frac{2}{\pi}\int_b^\infty \frac{\tau\Psi_{ebD}(\tau)d\tau}{\tau^2-t^2}\ . \tag{10.7.28}$$

These are the equations obtained by Gubenko and Mossakovskii (1960); they obtained a series solution of (10.7.27) valid for $b \ll a$.

Cooke (1963a) obtained a different, but equivalent, formulation by assuming $\psi(r,z)$ in the form (see Example 10.7.1)

$$\psi(r,z) = H_0[\xi^{-1}\Phi(\xi)\exp(-\xi z);r] , \tag{10.7.29}$$

and noting that equation (10.7.6) demands that

$$\Phi(\xi) = \int_b^a g(y)J_0(\xi y)dy . \tag{10.7.30}$$

A simple approximate solution of the punch problem may be found by expressing ψ_{ibN}, ψ_{eaN} in terms of the Legendre function solutions of 10.5. Thus if ξ_b, η_b denote the coordinates ξ, η of (10.5.1) for the circle $r = b$, and σ_a, τ_a denote the coordinates σ, τ of (10.5.29) for the circle $r = a$ then the assumed forms of ψ_{ib}, ψ_{ea} are

$$\psi_{ibN}(\xi_b,\eta_b) = \sum_{k=1}^{N} b_k Q_{2k-1}(i\xi_b)P_{2k-1}(\eta_b) \tag{10.7.31}$$

$$\psi_{eaN}(\sigma_a,\tau_a) = (1-\sigma_a^2+\tau_a^2)^{1/2} \sum_{k=1}^{N} a_k Q_{2k-1}(i\tau_a)P_{2k-1}(\sigma_a) \tag{10.7.32}$$

and the a_k, b_k may be found by supposing that equations (10.7.18) are satisfied *in some sense*, either momentwise, by making

$$\int_0^b I_i(r)m_k(r)dr = 0 = \int_a^\infty I_e(r)n_k(r)dr \tag{10.7.33}$$

for multipliers $m_k(r)$, $n_k(r)$, $k = 1,2,\ldots N$, or by equating the N leading coefficients of r in I_i and of r^{-1} in I_e to zero. The flat punch ($f(r) = 0$, $F(r) = d/\vartheta = \delta$) and $N = 1$ shows the working of the process. The potential ψ_{iaD} is

$$\psi_{iaD}(\xi_a,\eta_a) = c_0 Q_0(i\xi_a)P_0(\eta_a) . \tag{10.7.34}$$

The functions $iQ_{2k}(i\xi)$ are given by England and Shail (1977) as

$$iQ_{2k}(i\xi) = (-)^k\xi \int_0^1 \frac{x^{2k}(1-x^2)^k dx}{(\xi^2+x^2)^{k+1}} \tag{10.7.35}$$

so that $iQ_0(i\xi) = \tan^{-1}(\xi^{-1})$. Thus for $r < a$ $(\xi = 0)$ $\psi_{iaD}(0,\eta_a) = c_0Q_0(i0) = \delta$ so that $c_0 = 2i\delta/\pi$, and

$$\frac{\partial\psi_{ia}}{\partial z} = \frac{c_0}{a\eta_a} \left.\frac{dQ_0(i\xi_a)}{d\xi_a}\right|_0 = \frac{c_0 i}{a\eta_a} = -\frac{2\delta}{\pi a}\left(1 - \frac{r^2}{a^2}\right)^{-1/2}, \quad r < a\,. \tag{10.7.36}$$

For N = 1 the potentials ψ_{ib}, ψ_{ea} may be written

$$\psi_{ibN}(\xi_b,\eta_b) = \delta b_1 Q_1(i\xi_b)P_1(\eta_b)\;, \tag{10.7.37}$$

$$\psi_{eaN}(\sigma_a,\tau_a) = \delta a_1(1-\sigma_a^2+\tau_a^2)^{1/2}Q_1(i\tau_a)P_1(\sigma_a)\;, \tag{10.7.38}$$

and the odd Q's may be computed from (England and Shail (1977))

$$Q_{2k+1}(i\xi) = (-)^{k+1}\xi \int_0^1 \frac{x^{2k+1}(1-x^2)^{k+1/2}}{(\xi^2+x^2)^{k+3/2}}\,dx\;, \tag{10.7.39}$$

so that, in particular, $Q_1(i\xi) = -1+\xi\tan^{-1}(\xi^{-1})$. Thus, by making use of the equations linking ξ_b, η_b and σ_a, τ_a to r we find

$$\delta^{-1}\psi = \begin{cases} 1 - b_1\left(1 - \dfrac{r^2}{b^2}\right)^{1/2}, & 0 \le r < b \\ 1, & b \le r \le a \\ \dfrac{2}{\pi}\sin^{-1}\left(\dfrac{a}{r}\right) - a_1\dfrac{a}{r}\left(1 - \dfrac{a^2}{r^2}\right)^{1/2}, & r > a \end{cases} \tag{10.7.40}$$

$$-\delta^{-1}a\,\frac{\partial\psi}{\partial z} = \frac{2}{\pi}\,R\left(\frac{r}{a}\right) - a_1\,\frac{a^3}{r^3}\,S\left(\frac{r}{a}\right) - \frac{b_1}{\varepsilon}\,S\left(\frac{b}{r}\right), \tag{10.7.41}$$

where

$$R(x) = \begin{cases} (1-x^2)^{-1/2}, & 0 < x < 1\,, \\ 0, & x > 1\,, \end{cases} \qquad S(x) = \begin{cases} \sin^{-1}x - x(1-x^2)^{-1/2}, & 0 < x < 1, \\ \dfrac{\pi}{2}, & x > 1\,, \end{cases} \tag{10.7.42}$$

and it may be noted that $S(x) = -x^3/3+0(x^5)$ for small $|x|$. Notice that $\partial\psi/\partial z$, which gives the contact pressure, is finite outside the contact zone, and has the desired square-root singularities at $r = b+$, $r = a-$. It may be shown that if the coefficients a_1, b_1 are chosen according to (10.7.33) with $m_1(r) = r = n_1(r)$, or so that the constant term in $I_i(r)$ and the coefficient of r^{-3} in $I_e(r)$ are zero, then the resulting displacements and stresses agree very closely with those calculated by Shibuya et al (1974) for values of ε up to about 0.8. See Gladwell and Gupta (1979).

The torsion problem may be analyzed in a similar fashion. Now the individual potentials ψ_{iaD}, ψ_{eaN} etc. will all have the form (10.2.49); the equations corresponding to (10.7.25), (10.7.26) are given in Example 10.7.2. The approximate analysis using oblate spheroidal coordinates now makes use of solutions $P_n^1\ Q_n^1$ but otherwise proceeds on much the same lines. It is found now (Gladwell and Gupta (1979)) that the results obtained by using just one term in each of ψ_{eaN}, ψ_{ibN} agree extremely closely with the numerical results of Shibuya (1976).

Popov (1967c) and Kolybikhin and Popov (1969) have considered annulus problems for transversely isotropic and nonhomogeneous half-spaces.

Examples 10.7

1. The representation (10.7.29) for $\psi(r,z)$ leads to the integral equation

$$\int_b^a g(y)\left\{\int_0^\infty J_0(\xi r)J_0(\xi y)d\xi\right\}dy = F(r) .$$

Using the fact that

$$\int_0^\infty J_0(\xi r)J_0(\xi y)d\xi = \frac{2}{\pi}\int_0^{\min(r,y)} \frac{ds}{(r^2-s^2)^{1/2}(y^2-s^2)^{1/2}} ,$$

show that if

$$g(y) = -\frac{2}{\pi}\frac{d}{dy}\int_y^a \frac{sG(s)ds}{(s^2-y^2)^{1/2}} \text{ and } F(r) = 1,$$

then by changing the order of integration we may write the equation for G(s) as

$$G(s) + \frac{s}{(s^2-b^2)^{1/2}}\int_b^a \frac{tG(t)K(s,t)dt}{(t^2-b^2)^{1/2}} = \frac{s}{(s^2-b^2)^{1/2}},$$

where

$$K(s,t) = \frac{2}{\pi^2(s^2-t^2)}\left\{\frac{s^2-b^2}{s}\,\ell n\left[\frac{s+b}{s-b}\right] - \frac{t^2-b^2}{t}\,\ell n\left[\frac{t+b}{t-b}\right]\right\}.$$

2. Carry out an analysis similar to that in (10.7.19) - (10.7.26) for the Reissner-Sagoci problem for the annulus. The equations corresponding to (10.7.25), (10.7.26) are

$$\Psi_{ibN}(t) = -\frac{2}{\pi}\int_0^a \frac{\xi\Psi_{iaD}(\xi)d\xi}{\xi^2-t^2} - \frac{2}{\pi}\int_a^\infty \frac{\xi\Psi_{eaN}(\xi)d\xi}{\xi^2-t^2} + C, \quad 0 \le t < b,$$

$$\Psi_{eaN}(t) = \frac{2t}{\pi}\int_0^b \frac{\Psi_{ibN}(\xi)d\xi}{\xi^2-t^2} + \frac{2t}{\pi}\int_b^\infty \frac{\Psi_{ebD}(\xi)d\xi}{\xi^2-t^2}, \qquad t > a,$$

where C may be determined by equating both sides at t = 0.

CHAPTER XI - CONTACT PROBLEMS FOR PLATES AND LAYERS

"He goes about and sits on folk
That eat too much at night:
His duties are to pinch, and poke,
And squeeze them till they nearly choke."
(I said "It serves them right!")
-*Phantasmagoria*

11.1 *Foundation types*

Many hundreds of research papers and scores of books have been written on the subject of plates on elastic foundations. Many of these have been concerned with the approximate calculation of contact stresses and displacements for the type of plate - foundation configurations met in engineering practice, a subject which is not of primary interest here. An account of such theory and practice, particularly as it pertains to the extensive Russian literature, is given by Gorbunov-Posadov and Malikova (1973). Gorbunov-Posadov himself pioneered the use of (sometimes crude, but usually effective) approximate methods in foundation problems, starting from 1935 (see for example Gorbunov-Posadov (1939), (1940), (1941), (1949)). A more mathematically oriented view of the subject, based largely on a Galerkin (or variational) reduction of the governing partial differential equations to ordinary differential equations, has been provided by Vlasov and Leont'ev (1966). Both these treatises provide extensive tables and charts suitable for practical computations. A recent review of the Russian literature concerned with the mathematical analysis of the subject has been given by Popov (1972b); he has been responsible for the application of sophisticated analysis, particularly the use of orthogonal polynomials, to contact problems. Other reviews have been made by Korenev (1960b), Hetenyi (1964). Selvadurai (1979) has made an exhaustive study of soil-foundation interaction.

Much of the literature published before 1962 or so has little value today; the authors considered particular configurations, not realizing or being concerned with the fact that their analyses could easily be extended, without any essential complication, to more general situations. Advances in analytical techniques and the use of high-speed computers have also made many researches obsolete.

Consider a foundation, with properties as yet unspecified, occupying the half-space $z \geq 0$. For the mathematical formulation of any problem concerning contact between the foundation and any other body it is sufficient to know, for the half-space, the matrix of influence coefficients $R_{ij}(x,y,x_0,y_0)$. The typical coefficient $R_{32}(x,y,x_0,y_0)$ denotes the displacement $u_3(x,y,0) \equiv w(x,y,0)$ at the point (x,y) of the surface due to a unit concentrated force in the y-direction at $(x_0,y_0,0)$. Foundations may be categorized according to the sophistication of the influence coefficient matrix.

The simplest type of foundation is one characterized by the single influence coefficient $R_{33}(x,y,x_0,y_0) \equiv R(x,y,x_0,y_0)$, relating the normal displacement $w(x,y,0)$ on the surface $z = 0$ to a normal load at $(x_0,y_0,0)$. The simplest example is Winkler's foundation made up of independent springs in which

$$R(x,y,x_0,y_0) = k\delta(x-x_0)\delta(y-y_0) \tag{11.1.1}$$

where δ is the Dirac delta function. Numerous researchers have constructed modifications of the Winkler foundation to provide for shear effects by connecting the ends of the springs at the contact surface by a shear layer (Pasternak (1954)), a plate in bending (Hetenyi (1950)) or a stretched membrane (Filonenko-Borodich (1945)). These modifications and more general foundations may be unified by using the approach used by Vlasov and Leont'ev (1966); their underlying philosophy is that it is the governing equations which are of prime importance, not the model, mathematical or physical which is used to obtain the equations.

They use the following mathematical procedure:

(1) The foundation, of finite depth H, or of infinite depth, is assumed to be a homogeneous, isotropic elastic medium.

(2) The displacements u, v, w in the foundation are assumed to have the form

$$u(x,y,z) = \sum_{i=1}^{\ell} u_i(x,y)\phi_i(z) \ , \quad v(x,y,z) = \sum_{j=1}^{m} v_j(x,y)\chi_j(z) \ ,$$
$$w(x,y,z) = \sum_{k=1}^{n} w_k(x,y)\psi_k(z) \ , \tag{11.1.2}$$

where the functions $\phi_i(z)$, $\chi_j(z)$, $\psi_k(z)$ are assumed to be known, dimensionless, linearly independent functions.

(3) The equations governing the unknown functions u_i, v_j, w_k are obtained by applying the Principle of Virtual Work to the stresses obtained from these displacements, with the virtual displacements ϕ_i, χ_j, ψ_k. This process leads to $(\ell+m+n)$ second order partial differential equations for u_i, v_j and w_k.

The analysis runs as follows. The stress-strain relations give

$$\tau_{xx} = \frac{2\mu}{(1-2\nu)} [(1-\nu)e_{xx}+\nu(e_{yy}+e_{zz})] \ ,$$

$$= \frac{2\mu}{(1-2\nu)} \left[(1-\nu) \sum_{i=1}^{\ell} \frac{\partial u_i}{\partial x} \phi_i(z)+\nu \sum_{j=1}^{m} \frac{\partial v_j}{\partial y} \chi_j(z)+\nu \sum_{k=1}^{n} w_k\psi_k'(z)\right] \ , \tag{11.1.3}$$

$$\tau_{yy} = \frac{2\mu}{(1-2\nu)} \left[(1-\nu) \sum_{j=1}^{m} \frac{\partial v_j}{\partial y} \chi_j(z)+\nu \sum_{i=1}^{\ell} \frac{\partial u_i}{\partial x} \phi_i(z)+\nu \sum_{k=1}^{n} w_k\psi_k'(z)\right] \ , \tag{11.1.4}$$

$$\tau_{zz} = \frac{2\mu}{(1-2\nu)} \left[(1-\nu) \sum_{k=1}^{n} w_k\psi_k'(z)+\nu \sum_{i=1}^{\ell} \frac{\partial u_i}{\partial x} \phi_i(z)+\nu \sum_{j=1}^{m} \frac{\partial v_j}{\partial y} \chi_j(z)\right] \ , \tag{11.1.5}$$

$$\tau_{yz} = 2\mu e_{yz} = \mu\left[\sum_{j=1}^{m} v_j\chi_j'(z) + \sum_{k=1}^{n} \frac{\partial w_k}{\partial y} \psi_k(z)\right] \ , \tag{11.1.6}$$

$$\tau_{zx} = \mu \left[\sum_{k=1}^{n} \frac{\partial w_k}{\partial x} \psi_k(z) + \sum_{i=1}^{\ell} u_i \phi_i'(z) \right] , \tag{11.1.7}$$

$$\tau_{xy} = \mu \left[\sum_{i=1}^{\ell} \frac{\partial u_i}{\partial y} \phi_i(z) + \sum_{j=1}^{m} \frac{\partial v_j}{\partial x} \chi_j(z) \right] . \tag{11.1.8}$$

The virtual work equations may be derived by integrating the product of the equilibrium equations (1.9.5) - (1.9.7) and the displacements ϕ_i, χ_j, ψ_k over the depth of the foundation, and performing an integration by parts of the terms containing z-derivatives. Thus

$$\int_0^H \left\{ \frac{\partial \tau_{xx}}{\partial x} + \frac{\partial \tau_{xy}}{\partial y} + \frac{\partial \tau_{xz}}{\partial z} + F_x \right\} \phi_i(z) dz = 0 \tag{11.1.9}$$

gives

$$\int_0^H \left\{ \left[\frac{\partial \tau_{xx}}{\partial x} + \frac{\partial \tau_{xy}}{\partial y} + F_x \right] \phi_i(z) - \tau_{xz} \phi_i'(z) \right\} dz + [\tau_{xz} \phi_i(z)]_0^H = 0 , \tag{11.1.10}$$

while the remaining two equations give

$$\int_0^H \left\{ \left[\frac{\partial \tau_{yx}}{\partial x} + \frac{\partial \tau_{yy}}{\partial y} + F_y \right] \chi_j(z) - \tau_{yz} \chi_j'(z) \right\} dz + [\tau_{yz} \chi_j(z)]_0^H = 0 , \tag{11.1.11}$$

$$\int_0^H \left\{ \left[\frac{\partial \tau_{zx}}{\partial x} + \frac{\partial \tau_{zy}}{\partial y} + F_z \right] \psi_k(z) - \tau_{zz} \psi_k'(z) \right\} dz + [\tau_{zz} \psi_k(z)]_0^H = 0 . \tag{11.1.12}$$

The equations governing the generalized displacements u_i, v_j, w_k may now be obtained by substituting the expressions (11.1.3) - (11.1.8) for the stresses and carrying out the integrations in (11.1.10) - (11.1.12), all of which involve the integration of known functions.

In the simplest case, in which

$$u(x,y,z) \equiv 0 \equiv v(x,y,z) , \quad w(x,y,z) = w_2(x,y)\psi_2(z) , \qquad (11.1.13)$$

there is just one equation, derived from (11.1.12), namely

$$r_{22}(\mu\nabla_1^2 w_2+F_2) - s_{22}[2\mu(1-\nu)/(1-2\nu)]w_2 + [\tau_{zz}\psi_2(z)]_0^H = 0 , \qquad (11.1.14)$$

where

$$r_{22} = \int_0^H \psi_2^2(z)dz , \quad s_{22} = \int_0^H \psi_2'(z)^2 dz . \qquad (11.1.15)$$

The function $\psi_2(z)$ governs the z-variation of $w(x,y,z)$. If the foundation is a shallow layer resting on a rigid base we may take

$$\psi_2(z) = (H-z)/H , \qquad (11.1.16)$$

while the function

$$\psi_2(z) = \sinh\gamma(H-z)/\sinh\gamma H \qquad (11.1.17)$$

is more appropriate to a deep (or infinite) layer. In either case $\psi_2(1) = 1$, $\psi_2(0) = 0$. If there is no body force, i.e., $F_z = 0$, and $\tau_{zz}(x,y,0) = -p(x,y)$ then equation (11.1.14) becomes

$$r_{22}\mu\nabla_1^2 w_2 - s_{22}[2\mu(1-\nu)/(1-2\nu)]w_2 + p(x,y) = 0 , \qquad (11.1.18)$$

which has the form of the equation obtained from a foundation consisting of a set of springs joined at their tops by a stretched membrane, or by a shear layer. Higher, fourth and eighth order equations corresponding to a plate attached to Winkler springs, etc., may be obtained by suitable choice of more shape functions.

Popov (1972b) has proposed a quite different unifying theory based on Fourier transforms. He considers a general foundation in which the influence coefficients giving the surface displacements due to loads applied on the surface have the form

$$R_{jk}(x,y,x_0,y_0) = H_{jk}(x-x_0,y-y_0) \ . \tag{11.1.19}$$

This class includes, as a particular case, an anisotropic elastic half-space in which all 21 elastic constants depend arbitrarily on z. Put

$$H_{jk}(x,y) = F^*_{(2)}[\tilde{H}_{jk}(\xi_1,\xi_2);(\xi_1,\xi_2) \rightarrow (x,y)] \tag{11.1.20}$$

and use the abbreviations $H(x,y)$, $\tilde{H}(\xi_1,\xi_2)$ for the normal coefficients $H_{33}(x,y)$, $\tilde{H}_{33}(\xi_1,\xi_2)$ respectively. The displacement $u_j(x,y,0)$ on the surface of the half-space due to a distribution

$$\tau_{k3}(x_0,y_0,0) = -g_k(x_0,y_0) \tag{11.1.21}$$

over the plane z = 0, may be written

$$u_j(x,y) = \int_{-\infty}^{\infty}\int_{-\infty}^{\infty} H_{jk}(x-x_0,y-y_0)g_k(x_0,y_0)dx_0dy_0 \ . \tag{11.1.22}$$

The convolution theorem (5.7.4) for $F_{(2)}$ transforms allows this to be written

$$u_j(x,y) = \int_{-\infty}^{\infty}\int_{-\infty}^{\infty} \tilde{H}_{jk}(\xi_1,\xi_2)G_k(\xi_1,\xi_2)\exp(-i\xi_1x-i\xi_2y)d\xi_1d\xi_2 \tag{11.1.23}$$

where G_k is the $F_{(2)}$ transform of g_k. In particular, if there is only normal loading,

$$w(x,y) = \int_{-\infty}^{\infty}\int_{-\infty}^{\infty} \tilde{H}(\xi_1,\xi_2)P(\xi_1,\xi_2)\exp(-i\xi_1x-i\xi_2y)d\xi_1d\xi_2 \ , \tag{11.1.24}$$

where $P(\xi_1,\xi_2) = \mathcal{F}_{(2)}[p(x,y);(x,y) \to (\xi_1,\xi_2)]$. Various particular cases may now be recognized. If $\tilde{H}(\xi_1,\xi_2)$ is a function of $\xi = (\xi_1^2+\xi_2)^{1/2}$ then (5.7.10) shows that the kernel $H(x,y)$ is

$$H(x,y) = \mathcal{F}^*_{(2)}[\tilde{H}(\xi);(\xi_1,\xi_2) \to (x,y)] = \mathcal{H}_0[\tilde{H}(\xi);r] \qquad (11.1.25)$$

where $r = (x^2+y^2)^{1/2}$; in this case therefore $H(x-x_0,y-y_0)$ depends only on the distance $R = \{(x-x_0)^2+(y-y_0)^2\}^{1/2}$ between (x,y) and (x_0,y_0). It is convenient to use polar coordinates, as in Figure 11.1.1.

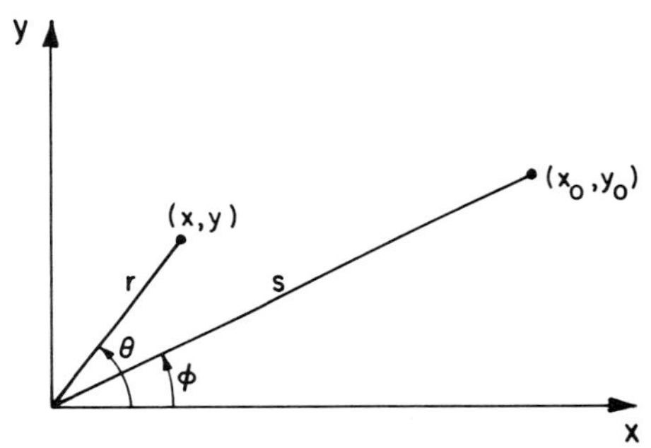

Figure 11.1.1 - Polar coordinates used for (x,y) and (x_0,y_0)

If the (normal) loading $p(x_0,y_0)$ is axisymmetric, so that $p(x_0,y_0) = p(s)$, then w will be axisymmetric also and

$$w(r) = \int_0^{2\pi} d\phi \int_0^{\infty} sH(r\cos\theta-s\cos\phi,\ r\sin\theta-s\sin\phi)p(s)ds\ , \qquad (11.1.26)$$

but

$$\int_0^{2\pi} H(r\cos\theta-s\cos\phi\ ,\ r\sin\theta-s\sin\phi)d\phi = \int_0^{2\pi} d\phi \int_0^{\infty} \xi\tilde{H}(\xi)J_0(\xi R)d\xi \qquad (11.1.27)$$

where $R = \{r^2+s^2-2rs\cos(\theta-\phi)\}^{1/2}$ and

$$(2\pi)^{-1} \int_0^{2\pi} J_0(\xi R) d\phi = J_0(\xi r) J_0(\xi s) \ , \qquad (11.1.28)$$

so that

$$w(r) = \int_0^{\infty} s k(r,s) p(s) ds \ , \qquad (11.1.29)$$

where

$$k(r,s) = 2\pi \int_0^{\infty} \xi \tilde{H}(\xi) J_0(\xi r) J_0(\xi s) d\xi \qquad (11.1.30)$$

For the isotropic elastic half-space $\tilde{H}(\xi_1,\xi_2) = \vartheta/(2\pi\xi)$ so that $H(x,y) = \vartheta/(2\pi r)$ and

$$H(x-x_0,y-y_0) = \vartheta/[2\pi\{(x-x_0)^2+(y-y_0)^2\}^{1/2}] \qquad (11.1.31)$$

in agreement with (2.5.27). In this case the kernel $k(r,s)$ is

$$k(r,s) = \vartheta \int_0^{\infty} J_0(\xi r) J_0(\xi s) d\xi \ , \qquad (11.1.32)$$

in agreement with (10.2.28).

If

$$\tilde{H}_\lambda(\xi) = (2\pi k)^{-1} \exp(-\lambda\xi) \ , \qquad (11.1.33)$$

where $\lambda > 0$ then

$$H_\lambda(x,y) = (2\pi k)^{-1} \lambda (r^2+\lambda^2)^{-3/2} \qquad (11.1.34)$$

The limiting value of this kernel as $\lambda \to 0$ has a special interest. The limiting value is in fact the kernel for the Winkler foundation i.e.,

$$\lim_{\lambda\to 0} H_\lambda(x,y) = k^{-1}\delta(x)\delta(y) \ . \qquad (11.1.35)$$

To show this it is sufficient to prove that

$$\lim_{\lambda\to 0} H_\lambda(x,y) = 0 \quad \text{if} \quad x^2+y^2 \neq 0 \;, \tag{11.1.36}$$

and that the integral of $H_\lambda(x,y)$ over all x,y is k^{-1}. Equation (11.1.36) follows immediately from (11.1.34), while

$$\int_{-\infty}^{\infty}\int_{-\infty}^{\infty} H_\lambda(x,y)dxdy = k^{-1}\int_0^\infty \lambda r(r^2+\lambda^2)^{-3/2}dr = k^{-1} \;. \tag{11.1.37}$$

Thus for a Winkler foundation

$$\tilde{H}_W(\xi) = (2\pi k)^{-1} \;, \quad H_W(x,y) = k^{-1}\delta(x)\delta(y) \;. \tag{11.1.38}$$

If

$$\tilde{H}(\xi) = (2\pi k)^{-1}+\vartheta/(2\pi\xi) \;, \tag{11.1.39}$$

then

$$H(x,y) = k^{-1}\delta(x)\delta(y)+\vartheta/(2\pi r) \;, \tag{11.1.40}$$

which is the kernel for an elastic half-space covered by a Winkler foundation, as discussed by Zhemochkin and Sinitsyn (1948).

In §12.5 nonhomogeneous elastic media are considered and it is shown in particular that a half-space $z \geq 0$ in which the shear modulus is directly proportional to z, and in which $\nu = 1/2$, is equivalent to a Winkler foundation in the sense that

$$H(x,y) = k^{-1}\delta(x)\delta(y) \;. \tag{11.1.41}$$

In §11.2 we shall discuss the contact between plates and foundations. The term 'thin plate' is used there. This refers to a plate governed by the classical Poisson-Kirchhoff theory, as described for instance in Selvadurai (1979, pp. 165-169). This theory, by using a number of simplifying assumptions, concentrates on the displacement w normal to the surface, $z = 0$, of the plate, assumes that

$$u = 0 = v \quad , \quad w = w_1(x,y) \tag{11.1.42}$$

and deduces that the equation governing the deflection due to a pressure $p(x,y)$ is

$$D_1 \nabla_1^4 w_1(x,y) = p(x,y) \tag{11.1.43}$$

where $\nabla_1^2 = \partial^2/\partial x^2 + \partial^2/\partial y^2$ and D_1 is the flexural rigidity of the plate.

Examples 11.1

1. Show that if $\tilde{H}(\xi_1,\xi_2)$ is a function of $(\xi_1^2+\gamma^2\xi_2^2)^{1/2}$, then $H(x,y)$ is a function of $(x^2+\gamma^2 y^2)^{1/2}$.
2. Show that if $\tilde{H}(\xi_1,\xi_2) = [P_n(\xi_1,\xi_2)]^{-1}$, where P_n is a polynomial in the ξ_1, ξ_2 and of arbitrary degree n, and p, w are linked by equation (11.1.22), then

   $$P_n(i\partial/\partial x \quad , \quad i\partial/\partial y) w(x,y) = p(x,y)$$

   so that equation (11.1.22) may be transformed into a single differential equation in w. Derive the differential equation for the two-parameter foundation specified by $P_n(\xi_1,\xi_2) = k_1\xi^2 + k_2$. This model occurs in the papers by Vlasov and Leont'ev (1966), Filonenko-Borodich (1940), (1945) and Pasternak (1954).

11.2 *Complete contact between infinite plates and foundations*

Thin plate on a Vlasov-Leont'ev foundation

In the simplest model of a thin plate (1) in complete contact with a foundation (2) over the whole of the plane $z = 0$ under the action of normal forces, which may be tensile or compressive, the displacement boundary condition is

$$w_1(x,y,0) = w_2(x,y,0) \quad \text{for all } x,y \ .$$

The equations governing the deflection of a thin plate of flexural rigidity D_1 in contact with a two-parameter foundation of Vlasov and Leont'ev type (equation (11.1.18)) are

$$\left.\begin{aligned} D_1\nabla_1^4 w_1(x,y) &= p_1(x,y) - p_2(x,y)\ , \\ r_{22}\mu\nabla_1^2 w_2(x,y) - s_{22}[2\mu(1-\nu)/(1-2\nu)]w_2(x,y)+p_2(x,y) &= 0\ , \end{aligned}\right\} \tag{11.2.1}$$

$$w_1(x,y) = w_2(x,y,0)\ , \tag{11.2.2}$$

where suffices 1, 2 refer to the plate and foundation respectively and r_{22}, s_{22} are defined in (11.1.15). These equations yield

$$\{D_1\nabla_1^4 - r_{22}\mu\nabla_1^2 + s_{22}[2\mu(1-\nu)/(1-2\nu)]\}w_1(x,y) = p_1(x,y)\ , \tag{11.2.3}$$

which may conveniently be written

$$(\nabla_1^4 - 2r^2\nabla_1^2 + s^4)w_1(x,y) = p_1(x,y)/D_1\ . \tag{11.2.4}$$

This equation may be solved by using the two-dimensional Fourier transforms. Putting

$$w_1(x,y) = F^*_{(2)}[W_1(\xi_1,\xi_2);(x,y)]\ , \tag{11.2.5}$$

$$p_1(x,y) = F^*_{(2)}[P_1(\xi_1,\xi_2);(x,y)] \ , \qquad (11.2.6)$$

and writing $\xi^2 = \xi_1^2+\xi_2^2$ we obtain

$$(\xi^4-2r^2\xi^2+s^4)W_1 = P_1/D_1 \ , \qquad (11.2.7)$$

from which particular solutions may be obtained.

Vlasov-Leont'ev Plate-Foundation System

Vlasov and Leont'ev's method lends itself naturally to the approximate treatment of thick plates. Thus consider a thick plate, of thickness h lying on a two-parameter foundation of thickness H, as shown in Figure 11.2.1. The plate may be modelled as a Vlasov-Leont'ev layer of shear modulus μ_1, Poisson's ratio ν_1 and the foundation as a layer with parameters μ_2, ν_2. The displacements u, v, w may be taken in the form

$$u(x,y,z) = u_1(x,y)\phi_1(z) \ , \quad v(x,y,z) = v_1(x,y)\phi_1(z) \ , \qquad (11.2.8)$$

$$w(x,y,z) = w_1(x,y)\psi_1(z) + w_2(x,y)\psi_2(z) \ , \qquad (11.2.9)$$

where for $0 < z < h$

$$\phi_1(z) = (h-2z)/2 \ , \quad \psi_1(z) = (h-z)/h \ , \quad \psi_2(z) = 1 \ , \qquad (11.2.10)$$

and for $h < z < h+h$

$$\phi_1(z) = 0 \ , \quad \psi_1(z) = 0 \ , \quad \psi_2(z) = (H+h-z)/H \ . \qquad (11.2.11)$$

The function $\phi_1(z)$ governs shear distortion of the plate while $\psi_1(z)$ describes its vertical compression. It is assumed that there is no friction between plate and foundation. The equations governing the deformation of the system under a distributed normal load may be obtained by using the general theory (11.1.3) - (11.1.12) with no body forces and with

$$\tau_{xz} = 0 = \tau_{yz} , \quad \tau_{zz} = -q(x,y) \text{ on } z = 0 , \tag{11.2.12}$$

and with the integration carried out over the total depth H+h.

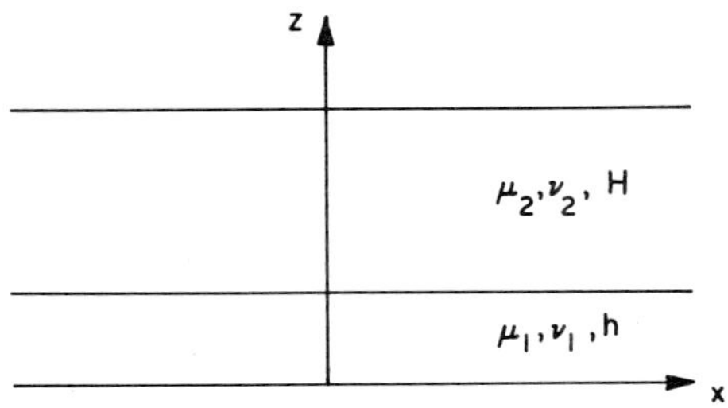

Figure 11.2.1 - A plate of thickness h lies on a layer of thickness H

Thin Plate on a Popov Foundation

If Popov's model of the foundation is used then the equations governing the deflection of a thin plate of flexural rigidity D_1 in contact with the foundation over Ω, the entire plane $z = 0$, are

$$D_1 \nabla_1^4 w_1(x,y) = p_1(x,y) - p_2(x,y) , \tag{11.2.13}$$

$$w_2(x,y,0) = \iint_{\Omega} H(x-x_0, y-y_0) p_2(x_0,y_0) dx_0 dy_0 , \tag{11.2.14}$$

where $p_1(x,y)$ is the external load and $p_2(x,y)$ is the contact pressure. Let $W_1(\xi_1,\xi_2)$, $P_1(\xi_1,\xi_2)$, $P_2(\xi_1,\xi_2)$ be the $F_{(2)}$ transforms of $w_1(x,y)$, $p_1(x,y)$, $p_2(x,y)$ respectively; then the $F_{(2)}$ convolution theorem (5.7.4) applied to (11.2.14) gives

$$w_2(x,y) = 2\pi(H \circ p_2)(x,y) , \tag{11.2.15}$$

and therefore (5.7.5) yields

$$W_2(\xi_1,\xi_2) = 2\pi\tilde{H}(\xi_1,\xi_2)P_2(\xi_1,\xi_2) , \tag{11.2.16}$$

or

$$P_2(\xi_1,\xi_2) = f_2(\xi_1,\xi_2)W_2(\xi_1,\xi_2) . \tag{11.2.17}$$

On the other hand, Example 11.1.2 shows that

$$D_1\xi^4W_1(\xi_1,\xi_2) = P_1(\xi_1,\xi_2)-P_2(\xi_1,\xi_2) \tag{11.2.18}$$

so that, on eliminating $P_2(\xi_1,\xi_2)$, we obtain

$$W_1(\xi_1,\xi_2) = P_1(\xi_1,\xi_2)/[D_1\xi^4+f_2(\xi_1,\xi_2)] \tag{11.2.19}$$

and

$$w_1(x,y) = F^*_{(2)}\{P_1(\xi_1,\xi_2)/[D_1\xi^4+f_2(\xi_1,\xi_2)];(x,y)\} . \tag{11.2.20}$$

This expression was essentially obtained by Korenev (1954), (1960).

Leonov (1939a) considered equations (11.2.13), (11.2.14) for a half-space, for which $H(x,y) = \vartheta_2/(2\pi r)$, and reduced the equations to a single partial differential equation, of the sixth order, in $w_1(x,y)$; when $\tilde{H}(\xi_1,\xi_2) = \vartheta_2\xi^{-1}$ equation (11.2.20) gives the transform solution of this equation. In a subsequent paper (1939b), Leonov considered the case in which the foundation is an elastic half-space, but now the displacements $u(x,y,0)$, $v(x,y,0)$ are constrained to be zero. By using the appropriate Papkovich-Neuber solution for the half-space he showed that $H(x,y) = \vartheta_2^*/(2\pi r)$, where $\vartheta_2^* = (3-4\nu_2)/[4\mu_2(1-\nu_2)]$ so that the above analysis holds with ϑ_2 replaced by ϑ_2^*.

When the loading is axisymmetric all the $F_{(2)}$ transforms may be replaced by H_0 transforms and equation (11.2.20), for instance, becomes

$$w(r) = \mathcal{H}_0\{P_1(\xi)/[D_1\xi^4+f_2(\xi)];r\} . \tag{11.2.21}$$

This equation has a long history. The particular case of a Winkler foundation was given by Hertz (1884). Other solutions were given by Marguerre (1937) for Winkler half-space; by Holl (1938), (1939) for Winkler, half-space and layer, the latter by using Marguerre's (1933) analysis for the layer; by Woinowsky-Krieger (1932), (1949) for Winkler and half-space; by Hogg (1938) for half-space and Hogg (1944) for layer. An illuminating discussion of various theoretical questions which arise in connection with the load carrying capacity of foundations is given by Reissner (1955). Further details of the theory and various graphical results are given in Sneddon, Gladwell and Coen (1975).

For a Winkler foundation equations (11.1.36), (11.2.16) yield $f_2(\xi) = k_2$,

$$w(r) = D_1^{-1}\mathcal{H}_0\{P_1(\xi)/[\xi^4+a^4];r\} , \quad a^4 = k_2/D_1 . \tag{11.2.22}$$

For a point load P_0 at the origin $P_1(\xi) = P_0/(2\pi)$ and

$$w(r) = \frac{P_0}{2\pi D_1}\int_0^\infty \frac{\xi J_0(\xi r)d\xi}{\xi^4+a^4} , \tag{11.2.23}$$

so that

$$w(0) = \frac{P_0}{2\pi D_1}\int_0^\infty \frac{\xi d\xi}{\xi^4+a^4} = \frac{P_0}{8D_1a^2} , \tag{11.2.24}$$

and

$$w(r) = -\frac{P_0}{2\pi D_1a^2}\,\mathrm{kei}_0(ar) , \tag{11.2.25}$$

where kei_0 is Kelvin's function defined by

$$K_0(z\exp(\pi i/4)) = \mathrm{ker}_0(z)+\mathrm{kei}_0(z) . \tag{11.2.26}$$

Complete contact between infinite plates and foundations

Figure 11.2.2 shows

$$\frac{w(r)}{w(0)} = -\frac{4}{\pi}\,\mathrm{kei}(ar) \tag{11.2.27}$$

plotted as a function of ar.

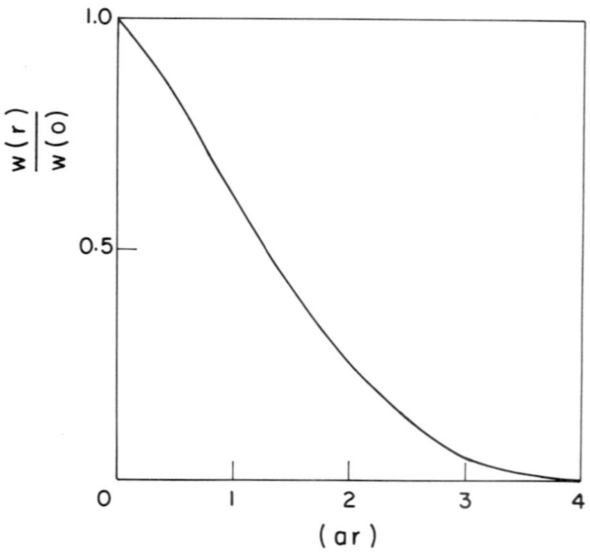

Figure 11.2.2 - Dimensionless displacement of a plate on a Winkler foundation due to a point load

The deflection is zero at $r = 3.91/a$. If it is assumed (Weitsman (1970)) that only compressive stresses are allowed between the plate and the foundation, then the plate and the foundation will lose contact at this radius, but then the theory has to be modified *ab initio* to take such separation into account. This is discussed in §11.5. Notice that the deflection (11.2.27) oscillates as r increases, the second zero being at $r = 8.34/a$, but that the deflection remains very small, i.e., $|w(r)/w(0)| < 0.025$, for $ar > 4$.

The extension of Hertz's analysis to annular plates on a Winkler foundation and to non-axisymmetric loading has been made by Reismann (1954).

For an elastic half-space, for which $\tilde{H}(\xi) = \vartheta_2/(2\pi\xi)$, equation (11.2.21) gives

$$w(r) = \frac{1}{D_1}\int_0^\infty \frac{P_2(\xi)J_0(\xi r)d\xi}{\xi^3+c^3}, \qquad c^3 = 1/(D_1\vartheta_2), \tag{11.2.28}$$

so that for a concentrated load at the origin

$$\frac{w(r)}{w(0)} = \frac{3\sqrt{3}}{2\pi}\,\omega(cr), \qquad \omega(x) = \int_0^\infty \frac{J_0(xt)dt}{1+t^2}. \tag{11.2.29}$$

Figure 11.2.3 shows $w(r)/w(0)$ as a function of r/h for five values of the dimensionless ratio $K = \vartheta_1/\vartheta_2$ where the flexural rigidity D_1 of the plate has been written

$$D_1 = \frac{E_1h^3}{12(1-\nu_1^2)} = \frac{h^3}{6\vartheta_1}. \tag{11.2.30}$$

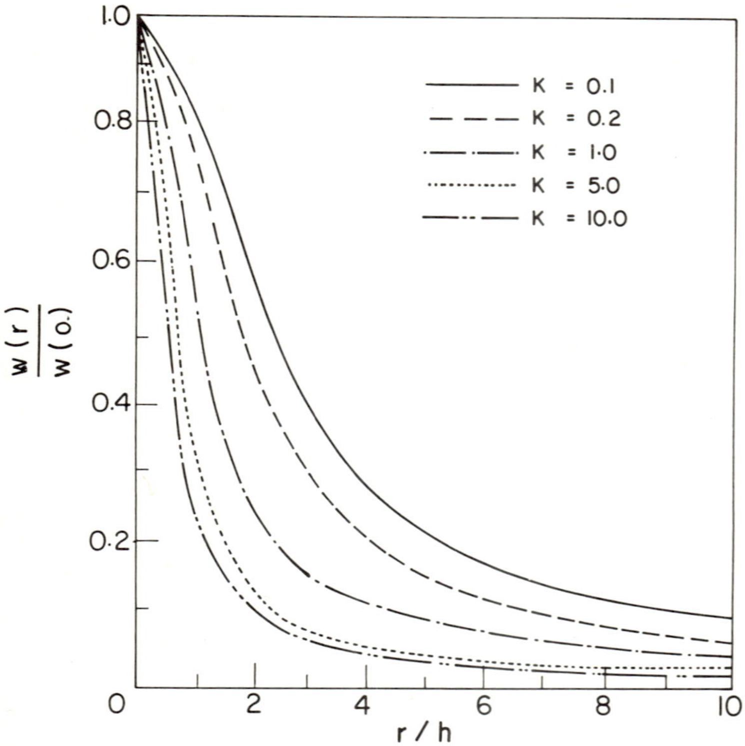

Figure 11.2.3 - Dimensionless displacement of a plate on an elastic foundation due to a point load

The foregoing analysis has related to a *thin* plate of infinite extent in complete contact with a foundation. Naghdi and Rowley (1953) use the thick plate theory derived by Reissner (1944), (1945), (1950) to analyze a plate in complete contact with a Winkler foundation; they use the so-called Schleicher (1926a), (1926b), (1926c) functions to solve the fourth order equation

$$\nabla^4 f - b^2 \nabla^2 f + f = 0 \tag{11.2.31}$$

which arises in the analysis. Frederick (1956) made a similar, more detailed study, considering both axisymmetric and asymmetric loading. Bardanov and Popov (1966) analyze the general equations governing thick plates, of infinite extent, in complete contact with an elastic half-space.

Popov (1961c) has considered the Wiener-Hopf problem of the bending of a *semi-infinite* plate resting on a linearly-deformable foundation; further references may be found in that paper and in Paskalenko and Popov (1972), and Vorob'ev, Popov and Serdyuk (1977).

Examples 11.2

1. Show by using $F_{(2)}$ transforms that the inverse of the operator L defined by

$$L[p(x_0,y_0);(x,y)] = (2\pi)^{-1}\iint_\Omega \{(x-x_0)^2+(y-y_0)^2\}^{-1/2}p(x_0,y_0)dx_0d$$

is $L^{-1} = -L\nabla_1^2$. Show also that $L\nabla_1^2 = \nabla_1^2 L$. Hence show that for the half-space, for which $H(x,y) = \vartheta_2/(2\pi r)$, equations (11.2.13 (11.2.14) may be reduced to

$$\nabla_1^6 w + (\vartheta_2 D_1)^{-2} w = D_1^{-1}\nabla_1^2 p_1 + (\vartheta_2 D_1^2)^{-1} L p_1 \ .$$

2. Show that if the displacements $u(x,y,0)$, $v(x,y,0)$ are constrained to be zero then the relation between $w(x,y,0)$ and $\tau_{zz}(x,y,0) = -p_2(x,y)$ is given by (11.2.14), where now $H(x,y) = \vartheta_2^*/(2\pi r)$, $\vartheta_2^* = (3-4\nu_2)/[4\mu_2(1-\nu_2)]$. (The boundary conditions on $z = 0$ are $u(x,y,0) = 0 = v(x,y,0)$, $\tau_{zz}(x,y,0) = -p_2$.)

11.3 *Complete contact of finite plates with foundations*

The solution for a finite circular thin plate in complete contact with a Winkler foundation was given by Hertz (1884). The governing equation is

$$D_1\nabla_1^4 w_1(r) + kw_1(r) = p_1(r) \ , \qquad (11.3.1)$$

together with the boundary conditions

$$w_1'' + \nu_1 w_1/r = 0 = w_1''' + w_1''/r - w_1'/r^2 \text{ at } r = a . \tag{11.3.2}$$

The general solution of the homogeneous equation corresponding to $p_1 = 0$ is

$$w_1(r) = A\mathrm{ber}(x) + B\mathrm{bei}(x) + C\mathrm{ker}(x) + D\mathrm{kei}(x) \tag{11.3.3}$$

where $x = (k/D)^{1/4} r$ and ber, bei, kei are Kelvin functions (Abramowitz and Stegun (1972, p. 379)). At the origin ber and bei are finite, ker has a $\ell n x$ singularity while keix behaves like $x^2 \ell n x$. If $w_1(r)$ is to be finite at the origin, then C must be zero; D will be zero unless there is a concentrated force at the origin; A, B will be determined so that the total $w_1(r)$ made up of the complementary solution (11.3.3) and the particular integral related to $p_1(r)$ satisfies the boundary conditions at the edge of the plate. Various examples may be found in Schleicher (1926a), (1926b), (1926c) and Timoshenko and Woinowsky-Krieger (1940) and Kerr (1963).

Borowicka (1936), (1939), (1943a) analyzed a finite circular thin plate lying on an elastic half-space. The governing equations are (11.1.41), (11.3.2) for the plate and (11.1.27), namely

$$w_2(r) = \int_0^a s k(r,s) p_2(s) ds , \tag{11.3.4}$$

for the half-space. Contact between plate and half-space demands

$$w_1(r) = w_2(r) , \quad 0 \leq r \leq a . \tag{11.3.5}$$

His method is as follows. The equation governing the plate is

$$D_1 \nabla_1^4 w_1(r) = p_1(r) - p_2(r) \tag{11.3.6}$$

together with the boundary conditions (11.3.2). The complementary function for equation (11.3.6) is

$$w_1(r) = c_0+c_1\ln r+c_2r^2+c_3r^2\ln r \tag{11.3.7}$$

and, in the absence of concentrated loads, $c_1 = 0 = c_3$. Assume that the contact pressure $p_2(r)$ has the form

$$p_2(r) = \sum_{n=0}^{\infty} b_{2n}r^{2n}, \qquad 0 \leq r < a, \tag{11.3.8}$$

and hence find the deflection $w_1(r)$ in the form

$$w_1(r) = c_0+c_2r^2+\sum_{m=2}^{\infty} a_{2m}r^{2m}. \tag{11.3.9}$$

The coefficients c_2, a_{2m} may be expressed in terms of $p_1(r)$, assumed constant, and the b_{2n} by using (11.3.2), (11.3.6). Substitution of (11.3.8) into (11.3.4) leads to an expression for $w_2(r)$, which, like that for $w_1(r)$, is a series in the a_{2n} and powers of r^2. Equating coefficients of r^{2n} in the series for w_1, w_2 we obtain the solution. Borowicka plots the central contact pressure as a function of

$$k = \frac{D_1\vartheta_2}{a^3} = \frac{1}{6}\cdot\frac{\vartheta_2}{\vartheta_1}\frac{h^3}{a^3} \tag{11.3.10}$$

and the contact stress distribution as a function of k and r. He shows, moreover, that as $k \to \infty$, the series for the contact pressure reduces to that for a flat punch.

Borowicka's solution suffers from the defect that it does not exhibit the correct $(a^2-r^2)^{-1/2}$ singularity in the contact pressure. Ishkova (1947) carried out an analysis for the same problem, but assumed that

$$p(r) = A(a^2-r^2)^{-1/2} + \sum_{n=0}^{\infty} B_{2n}r^{2n} \tag{11.3.11}$$

and proved the convergence of the series. Ishkova's solution may be called the 'exact' solution for the problem, but it did not yield a convenient method for calculating stresses etc. In later work, Ishkova (1957), (1958), (1960) generalized this work to obtain an exact solution for non-axisymmetric loads, and an approximate solution for a combined foundation (1962). These analyses are long and complicated.

Klubin (1952) derived an effective numerical method for the problem by using the expansion (2.6.16), namely

$$p_2(r) = (1-r^2/a^2)^{-1/2} \sum_{n=0}^{\infty} a_n P_{2n}[(1-r^2/a^2)^{1/2}] \tag{11.3.12}$$

for the contact pressure. It was shown in (2.6.15) that the displacement of the surface of the elastic half-space due to this pressure may be explicitly determined, and is

$$w_2(r) = \vartheta_2 a \sum_{n=0}^{\infty} \lambda_n a_n P_{2n}[(1-r^2/a^2)^{1/2}] \tag{11.3.13}$$

where $\lambda_n = \pi[P_{2n}(0)]^2/2$. He now solved the plate equation for the pressure (11.3.12) and equated powers of r/a in the polynomial for $w_2(r)$ and the expression for $w_1(r)$ which involves both polynomials, $(a^2-r^2)^{-1/2}$ and logarithmic terms. This latter step naturally leads to extensive arithmetic and algebraic analysis, but he proceeds to a final solution and gives tables and graphs of contact stresses etc. for a number of loading configurations.

Perhaps the most useful basis for a computational approach is to express the relations between displacement and pressure for both foundation and plate in *integral* form. This was the approach followed by Leonov (1939a) who used

$$w_2(r) \equiv w_2(x,y) = \frac{\vartheta_2}{2\pi} \iint_S \frac{p_2(x_0,y_0)\,dx_0dy_0}{\{(x-x_0)^2+(y-y_0)^2\}^{1/2}} \tag{11.3.14}$$

for the half-space and

$$D_1 w_1(r) = \int_0^a sG_1(r,s)[p_1(s)-p_2(s)]ds+c_0 \qquad (11.3.15)$$

for the plate. Here G(r,s) denotes the Green's function for the plate under the specified edge conditions. Equating (11.3.14) and (11.3.15) we obtain an integral equation for p(s) which Leonov suggested could be solved iteratively in powers of $2\pi/(D_1\vartheta_2)$; see Leonov (1939b), (1940) for other attempts at solution. If the problem is in fact axisymmetric it is best to write (11.3.14) in the axisymmetric form

$$w_2(r) = \vartheta_2 \int_0^a sk(r,s)p_2(s)ds \qquad (11.3.16)$$

to obtain the integral equation

$$\int_0^a s[k(r,s)+cG(r,s)]p_2(s)ds = c\left[\int_0^a sG(r,s)p_1(s)ds+c_0\right] \qquad (11.3.17)$$

where $c = (D_1\vartheta_2)^{-1}$. Clearly this equation may be derived for any foundation for which w_2 and p_2 are related by an equation such as (11.3.16).

The following is a brief list of other papers on plates on elastic foundations. It should be noted that different authors give different meanings to the terms used; a somewhat fuller list may be found in Iyer (1973), and an extensive one in Selvadurai (1979).

Szabo (1951a), (1951b), (1952a), (1952b) considered a finite circular plate governed by higher-order plate theory, in frictionless or frictional contact with an isotropic elastic half-space. Pickett el al (1951) and Pickett and McCormick (1952) considered infinite or circular thin plates on a Winkler foundation, a layer or a half-space. Reismann (1954) also considers such plates on a Winkler foundation. Yi-Yuan Yu (1957) considered a circular thin plate on a modified Winkler foundation that can resist rotation also. Palmov (1960) considers an infinite thin plate on an elastic layer. Golub and Mossakovskii (1960) consider a plate in which the shear effects are included, on a half-space; see also Golub (1964). Essenburg and Gulati (1966) consider the contact of *two* axisymmetric Reissner plates.

Examples 11.3

1. Integrate equation (11.3.6) under the boundary conditions (11.3.2), and the appropriate regularity conditions at $r = 0$, to give (11.3.15).

11.4 *Equations for the layer*

The equations governing *axisymmetric* layer problems may be found by using the Papkovich-Neuber potentials as in §10.2 for the half-space, and including the $\exp(\xi z)$ terms. Alternatively, a short cut to the solution may be found by using the relations between axisymmetric and x-symmetric plane solutions set forth in §10.6 together with the solution for the plane strip in §6.3. Thus Example 10.6.5 and Example 6.3.4 give

$$u(r,z) = A_2^{-1}[x^{-1}u^p(x,z);r] =$$

$$= A_2^{-1}[x^{-1}F_s\{U(\xi,z);x\};r] = H_1[U(\xi,z);r] \quad , \tag{11.4.1}$$

where U is given by the first line of (6.3.12) with i omitted. In a similar way we find the other three parts of equation (10.2.8), (10.2.9) where W, T_{13} and T_{33} are given by lines 2, 4, 5 of equation (6.3.12), again with i omitted. This means that the equations governing a layer subject to certain boundary conditions on the faces $z = 0,h$ may be obtained by replacing the F operators of the plane equations by the appropriate H operators.

For an example, consider a layer lying on a frictionless rigid foundation at $z = h$ and indented by an axisymmetric punch, rigid or elastic, over the circle $r < a$, $z = 0$. The boundary conditions

$$\tau_{rz}(r,0) = 0 = \tau_{rz}(r,h) \; ; \quad w(r,h) = 0 \, , \quad r > 0 \, , \tag{11.4.2}$$

$$w(r,0) = f(r) \, , \quad r < a \; ; \quad \tau_{zz}(r,0) = -p(r)H(a-r) \, , \tag{11.4.3}$$

are the analogues of equations (6.3.4) - (6.3.6) for the strip, so that the analogues of equations (6.3.15) - (6.3.16) are

$$\tau_{zz}(r,0) = -\mathcal{H}_0[P(\xi);r] = -p(r)H(a-r) \, , \tag{11.4.4}$$

$$w(r,0) = \vartheta \mathcal{H}_0[\xi^{-1}L(\omega)P(\xi);r] = f(r) \, , \quad r < a \, , \tag{11.4.5}$$

where $\omega = \xi h$ and $L(\omega)$ is given by (6.7.20); these equations appear to have been given first by Marguerre (1933). They may be reduced to dimensionless form as in (6.3.17) - (6.3.20) to give

$$aP_0^{-1}\tau_{zz}(r,0) = -\mathcal{H}_0[\Phi(s);\xi] = -\phi(\xi)H(1-\xi) \, , \tag{11.4.6}$$

$$(\vartheta P_0)^{-1}w(r,0) = \mathcal{H}_0[s^{-1}L(\omega)\Phi(s);\xi] = \chi(\xi) \, , \quad \xi < 1 \, , \tag{11.4.7}$$

and these may be manipulated in many ways. See Tranter (1950) for an early study of such equations. Thus the analogue of the method used in (6.3.38) is to operate on equation (11.4.6) by entry 10 of Table 5.7.1 to obtain

$$\mathcal{F}_c[\Phi(s);\tau] = \mathcal{A}_2[\xi\phi(\xi)H(1-\xi);\tau] = \psi(\tau) \, , \tag{11.4.8}$$

where $\psi(\tau) = 0$ for $\tau > 1$. Thus

$$\Phi(s) = \mathcal{F}_c[\psi(\tau);s] = (2/\pi)^{1/2}\int_0^1 \psi(\tau)\cos s\tau \, d\tau \, , \tag{11.4.9}$$

so that entry 1 of Table 5.7.1 transforms (11.4.7) into

$$\psi(\tau) + \int_0^1 H(t,\tau)\psi(t)dt = \theta(\tau) , \tag{11.4.10}$$

where

$$H(t,\tau) = \frac{2}{\pi}\int_0^\infty N(\omega)\cos st\cos s\tau ds \tag{11.4.11}$$

and

$$L(\omega) = 1+N(\omega) , \quad \theta(\tau) = A_1^{-1}[\chi(s);\tau] . \tag{11.4.12}$$

These equations were obtained by Lebedev and Ufliand (1958), who obtained a numerical solution for the case of a punch with a plane base. Pupyrev and Ufliand (1960) consider the general problem for the elastic layer when a circle separates the boundary conditions on one of the faces. Keer (1964a) considers the indentation of an elastic layer by an elastic sphere. Here Hertz' approximation is used, so that the sphere is treated as a half-space with a parabolic surface. The governing integral equations have exactly the same form as (11.4.4), (11.4.5), with a slightly changed $L(\omega)$ (see Example 11.4.4). Keer solves them for large h/a.

When the contact region is not necessarily circular, the action of a stamp upon an elastic layer may be found by using double Fourier transforms combined with the Papkovich-Neuber potentials. Thus suppose, for example, that the layer lies, without lifting, on a smooth rigid base $z = h$ while its upper surface $z = 0$ is subject to a normal load $p(x,y)$. The boundary conditions are

$$\tau_{xz}(x,y,0) = 0 = \tau_{yz}(x,y,0) , \tag{11.4.13}$$

$$\tau_{zz}(x,y,0) = -p(x,y) , \tag{11.4.14}$$

$$\tau_{xz}(x,y,h) = 0 = \tau_{yz}(x,y,h) = w(x,y,h) \ . \qquad (11.4.15)$$

It may be shown (Example 11.4.2) (Lur'e (1964)) that the $\mathcal{F}_{(2)}$ transforms of w, τ_{xz}, τ_{yz}, τ_{zz} respectively are given by

$$2\mu\xi\Delta W(\xi_1,\xi_2) = 2P(\xi_1,\xi_2)[-2(1-\nu)\sinh\xi(z-h)\sinh\xi h +$$

$$+ \xi z\cosh\xi(z-h)\sinh\xi h - \xi h\sinh\xi z] \ , \qquad (11.4.16)$$

$$\xi\Delta T_{xz}(\xi_1,\xi_2) = -2i\xi_1 P(\xi_1,\xi_2)[\xi z\cosh\xi(z-h)\sinh\xi h - \xi h\sinh\xi z] \ , \qquad (11.4.17)$$

$$\xi\Delta T_{yz}(\xi_1,\xi_2) = -2i\xi_2 P(\xi_1,\xi_2)[\xi z\cosh\xi(z-h)\sinh\xi h - \xi h\sinh\xi z] \ , \qquad (11.4.18)$$

$$\Delta T_{zz}(\xi_1,\xi_2) = -2P(\xi_1,\xi_2)[\sinh\xi h\cosh\xi(z-h) - \xi z\sinh\xi h\sinh\xi(z-h) +$$

$$+ \xi h\cosh\xi z] \ , \qquad (11.4.19)$$

where $\xi = (\xi_1^2+\xi_2^2)^{1/2}$ and $\Delta = 2\xi h+\sinh(2\xi h)$. Thus if a frictionless punch is in contact with the layer over a region Ω of the plane $z = 0$, so that the boundary condition there is

$$w(x,y,0) = \delta+\alpha x+\beta y+f(x,y) \ , \quad (x,y) \text{ in } \Omega \ , \qquad (11.4.20)$$

then the integral relation linking $w(x,y,0)$ and $p(x,y)$ is

$$2\mu\xi\Delta W(\xi_1,\xi_2) = 4(1-\nu)\sinh^2\xi h P(\xi_1,\xi_2) \ , \qquad (11.4.21)$$

or, in the notation of (11.4.5),

$$W(\xi_1,\xi_2) = \vartheta\xi^{-1}L(\omega)P(\xi_1,\xi_2) \ , \qquad (11.4.22)$$

i.e.,

$$w(x,y,0) = \vartheta\mathcal{F}^{*}_{(2)}[\xi^{-1}L(\omega)P(\xi_1,\xi_2);(\xi_1,\xi_2) \rightarrow (x,y)] \ , \qquad (11.4.23)$$

and the convolution theorem, together with the fact that $p(x,y)$ is zero outside Ω, allows this to be written

$$w(x,y,0) = \iint H_{33}(x-x_0,y-y_0)p(x_0,y_0)dx_0dy_0 \qquad (11.4.24)$$

for (x,y) in Ω; here

$$H_{33}(x,y) = \frac{\vartheta}{2\pi} F^*_{(2)}[\xi^{-1}L(\omega);(\xi_1,\xi_2) \to (x,y)] , \qquad (11.4.25)$$

and since the operand is a function of ξ the $F^*_{(2)}$ operator may be replaced by H_0 ((5.7.10)). If $p(x,y)$ and the region are axisymmetric the $F^*_{(2)}$ operator in (11.4.3) may be replaced by H_0 to yield (Example 11.4.2)

$$w(r) = \int_0^a sk_{33}(r,s)p(s)ds , \qquad (11.4.26)$$

where

$$k_{33}(r,s) = \vartheta \int_0^\infty L(\omega)J_0(\xi r)J_0(\xi s)d\xi . \qquad (11.4.27)$$

For the general, non-axisymmetric case see Dovnorovich (1964).

Equations (11.4.10) and (11.4.26) are very similar to the equations obtained in §6.3 and §6.7 for the corresponding problem of the plane strip. In face, since the axisymmetric and x-symmetric quantities are linked by the equations

$$w(r,z) = r^{-1}A_2^{-1}[w^p(x,z);r]; \ \tau_{zz}(r,z) = r^{-1}A_2^{-1}\{\tau^p_{zz}(x,z);r] , \qquad (11.4.28)$$

equation (11.4.26), for instance, may be obtained by operating on (6.3.31) by the appropriate A_2 operators. This means also that the methods described in §6.7 for solving the plane equations, for small or large h/a, may be used, with the required changes, for the corresponding space problems. We shall therefore give only a brief account of the research into the solution procedures.

V.M. Aleksandrov (1963b) considered equation (11.4.24), wrote $L(\omega) = 1 + N(\omega)$ and

$$H_{33}(x,y) = \frac{\vartheta}{2\pi r} + \frac{1}{h}\int_0^\infty N(\omega) J_0\left(\frac{\omega r}{h}\right) d\omega = \frac{\vartheta}{2\pi r} + \frac{f(r/h)}{h} , \qquad (11.4.29)$$

so that equation (11.4.24) becomes, for (x,y) in Ω,

$$w(x,y,0) = \frac{\vartheta}{2\pi}\int\int_\Omega \frac{p(x_0,y_0)}{R} dx_0 dy_0 + \frac{1}{h}\int\int_\Omega f(R/h) p(x_0,y_0) dx_0 dy_0 , \qquad (11.4.30)$$

where $R^2 = (x-x_0)^2+(y-y_0)^2$. For large h the second term may be ignored and the equation reduces to that for the half-space, for which many solutions are known. Higher approximations, valid for h large compared to the size of the contact region, are given in Vorovich and Ustinov (1959), V.M. Aleksandrov and Vorovich (1960), V.M. Aleksandrov (1963b). For very small h the argument used in §6.7 may be modified to show that H_{33} may be approximated by

$$H_{33}(x,y) = \frac{\vartheta A}{h} \delta\left(\frac{R}{h}\right) \qquad (11.4.31)$$

where δ is the Dirac delta function. Thus

$$p(x,y) = w(x,y,0)/[\vartheta A h] . \qquad (11.4.32)$$

V.M. Aleksandrov (1964b) noted that although the $p(x,y)$ found from equation (11.4.30) will have the same square root singularity at the edge of the contact region that is a characteristic of half-space contact problems, yet that found from (11.4.32) will not possess such a singularity. He then proceeds to obtain a solution that does have such a singularity by replacing $L(\omega)$ by the $L_0(\omega)$ of equation (6.7.22). See also V.M. Aleksandrov, Babeshko and Kucherov (1966).

V.M. Aleksandrov (1967c) considers the formulation given by equations (11.4.26), (11.4.27) and writes

$$k_{33}(r,s) = k_{33}^{0}(r,s) + k_{33}^{1}(r,s) , \qquad (11.4.33)$$

where $k_{33}^{0}(r,s)$ is the kernel (2.6.8) for the half-space; thus

$$k_{33}^{0}(r,s) = \vartheta \int_0^\infty J_0(\xi r)J_0(\xi s)d\xi , \qquad (11.4.34)$$

$$k_{33}^{1}(r,s) = \vartheta \int_0^\infty [L(w)-1]J_0(\xi r)J_0(\xi s)d\xi , \qquad (11.4.35)$$

and the latter may be expanded in a series

$$k_{33}^{1}(r,s) = \vartheta a^{-1} \sum_{m,n=0}^{\infty} e_{m,n}P_{2m}[(1-r^2/a^2)^{1/2}]P_{2n}[(1-s^2/a^2)^{1/2}] . \qquad (11.4.36)$$

If therefore

$$w(r) = a \sum_{n=0} b_n P_{2n}[(1-r^2/a^2)^{1/2}] , \qquad (11.4.37)$$

$$p(r) = (1-r^2/a^2)^{-1/2} \sum_{n=0}^{\infty} a_n P_{2n}[(1-r^2/a^2)^{1/2}] , \qquad (11.4.38)$$

then the analysis of §2.6 or §10.5 shows that

$$\int_0^a sk_{33}^{0}(r,s)p(s)ds = \frac{\pi}{a}\vartheta a \sum_{n=0}^{\infty} [P_{2n}(0)]^2 a_n P_{2n}[(1-r^2/a^2)^{1/2}] , \qquad (11.4.39)$$

while the orthogonality of the $P_{2n}[(1-r^2/a^2)^{1/2}]$ over $(0,a)$ gives

$$\int_0^a sk_{33}^{1}(r,s)p(s)ds = \vartheta a \sum_{m,k=0}^{\infty} (4k+1)^{-1} e_{mk} a_k P_{2m}[(1-r^2/a^2)^{1/2}] , \qquad (11.4.40)$$

so that on equating coefficients of P_{2m} on either side of (11.4.26) we obtain

$$\frac{\pi}{2}\vartheta[P_{2n}(0)]^2 a_n + \vartheta \sum_{k=0}^{\infty}(4k+1)^{-1} e_{nk} a_k = b_n \ , \quad (n = 0,1,\ldots) \ . \quad (11.4.41)$$

V.M. Aleksandrov (1967c) discusses the convergence of the series (11.4.36) and the regularity of the system (11.4.41). V.M. Aleksandrov (1969) considers more sophisticated asymptotic solutions. R.S. Rao and Shadley (1970) solve equation (11.4.10) by the method of invariant imbedding. See also Kalaba, Yakush and Zagustin (1972), (1976), Kalaba and Zagustin (1972a), (1972b), (1973), (1975) and Buell et al (1972). Potelezhko and Filippov (1967), Dhaliwal (1970), I.S. Rau and Dhaliwal (1972) and Civelek and Erdogan (1974) analyze axisymmetric contact problems for a punch indenting an elastic layer, which itself lies in complete contact with an isotropic elastic half-space: the integral equations obtained are solved iteratively or numerically. Gecit and Erdogan (1978) consider separation of a layer and a frictionless rigid foundation. Kizima (1964) obtained an approximate solution for a flat punch in adhesive contact with a layer. The action of an annular punch on a layer, which may itself be lying on a half-space, is considered by Valov (1968) and Gubenko, Kiselev and Lamzyuk (1975) and by others to whom reference is made in these papers. Shibuya, Koizumi, Nakahara and Tanaka (1975), (1977) consider a layer under normal or torsional displacement by two identical annular punches opposing each other, one on each face of the layer. Low (1964a), (1964b) considers a rigid flat-ended circular punch of radius a indenting a layer which rests on a rigid foundation from which the circular portion $0 \leq r \leq a$ has been deleted. The resulting pair of Fredholm integral equations is solved iteratively for large h/a. Bielak and Martinez (1974) and Keer and Luk (1976) have considered asymmetric problems for elastic layers. Goodman and Keer (1975) consider the influence of an elastic layer on the tangential compliance (§12.4) of bodies in contact. England (1962) considered the axisymmetric punch problem for a transversely isotropic layer, including a discussion of incomplete contact; the general formulation for such a layer has been given by Lekhnitskii (1962b). Zlatin and Ufliand (1976) study the axisymmetric contact problem for a finite elastic cylinder indenting an elastic layer. Kuz'min and Ufliand (1967) analyze a layer compressed by two rigid punches indenting opposite faces of the layer. See also Greenwood (1964) and Zakorko (1974). Hooper (1974), (1975), (1976) has carried out an extensive analysis of various problems for flexible rafts in adhesive contact with isotropic and transversely isotropic media. He analyzes the various available solutions, gives extensive bibliographies and presents many numerical results. Büfler (1971) reviews research into the theory of a multi-layered medium. See also Naumov and Shevlyakov (1967), Wu and Chiu (1967), Pao, Wu and Chiu (1971). Robinson (1973) presents a method for computing the behaviour of a layer under arbitrary normal loading and reviews similar practically-oriented research.

Examples 11.4

1. Derive equation (11.4.10) from equations (11.4.6), (11.4.7).
2. Take the Papkovich-Neuber solution (1.10.11) with $\psi_1 = 0 = \psi_2$, $\psi_3 = \psi$ and show that the $\mathcal{F}_{(2)}$ transforms of τ_{xz}, τ_{yz}, τ_{zz}, w are given by

$$T_{xz} = i\xi_1 T \ , \quad T_{yz} = i\xi_2 T \ , \quad T_{zz} = 2(1-\nu)\frac{\partial\Psi}{\partial z} - \xi^2\Phi \ ,$$

$$2\mu W = 4(1-\nu)\Psi - \frac{\partial}{\partial z}(z\Psi+\Phi) \ , \quad T = -2(1-\nu)\Psi + \frac{\partial}{\partial z}(z\Psi+\Phi) \ .$$

3. Show that if $p(x,y) = f(r)\sin\theta$, where $x = r\cos\theta$, $y = r\sin\theta$, then

$$P(\xi_1,\xi_2) = \mathcal{F}_{(2)}[p(x,y);(x,y) \to (\xi_1,\xi_2)] = i\sin\phi\mathcal{H}_1[f(r);\xi] \ ,$$

where $\xi_1 = \xi\cos\phi$, $\xi_2 = \xi\sin\phi$.

4. Use the distribution in (11.4.16) - (11.4.19) to show that the normal stress under a rigid flat-ended punch of radius a tipped through an angle γ and lying on a layer of thickness h, so that $w(r,\theta,0) = \gamma r\sin\theta$, is given by the pair of dual integral equations (Florence (1961))

$$w(r,\theta,0) \equiv \vartheta \sin\theta \mathcal{H}_1[\xi^{-1}L(w)F(\xi);r] = \gamma r \sin\theta \ , \quad r < a \ ,$$

$$\tau_{zz}(r,\theta,0) \equiv -\mathcal{H}_1[F(\xi);r] = 0 \ , \quad r > a \ .$$

Write $F(\xi) = \mathcal{F}_s[g(t);\xi]$ and obtain a Fredholm integral equation for $g(t)$.

5. Show that the integral equations governing the contact between an elastic 'sphere' (body 2) and an elastic layer (body 1) under Hertz's approximation are

$$\mathcal{H}_0[P(\xi);r] = 0 \ , \quad r > a \ ,$$

$$\mathcal{H}_0[\xi^{-1}(\vartheta_1 L(\omega)+\vartheta_2)P(\xi);r] = f(x) \ , \quad r < a.$$

11.5 *Receding contact between circular plates and foundations*

The analysis of 11.3 was based on the assumption that the plate remains in continuous contact with the foundation. However it may be easily shown (see Privarnikov (1972) for the case of a layer lying on a half-space) that, under a general distribution of load over a finite region, the contact stress between plate and foundation will become negative (i.e., tensile) over part of the supposed contact region and, unless some means is used to prevent it, separation between plate and foundation will occur. In this section we describe a number of analyses of plate-foundation configurations in which it is assumed that contact will be maintained only if the contact pressure is positive (i.e., compressive). One of the primary unknowns in such analyses is the extent of the contact area, at the boundary of which the contact pressure is zero.

The problem may be specified as follows. A circular plate of radius a and flexural rigidity D_1 rests on an axisymmetric foundation and is deformed under its own weight and some other axisymmetric load, so that it is in contact with the foundation over some (axisymmetric) region S. It is required to find the extent of S, the contact pressure and the displacements of the plate and foundation.

Receding contact between circular plates and foundations

The formulation which has been followed, more or less, by most investigators is to express the displacements of the plate and the foundation in terms of the contact pressure and external loading, in integral forms, and so to obtain an integral equation for the contact pressure. This equation is then solved under the restriction that the contact pressure is zero at the boundary of the contact region - a restriction which thus yields the extent of this region.

The most important case is that in which the contact region S is a circle of radius c. The governing equations are then

$$w_1(r) = \int_0^a sG(r,s)\{p_1(s)-p_2(s)\}ds \tag{11.5.1}$$

for the plate and

$$w_2(r) = \int_0^c sk(r,s)p_2(s)ds \tag{11.5.2}$$

for the foundation. Since contact occurs over the circle of radius c, $p_2(s)$ in (11.5.1) may be integrated over this circle only, yielding the equation

$$\int_0^c s\{k(r,s)+G(r,s)\}p_2(s)ds = \int_0^a sG(r,s)p_1(s)ds = f(s) \ . \tag{11.5.3}$$

The radius c must be found so that $p_2(c) = 0$. Before considering the solution of this equation we give a brief historical review of attempts to solve the problem.

Jung (1952) considered the foundation to be a half-space and used equations (10.2.14), (10.2.15), i.e.,

$$w_2(r) = \vartheta_2 \mathcal{H}_0[\xi^{-1}P_2(\xi);r] \ , \tag{11.5.4}$$

$$p_2(r) = \mathcal{H}_0[P_2(\xi);r] \ . \tag{11.5.5}$$

He then took the expansion

$$P_2(\xi) = \sum_{n=0}^{\infty} a_n \xi^{-(n+1/2)} J_{n+1/2}(c\xi) , \tag{11.5.6}$$

which yields a contact pressure $p_2(r)$ that is automatically zero for $r > c$, and gives a displacement which is a polynomial for $r \leq c$. However the use of the pressure in (11.5.4) leads to extremely complicated integrals. In addition, since the terms in (11.5.6) are very similar to each other, the equations obtained by equating the displacements $w_1(r)$, $w_2(r)$ in some approximate fashion, are very ill-conditioned. Jung's method was reconsidered in detail by Iyer (1973).

Weitsman (1969) considered an infinite weightless plate loaded by a concentrated load at the centre; he carried out integrations by parts in the integrals (11.5.1) and (11.5.2) and so expressed w_1, w_2 and the integral equation (11.5.3) in terms of $p_2'(s)$; the kernel has a $\ln|r^2-s^2|$ singularity. He then used a variational argument due to Noble and Hussain (1967), applicable to integral equations with a $\ln|r^2-s^2|$ singularity in the kernel, to obtain an approximate, one-term solution for $p_2'(s)$, and a resulting approximation to the contact radius. He assumes that $p_2'(s)$ has the form

$$p_2'(s) = B(c)(c^2-s^2)^{-1/2} + g(s) \tag{11.5.7}$$

where $g(s)$ is continuous at $s = c$, and uses the variational process to find c, taking $g(s) = A(c)(c^2-s^2)^{1/2}$ and choosing c so that $B(c) = 0$. However, it would appear that this way of choosing c is not correct since, *even when the correct value of c is found*, we expect $p_2(s)$ to have the form $g(s)(c^2-s^2)^{1/2}$ near $s = c$ so that $p_2'(s)$ would still have a $(c^2-s^2)^{-1/2}$ singularity. Weitsman's problem was solved by Gladwell and Iyer (1974) and by Gladwell (1975), for which see below.

Receding contact between circular plates and foundations

For an infinite, weightless plate loaded over a circular region, of radius b, and resting on a half-space, the general theory of §4.10 shows that the contact radius will be independent of the level of loading, although it will of course depend on b. For a concentrated load at the centre the problem will contain only three parameters, the contact radius c, the plate rigidity $D_1 \equiv h^3/(12\vartheta_1)$, and the foundation constant ϑ_2; dimensional analysis shows that these three parameters must be linked by a relation

$$(c/h)^3 = k(\vartheta_2/\vartheta_1) \ , \qquad (11.5.8)$$

where k is a numerical constant. Weitsman (1969) obtained the value k = 4.18 whereas Gladwell (1975), in analyzing a plate of finite radius a, found that, for large (a/h), k was almost independent of (a/h), and that for a/h = 50, k = 4.15. Weitsman's agreement with this result may well be fortuitous in that the effects of the wrong assumption concerning the nature of $p_2(s)$ near s = c, and the one term approximation used in the variational argument may have combined to produce this result. It should be noted that in the problem of the incomplete contact of a *beam*, Weitsman's assumption that $p_2'(y) \propto (c^2-y^2)^{1/2}$ near y = c led to a completely wrong numerical value for the constant corresponding to k (see equation (6.7.32)).

It should be noted that if the half-space is *rigid*, so that $\vartheta_2 = 0$, then equation (11.5.8) yields c = 0, which means that the plate has only point contact with the foundation. This physically unrealistic result is due to the use of thin plate theory; such a theory can be used only when the radius of contact is *large* compared to the thickness of the plate. An analysis in which the plate is treated as a layer is given in §11.7.

The unbonded contact of a finite circular plate, pressed by an axisymmetric load and its own weight against an elastic half-space, was considered by Gladwell (1975). The analysis runs as follows.

The plate equation (11.3.6) may be integrated to give

$$w_1'(r) = \int_0^a sf_1(r,s)\{p_1(s)-p_2(s)\}ds \;, \tag{11.5.9}$$

where

$$4D_1f_1(r,s) = \begin{cases} \kappa_1 rs^2/a^2+s^2/r \;, & s < r \;, \\ \kappa_1 rs^2/a^2+r+2r\ln(s/r) \;, & s \geq r \;. \end{cases} \tag{11.5.10}$$

The half-space equations (11.5.4), (11.5.5) may be solved by writing

$$P_2(\xi) = \int_0^c g(t)\cos\xi t\,dt \;, \tag{11.5.11}$$

and then

$$w_2(r) = \vartheta_2 \int_0^{\min(c,r)} (r^2-t^2)^{-1/2} g(t)dt \;, \tag{11.5.12}$$

and the condition for unbonded contact implies $g(c) = 0$. Equations (11.5.12) may be inverted to give

$$g(t) - g(0) = \frac{2}{\pi\vartheta_2} t \int_0^t \frac{w_2'(r)dt}{(t^2-r^2)^{1/2}} \;, \quad t \leq c \;, \tag{11.5.13}$$

and equations (11.5.9), (11.5.13) may be combined by substituting the expression for w_1' into (11.5.13), expressing $p_2(r)$ in terms of $g(t)$ to obtain a Fredholm integral equation

$$2\pi\vartheta_2 D_1 g(t) = \int_0^c H(t,u)g(u)du \;. \tag{11.5.14}$$

This equation is best solved by specifying c and finding the value of the eigenvalue $2\pi\vartheta_2 D_1$, which may be expressed in terms of Dundurs' mismatch parameter α, for which c is the correct contact radius. Further details and graphical results may be found in Gladwell (1975).

Examples 11.5

1. Show that if $P_2(\xi)$ is given by (11.5.11) then $p_2(r) = 0$ for $r > c$; show also that if $p_2(c) = 0$ then $g(c) = 0$ and that $w_2(r)$ is given by (11.5.12).
2. Derive equation (11.5.14) for $g(t)$.

11.6 *Torsional contact problems*

The analysis of torsional problems follows the same lines as that for the half-space. Thus, with the Papkovich-Neuber potential (10.2.45), the displacements and stresses are given by equation (10.2.46) and now

$$\psi(r,z) = H_1[\xi^{-1}\{A(\xi)\exp(-\xi z)+B(\xi)\exp(\xi z)\};\xi \to r] \ , \tag{11.6.1}$$

so that if the layer occupies $0 \leq z \leq h$ and if

$$v(r,h) = 0 \ , \quad \tau_{\theta z}(r,0) = -q(r) \ , \tag{11.6.2}$$

then

$$\mu v(r,z) = H_1\left[\xi^{-1}\ \frac{\sinh\xi(h-z)}{\cosh\xi h}\ Q(\xi);\xi \to r\right] , \tag{11.6.3}$$

$$\tau_{\theta z}(r,z) = -H_1\left[\frac{\cosh\xi(h-z)}{\cosh\xi h}\ Q(\xi);\xi \to r\right] \ . \tag{11.6.4}$$

where $Q(\xi) = H_1[q(r);\xi]$. Thus the dual integral equations relating to a rigid disc of radius a in contact with the layer on $z = 0$ and twisted through an angle γ are

$$\mu v(r,0) \equiv H_1[\xi^{-1}\tanh\xi h Q(\xi);\xi \to r] = \gamma r \ , \quad r \leq a \ , \tag{11.6.5}$$

$$\tau_{\theta z}(r,0) = -H_1[Q(\xi);\xi \to r] = 0 \ , \quad r > a \ . \tag{11.6.6}$$

Applying operator 11 of Table 5.7.1 to equation (11.6.6) we find

$$xA_2[\tau_{\theta z}(r,0);x] \equiv -F_s[Q(\xi);x] \equiv -g(x) = 0 \ , \quad x > a \ , \tag{11.6.7}$$

so that

$$Q(\xi) = F_s[g(x);\xi] = (2/\pi)^{1/2} \int_0^a g(x)\sin x\xi dx \ , \tag{11.6.8}$$

and applying operator 2 of Table 5.7.1 to equation (11.6.5) we obtain

$$\mu x^{-1}A_1^{-1}[rv(r,0);x] = F_s[Q(\xi);x]+F_s[(\tanh\xi h-1)Q(\xi);x] \ , \tag{11.6.9}$$

so that (Florence (1961), Gladwell (1969))

$$g(x) + \frac{2}{\pi}\int_0^a M(x,y)g(y)dy = (2/\pi)^{1/2}2\mu\gamma x \ , \tag{11.6.10}$$

where

$$M(x,y) = \int_0^\infty (\tanh\xi h-1)\sin x\xi \sin y\xi d\xi \ . \tag{11.6.11}$$

For large h/a this equation may be solved by expanding M(x,y) in the form

$$M(x,y) = \frac{1}{h}\sum_{m=1}^{\infty} a_m\left\{\left(\frac{x+y}{h}\right)^{2m} - \left(\frac{x-y}{h}\right)^{2m}\right\} \tag{11.6.12}$$

and seeking a solution in the form (Example 11.6.2)

$$g(x) = \left(\frac{2}{\pi}\right)^{1/2} 2\mu\gamma \sum_{n=0}^{\infty}\left(\frac{a}{h}\right)^n \theta_n(x) \ . \tag{11.6.13}$$

O'Connor (1966) performed a similar analysis for a layer of thickness h pressed between two identical spheres, which were so large as to be considered half-spaces. Jabali (1975) analyzed the torsional contact problem for a rigid disc of radius a attached to a layer of thickness h and shear modulus μ, which itself is in contact with a half-space of shear modulus μ_2. Dhaliwal and Singh (1978c) analyzed a non-homogeneous elastic version of this problem. See also Hassan (1979).

The contact stress in each of the problems listed above has a square root singularity at the edge (r = a) of the contact region; in practice, the adhesion will break down near this boundary and there will be slip. Various attempts have been made to construct a model in which perfect adhesion is maintained over an inner circle of radius b, while slip occurs over the annular region $b < r \leq a$. An account of these attempts for the half-space is given in §12.4; Keer (1964b) analyzed the problem for the layer and supposed that the 'friction-law', i.e., the shear stress in the slip region, was arbitrary. This includes Lubkin's (1951) assumption, that the shear stress is a multiple of the normal stress between two frictionless elastic spheres, as a special case. The boundary conditions for Keer's problem are therefore

$$v(r,\theta,0) = \gamma r \ , \quad 0 \leq r \leq b \ ; \ \tau_{\theta z}(z,\theta,0) = -\mu f(r) \ , \ b < r \leq a \ , \tag{11.6.14}$$

$$\tau_{\theta z}(r,\theta,0) = 0 \ , \quad r > a \ ; \ v(r,\theta,h) = 0 \ , \quad r \geq 0 \ . \tag{11.6.15}$$

The solution may be sought as the sum of two parts: (i) Florence's problem (11.6.10) - (11.6.11) for radius b and (ii) the solution for the boundary conditions

$$v(r,\theta,0) = 0 \ , \quad 0 \leq r \leq b \ ; \ \tau_{\theta z}(r,\theta,0) = -\mu f(r)H(a-r) \ , \quad r > b \ , \tag{11.6.16}$$

$$v(r,\theta,h) = 0 \quad , \quad r \geq 0 \ . \tag{11.6.17}$$

Keer solves this second problem using representations of the form (10.6.11), which had earlier been used in England (1962), but the problem may equally well be solved by using (11.6.5) - (11.6.6). The solution to each problem will exhibit a $(b^2-r^2)^{-1/2}$ singularity in $\tau_{\theta z}(r,\theta,0)$, as $r \to b-$; the solution is completed by finding the relation between a, b, γ and f(r) which will remove this singularity.

Thus the shear stress $\tau_{\theta z1}$ for problem (i) is

$$\tau_{\theta z1} = -A_2^{-1}[x^{-1}g_1(x);r] = \left(\frac{2}{\pi}\right)^{1/2}\frac{d}{dr}\int_r^b \frac{g_1(x)dx}{(x^2-r^2)^{1/2}} , \qquad (11.6.18)$$

where $g_1(x)$ satisfies

$$g_1(x) + \frac{2}{\pi}\int_0^b M(x,y)g_1(y)dy = \left(\frac{2}{\pi}\right)^{1/2} 2\mu\gamma x . \qquad (11.6.19)$$

Problem (ii) may be solved by writing $\tanh\xi h Q(\xi) = G_2(\xi)$ in (11.6.5) to yield

$$\mu v_2(r) \equiv H_1[\xi^{-1}G_2(\xi);r] = 0 , \qquad r \le b , \qquad (11.6.20)$$

$$\tau_{\theta z2} \equiv -H_1[\coth\xi h G_2(\xi);r] = -\mu f(r)H(a-r) , r > b , \qquad (11.6.21)$$

and operating on these equations by entries 2, 11 of Table 5.7.1 respectively we find

$$\mu x^{-1}A_1^{-1}\{rv_2(r);x\} = F_s[G_2(\xi);x] \equiv g_2(\xi) = 0 , \qquad x \le b , \qquad (11.6.22)$$

$$-xA_2\{\tau_{\theta z2};x\} \equiv g_2(x) + \frac{2}{\pi}\int_0^\infty N(x,y)g_2(y)dy = \mu F(x) , \qquad (11.6.23)$$

where

$$N(x,y) = \int_0^\infty (\coth\xi h-1)\sin x\xi \sin y\xi d\xi , \quad F(x) = \left(\frac{2}{\pi}\right)^{1/2} x\int_x^a \frac{f(r)dr}{(r^2-x^2)^{1/2}} . \qquad (11.6.24)$$

Both (11.6.19), (11.6.23) are standard Fredholm equations which may be solved for g_1, g_2. It may easily be verified that

$$\tau_{\theta z_1} = -(2/\pi)^{1/2} g_1(b)(b^2-r^2)^{-1/2} + 0(b^2-r^2)^{1/2} \text{ as } r \to b- . \qquad (11.6.25)$$

Equation (11.6.23) gives

$$\tau_{\theta z 2} = -A_2^{-1}\left\langle x^{-1}\left[g_2(x) + \frac{2}{\pi}\int_b^\infty N(x,y)g_2(y)dy\right];r\right\rangle =$$

$$= \left(\frac{2}{\pi}\right)^{1/2}\frac{d}{dr}\int_b^\infty \frac{g_2(x)dx}{(x^2-r^2)^{1/2}} - \left(\frac{2}{\pi}\right)^{3/2}\int_b^\infty N^*(r,y)g_2(y)dy , \qquad (11.6.26)$$

where

$$N^*(r,y) = \int_0^\infty \xi(\coth\xi h-1)J_1(r\xi)\sin y\xi d\xi . \qquad (11.6.27)$$

The first term in (11.6.26) may be rewritten by using equation (11.6.23) to yield

$$\tau_{\theta z 2} = r(b^2-r^2)^{-1/2}\left\{\frac{2\mu}{\pi}\int_b^a \frac{(s^2-b^2)^{1/2}f(s)ds}{s^2-r^2} - \right.$$

$$\left. - \left(\frac{2}{\pi}\right)^{3/2}\int_b^\infty g_2(y)dy\int_b^\infty \frac{(t^2-b^2)^{1/2}}{(t^2-r^2)}N^*(t,y)dt\right\} - \left(\frac{2}{\pi}\right)^{3/2}\int_b^\infty N^*(r,y)g_2(y)dy , \qquad (11.6.28)$$

so that the condition for there to be no stress singularity as $r \to b-$ is

$$g_1(b) + \frac{2}{\pi}b\int_b^\infty g_2(y)dy\int_b^\infty (t^2-b^2)^{-1/2}N^*(t,y)dt =$$

$$= \left(\frac{2}{\pi}\right)^{1/2}\mu b\int_b^a (s^2-b^2)^{-1/2}f(s)ds . \qquad (11.6.29)$$

For the half-space this agrees with Lubkin's result (Example 10.2.12)

There is a class of torsional contact problems involving a cylindrical inclusion of radius a in a half-space or layer, or problems relating to an elastic rod attached to a half-space, etc. These problems, which invariably involve a consideration of the roots of Bessel function equations, such as $J_0(\mu_k a) = 0$, and the use of dual series equations, are not considered here. The reader should refer to: Srivastav (1964), Arutiunian and Babloian (1966), Sneddon (1966a), Freeman and Keer (1967a), (1967b), Abramian and Arutuinian (1967), Grilitskii and Kizyma (1967), and Dhaliwal, Singh and Sneddon (1979). Further references may be found in these papers and in the book by Sneddon (1966a).

Examples 11.6

1. Show that for h/a > 1 the kernel M(x,y) may be expanded in the form

$$M(x,y) = \sum_{m=1}^{\infty} (-)^m \frac{(1-2^{-2m})\zeta(2m+1)}{2^{2m+1}h^{2m+1}} \{(x+y)^{2m}-(x-y)^{2m}\},$$

where $\zeta(2m+1) = \sum_{n=1}^{\infty} n^{-(2m+1)}$ is the Riemann Zeta function.

2. Use Example 11.6.1 to obtain a solution

$$g(x) = \left(\frac{2}{\pi}\right)^{1/2} 2\mu\gamma x \left\{1 + \frac{\zeta(3)}{4\pi}\left(\frac{a}{h}\right)^3 - \frac{\zeta(5)}{32\pi}\frac{(5x^2+3)}{a^2}\left(\frac{a}{h}\right)^5 + \cdots\right\},$$

and show that the total torque applied by the disc is

$$T = -2\pi\int_0^a r^2\tau_{\theta z}dr = 4\pi\left(\frac{2}{\pi}\right)^{1/2}\int_0^a tg(t)dt =$$

$$= \frac{16\mu a^3\gamma}{3}\left\{1 + \frac{\zeta(3)}{4}\left(\frac{a}{h}\right)^3 - \frac{3\zeta(5)}{16\pi}\left(\frac{a}{h}\right)^5 + \frac{\zeta^2(3)}{16\pi^2}\left(\frac{a}{h}\right)^6 + \cdots\right\}.$$

11.7 *Receding contact between layers and foundations*

The analysis of *receding* contact (see §4.10) given in §11.5 was based on the assumption that the layer could be treated as a plate governed by Kirchhoff theory; one of the physically unrealistic results of this analysis was that a plate pressed against a rigid half-space by a point load had only a point contact with the foundation. In this section the plate will be considered to be a layer governed by the theory of §11.4.

Of the many problems that fall into the title category the first that will be considered is that studied by Keer, Dundurs and Tsai (1972), the axisymmetric analogue of that analyzed in §6.8. A normal pressure distributed over a circle of radius b on the surface $z = -h$ presses a layer into a half-space occupying the region $z \geq 0$. Contact between layer and half-space is maintained over the circle $r \leq a$, $z = 0$. The boundary conditions are

$$\tau_{rz}^{(1)}(z,-h) = 0 , \quad \tau_{zz}^{(1)}(r,-h) = -q(r)H(b-r) , \tag{11.7.1}$$

$$\tau_{rz}^{(1)}(r,0) = 0 = \tau_{rz}^{(2)}(r,0) , \tag{11.7.2}$$

$$\tau_{zz}^{(1)}(r,0) = \tau_{zz}^{(2)}(r,0) = -p(r)H(a-r) , \tag{11.7.3}$$

$$w^{(1)}(r,0) = w^{(2)}(r,0) , \quad 0 \leq r \leq a . \tag{11.7.4}$$

The governing equations may be derived from equations (6.8.1) - (6.8.32) by using the relations between plane and axisymmetric states given in §10.6, in particular (Example 10.6.5)

$$[\tau_{zz}(r,z),w(r,z)] = r^{-1}A_2^{-1}[\tau_{zz}^p(x,z),w^p(x,z);r] . \tag{11.7.5}$$

Systematic use of these relations shows that

$$p(r) = \mathcal{H}_0[P(\xi);r], \quad P(\xi) = \mathcal{F}_c[\phi(x);\xi] ,$$

where $\phi(x) = 0$ for $x > a$, so that

$$p(r) = r^{-1}A_2^{-1}[\phi(x);r] =$$

$$= \left(\frac{2}{\pi}\right)^{1/2}\{(a^2-r^2)^{-1/2}\phi(a) - \int_r^a (x^2-r^2)^{-1/2}\phi'(x)dx\} , \tag{11.7.7}$$

and the condition for receding contact is that $\phi(a) = 0$. In addition the total applied force is

$$P_0 = 2\pi\int_0^a rp(r)dz = 2\pi P(0) = 2\pi\left(\frac{2}{\pi}\right)^{1/2}\int_0^a \phi(x)dx . \tag{11.7.8}$$

The displacements $w^{(1)}(r,0)$, $w^{(2)}(r,0)$ are given by

$$w^{(i)}(r,0) = H_0[\xi^{-1}V^{(i)}(\xi);r] , \tag{11.7.9}$$

and equation (6.3.12), together with (11.7.4), yields

$$(\vartheta_1+\vartheta_2)\phi(s) + \frac{2}{\pi}\vartheta_1\int_0^a K_1(s,t)\phi(t)dt = C , \tag{11.7.10}$$

where C is a constant referring to an arbitrary relative rigid-body displacement of layer and half-space and

$$K_1(s,t) = \int_0^\infty [e(u)\cos\xi t-f(u)\hat{Q}(\xi)]\cos\xi s d\xi , \tag{11.7.11}$$

where $e(u)$, $f(u)$ are given in (6.8.20) and

$$\hat{Q}(\xi) = (2\pi/P_0)H_0[q(r);\xi] . \tag{11.7.12}$$

Integration by parts applied to the integral in (11.7.10), followed by differentiation of the resulting equation, now leads to

$$\int_0^a K_2(s,t)\phi'(t) = \frac{\pi}{1+\alpha}\phi'(s) , \tag{11.7.13}$$

where α is the mismatch parameter (6.8.31) and

$$K_2(s,t) = \int_0^\infty [t\xi f(u)\hat{Q}(\xi) - e(u)\sin\xi t]\sin\xi s d\xi \ . \tag{11.7.14}$$

Keer et al (1972) investigated equation (11.7.13) for two cases: a concentrated load at $r = 0$ and a uniform load distributed over the circle $r = b$. For numerical purposes it is convenient to choose a and find the value of α ($-1 \leq \alpha \leq 1$) for which equation (11.7.13) has a solution corresponding to a purely compressive contact stress. In each case it was found that this value of α was unique. It may easily be verified that a/h is independent of the total load and is thus a function of b/h and α alone. Figure 11.7.1 shows a/h as a function of α for different values of b/h. Also shown are the values obtained by Weitsman (1969) which are described in §11.5, and which yield a = 0 for the rigid half-space (α = 1).

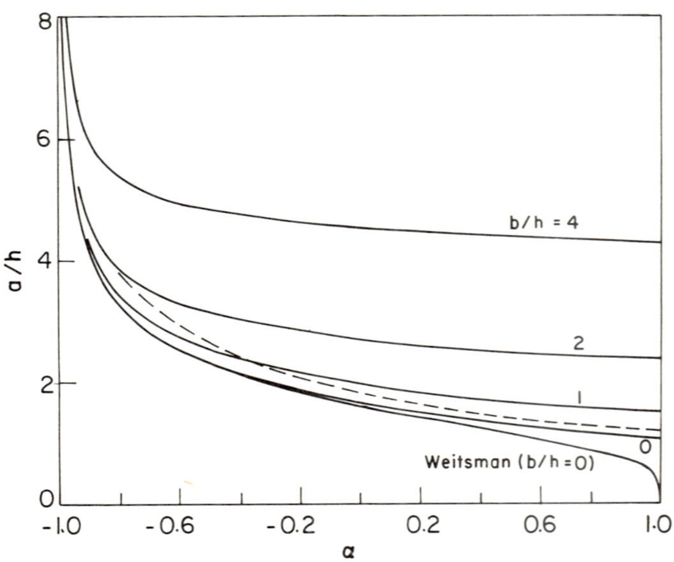

Figure 11.7.1 - The extent of contact for uniformly distributed and concentrated loads

The title problem had been studied earlier by a number of other researchers. Thus Hussain and Anderson (1969) formulated the problem of receding contact between a layer and a rigid foundation, and solved the governing integral equation by using a variational method. For a point load they found the value $a/h = 1.16$; this is shown in Figure 11.7.1 and is slightly higher than the value obtained by Keer et al. Pu and Hussain (1970) formulated the problem for a layer on a half-space and again solved the governing equation by using a one-term variational approximation, obtaining the results shown by the broken line in Figure 11.7.1. Lamziuk and Privarnikov (1971) consider a layer bounded by $z = -h$, and $z = 0$, pressed by a concentrated normal force at $z = -h$, $r = 0$ against an axisymmetric rigid convex punch. The convex surface of the punch is given by $z = f(r)$ while the lateral surface is a cylinder of radius R. The punch is in incomplete contact with the layer over a circle of radius $a < R$. The authors show that, of all possible surfaces, the flat surface is the one for which the radius of contact is largest, and that in this case $a/h = 1.1$. This value agrees with that given by Keer et al for $\alpha = 1$. Gecit and Erdogan (1978) consider separation between a layer and a smooth foundation under axisymmetric loading.

Among other papers we mention those by Tu and Gazis (1964) and Tu (1967); both concern a layer pressed between two spheres; the former gives a method suitable for large h/a, the latter, a simple numerical method for small h/a. Keer and Luk (1976) analyze a layer attached (i.e., bonded) to a half-space of the same material.

CHAPTER XII - THREE-DIMENSIONAL CONTACT PROBLEMS

"The time has come", the Walrus said,
"To talk of many things:
Of shoes - and ships - and sealing wax -
Of cabbages - and kings -
And why the sea is boiling hot -
And whether pigs have wings."
-*Through the Looking Glass*

12.1 *The anisotropic half-space*

The basic tool for the solution of contact problems for the elastic half-space is the influence function H_{jk} giving the displacements $u_j(x_1,x_2,x_3)$, especially those on the surface $x_3 = 0$, arising from a concentrated unit traction τ_{k3} at a point $(x_1',x_2',0)$. For the isotropic half-space, the use of, say, the Papkovich-Neuber representation allows contact problems to be reduced to potential problems, as discussed earlier in Chapter X. For the general anisotropic half-space this reduction is not possible, but Willis (1966) has shown, by using the two-dimensional Fourier transform (§5.7), that enough information may be found about the influence function so that many important contact problems, including that of Hertz, may be reduced to the computation of one-dimensional integrals. In this section Willis' analysis is introduced and its particularization to a transversely isotropic medium, for which a potential solution is possible, is discussed.

In the absence of body-forces the equations governing the static elastic field in a homogeneous elastic half-space are the equilibrium equations (1.4.11) - (1.4.13) which may be written

$$\tau_{jk,k} = 0 , \qquad (12.1.1)$$

where the comma denotes differentiation, and the stress-strain relations (1.8.1), namely

$$\tau_{jk} = c_{jk\ell m}e_{\ell m} \; . \tag{12.1.2}$$

The notation is as given in §1.8, so that a repeated suffix implies summation over 1, 2, 3. It is convenient to introduce the convention that Latin suffices take the values 1, 2, 3 while Greek suffices take the values 1, 2. Then by eliminating τ_{ij} between (12.1.1) and (12.1.2), and following Willis (1966), we obtain

$$c_{j\alpha k\beta}u_{k,\alpha\beta} + (c_{j\alpha k3}+c_{j3k\alpha})u_{k,\alpha 3} + c_{j3k3}u_{k,33} = 0 \; . \tag{12.1.3}$$

Now define the two-dimensional Fourier transform (§5.7) $F_{(2)}$ to be

$$\tilde{u}_j(\xi_1,\xi_2,x_3) = F_{(2)}[u_j(x_1,x_2,x_3);\; (x_1,x_2) \to (\xi_1,\xi_2)]$$

$$= \frac{1}{2\pi}\int_{-\infty}^{\infty}\int_{-\infty}^{\infty} u_j(x_1,x_2,x_3)\exp i(x_\alpha\xi_\alpha)dx_1dx_2 \; , \tag{12.1.4}$$

where $x_\alpha\xi_\alpha = x_1\xi_1+x_2\xi_2$, and take the $F_{(2)}$ transform of (12.1.3) to obtain

$$-\xi_\alpha\xi_\beta c_{j\alpha k\beta}\tilde{u}_k - i\xi_\alpha(c_{j\alpha k3}+c_{j3k\alpha})\tilde{u}_{k,3}+c_{j3k3}\tilde{u}_{k,33} = 0 \; . \tag{12.1.5}$$

This system of three, second-order ordinary differential equations has the general solution

$$\tilde{u}_j(\xi_1,\xi_2,x_3) = \sum_{A=1}^{6} a_j^A(\xi_1,\xi_2)\exp[im_A(\xi_1,\xi_2)x_3]b_A(\xi_1,\xi_2) \; , \tag{12.1.6}$$

where the m_A are the six roots of the determinantal equation

$$\left| -\xi_\alpha\xi_\beta c_{j\alpha k\beta}+m\xi_\alpha(c_{j\alpha k3}+c_{j3k\alpha})-m^2c_{j3k3}\right| = 0 \tag{12.1.7}$$

and the a_k^A are suitably normalized solutions of the equations

$$[-\xi_\alpha \xi_\beta c_{j\alpha k\beta} + m_A \xi_\alpha (c_{j\alpha k3} + c_{j3k\alpha}) - m_A^2 c_{j3k3}] a_k^A = 0 , \qquad (12.1.8)$$

where $j,k = 1, 2, 3$, $A = 1, 2, \ldots, 6$, and the $b_A(\xi_1,\xi_2)$ are arbitrary.

Since equations (12.1.3) are elliptic, the roots m_A will be non-real, and, since the coefficients in (12.1.7) are all real, the m_A will appear in complex conjugate pairs. Denoting the three roots with positive imaginary parts by m_1, m_2, m_3, we note that the solution appropriate to the half-space $x_3 \geq 0$ will be

$$\tilde{u}_j(\xi_1,\xi_2,x_3) = \sum_{n=1}^{3} a_j^n(\xi_1,\xi_2)\exp[im_n(\xi_1,\xi_2)x_3]b_n(\xi_1,\xi_2) . \qquad (12.1.9)$$

When two of the roots m_n coincide, as happens for an isotropic half-space, the theory must be modified, as will be discussed below.

The unknown $b_n(\xi_1,\xi_2)$ must be determined from the boundary conditions. If these relate to specified surface tractions on $x_3 = 0$ then it is sufficient to consider a concentrated point load at the origin; any distributed load may then be obtained by superimposing point loads. Thus suppose that

$$\tau_{j3}(x_1,x_2,0) = -\delta_{jk}\delta(x_1)\delta(x_2) , \qquad (12.1.10)$$

where k can take any value 1, 2, 3, δ_{jk} is the Kronecker delta and $\delta(x_1)$ is the Dirac delta function. Then

$$\tilde{\tau}_{j3}(\xi_1,\xi_2,0) = -\delta_{jk}/(2\pi) . \qquad (12.1.11)$$

But

$$\tau_{j3}(x_1,x_2,x_3) = c_{j3\ell\alpha}u_{\ell,\alpha} + c_{j3\ell3}u_{\ell,3} , \qquad (12.1.12)$$

so that

$$\tilde{\tau}_{j3}(\xi_1,\xi_2,x_3) = -i\xi_\alpha c_{j3\ell\alpha}\tilde{u}_\ell + c_{j3\ell3}\tilde{u}_{\ell,3} , \tag{12.1.13}$$

and combining (12.1.9), (12.1.11) and (12.1.13) we obtain

$$\sum_{n=1}^{3} \{-i\xi_\alpha c_{j3\ell\alpha} + im_n c_{j3\ell3}\} a_\ell^n b_n(\xi_1\xi_2) = -\delta_{jk}/(2\pi) . \tag{12.1.14}$$

Equations (12.1.7) show that the m_n are homogeneous functions of degree +1 in ξ_1, ξ_2 so that the b_n, and hence the $\tilde{u}_j$, are homogeneous of degree -1 in ξ_1, ξ_2. The $\tilde{u}_j$ due to a unit load applied to the surface $x_3 = 0$ in direction k at the origin will be denoted $\tilde{H}_{jk}(\xi_1,\xi_2)$; it is a homogeneous function of ξ_1, ξ_2 of degree -1. By superposition, the surface displacements at x_1, x_2 due to a distribution

$$\tau_{k3}(x_1',x_2',0) = -g_k(x_1',x_2') \tag{12.1.15}$$

over a region S of the plane $x_3 = 0$ may be written

$$u_j(x_1,x_2) = \iint_S H_{jk}(x_1-x_1',x_2-x_2')g_k(x_1',x_2')dx_1',dx_2' , \tag{12.1.16}$$

where

$$H_{jk}(x_1,x_2) = F^*_{(2)}[\tilde{H}_{jk}(\xi_1,\xi_2);(\xi_1,\xi_2) \to (x_1,x_2)] \tag{12.1.17}$$

will be homogeneous of degree -1 in x_1, x_2. The convolution theorem for $F_{(2)}$ transforms (5.7.4) allows (12.1.16) to be written

$$u_j(x_1,x_2) = \int_{-\infty}^{\infty}\int_{-\infty}^{\infty} \tilde{H}_{jk}(\xi_1,\xi_2)G_k(\xi_1,\xi_2)\exp(-i\xi_1x_1-i\xi_2x_2)d\xi_1 d\xi_2 , \tag{12.1.18}$$

where

$$G_k(\xi_1,\xi_2) = F_{(2)}[g_k(x_1',x_2');(x_1',x_2') \to (\xi_1,\xi_2)] \ . \qquad (12.1.19)$$

Willis (1966) gives expressions for the displacements within the half-space. Svelko (1964) shows that some simple results may be found for orthotropic media particularly if the loading is axisymmetric. See also Kim and Suncheleev (1970) and Svelko (1970), (1974).

When the medium is transversely isotropic (Example 1.8.3) with the x_3-axis along the axes of symmetry the stress-strain equations yield

$$\left.\begin{aligned} \tau_{11} &= c_{11}u_{1,1}+c_{12}u_{2,2}+c_{13}u_{3,3} \ , & \tau_{23} &= c_{44}(u_{2,3}+u_{3,2}) \ , \\ \tau_{22} &= c_{12}u_{1,1}+c_{11}u_{2,2}+c_{13}u_{3,3} \ , & \tau_{31} &= c_{44}(u_{3,1}+u_{1,3}) \ , \\ \tau_{33} &= c_{13}u_{1,1}+c_{13}u_{2,2}+c_{33}u_{3,3} \ , & \tau_{12} &= c_{66}(u_{1,2}+u_{2,1}) \ , \end{aligned}\right\} \quad (12.1.20)$$

where $c_{66} = (c_{11}-c_{22})/2$, and equations (12.1.8) reduce to

$$\begin{bmatrix} c_{11}\xi_1^2+c_{66}\xi_2^2+c_{44}m^2 \ , & \frac{1}{2}(c_{11}+c_{12})\xi_1\xi_2 \ , & -(c_{13}+c_{44})\xi_1 m \\ \frac{1}{2}(c_{11}+c_{12})\xi_1\xi_2 \ , & c_{66}\xi_1^2+c_{11}\xi_2^2+c_{44}m^2 \ , & -(c_{13}+c_{44})\xi_2 m \\ -(c_{13}+c_{44})\xi_1 m \ , & -(c_{13}+c_{44})\xi_2 m \ , & c_{44}(\xi_1^2+\xi_1^2)+c_{33}m^2 \end{bmatrix} \begin{bmatrix} a_1 \\ a_2 \\ a_3 \end{bmatrix} = 0$$

(12.1.21)

These equations have the solution $a_1 = \xi_1$, $a_2 = \xi_2$, $a_3 = \beta\xi$ where $\xi^2 = \xi_1^2+\xi_2^2$, provided that

$$\left.\begin{aligned} & c_{11}\xi^2+c_{44}m^2-(c_{13}+c_{44})\beta\xi m = 0 \ , \\ & -(c_{13}+c_{44})\xi m+(c_{44}\xi^2+c_{33}m^2)\beta = 0 \ , \end{aligned}\right\} \quad (12.1.22)$$

i.e., provided that $m^2 = \kappa^2\xi^2$ where κ^2 satisfies

$$c_{33}c_{44}\kappa^4+(c_{11}c_{33}-2c_{13}c_{44}-c_{13}^2)\kappa^2+c_{11}c_{44} = 0 \ . \tag{12.1.23}$$

Thus the solutions of (12.1.21) corresponding to the first two roots may be written

$$a_\alpha^A = \xi_\alpha \ , \quad \beta_A = \frac{(c_{11}+c_{44}\kappa_A^2)}{(c_{13}+c_{44})\kappa_A} \ , \quad A = 1,\ 2, \tag{12.1.24}$$

where κ_1, κ_2 are the two roots of (12.1.23) with positive imaginary parts; the only important case in which the roots of (12.1.23) coincide occurs when the medium is isotropic. The third solution of (12.1.21) may be found by writing

$$a_1^3 = \xi_2 \ , \quad a_2^3 = -\xi_1 \ , \quad a_3^3 = 0 \ . \tag{12.1.25}$$

This will satisfy (12.1.21) provided that $m_3 = \kappa_3\xi$ and

$$\kappa_3^2 = -(c_{11}-c_{12})/(2c_{44}) \ . \tag{12.1.26}$$

The stresses τ_{j3} are now

$$\left.\begin{aligned} &\tau_{13} = c_{44}(u_{3,1}+u_{1,3}) \ , \quad \tau_{23} = c_{44}(u_{3,2}+u_{2,3}) \ , \\ &\tau_{33} = c_{13}(u_{1,1}+u_{2,2})+c_{33}u_{3,3} \ , \end{aligned}\right\} \tag{12.1.27}$$

so that the stress transforms on $x_3 = 0$ are

$$\left.\begin{aligned} \tilde{\tau}_{13} &= c_{44}i\{b_1\xi_1(m_1-a_3^1)+b_2\xi_1(m_2-a_3^2)+b_3\xi_2m_3\} \ , \\ \tilde{\tau}_{23} &= c_{44}i\{b_1\xi_2(m_1-a_3^1)+b_2\xi_2(m_2-a_3^2)-b_3\xi_1m_3\} \ , \\ \tilde{\tau}_{33} &= -i\{b_1(c_{13}\xi^2-c_{33}m_1a_3^1)+b_2(c_{13}\xi^2-c_{33}m_2a_3^2)\} \ . \end{aligned}\right\} \tag{12.1.28}$$

When therefore k = 3 in (12.1.11) we have

$$b_1 = -i(\kappa_2-\beta_2)C\xi^{-2}\ , \quad b_2 = i(\kappa_1-\beta_1)C\xi^{-2}\ , \quad b_3 = 0\ , \tag{12.1.29}$$

and

$$2\pi C\{c_{13}(\kappa_2-\beta_2-\kappa_1+\beta_1)+c_{33}[\kappa_2\beta_2(\kappa_1-\beta_1)-\kappa_1\beta_1(\kappa_2-\beta_2)]\} = 1\ ,$$

i.e.,

$$2\pi C\{(c_{13}-c_{33}\beta_1\beta_2)(\kappa_2-\kappa_1) - (c_{13}-c_{33}\kappa_1\kappa_2)(\beta_2-\beta_1)\} = 1\ . \tag{12.1.30}$$

Now equation (12.1.24) shows that

$$\beta_2-\beta_1 = -\frac{(c_{11}-c_{44}\kappa_1\kappa_2)}{(c_{13}+c_{44})\kappa_1\kappa_2}(\kappa_2-\kappa_1)\ , \quad \beta_1\beta_2 = \frac{c_{11}}{c_{33}\kappa_1\kappa_2}\ , \tag{12.1.31}$$

so that

$$2\pi C\ \frac{(\kappa_2-\kappa_1)}{\kappa_1\kappa_2}\ (c_{13}\kappa_1\kappa_2-c_{11})+(c_{13}-c_{33}\kappa_1\kappa_2)\ \frac{(c_{11}-c_{44}\kappa_1\kappa_2)}{(c_{13}+c_{44})} = 1\ ,$$

which reduces to

$$2\pi C(\kappa_2-\kappa_1)(c_{13}^2-c_{11}c_{33}) = (c_{13}+c_{44})\ . \tag{12.1.32}$$

The roots κ_1^2, κ_2^2 will be *either* (1) real and negative so that $\kappa_1^2 = -K_1^2$, $\kappa_2^2 = -K_2^2$, $\kappa_1 = iK_1$, $\kappa_2 = iK_2$ in which case C will be purely imaginary *or* (2) complex conjugates so that $\kappa_1^2 = K^2(\cos 2\gamma+i\sin 2\gamma)$, $\kappa_2^2 = \overline{\kappa}_1^2$, $\kappa_1 = K(\cos\gamma+i\sin\gamma)$, $\kappa_2 = -\overline{\kappa}_1$, $\kappa_2-\kappa_1 = -2\kappa\cos\gamma$, $\kappa_1\kappa_2 = -K^2$, so that C will be purely real.

The quantity $\tilde{H}_{jk}(\xi_1,\xi_2)$ relates to the transform $\tilde{u}_j(\xi_1,\xi_2,0)$. Taking j = k = 3 we have

$$\tilde{u}_3(\xi_1,\xi_2,0) = a_3^1b_1+a_3^2b_2+a_3^3b_3 = -iC\xi^{-1}\{\beta_1(\kappa_2-\beta_2)-\beta_2(\kappa_1-\beta_1)\} =$$

$$= \frac{ic_{11}}{2\pi(c_{11}c_{33}-c_{13}^2)}\cdot\frac{\kappa_1+\kappa_2}{\kappa_1\kappa_2}\cdot\frac{1}{\xi} = \frac{\vartheta}{2\pi\xi}\ , \tag{12.1.33}$$

and in both cases (1) and (2) the coefficient of ξ^{-1} is *real;* ϑ is a combination of the elastic constants which reduces to $(1-\nu)/\mu$ for isotropy. At points within the half-space

$$\tilde{u}_3(\xi_1,\xi_2,x_3) = -iC\xi^{-1}\{\beta_1(\kappa_2-\beta_2)\exp(i\kappa_1 x_3\xi)-\beta_2(\kappa_1-\beta_1)\exp(i\kappa_2 x_3\xi)\} , \tag{12.1.34}$$

which gives the normal displacement due to a unit normal load at the origin to be

$$u_3(x_1,x_2,x_3) = -iC\beta_1(\kappa_2-\beta_2)(x_1^2+x_2^2-\kappa_1^2x_3^2)^{-1/2}+iC\beta_2(\kappa_1-\beta_1)(x_1^2+x_2^2-\kappa_2^2x_3^2)^{-1/2} \tag{12.1.35}$$

and again it may be verified that this is always real. This shows in particular that the displacement on the plane $x_3 = 0$ due to a unit normal concentrated force at the origin is proportional to r^{-1}, just as in the isotropic case (Example 12.1.7). Further results may be found in Lekhnitskii (1962b), (1963a); practical examples of the analysis of a transversely isotropic half-space may be found in Conway and Farnham (1967b), Conway, Farnham and Ku (1967), Fabrikant (1971), Dahan and Zarka (1977) and Pouyet and Lataillade (1979). These results, and especially (12.1.33), will be used in some contact problems in §12.2.

The analysis given above for a transversely isotropic medium may be rewritten by using the representation

$$\left.\begin{aligned} u_1(x_1,x_2,x_3) &= \phi^1_{;1}+\phi^2_{;1}-\phi^3_{;2} , \\ u_2(x_1,x_2,x_3) &= \phi^1_{;2}+\phi^2_{;2}+\phi^3_{;1} , \\ u_3(x_1,x_2,x_3) &= -\frac{\beta_1}{\kappa_1}\phi^1_{;3}-\frac{\beta_2}{\kappa_2}\phi^2_{;3} , \end{aligned}\right\} \tag{12.1.36}$$

where the $\phi^k(x_1,x_2,x_3)$ are potentials satisfying

$$\left(\frac{\partial^2}{\partial x_1^2}+\frac{\partial^2}{\partial x_2^2}-\frac{1}{\kappa_k^2}\frac{\partial^2}{\partial x_3^2}\right)\phi^k = 0 , \quad k = 1, 2, 3 . \tag{12.1.37}$$

This representation is considered further in Example 12.1.6 and has been used for the solution of crack and punch problems by H.A. Elliott (1948), (1949) and Shield (1951); see also Green and Zerna (1954, p. 183) and Deutsch (1963). Svelko (1969) has developed another analysis based on the representation

$$u_1 = \phi^1_{;1}+\phi^2_{;2}\ , \quad u_2 = \phi^1_{;2}-\phi^2_{;1}\ , \quad u_3 = \phi^3_{;3}\ . \tag{12.1.38}$$

Bickford (1977) has generalized the Papkovich-Neuber solution to transversely isotropic media. When the medium is isotropic

$$c_{33} = c_{11} = \frac{2\mu(1-\nu)}{1-2\nu}\ , \quad c_{13} = c_{12} = \frac{2\mu\nu}{1-2\nu}\ , \quad c_{44} = \mu\ , \tag{12.1.39}$$

so that the roots of (12.1.23) are $\kappa_1^2 = \kappa_2^2 = \kappa_3^2 = -1$. In this degenerate case equation (12.1.9) becomes

$$\left.\begin{aligned} \tilde{u}_1(\xi_1,\xi_2,x_3) &= \exp(-\xi x_3)\{b_1\xi_1+b_2\xi_1(\xi x_3-\kappa)-2b_3\xi_2\}\ ,\\ \tilde{u}_2(\xi_1,\xi_2,x_3) &= \exp(-\xi x_3)\{b_1\xi_2+b_2\xi_2(\xi x_3-\kappa)+2b_3\xi_1\}\ ,\\ \tilde{u}_3(\xi_1,\xi_2,x_3) &= \exp(-\xi x_3\{-ib_1\xi-ib_2\xi^2x_3\}\ , \end{aligned}\right\} \tag{12.1.40}$$

and

$$\left.\begin{aligned} \tilde{\tau}_{13}(\xi_1,\xi_2,x_3) &= 2\mu\xi\exp(-\xi x_3)\{-b_1\xi_1+b_2\xi_1(2-2\nu-\xi x_3)+b_3\xi_2\}\ ,\\ \tilde{\tau}_{23}(\xi_1,\xi_2,x_3) &= 2\mu\xi\exp(-\xi x_3)\{-b_1\xi_2+b_2\xi_2(2-2\nu-\xi x_3)-b_3\xi_1\}\ ,\\ \tilde{\tau}_{33}(\xi_1,\xi_2,x_3) &= 2\mu i\xi^2\exp(-\xi x_3)\{b_1-b_2(1-2\nu-\xi x_3)\}\ . \end{aligned}\right\} \tag{12.1.41}$$

Equations (12.1.40) - (12.1.41) hold the same information as (2.5.25) - (2.5.27), as shown in Example 12.1.3.

Other punch problems for anisotropic, and particularly hexagonally elastic, half-spaces have been considered by Guidera (1975) and Guidera, Lardner and Tupholme (1978).

Examples 12.1

1. Show that for an isotropic medium the displacement transforms due to a unit concentrated normal load at the origin are given by

$$4\pi\mu\tilde{u}_1(\xi_1,\xi_2,x_3) = i\xi^{-2}\xi_1\exp(-\xi x_3)[-1+2\nu+\xi x_3] ,$$

$$4\pi\mu\tilde{u}_2(\xi_1,\xi_2,x_3) = i\xi^{-2}\xi_2\exp(-\xi x_3)[-1+2\nu+\xi x_3] ,$$

$$4\pi\mu\tilde{u}_3(\xi_1,\xi_2,x_3) = \xi^{-1}\exp(-\xi x_3)[2-2\nu+\xi x_3) ,$$

where $\xi^2 = \xi_1^2+\xi_2^2$.

2. Use Example 12.1.1 to show that for an isotropic medium

$$\tilde{H}_{13}(\xi_1,\xi_2) = \frac{-i(1-2\nu)\xi_1}{4\pi\mu\xi^2} , \quad \tilde{H}_{23}(\xi_1,\xi_2) = \frac{-i(1-2\nu)\xi_2}{4\pi\mu\xi^2} ,$$

$$\tilde{H}_{33}(\xi_1,\xi_2) = \frac{\vartheta}{2\pi\xi} ,$$

and show that the same value of $\tilde{H}_{33}$ is obtained by taking the limiting value of (12.1.33) as the medium becomes isotropic. Show also that

$$\tilde{H}_{11}(\xi_1,\xi_2) = \frac{(1-\nu)\xi_1^2+\xi_2^2}{2\pi\mu\xi^3} , \quad \tilde{H}_{21}(\xi_1,\xi_2) = \frac{\nu\xi_1\xi_2}{2\pi\mu\xi^3} ,$$

$$\tilde{H}_{31}(\xi_1,\xi_2) = \frac{i(1-2\nu)\xi_1}{4\pi\mu\xi^2} .$$

3. Evaluate the displacements $u_k(x_1,x_2,x_3)$ by taking the $F^*_{(2)}$ transforms of the $\tilde{u}_k$ in Example 12.1.1. Use

$$F^*_{(2)}[\xi^{-1}\exp(-\xi x_3);(\xi_1,\xi_2) \to (x_1,x_2)] = H_0[\xi^{-1}\exp(-\xi x_3);\xi \to r] =$$

where $r^2 = x_1^2+x_2^2$, $R^2 = x_1^2+x_2^2+x_3^2$ to derive the parts of (2.5.25) - (2.5.27) relating to the normal load.

4. Show that if the $c_{ijk\ell}$ satisfy $c_{ijk\ell} = c_{jik\ell} = c_{k\ell ij}$, then $\tilde{H}_{jk}(\xi_1,\xi_2) = \bar{\tilde{H}}_{kj}(\xi_1,\xi_2)$, where the bar denotes complex conjugate, and this implies

$$H_{jk}(x_1-x_1',x_2-x_2') = H_{kj}(x_1'-x_1,x_2'-x_2),$$

which is a particular case of the reciprocal theorem.

5. Show that for a transversely isotropic medium

$$\tilde{H}_{\alpha 3}(\xi_1,\xi_2) = \frac{i\xi_\alpha \xi^{-2}}{2\pi} \cdot \frac{(c_{13}\kappa_1\kappa_2+c_{11})}{(c_{11}c_{33}-c_{13}^2)} \cdot \frac{1}{\kappa_1\kappa_2}, \qquad \alpha = 1,\ 2.$$

6. Find the stresses $\tau_{jk}(x_1,x_2,x_3)$ corresponding to (12.1.36) and show that if $\tau_{13} = 0 = \tau_{23}$ on $x_3 = 0$ then $\phi^3 \equiv 0$, $(1-\beta_1/\kappa_1)\phi^1+(1-\beta_2/\kappa_2)\phi^2 = 0$ when $x_3 = 0$. Put

$$\phi^1 = \frac{\kappa_1}{\kappa_1-\beta_1}\,\Phi(x_1,x_2,z_1), \quad \phi^2 = \frac{-\kappa_2}{\kappa_2-\beta_2}\,\Phi(x_1,x_2,z_2),$$

where $z_\alpha = -i\kappa_\alpha x_3$ and $\Phi(x_1,x_2,x_3)$ satisfies $(\partial^2/\partial x_1^2+\partial^2/\partial x_2^2+\partial^2/\partial x_3^2)\Phi = 0$ and show that

$$u_3 = \frac{i\beta_1\kappa_1}{\kappa_1-\beta_1}\frac{\partial\Phi}{\partial z_1}(x_1,x_2,z_1) - \frac{i\beta_2\kappa_2}{\kappa_2-\beta_2}\frac{\partial\Phi}{\partial z_2}(x_1,x_2,z_2),$$

$$\tau_{33} = -(c_{13}-c_{33}\beta_1\kappa_1)\frac{\kappa_1}{\kappa_1-\beta_1}\frac{\partial^2\Phi}{\partial z_1^2}(x_1,x_2,z_1) +$$

$$+ (c_{13}-c_{33}\beta_2\kappa_2)\frac{\kappa_2}{\kappa_2-\beta_2}\frac{\partial^2\Phi}{\partial z_2^2}(x_1,x_2,z_2).$$

On $x_3 = 0$ therefore

$$u_3 = \left\{\frac{i\beta_1\kappa_1}{\kappa_1-\beta_1} - \frac{i\beta_2\kappa_2}{\kappa_2-\beta_2}\right\}\frac{\partial\Phi}{\partial x_3}(x_1,x_2,x_3)\bigg|_{x_3 = 0},$$

$$\tau_{33} = \left\{-(c_{13}-c_{33}\beta_1\kappa_1)\frac{\kappa_1}{\kappa_1-\beta_1} + \right.$$

$$\left. + (c_{13}-c_{33}\beta_2\kappa_2)\frac{\kappa_2}{\kappa_2-\beta_2}\right\}\frac{\partial^2\Phi}{\partial x_3^2}(x_1,x_2,x_3)\bigg|_{x_3 = 0},$$

so that the methods developed for the Boussinesq problem for isotropic bodies may immediately be extended to transversely isotropic bodies.

7. Show that $G_3(\xi_1,\xi_2)$ due to a unit load uniformly distributed over the circle $x_1^2+x_2^2 = a^2$ is

$$G_3(\xi_1,\xi_2) = \frac{J_1(\xi a)}{\pi\xi a},$$

so that, for a unit concentrated load at the origin $G_3(\xi_1,\xi_2) = 1/(2\pi)$. Equation (12.1.18) states that the normal displacement transform due to a distribution of normal load is given by $\tilde{u}_3(\xi_1,\xi_2,0) = 2\pi\tilde{H}_{33}(\xi_1,\xi_2)G_3(\xi_1,\xi_2)$. Use this to show that the normal displacement $u_3(x_1,x_2,0)$ due to a unit normal concentrated load at the origin is

$$u_3(x_1,x_2,0) = \vartheta/(2\pi r)\ .$$

[Use the integrals $\int_0^a rJ_0(\xi r)dr = aJ_1(\xi a)/\xi$; $\int_0^\infty J_0(\xi r)d\xi = 1/r$.]

8. Use $\tilde{H}_{33}$ from Example 12.1.2 and the integral

$$\int_0^{2\pi} \exp(ir\cos(\theta-\phi))d\theta = 2\pi J_0(r)$$

to show that the displacement $u_3(x_1,x_2,0)$ due to a load P_0 uniformly distributed over the circle $r^2 \equiv x_1^2+x_2^2 \leq a^2$ is

$$u_3(x_1,x_2,0) = \vartheta P/(\pi a) \int_0^\infty \xi^{-1}J_1(a\xi)J_0(x\xi)d\xi\ .$$

Compare this with Example 2.6.1.

12.2 *Loading over an elliptical region*

Klubin (1952) was perhaps the first to note that a normal pressure

$$p(r) = p_0(1-r^2/a^2)^{-1/2}P_{2n}[(1-r^2/a^2)^{1/2}] \tag{12.2.1}$$

applied over a circle $r \le a$ on the surface of an isotropic elastic half-space gives rise to a normal displacement $w(r)$ on the surface which, within the circle, is proportional to $P_{2n}[(1-r^2/a^2)^{1/2}]$, a polynomial in r/a. This result, which was stated in §2.6, and which appeared in §10.5 as a special case of a separated solution of the elasticity equations in oblate spheroidal co-ordinates, was rederived by Popov (1962) who found the explicit expression

$$w(r) = \lambda_n \vartheta a p_0 P_{2n}[(1-r^2/a^2)^{1/2}] \ , \tag{12.2.2}$$

where

$$\lambda_n = \pi[P_{2n}(0)]^2/2 \ . \tag{12.2.3}$$

A number of special cases of loading over an elliptical region S: $x^2/a^2+y^2/b^2 \le 1$, $z = 0$ were stated and used in §2.9. For example, it has been known since the time of Hertz that the loading

$$p(x,y) = p_0(1-x^2/a^2-y^2/b^2)^{-1/2} \ , \quad (x,y) \text{ within } S \tag{12.2.4}$$

gives a constant displacement

$$w(x,y) = \vartheta b p_0 K(e) \ , \quad (x,y) \text{ within } S \tag{12.2.5}$$

where $e^2 = 1-b^2/a^2$ and $K(\cdot)$ is the complete elliptic integral of the first kind. This and the other examples given in §2.9 were known to Mindlin (1949) and Lur'e (1964) who obtained them from the Boussinesq integrals over the loaded region which are given in §2.5. They compute the integrals by transforming to a set of polar coordinates with the field point (x,y) as origin; their analysis, besides being complicated, has a flaw, which, however, does not affect the final results. (Many exact and approximate relations, together with numerical results, were given by Deresiewicz (1960).) Galin (1947b), (1961) showed that any loading over *S* which has the form

$$p(x,y) = (1-x^2/a^2-y^2/b^2)^{-1/2}p_n(x,y) , \quad (x,y) \text{ within } S, \quad (12.2.6)$$

where $p_n(x,y)$ is a polynomial of combined degree n in x,y yields a displacement which, within *S*, is a polynomial $q_n(x,y)$; this is often called Galin's theorem. Galin did not give the explicit relationship between p_n and q_n. Willis (1966) showed that the loading

$$p(x,y) = p_0(1-x^2/a^2-y/b^2)^{1/2} , \quad (x,y) \text{ within } S , \quad (12.2.7)$$

on the surface of an *anisotropic* elastic half-space yields a quadratic polynomial displacement within *S*, and he applied this result to the problem of Hertzian contact between two anisotropic bodies. Shield (1951) had earlier considered elliptical punch and crack problems for a hexagonally aeolotropic half-space. Using the notation of §2.9 we suppose that the shapes of the two bodies, near o, their contact point in the unstressed state, are given by (2.9.1). Hertz' basic assumption is that the two bodies may be approximated by half-spaces, and that the pressure over the contact region *S* gives rise to displacement w_1, w_2 of the surfaces of the half spaces satisfying

$$w_1+w_2 = \delta-Ax^2-By^2-2Hxy , \quad (x,y) \text{ in } S . \quad (12.2.8)$$

(In the anisotropic case it is convenient to choose the x and y axes along the principal axes of the contact ellipse, not so that the term H is zero.) Willis showed that the contact pressure between the bodies will have the form (12.2.7); in the general case the relationship between p_0, a, b and δ, A, B, H will involve the evaluation of some complicated integrals, but when the medium is transversely isotropic these integrals may easily be found, as shown below.

Willis (1967) showed that the loading (12.2.6) applied to an anisotropic half-space produces a polynomial displacement $w = q_n(x,y)$ within S; again the relationships between the coefficients of the polynomials p_n, q_n involve complicated contour integrals. Willis (1968), (1970) found comparable results for an elliptical crack. Gladwell (1978) obtained explicit relationships between the polynomial p_n, q_n for a transversely isotropic medium. Shail (1978) gave a comprehensive treatment of elliptical punch and crack problems for an isotropic half-space, using ellipsoidal coordinates; Lur'e (1941) and Galin (1961) derived similar, but less extensive, results. Shibuya (1977) analyzed contact problems for an elliptic punch using a method essentially similar to that given below.

Equation (12.1.18) is a convenient starting point, and here we revert to the notation x_1, x_2 for x, y, as in §12.1. This gives the surface displacement $u_j(x_1,x_2)$ due to a distribution of surface stresses

$$\tau_{k3}(x_1',x_2',0) = -g_k(x_1',x_2'), \quad (x_1',x_2') \text{ in } S \tag{12.2.9}$$

in the form

$$u_j(x_1,x_2) = \int_{-\infty}^{\infty}\int_{-\infty}^{\infty} \tilde{H}_{jk}(\xi_1,\xi_2)G_k(\xi_1,\xi_2)\exp(-i\xi_1x_1-i\xi_2x_2)d\xi_1d\xi_2 \,, \tag{12.2.10}$$

where $G_k(\xi_1,\xi_2)$ is the $F_{(2)}$ transform of g_k. In order to evaluate $G_k(\xi_1,\xi_2)$ we use the coordinates

$$x_1' = a\rho\cos\phi \,, \quad x_2' = b\rho\sin\phi \,, \tag{12.2.11}$$

in which S is the region $|\rho| \leq 1$, $0 \leq \phi \leq 2\pi$. Suppose that in these coordinates g_k has the form

$$g_k(x_1', x_2') = F(\rho)\cos m(\phi+\phi_0) \ , \qquad (12.2.12)$$

then

$$G_k(\xi_1,\xi_2) = \frac{ab}{2\pi}\int_0^1 \rho F(\rho)d\rho \int_0^{2\pi} \cos m(\phi+\phi_0)\exp[i(a\xi_1\cos\phi+b\xi_2\sin\phi)\rho]d\phi \ . \qquad (12.2.13)$$

Put

$$a\xi_1 = c\cos\psi \ , \quad b\xi_2 = c\sin\psi \ , \qquad (12.2.14)$$

then the ϕ-integral in (12.2.13) becomes

$$I_m = \frac{1}{2\pi}\int_0^{2\pi} \cos m(\phi+\phi_0)\exp[ic\rho\cos(\phi-\psi)]d\phi \ , \qquad (12.2.15)$$

which can be shown to be

$$I_m = i^m \cos m(\psi+\phi_0)J_m(c\rho) \ , \qquad (12.2.16)$$

so that

$$G_k(\xi_1,\xi_2) = abi^m\cos m(\psi+\phi_0)\int_0^1 \rho F(\rho)J_m(c\rho)d\rho \ . \qquad (12.2.17)$$

It is required to find the surface displacements due to a loading (normal or shear) of the form (12.2.6). A typical term in $g_k(x_1', x_2')$ is therefore

$$(1-x_1'^2/a^2-x_2'^2/b^2)^{-1/2}(x_1')^p(x_2')^q = (1-\rho^2)^{-1/2}a^p b^q \rho^{p+q}\cos^p\phi\sin^q\phi \ , \qquad (12.2.18)$$

and $\cos^p\phi \ \sin^q\phi$ may be expressed as a linear combination of terms $\cos m(\phi+\phi_0)$, where the values of m have the same parity as p+q. Each term $\rho^{p+q}\cos m(\phi+\phi_0)$, with $p+q \geq m$, may itself be expressed as a combination of terms $P_n^m[(1-\rho^2)^{1/2}]\cos m(\phi+\phi_0)$ where m, n have the same parity and $n \geq m$. Thus it is sufficient to take

$$F(\rho) = (1-\rho^2)^{-1/2}P_n^m[(1-\rho^2)^{1/2}] , \tag{12.2.19}$$

where m, n have the same parity and $n \geq m$. In this case the integral in (12.2.17) is

$$I_{n,m}(c) = \int_0^1 \rho(1-\rho^2)^{-1/2}P_n^m[(1-\rho^2)^{1/2}]J_m(c\rho)d\rho , \tag{12.2.20}$$

and (Example 12.2.1)

$$I_{n,m}(c) = (-)^{(n-m)/2}P_n^m(0)j_n(c) , \tag{12.2.21}$$

where $j_n(c)$ is the spherical Bessel function of the first kind. Thus

$$G_k(\xi_1,\xi_2) = ab(-)^{n/2}P_n^m(0)\cos m(\psi+\phi_0)j_n(c) . \tag{12.2.22}$$

Now substitute this into (12.2.10), make the substitution (12.2.14) and use the fact, proved in §12.1, that $\tilde{H}_{jk}(\xi_1,\xi_2)$ is homogeneous of degree -1, so that

$$\tilde{H}_{jk}(\xi_1,\xi_2) = c^{-1}\tilde{H}_{jk}(a^{-1}\cos\psi, b^{-1}\sin\psi) = c^{-1}\omega_{jk}(\psi) . \tag{12.2.23}$$

Then

$$u_j(x_1,x_2) = P_n^m(0)\left\{(-)^{n/2}\int_0^\infty \exp(-i\zeta c)j_n(c)dc\right\}\int_0^{2\pi}\omega_{jk}(\psi)\cos m(\psi+\phi_0)d\psi , \tag{12.2.24}$$

where

$$\zeta = a^{-1}x_1\cos\psi + b^{-1}x_2\sin\psi . \tag{12.2.25}$$

The crucial matter is whether ω_{jk}, i.e., $\tilde{H}_{jk}$, is real or not. When the half-space has Green's type of anisotropy then Example 12.1.4 shows that $\tilde{H}_{jk}$ is real when $j = k$; it is real in other cases also; see Example 12.1.2. In this case it is sufficient to consider only the real part of the bracketed term in (12.2.24); this is (Abramowitz and Stegun (1972, 11.4.26))

$$\mathrm{Re}\left\{(-)^{n/2}\int_0^\infty \exp(-i\zeta c)j_n(c)dc\right\} = \begin{cases}(\pi/2)P_n(\zeta) , & |\zeta| \le 1 ,\\ 0 , & |\zeta| > 1\end{cases} \tag{12.2.26}$$

Within the ellipse, x_1, x_2 may be written

$$x_1 = ar\cos\theta , \qquad x_2 = br\sin\theta , \qquad |r| \le 1 , \tag{12.2.27}$$

so that $\zeta = r\cos(\theta-\psi)$ and $|\zeta| \le 1$. Thus, within S,

$$u_j(x_1,x_2) = (\pi/2)P_n^m(0)\int_0^{2\pi} \omega_{jk}(\psi)P_n[r\cos(\theta-\psi)]\cos m(\psi+\phi_0)d\psi \tag{12.2.28}$$

which is a polynomial in $r\cos\theta$, $r\sin\theta$, i.e., a polynomial in x_1, x_2. When $j = k = 3$, $\omega_{jk}(\psi)$ is real, so that this shows that the loading (12.2.19) gives rise to a normal surface displacement $u_3(x_1,x_2)$ which is a polynomial in x_1, x_2. This proves Galin's theorem for an anisotropic half-space.

The coefficients in the polynomial $q_n(x,y)$ may be determined only when $\omega_{jk}(\psi)$ is known. When the half-space is transversely isotropic and $j = k = 3$ then

$$\tilde{H}_{33}(\xi_1,\xi_2) = (2\pi)^{-1}\vartheta(\xi_1^2+\xi_2^2)^{-1/2} \tag{12.2.29}$$

where ϑ is a combination of the elastic constants equal to $(1-\nu)/\mu$ when the half-space is isotropic. Then

$$\omega_{33}(\psi) = (2\pi)^{-1}\vartheta[a^{-2}\cos^2\psi+b^{-2}\sin^2\psi]^{-1/2} , \tag{12.2.30}$$

$$u_3(x_1,x_2) = \vartheta b P_n^m(0) J_{n,m} , \tag{12.2.31}$$

$$J_{n,m} = \frac{1}{4}\int_0^{2\pi} (1-e^2\cos^2\psi)^{-1/2} P_n[r\cos(\theta-\psi)]\cos m(\psi+\phi_0)\,d\psi . \tag{12.2.32}$$

The Legendre functions satisfy the addition theorem (Gradshteyn and Ryhzik (1965, 8.794))

$$P_n[r\cos(\theta-\psi)] = 2 \sum_{\ell=0}^{n}{}' \frac{(n-\ell)!}{(n+\ell)!} P_n^\ell(0) P_n^\ell[(1-r^2)^{1/2}]\cos\ell(\theta-\psi) , \tag{12.2.33}$$

where ' indicates that the first term is halved. Now since $P_n^\ell(0)$ is zero when ℓ, n have opposite parity, we may consider only those terms in the integral (12.2.33) in which ℓ, m, n have the same parity. The typical term to be integrated is

$$\int_0^{2\pi} (1-e^2\cos^2\psi)^{-1/2}\cos\ell(\theta-\psi)\cos m(\psi+\theta_0)\,d\psi =$$

$$= 2\cos(m\gamma-\ell\theta)K_M(e) + 2\cos(m\gamma+\ell\theta)K_L(e) , \tag{12.2.34}$$

where $M = (m+\ell)/2$, $L = |(m-\ell)/2|$, and

$$K_p(e) = \int_0^{\pi/2} (1-e^2\cos^2\psi)^{-1/2}\cos 2p\psi\,d\psi . \tag{12.2.35}$$

Thus

$$J_{n,m} = \sum_{\ell=0}^{n}{}' \frac{(n-\ell)!}{(n+\ell)!} P_n^\ell(0) P_n^\ell[(1-r^2)^{1/2}]\{\cos(m\phi_0-\ell\theta)K_M(e) +$$

$$+ \cos(m\phi_0+\ell\theta)K_L(e)\} . \tag{12.2.36}$$

The integrals $K_p(e)$ may be expressed as combinations of the complete elliptic integrals K(e), E(e) with coefficients that are polynomials in e^{-2}. Thus by comparing items 560.04 and 561.03 of Byrd and Friedman (1971) we may deduce that

$$K_p(e) = Q_{p-1/2}(z) \ , \qquad z = 2e^{-2}-1 \ , \tag{12.2.37}$$

where $Q_{p-1/2}$ is a Legendre function of the second kind. This means that the K_p may be determined from the initial values

$$K_0(e) = K(e) \ , \quad K_1(e) = (2e^{-2}-1)K_0(e)-2e^{-2}E(e) \ , \tag{12.2.38}$$

and the recurrence

$$(2n+1)K_{n+1}(e) = 4n(2e^{-2}-1)K_n(e)-(2n-1)K_{n-1}(e) \ , \tag{12.2.39}$$

which may be deduced from the recurrence

$$(\nu-\mu+1)P^{\mu}_{\nu+1}(z) = (2\nu+1)zP^{\mu}_{\nu}(z)-(\nu+\mu)P^{\mu}_{\nu-1}(z) \tag{12.2.40}$$

satisfied by both the P^{μ}_{ν} and the Q^{μ}_{ν}. As an application of these results consider the loading (12.1.7). Here m = 0 and

$$F(\rho) = 1-x_1'^2/a^2-x_2'^2/b^2 = 1-\rho^2 = a_0P_0+a_1P_2[(1-\rho^2)^{1/2}] =$$

$$= a_0+a_1(2-3\rho^2)/2 \ , \tag{12.2.41}$$

so that $a_0 = 1/3$, $a_1 = 2/3$. Equations (12.1.31), (12.1.36) now give the normal displacement within the ellipse as

$$u_3(x_1,x_2,0) = \vartheta b\{P_0(0)J_{0,0}+2P_2(0)J_{2,0}\} \ , \tag{12.2.42}$$

where $J_{2,0} = K_0(e)$ and

$$J_{2,0} = P_2(0)P_2[(1-r^2)^{1/2}]K_0(e)+2P_2^2(0)P_2^2[(1-r^2)^{1/2}]\cos2\theta K_1(e)/4!$$

$$= -\frac{(2-3r^2)}{4}K(e) + \frac{3}{4}r^2K_1(e)\cos2\theta \ . \tag{12.2.43}$$

Now writing $r^2 = x^2/a^2+y^2/b^2$, $r^2\cos2\theta = x^2/a^2-y^2/b^2$ we recover (2.7.6)

The analysis of this section yields the required generalization of Galin's theorem for anisotropic media and, in particular, gives the normal displacements within the contact area due to loading of the form (12.1.6). Various cases of these results may be found in the literature, e.g., Thomas and Hoersch (1930), Cattaneo (1938), Way (1940), Galin (1947b), (1961), Mindlin (1949), Lur'e (1964), V.M. Aleksandrov and Solov'ev (1966) and Segedin (1968). It should be emphasized that the generalization lies in equation (12.1.28) and states that the displacement $u_j(x_1,x_2,0)$ within S, due to loading $\tau_{k3}(x_1,x_2,0)$ within S which has the form (12.1.12), (12.1.19), will be a polynomial of degree n in x_1, x_2, whenever $\tilde{H}_{jk}$ is *real*. It is known that $\tilde{H}_{jk}$ is real when j = k; it is real also when j = 1, k = 2 (or j = 2, k = 1) for an isotropic medium (Example 12.1.2). The method of this section is not well suited to the computation of displacements and stresses *off* the contact region, even for an isotropic medium; for this it is best to use ellipsoidal coordinates, as described in the following section.

Examples 12.2

1. Show that when m, n are even, (12.2.21) follows from 7.3331 of Gradshteyn and Ryzhik (1965). When m, n are both odd the integral is given in Gladwell (1978).
2. Use the values of $\tilde{H}_{11}$, $\tilde{H}_{21}$ given in Example 12.1.2 to show that the displacements inside the ellipse S due to the shear-stress

$$\tau_{13}(x_1,x_2,0) = \frac{-P}{2\pi ab}\left(1 - \frac{x_1^2}{a^2} - \frac{x_2^2}{b^2}\right)^{-1/2}, \qquad (x_1,\ x_2) \text{ in } S$$

and $\tau_{13}(x_1,x_2,0) = 0$ for (x_1,x_2) outside S are given by

$$u_1(x_1,x_2,0) = \frac{P}{2\pi\mu a}\,[K(e)-\nu B(e)]\ , \quad u_2(x_1,x_2,0) = 0\ ,$$

where B(e) is given in (2.7.7). What are the displacements on $x_3 = 0$ outside S?

12.3 *The use of ellipsoidal coordinates*

Just as the use of oblate spheroidal coordinates in §10.5 led to a unified treatment of problems in which a circle divides the boundary conditions, so the use of ellipsoidal coordinates leads to a unified treatment of problems relating to an ellipse. The treatment in §12.2, while yielding general results for anisotropic and transversely isotropic media, does not lend itself to the determination of field quantities - displacements and stresses - off the contact region. Thus even in the comparatively simple problem of finding the normal displacement w(x,y,0) due to the pressure (12.2.4), when (x,y) lies outside *S*, it is necessary to compute the integral

$$\frac{1}{2}\int_{\theta-\beta}^{\theta+\beta} [1-k^2\sin^2\chi]^{-1/2}d\chi \ , \qquad \beta = \sin^{-1}(1/r) \ .$$

This may be written as $1/2[F(\theta+\beta,k) - F(\theta-\beta,k]$, and may be expressed in the form $F(\omega,k)$ where $\omega = \sin^{-1}(1/\rho)$ and ρ is a seemingly complicated function of θ, β, k. Only when ellipsoidal coordinates are used does it become apparent that ρ is in fact one of the coordinates in a system which is designed to fit the ellipse.

There are three ways of introducing ellipsoidal coordinates, using (1) algebraic functions, (2) Weierstrassian elliptic functions, (3) Jacobian elliptic functions. The first, used by Hobson (1931) in a general treatment and by Galin (1947b), (1961), Lur'e (1940), (1941), (1964) and Green and Sneddon (1950) in contact problems, is based on the coordinates τ, σ, ρ which lie in the intervals

$$0 \le \tau^2 \le k^2, \quad k^2 \le \sigma^2 \le 1 \ , \quad 1 \le \rho^2 < \infty,$$

and are related to x, y, z by

$$k^2x^2 = a^2\rho^2\sigma^2\tau^2 , \quad k'^2k^2y^2 = a^2(\rho^2-k^2)(\sigma^2-k^2)(k^2-\tau^2) , \tag{12.3.1}$$

$$k'^2z^2 = a^2(\rho^2-1)(1-\sigma^2)(1-\tau^2) , \tag{12.3.2}$$

where $k'^2 = 1-k^2$. This is the ellipsoidal analogue of (10.5.1). A development based on Weierstrassian elliptic functions may be found in Whittaker and Watson (1952). Here we shall use Jacobian elliptic functions, following Vol. 3 of Erdelyi (1955) and Arscott (1962); the application to contact problems is based on Shail (1978). Bibliographic notes may be found in the above-mentioned works.

We introduce ellipsoidal coordinates α, β, γ by the relations

$$x = ak^2\text{sn}\alpha\,\text{sn}\beta\,\text{sn}\gamma , \quad y = -ak^2k'^{-1}\text{cn}\alpha\,\text{cn}\beta\,\text{cn}\gamma , \quad a = iak'^{-1}\text{dn}\alpha\,\text{dn}\beta\,\text{dn}\gamma , \tag{12.3.3}$$

where, for the sake of definiteness, α, β, γ are taken in the following ranges: α from $-2K$ to $2K$; β from K to $K+iK'$; γ from iK' to $K+iK'$, where $K = K(k)$ and $K(k')$ are the complete integrals of the first kind. For these ranges x, y, z will be real.

The elliptic functions are defined as in standard texts (e.g., Byrd and Friedman (1971)). Thus the basic equation is

$$\zeta(y,k) \equiv \zeta = \int_0^y \frac{dt}{[(1-t^2)(1-k^2t^2)]^{1/2}} = \int_0^\phi \frac{d\theta}{[1-k^2\sin^2\theta]^{1/2}} = F(\phi,k) . \tag{12.3.4}$$

For a given value of y, the first integral defines ζ. The elliptic function snζ gives the inverse of this equation; snζ is the value of y for which the integral yields ζ. We write

$$y = \sin\phi = \text{sn}(\zeta,k) = \text{sn}\zeta , \quad \phi = \text{am}(\zeta,k) = \text{am}\zeta , \tag{12.3.5}$$

so that 'sn' may be considered as an abbreviation of sin(amζ). The functions cnζ, dnζ are

$$\text{cn}\zeta = \cos\phi = (1-\text{sn}^2\zeta)^{1/2} , \quad \text{dn}\zeta = (1-k^2\sin^2\phi)^{1/2} = (1-k^2\text{sn}^2\zeta)^{1/2} , \tag{12.3.6}$$

and note that (12.3.4) gives $d\zeta/d\phi = (1-k^2\sin^2\phi)^{-1/2}$, so that

$$\mathrm{dn}\zeta = (1-k^2\sin^2\phi)^{1/2} = \frac{d\phi}{d\zeta} = \frac{d}{d\zeta}(\mathrm{am}\zeta) . \tag{12.3.7}$$

The other derivatives are

$$\frac{d}{d\zeta}(\mathrm{sn}\zeta) = \mathrm{cn}\zeta\mathrm{dn}\zeta , \quad \frac{d}{d\zeta}(\mathrm{cn}\zeta) = -\mathrm{sn}\zeta\mathrm{dn}\zeta ,$$

$$\frac{d}{d\zeta}(\mathrm{dn}\zeta) = -k^2\mathrm{sn}\zeta\mathrm{cn}\zeta . \tag{12.3.8}$$

The functions $\mathrm{sn}\zeta$, $\mathrm{cn}\zeta$, $\mathrm{dn}\zeta$ are doubly periodic, with periods (2K, 2iK'), (4K, 2K+2iK'), (2K, 4iK') respectively. The functions also satisfy the following relations:

$$\mathrm{sn}(-\zeta) = -\mathrm{sn}(\zeta), \quad \mathrm{cn}(-\zeta) = \mathrm{cn}(\zeta), \quad \mathrm{dn}(-\zeta) = \mathrm{dn}(\zeta) , \tag{12.3.9}$$

$$\mathrm{sn}(0) = 0 , \quad \mathrm{sn}(K) = 1 , \quad \mathrm{cn}(0) = 1 , \quad \mathrm{cn}(K) = 0 , \quad \mathrm{dn}(0) = 1 , \quad \mathrm{dn}(K) = k' , \tag{12.3.10}$$

$$\mathrm{sn}(\zeta+K) = \frac{\mathrm{cn}\zeta}{\mathrm{dn}\zeta} , \quad \mathrm{cn}(\zeta+K) = -\frac{k'\mathrm{sn}\zeta}{\mathrm{dn}\zeta} , \quad \mathrm{dn}(\zeta+K) = \frac{k'}{\mathrm{dn}\zeta} , \tag{12.3.11}$$

$$\mathrm{sn}(\zeta+iK') = \frac{1}{k\mathrm{sn}\zeta} , \quad \mathrm{cn}(\zeta+iK') = -\frac{i\mathrm{dn}\zeta}{k\mathrm{sn}\zeta} , \quad \mathrm{dn}(\zeta+iK') = -\frac{i\mathrm{cn}\zeta}{\mathrm{sn}\zeta} , \tag{12.3.12}$$

$$\mathrm{sn}(i\zeta,k) = \frac{i\mathrm{sn}(\zeta,k')}{\mathrm{cn}(\zeta,k')} , \quad \mathrm{cn}(i\zeta,k) = \frac{1}{\mathrm{cn}(\zeta,k')} , \quad \mathrm{dn}(i\zeta,k) = \frac{\mathrm{dn}(\zeta,k')}{\mathrm{cn}(\zeta,k')} . \tag{12.3.13}$$

Many other relationships may be deduced from these (Examples 12.3.1, 12.3.2). The usual abbreviations are used, e.g.,

$$\mathrm{cd}(\zeta) = \frac{\mathrm{cn}(\zeta)}{\mathrm{dn}(\zeta)} , \quad \mathrm{nd}(\zeta) = \frac{1}{\mathrm{dn}(\zeta)} , \quad \mathrm{sd}(\zeta) = \frac{\mathrm{sn}(\zeta)}{\mathrm{dn}(\zeta)} , \text{ etc.} \tag{12.3.14}$$

These relations show that if α, β, γ lie in the stated ranges, so that $\beta = K+i\xi$, $\gamma = \eta+iK'$ where ξ, η are real, then

$$x = ka\,\mathrm{sn}\alpha\,\mathrm{nd}(\xi,k')\,\mathrm{ns}\eta\ , \qquad y = ka\,\mathrm{cn}\alpha\,\mathrm{sd}(\xi,k')\,\mathrm{ds}\eta\ , \tag{12.3.15}$$

$$z = a\,\mathrm{dn}\alpha\,\mathrm{cd}(\xi,k')\,\mathrm{cs}\eta\ , \tag{12.3.16}$$

so that if $\eta = K$ then

$$x = ka\,\mathrm{sn}\alpha\,\mathrm{nd}(\xi,k')\ , \quad y = k'ka\,\mathrm{cn}\alpha\,\mathrm{sd}(\xi,k')\ , \quad z = 0\ , \tag{12.3.17}$$

and x, y lies on the ellipse

$$\frac{x^2}{[ka\,\mathrm{nd}(\xi,k')]^2} + \frac{y^2}{[k'ka\,\mathrm{sd}(\xi,k')]^2} = 1\ . \tag{12.3.18}$$

This ellipse has major and minor axes $ka\,\mathrm{nd}(\xi,k')$ and $kk'a\,\mathrm{sd}(\xi,k')$ and Example 12.3.3 shows that $|k\,\mathrm{nd}(\xi,k')| \le 1$, $|ka\,\mathrm{sd}(\xi,k')| \le 1$ so that the ellipse (12.3.18) lies within the ellipse S_0: $x^2/a^2+y^2/k'^2a^2 = 1$. This situation is described by saying that $\gamma = K+iK'$ is the region inside S_0, actually described twice. A similar argument shows that $\beta = K+iK'$ is the region of the plane $z = 0$ exterior to S_0.

The functions sn, cn, dn all satisfy second-order differential equations; these equations may be deduced from (12.3.8) and all have the form

$$\frac{d^2Y}{d\zeta^2} + (h-2k^2\mathrm{sn}^2\zeta)Y = 0\ , \tag{12.3.19}$$

where h takes the values $1+k^2$, 1, k^2 respectively.

Ellipsoidal coordinates are *orthogonal*; when Laplace's equation $\nabla^2W(x,y,z) = 0$ is expressed in terms of them, it becomes

$$(\mathrm{sn}^2\gamma-\mathrm{sn}^2\beta)\,\frac{\partial^2W}{\partial\alpha^2} + (\mathrm{sn}^2\alpha-\mathrm{sn}^2\gamma)\,\frac{\partial^2W}{\partial\beta^2} + (\mathrm{sn}^2\beta-\mathrm{sn}^2\alpha)\,\frac{\partial^2W}{\partial\gamma^2} = 0\ . \tag{12.3.20}$$

This equation will have separated solutions $W = A(\alpha)B(\beta)C(\gamma)$ if

$$(\mathrm{sn}^2\gamma-\mathrm{sn}^2\beta)\frac{A''}{A} + (\mathrm{sn}^2\alpha-\mathrm{sn}^2\gamma)\frac{B''}{B} + (\mathrm{sn}^2\beta-\mathrm{sn}^2\alpha)\frac{C''}{C} = 0\ , \qquad (12.3.21)$$

and since this must be an identity in α, β, γ, there must be constants h, ℓ such that

$$\frac{A''}{A} = \ell\mathrm{sn}^2\alpha-h\ , \quad \frac{B''}{B} = \ell\mathrm{sn}^2\beta-h\ , \quad \frac{C''}{C} = \ell\mathrm{sn}^2\gamma-h\ . \qquad (12.3.22)$$

We write $\ell = k^2n(n+1)$ and see that A, B, C must satisfy *Lamé's equation*

$$\frac{d^2Y}{d\zeta^2} + \{h-k^2n(n+1)\mathrm{sn}^2\zeta\}Y = 0\ . \qquad (12.3.23)$$

It is proved in the standard treatises that if one solution of (12.3.23) is to be doubly periodic, like sn, cn and dn, then n must be an integer, which may be taken as non-negative, and h must take one of 2n+1 eigenvalues; in that case the second solution of (12.3.23) is not periodic at all.

Clearly $Y = 1$ is a solution of (12.3.23) corresponding to $h = 0 = n$; this gives the harmonic function $W = 1$. By comparing equations (12.3.19), (12.3.23) we see that $\mathrm{sn}\zeta$, $\mathrm{cn}\zeta$, $\mathrm{dn}\zeta$ are all solutions corresponding to $n = 1$, with eigenvalues $h = 1+k^2, 1, k^2$ respectively. These give the W functions $\mathrm{sn}\alpha\mathrm{sn}\beta\mathrm{sn}\gamma$, $\mathrm{cn}\alpha\mathrm{cn}\beta\mathrm{cn}\gamma$, $\mathrm{dn}\alpha\mathrm{dn}\beta\mathrm{dn}\gamma$, which equation (12.3.3) shows are merely multiples of x, y, z respectively.

Before considering the solutions of (12.3.23) corresponding to higher values of n, we consider how the knowledge of one solution Y_1 of (12.3.23) can lead to a second solution Y_2. As in Sturm-Liouville theory, we note that if Y_1, Y_2 are solutions, then $Y_1Y_2''-Y_2Y_1'' = 0$, giving $Y_1Y_2'-Y_2Y_1' = \text{const.} = H$. Now putting $Y_2 = Y_1Z$ we see that $Y_1^2Z' = H$ so that

$$Y_2 = HY_1 \int_{\zeta_0}^{\zeta} [Y_1(w)]^{-2} dw \,. \qquad (12.3.24)$$

In normalizing the second solution one needs to choose H and the datum ζ_0. In spheroidal coordinates $\gamma = iK'$ corresponds to the sphere at infinity so that it is convenient to take $\zeta_0 = iK'$; in this case $Y_2(iK') = 0$.

If $Y_1(\zeta) = \mathrm{sn}\zeta$, then

$$Y_2(\zeta) = \mathrm{sn}\zeta I(\zeta) \,, \quad I(u) = \int_{iK'}^{u} (\mathrm{sn}w)^{-2} dw \,. \qquad (12.3.25)$$

The integral may be calculated as follows. Put $\zeta = iK'+\eta$, $w = iK'+t$; then (12.3.12) gives $\mathrm{sn}w = 1/(k\,\mathrm{sn}t)$. Put $\mathrm{sn}t = \sin\phi$, $\mathrm{sn}\eta = \sin\psi$; then $\mathrm{dn}t\,dt = d\phi$, so that $dw = dt = d\phi/\mathrm{dn}t = (1-k^2\sin^2\phi)^{-1/2}d\phi$, and

$$I(\zeta) = \int_0^{\psi} (1-k^2\sin^2\phi)^{-1/2} k^2\sin^2\phi\, d\phi = F(\psi,k)-E(\psi,k) \,. \qquad (12.3.26)$$

But $F(\psi,k) = \eta = \zeta-iK'$ and $E(\psi,k) = E(\mathrm{am}\eta,k)$, so that

$$Y_2(\zeta) = \mathrm{sn}\zeta\{\zeta-iK'-E(\mathrm{am}(\zeta-iK'),k)\} \,. \qquad (12.3.27)$$

The (first) solutions of equation (12.3.23) may all be written as the product of up to three of the terms 1, $\mathrm{sn}\zeta$, $\mathrm{cn}\zeta$, $\mathrm{dn}\zeta$, and a polynomial in $\mathrm{sn}^2\zeta$. These solutions, called *Lamé polynomials*, are labelled ${}_qE_n^m(\zeta)$ where m, n are non-negative integers, of which n is the total order of the polynomial and q is a symbol taking the values u, s, c, d, sc, sd, cd, scd to indicate which product appears before the polynomial. Thus

$$n = 0: \quad {}_uE_0^0(\zeta) = 1 \,, \quad h = 0 \,, \qquad (12.3.28)$$

$$\left.\begin{aligned} n = 1: \quad & {}_sE_1^0(\zeta) = \mathrm{sn}\zeta\,, \quad && h = 1+k^2\,, \\ & {}_cE_1^0(\zeta) = \mathrm{cn}\zeta\,, \quad && h = 1\,, \\ & {}_dE_1^0(\zeta) = \mathrm{dn}\zeta\,, \quad && h = k^2\,, \end{aligned}\right\} \tag{12.3.29}$$

$$\left.\begin{aligned} n = 2: \quad & {}_{sc}E_2^0(\zeta) = \mathrm{sn}\zeta\mathrm{cn}\zeta\,, \quad && h = 4+k^2\,, \\ & {}_{sd}E_2^0(\zeta) = \mathrm{sn}\zeta\mathrm{dn}\zeta\,, \quad && h = 1+4k^2\,, \\ & {}_{cd}E_2^0(\zeta) = \mathrm{cn}\zeta\mathrm{dn}\zeta\,, \quad && h = 1+k^2\,, \end{aligned}\right\} \tag{12.3.30}$$

$${}_uE_2^m(\zeta) = \mathrm{sn}^2\zeta-\lambda_m\,, \quad h_m = 3(1+k^2-k^2\lambda_m)\,, \tag{12.3.31}$$

$$\lambda_m = [1+k^2\pm(1-k^2k'^2)^{1/2}]/3k^2\,, \quad m = 1,\ 2\,. \tag{12.3.32}$$

The corresponding second solutions for $n = 0,\ 1$ and the constant H in (12.3.24) equal to 1, are:

$$n = 0: \quad {}_uF_0^0(\zeta) = \eta\,, \tag{12.3.33}$$

$$\left.\begin{aligned} n = 1: \quad & {}_sF_1^0(\zeta) = \mathrm{sn}\zeta[\eta-E(\eta)]\,, \\ & {}_cF_1^0(\zeta) = k'^{-2}\{\mathrm{sn}\zeta[k'^2\eta-E(\eta)]-ik\mathrm{cn}\eta\}\,, \end{aligned}\right\} \tag{12.3.34}$$

$$\begin{aligned} n = 2: \quad {}_uF_2^n(\zeta) = a_m\{&(\mathrm{sn}^2\zeta-\lambda_m)[(1-\lambda_m k^2)\eta-E(\eta)] \\ &+ \lambda_m\mathrm{cn}\eta\mathrm{dn}\eta\}\,, \end{aligned} \tag{12.3.35}$$

where $\zeta = iK'+\eta$ and $E(\eta)$ denotes $E(\mathrm{am}\eta,k)$, and

$$a_m = 1/[2\lambda_m(1-\lambda_m)(1-\lambda_m k^2)]\,. \tag{12.3.36}$$

Now consider the frictionless punch problem for the region S_0. According to §2.5 this is equivalent to finding harmonic ϕ, ψ so that $\partial\phi/\partial z = (1-2\nu)\psi$ and

$$2\mu u = -\frac{z\partial\psi}{\partial x} - \frac{\partial\phi}{\partial x}, \quad 2\mu v = -\frac{z\partial\psi}{\partial y} - \frac{\partial\phi}{\partial y}, \tag{12.3.37}$$

$$2\mu w = 2(1-\nu)\psi - z\frac{\partial\psi}{\partial z}, \quad \tau_{zz} = \frac{\partial\psi}{\partial z} - z\frac{\partial^2\psi}{\partial z^2}. \tag{12.3.38}$$

The boundary conditions for ψ are therefore

$$\vartheta\psi(x,y,0) = f(x,y), \quad (x,y) \text{ in } S_0, \tag{12.3.39}$$

$$\frac{\partial\psi}{\partial z}(x,y,0) = 0, \quad (x,y) \text{ outside } S_0. \tag{12.3.40}$$

In terms of ellipsoidal coordinates, and with the aid of Example 12.3.7, these become

$$\vartheta\psi(\alpha,\beta,K+iK') = f(\alpha,\beta), \quad \frac{\partial\psi}{\partial\beta}(\alpha,K+iK',\gamma) = 0. \tag{12.3.41}$$

For the flat elliptic punch $f(\alpha,\beta) = d_0$ and equations (12.3.41) have the immediate solution

$$\psi(\alpha,\beta,\gamma) = \vartheta^{-1}d_0(\gamma-iK')/K = \vartheta^{-1}d_0\eta/K, \tag{12.3.42}$$

and note that this is a multiple of ${}_uF_0^0(\gamma)$. The displacement on $z = 0$ outside S_0 is

$$w(\alpha,K+iK', iK'+\eta) = \vartheta\psi = d_0\eta/K. \tag{12.3.43}$$

Example 12.3.6 shows that $\mathrm{sn}\eta = 1/\rho$ so that $\eta = F(\omega,k)$ where $\sin\omega = 1/\rho$. Thus

$$w(x,y,0) = d_0F(\omega,k) \text{ outside } S_0. \tag{12.3.44}$$

The contact pressure is given by (12.3.36) as

$$p = \frac{-\partial\psi}{\partial z} = -\frac{d_0}{\vartheta K}\frac{\partial\gamma}{\partial z} = \frac{d_0}{\vartheta K}\cdot\frac{\mathrm{dn}(\xi,k')}{k'a\,\mathrm{dn}\alpha\,\mathrm{cn}(\xi,k')} = \frac{d_0}{\vartheta Kk'a}\left(1 - \frac{x^2}{a^2} - \frac{y^2}{k'^2a^2}\right)^{-1/2} \tag{12.3.45}$$

and the total force applied by the punch is

$$P = \frac{2\pi d_0 a}{\vartheta K} . \tag{12.3.46}$$

For the flat punch one may obtain stresses and displacements off the plane $z = 0$ without difficulty. Thus equations (12.3.38) show that w and τ_{zz} involve $\partial\psi/\partial z$ and $\partial^2\psi/\partial z^2$. The derivative $\partial\gamma/\partial z$ is given in Example 12.3.7; it involves α, β, γ only through the elliptic functions, so that it may be expressed as an *elementary* function of the coordinates ρ, σ, τ; see Example 12.3.8.

For the general Boussinesq punch problem we note that if $f(x,y)$ is a polynomial defined over the interior $\gamma = K+iK'$ of the ellipse S_0 then $f(x,y)$ may be expressed as a polynomial in $\mathrm{sn}\alpha$, $\mathrm{sn}\beta$, $\mathrm{cn}\alpha$, $\mathrm{cn}\beta$ symmetrical in α and β. By using the orthogonality relations discussed by Arscott (1962) we may expand f as a linear combination of Lamé products ${}_qEp_n^m(\alpha,\beta) \equiv {}_qE_n^m(\alpha)\,{}_qE_n^m(\beta)$ where q is s, c or sc. Thus it is sufficient to consider a punch with a profile which in ellipsoidal coordinates is a single term ${}_qEp_n^m(\alpha,\beta)$. Then equations (12.3.41) reduce to

$$\vartheta\psi(\alpha,\beta,K+iK') = {}_qEp_n^m(\alpha,\beta) , \quad \frac{\partial\psi}{\partial\beta}(\alpha,K+iK',\gamma) = 0 , \tag{12.3.47}$$

with solution

$$\psi(\alpha,\beta,\gamma) = \vartheta^{-1}\,{}_qEp_n^m(\alpha,\beta)\,{}_qF_n^m(\gamma)/{}_qF_n^m(K+iK') . \tag{12.3.48}$$

The associated contact pressure has the form

$$p = \frac{-\partial\psi}{\partial z} = \frac{1}{\vartheta a}\,\frac{\mathrm{nd}\alpha\,\mathrm{nd}\beta}{{}_qF_n^m(K+iK')}\,{}_qEp_n^m(\alpha,\beta)\,{}_qF_n^{m\prime}(K+iK') , \tag{12.3.49}$$

and by Example 12.3.10 this may be written

$$p = \frac{1}{\vartheta ak'}\,\frac{{}_qF_n^{m\prime}(K+iK')}{{}_qF_n^m(K+iK')}\left\{1 - \frac{x^2}{a^2} - \frac{y^2}{k'^2a^2}\right\}^{-1/2}{}_qEp_n^m(\alpha,\beta) , \tag{12.3.50}$$

which is a constant, multiplied by $(1-x^2/a^2-y^2/k'^2a^2)^{-1/2}$ and by the original polynomial ${}_qEp_n^m(\alpha,\beta)$. The proves Galin's theorem once more. The normal displacement on z = 0 outside S_0 follows from (12.3.47) as

$$w(\alpha,K+iK',\gamma) = {}_qEp_n^m(\alpha,K+iK')\,{}_qF_n^m(\gamma)/{}_qF_n^m(K+iK') \ . \tag{12.3.51}$$

The computation of ψ and of w is thus reduced to the computation of the appropriate second solutions $F_n^m(\gamma)$; for n = 0,1,2 these are given in (12.3.33) - (12.3.35); see Example 12.3.12.

Calculations of the displacements and stresses for Hertz contact have been made by various authors; see Morton and Close (1922), Thomas and Hoersch (1930) and J.O. Smith and Liu (1953).

Examples 12.3

1. Show that

$$\text{sn}(\zeta+2K) = -\text{sn}\zeta,\quad \text{cn}(\zeta+2K) = -\text{cn}\zeta,\quad \text{dn}(\zeta+2K) = \text{dn}\zeta \ ,$$

$$\text{sn}(\zeta+2iK') = \text{sn}\zeta,\quad \text{cn}(\zeta+2iK') = -\text{cn}\zeta,\quad \text{dn}(\zeta+2iK') = -\text{dn}\zeta \ .$$

2. Show that $\text{sn}(K+iK') = k^{-1}$, $\text{cn}(K+iK') = -ik^{-1}k'$, $\text{dn}(K+iK') = 0$.

3. Show that if ζ is real then

$$-1 \le \text{sn}\zeta \le 1 \ ,\quad -1 \le \text{cn}\zeta \le 1 \ ,\quad k' \le \text{dn}\zeta \le 1 \ .$$

4. Show that the surfaces on which one of α, β, γ is constant are given by

$$\frac{x^2}{k^2a^2\text{sn}^2\zeta} - \frac{y^2}{k^2a^2\text{cn}^2\zeta} - \frac{z^2}{a^2\text{dn}^2\zeta} = 1 \ ,\quad \zeta = \alpha,\ \beta,\ \gamma \ ,$$

and that when α = real, $\beta = K+i\xi$, $\gamma = iK'+\eta$ then α = constant is a hyperboloid of two sheets; β = constant, one of one sheet; γ = constant, an ellipsoid.

5. If $\beta = K+i\xi$, $\gamma = iK'+\eta$ show that

$$\mathrm{sn}\beta = \mathrm{nd}(\xi,k'), \quad \mathrm{cn}\beta = -ik'\mathrm{sd}(\xi,k'), \quad \mathrm{dn}\beta = k'\mathrm{cd}(\xi,k'),$$

$$\mathrm{sn}\gamma = k^{-1}\mathrm{ns}\eta, \quad \mathrm{cn}\gamma = -ik^{-1}\mathrm{ds}\eta, \quad \mathrm{dn}\gamma = -i\mathrm{cs}\eta.$$

6. Show that correspondence between (12.3.1), (12.3.2), (12.3.3), (12.3.15), (12.3.16) may be obtained by putting

$$\tau = k\mathrm{sn}\alpha, \quad \sigma = k\mathrm{sn}\beta = k\mathrm{nd}(\xi,k'), \quad \rho = k\mathrm{sn}\gamma = \mathrm{ns}\eta.$$

7. By differentiating all three of equations (12.3.3) with respect to z show that

$$\frac{\partial\gamma}{\partial z} = \frac{i}{ak^2k'} \cdot \frac{\mathrm{dn}\alpha\mathrm{dn}\beta\mathrm{sn}\gamma\mathrm{cn}\gamma}{(\mathrm{sn}^2\beta-\mathrm{sn}^2\gamma)(\mathrm{sn}^2\gamma-\mathrm{sn}^2\alpha)}.$$

8. Use Example 12.3.5 to write the derivative $\partial\gamma/\partial z$ as

$$\frac{\partial\gamma}{\partial z} = \frac{\mathrm{dn}\alpha\mathrm{cd}(\xi,k')\mathrm{ns}\eta\mathrm{ds}\eta}{a[k^2\mathrm{nd}^2(\xi,k')-\mathrm{ns}^2\eta][\mathrm{ns}^2\eta-k^2\mathrm{sn}^2\alpha]},$$

and show that this may be written

$$\frac{\partial\gamma}{\partial z} = \frac{\rho}{ak'} \frac{[(1-\tau^2)(1-\sigma^2)(\rho^2-k^2)]^{1/2}}{(\sigma^2-\rho^2)(\rho^2-\tau^2)}.$$

9. Show by differentiating the third of equations (12.3.3) with respect to z and then putting $\beta = K+i\xi$, $\gamma = \eta+iK'$ that

$$\frac{\partial}{\partial z} = \frac{1}{a\mathrm{dn}\alpha\mathrm{dn}\beta}\frac{\partial}{\partial\gamma} = -\frac{\mathrm{dn}(\xi,k')}{k'a\mathrm{dn}\alpha\mathrm{cn}(\xi,k')}\frac{\partial}{\partial\eta}, \quad (x,y) \text{ in } S_0,$$

$$\frac{\partial}{\partial z} = \frac{-1}{a\mathrm{dn}\alpha\mathrm{dn}\gamma}\frac{\partial}{\partial\beta} = \frac{-\mathrm{sn}\eta}{a\mathrm{dn}\alpha\mathrm{cn}\eta}\frac{\partial}{\partial\xi}, \quad x,y \text{ outside } S_0.$$

10. Show that when x,y are given by (12.3.15), (12.3.16) then

$$1 - \frac{x^2}{a^2} - \frac{y^2}{k'^2a^2} = \mathrm{dn}^2\alpha\mathrm{cd}^2(\xi,k') = \frac{\mathrm{dn}^2\alpha\mathrm{dn}^2\beta}{k'^2}.$$

11. Use (12.3.3) to show that when $\gamma = K+iK'$ then the element of area dS is given by

$$dS = ik^2a^2 \mathrm{dn}\alpha\,\mathrm{dn}\beta(\mathrm{sn}^2\alpha-\mathrm{sn}^2\beta)d\alpha d\beta =$$

$$= -k^2k'^2a^2 \mathrm{dn}\alpha\,\mathrm{cd}(\xi,k')[\mathrm{sn}^2\alpha-\mathrm{nd}^2(\xi,k')]d\alpha d\xi \ .$$

If $p(\alpha,\xi)$ is given by (12.3.45), show that

$$\int_{-2K}^{2K}\int_{0}^{K'} dS = \pi k'a^2 \ , \qquad \int_{-2K}^{2K}\int_{0}^{K'} p(\alpha,\xi)dS = \frac{2\pi d_0 a}{K\vartheta} \ .$$

Hint: the second integral appears as a multiple of the left hand side of Legendre's indentity $EK'+E'K-KK' = \pi/2$.

12. For Hertzian contact over the ellipse S_0 the potential $\psi(\alpha,\beta,\gamma)$ may be sought in the form

$$\psi(\alpha,\beta,\gamma) = A_0 \frac{{}_uF_0(\gamma)}{{}_uF_0(K+iK')} + \sum_{m=1}^{2} A_{mu}Ep_2^m(\alpha,\beta) \frac{{}_uF_2^m(\gamma)}{{}_uF_2^m(K+iK')} \ .$$

The constants A_0, A_1, A_2 must be chosen so that the contact pressure has the form

$$p(\alpha,\beta,K+iK') = \frac{-\partial\psi}{\partial z} = \frac{1}{a\,\mathrm{dn}\alpha\,\mathrm{dn}\beta}\frac{\partial\psi}{\partial\gamma} = p_0\left(1 - \frac{x^2}{a^2} - \frac{y^2}{k'^2a^2}\right)^{1/2} =$$

$$= p_0(k')^{-1}\mathrm{dn}\alpha\,\mathrm{dn}\beta \ , \qquad \text{where } 3P = 2\pi abp_0 \ .$$

Show that A_0, A_1, A_2 have the values $A_0 = (\lambda_1-\lambda_2)KC$,

$$A_1 = {}_uF_2^1(K+iK')C \ , \qquad A_2 = -\,{}_uF_2^2(K+iK')C \ ,$$

where $C = k'ap_0/[3(\lambda_1-\lambda_2)]$. Hence show that on the plane $z = 0$ outside S_0

$$w(x,y) = \frac{3\vartheta P}{4\pi a}\left[\eta + \frac{x^2}{a^2k^2}\left\{E(\eta)-\eta\right\} - \frac{y^2}{a^2k^2}\left\{\frac{E(\eta)}{k'^2} - \eta - \right.\right.$$

$$\left.\left. - \frac{k^2}{k'^2}\frac{\mathrm{sn}\eta\,\mathrm{cn}\eta}{\mathrm{dn}\eta}\right\}\right] ,$$

where $E(\eta) \equiv E(am(\eta),k)$; i.e., $sn\eta = \sin\omega = 1/\rho$, $\eta = F(\omega,k)$. This result appears in Way (1940), who attributes it to Thomas and Hoersch (1930).

13. The $w(x,y)$ of Example 12.3.12 *on* the ellipse, $sn = 1$, is continuous with that given by equation (2.9.6) (with say $\vartheta_1 = \vartheta$, $\vartheta_2 = 0$). Show that

$$\frac{\partial w}{\partial \eta} = \frac{3\vartheta P}{4\pi a}(1-k^2\sin^2\omega)^{1/2}\left[1 - \frac{x^2}{a^2}\sin^2\omega - \frac{y^2}{a^2}\frac{\sin^2\omega}{(1-k^2\sin^2\omega)}\right]$$

and show that, when expressed in terms of ρ, τ, the term in brackets is $-(k^2-\tau^2)(\rho^2-1)/k^2$. Outside the ellipse η *decreases* from its value K which it has on the ellipse. This means that the separation condition (2.9.2), i.e., $w > \delta - Ax^2 - By^2$ is satisfied outside the ellipse. Construct a simpler proof of this result!

12.4 *Compliance of elastic bodies in contact*

We conclude this chapter by presenting the results obtained by Mindlin (1949) in one of the most important papers in the field. In the paper it is supposed that two bodies are in Hertzian contact, as described in §2.9. An additional system of forces is applied to the bodies, as in Figure 12.4.1, such that, across the contact surface, one body exerts on the other a small additional force in the direction Oz, a small force perpendicular to Oz and a couple with axis Oz. The compliances are defined to be the rates of change of the relative displacements of the bodies due to these small forces. In this section u, v, w will denote Cartesian displacements.

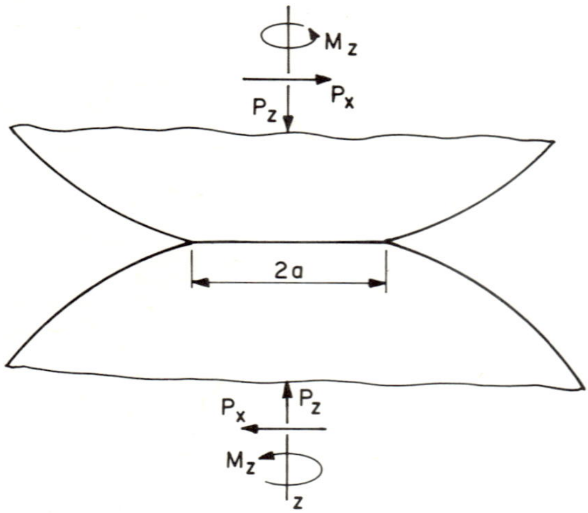

Figure 12.4.1 - An additional system of forces is applied to two bodies in contact

The normal compliance may be found directly from equations (2.9.6) - (2.9.11).

$$\delta_z = \frac{3(\vartheta_1+\vartheta_2)}{4\pi} \cdot K(e) \frac{P_z}{a} , \quad \frac{\partial \delta_z}{\partial P_z} = \frac{3(\vartheta_1+\vartheta_2)K(e)}{4\pi a} \left\{1 - \frac{P_z}{a} \frac{\partial a}{\partial P_z}\right\} , \tag{12.4.1}$$

and equation (2.9.10) shows that $\partial a/\partial P_z = a/(3P_z)$ so that the normal compliance is

$$C_z = \frac{\partial \delta_z}{\partial P_z} = \frac{(\vartheta_1+\vartheta_2)}{2\pi a} K(e) . \tag{12.4.2}$$

In order to find what he calls the *tangential* compliance, i.e., that referring to the relative x-displacement due to a small force applied by one body on the other in the x-direction, Mindlin first considers the case in which the system composed of the two bodies is elastically and geometrically symmetrical with respect to each of the three coordinate planes. If to this system is applied a (body - or exterior) force field P_x which has the property $P_x(x,y,z_1) = -P_x(x,y,z_2)$ then we should expect, by Saint Venant's Principle, that the displacements in each of the bodies would vanish at infinity, and

$$w_1(x,y,z_1) = -w_2(x,y,z_2) . \qquad (12.4.3)$$

Note however that this will hold only when the two bodies are *elastically and goemetrically similar,* as assumed. If in addition the force field has the properties

$$P_x(x,y,z_i) = P_x(x,-y,z_i) , \quad P_x(x,y,z_i) = -P_x(-x,y,z_i) , \qquad (12.4.4)$$

then the Cartesian displacement fields in the two bodies must satisfy

$$u_1(x,y,z_1) = -u_2(x,y,z_2) + \delta_x , \quad v_1(x,y,z_1) = -v_2(x,y,z_2) . \qquad (12.4.5)$$

If there is *no slip* over the region of contact, the ellipse S, then these conditions imply

$$u_1(x,y,0) = u_2(x,y,0) = \delta_x/2 , \quad v_1(x,y,0) = 0 = v_2(x,y,0) , \quad (x,y) \text{ in } S . \qquad (12.4.6)$$

The condition (12.4.3), together with the equilibrium of the normal stresses across S, yields

$$\tau_{zz}(x,y,0) = 0 , \quad (x,y) \text{ in } S . \qquad (12.4.7)$$

We conclude that in this case the displacement δ_x is proportional to the total force F_x, that the contact region S moves as a rigid body a distance δ_x, and that the compliance is

$$C_x = \delta_x/F_x . \qquad (12.4.8)$$

Compliance of elastic bodies in contact

When the two bodies are not *geometrically* identical then the contact surface will no longer be strictly the x-y plane but will be slightly warped (see equation (2.9.4)), but this warping can be (in fact, must be) ignored in small displacement, small rotation theory. If the bodies are not *elastically* identical then the continuity condition $w_1 = -w_2$ on S can be maintained only by an additional normal stress on S. This normal stress will be ignored on the grounds that the Hertz theory of §2.9 does not demand the continuity of tangential (i.e., u, v) displacements on S.

With this simplification each body may be treated as a separate half-space and it is sufficient to consider the following problem for the half-space $z \geq 0$

$$u(x,y,0) = \delta_x^i, \quad v(x,y,0) = 0, \quad \tau_{zz}(x,y,0) = 0, \quad (x,y) \text{ in } S, \tag{12.4.9}$$

$$\tau_{xz}(x,y,0) = 0 = \tau_{yz}(x,y,0) = \tau_{zz}(x,y,0), \quad (x,y) \text{ not in } S, \tag{12.4.10}$$

and to calculate

$$C_x^i = \delta_x^i/F_x, \tag{12.4.11}$$

where

$$F_x = -\iint_S \tau_{xz}(x,y,0)\,dxdy . \tag{12.4.12}$$

Then C_x is given by

$$C_x = C_x^1 + C_x^2 = (\delta_x^1 + \delta_x^2)/F_x . \tag{12.4.13}$$

The solution of (12.4.9), (12.4.10) for the case when S is a circle of radius a is given in (10.3.28) - (10.3.36) and Example 10.3.3 shows that $C_x^i = (2-\nu_i)8\mu_i a$, so that

$$C_x = \frac{1}{8a}\left[\frac{2-\nu_1}{8\mu_1} + \frac{2-\nu_2}{8\mu_2}\right] \quad . \tag{12.4.14}$$

In general S will be the ellipse $x^2/a^2+y^2/b^2 \leq 1$. In this case C_x^i is given by Example 12.2.2 and is

$$C_x^i = \frac{1}{2\pi\mu_i a}\,[K(e)-\nu_i B(e)] \quad , \tag{12.4.15}$$

where $B(e)$ is given by equation (2.7.7) and $b^2 = a^2(1-e^2)$. This reduces to the former value when $a = b$. (See also Example 12.4.1.)

The shear stress τ_{xz} in the foregoing solutions has a square-root singularity at the edge of the contact region. Mindlin therefore constructed a modified problem in which the shear stress never exceeds the product of the friction coefficient, k and the Hertzian normal stress p. Moreover he assumes that there is an outer slip region in which the shear stress remains at its maximum value kp. Thus the boundary conditions on $z = 0$ are

$$u = \delta \;,\quad v = 0 \;,\quad 0 \leq -\tau_{xz} \leq kp \;,\quad \tau_{yz} = 0 = \tau_{zz} \;,\quad (x,y) \text{ in } S' \;, \tag{12.4.16}$$

$$\tau_{xz} = -kp \;,\quad \tau_{yz} = 0 = \tau_{zz} \;,\quad (x,y) \text{ in } S \text{ and outside } S' \;, \tag{12.4.17}$$

$$\tau_{xz} = 0 = \tau_{yz} = \tau_{zz} \;,\qquad (x,y) \text{ outside } S \;. \tag{12.4.18}$$

Keer and Goodman (1976) consider the case when the resistance to slip is arbitrary.

First consider the case in which S is a circle of radius a and S' is a circle of radius $a' < a$. The Hertzian normal stress corresponding to a total normal force P_z and contact radius a is given in §2.9 and is

$$p = \frac{3P_z}{2\pi a^2}\left(1 - \frac{r^2}{a^2}\right)^{1/2} \quad . \tag{12.4.19}$$

Mindlin builds up a solution by first considering the displacements due to $\tau_{xz} = -kp$ acting over the *whole* of *S*. These displacements were obtained in Example 10.3.6 and are given by

$$64\mu a^3 u = 3kP_z[2(2-\nu)(2a^2-r^2)+\nu r^2\cos 2\theta] \ , \quad 64\mu a^3 v = 3kP_z \nu r^2 \sin 2\theta \qquad (12.4.20)$$

inside the circle $r \leq a$. Now add to this the displacements due to the shear stress

$$\tau'_{xz} = \frac{3kQ}{2\pi a'^2}\left(1 - \frac{r^2}{a'^2}\right)^{1/2} \ , \qquad 0 \leq r \leq a' \ . \qquad (12.4.21)$$

Inside the circle $r \leq a'$ these will be given by

$$64\mu a'^3 u' = -3kQ[2(2-\nu)(2a'^2-r^2)+\nu r^2\cos 2\theta] \ , \quad 64\mu a'^3 v' = -3kQ\nu r^2 \sin 2\theta \qquad (12.4.22)$$

The displacements $u+u'$, $v+v'$ will satisfy equation (12.4.5) within S' if $Q = P_z a'^3/a^3$. Then

$$16\mu a^3(u+u') = 3kP_z(2-\nu)(a^2-a'^2) \ , \quad v+v' = 0 \ , \quad 0 \leq r \leq a' \ , \qquad (12.4.23)$$

and the total tangential force P_x is

$$P_x = k(P_z-Q) = kP_z(1-a'^3/a^3) \ , \qquad (12.4.24)$$

so that

$$a' = a(1-P_x/kP_z)^{1/3} \ . \qquad (12.4.25)$$

The compliance C_x is now given in Example 12.4.2. The total shear stress $\tau_{xz}+\tau'_{xz}$ also satisfies the frictional constraint. Finally, since it is known that the problem, with certain regularity conditions appended, has a unique solution, it may be concluded that the solution found is this solution. The corresponding solution for an elliptic contact area was given by Cattaneo (1938), but it may also be found by using the method of §12.2; this analysis shows immediately that the shear stress distribution -kp over S yields (second-order) polynomial displacements u, v inside S (Example 12.4.3).

The compliance due to a torsional couple may be found in a similar way. Again it may be shown that the moment about the z-axis will induce no normal stress on the contact surface so that each body will rotate without distortion in the x, y-plane. The boundary conditions on z = 0 are therefore

$$u = -\gamma y\ , \quad v = \gamma x\ , \quad \tau_{zz} = 0\ , \quad (x,y) \text{ in } S\ , \tag{12.4.26}$$

$$\tau_{xz} = 0 = \tau_{yz} = \tau_{zz}\ , \quad (x,y) \text{ not in } S\ . \tag{12.4.27}$$

For the circular contact area this problem was solved in §10.2.5) and equation (10.2.58) gives the torsional compliance for one body to be

$$\frac{\gamma}{T} = \frac{3}{16\mu a^3}\ . \tag{12.4.28}$$

For the elliptic contact area the general theory of §10.9 shows that a shear stress distribution in which τ_{xz}, τ_{yz} have the form (12.2.12), (12.2.19), with m = n = 1, will give displacements within S that are linear in x, y. A detailed calculation shows that the shear stress distributions

$$\tau_{xz} = -T_1 y(1-x^2/a^2-y^2/b^2)^{-1/2}\ , \quad \tau_{yz} = T_2 x(1-x^2/a^2-y^2/b^2)^{-1/2} \tag{12.4.29}$$

within S give displacements in S such that

$$\mu u = by\{BT_1+\nu C[T_2-(1-e^2)T_1]\}\ , \tag{12.4.30}$$

$$\mu v = bx\{-DT_2+\nu C[T_2-(1-e^2)T_1]\}\ , \tag{12.4.31}$$

where B, D are given in equation (2.7.7) and $C = (D-B)/e^2$. If therefore equation (12.4.26) is to be satisfied then T_1, T_2 must have the values

$$T_1 = \frac{\mu\gamma}{b}\,\frac{(D-2\nu C)}{BD-\nu CE}\;, \qquad T_2 = \frac{\mu\gamma}{b}\,\frac{[B-2\nu(1-e^2)C]}{BD-\nu CE}\;. \tag{12.4.32}$$

The total moment applied to the half-space is

$$M = \iint_S (-y\tau_{yz}+x\tau_{yz})dxdy = \frac{2}{3}\,\mu a^3\gamma\pi\,\frac{[E-4\nu(1-e^2)C]}{BE-\nu CE}\;, \tag{12.4.33}$$

and the torsional compliance is

$$\frac{\gamma}{M} = \frac{3}{16\mu a^3}\cdot\frac{8(BD-\nu CE)}{\pi[E-4\nu(1-e^2)C]}\;. \tag{12.4.34}$$

When the ellipse is circular the second factor is unity.

Again it will be noted that the shear stress, given in equation (10.2.57) for the circular contact area, or by equation (12.4.29) for the ellipse, is singular at the boundary of the contact region, indicating that, in practice, the bodies will slip. Lubkin (1951) investigated this problem for a circular contact region, assuming that there was no slip over an inner circle of radius a', while the shear stress attained its maximum value k (Hertzian normal stress) over the annulus $a' < r < a$. Again the compliance for two bodies in contact can be found as the sum of the compliances for two separate half-spaces. For the latter it is necessary to solve the axisymmetric half-space problem subject to the conditions

$$u_\theta = \gamma r\;, \qquad |\tau_{\theta z}| \le kp(r)\;, \qquad 0 \le r \le a'\;, \tag{12.4.35}$$

$$\tau_{\theta z} = -kp(r)\;, \qquad a' \le r \le a\;, \tag{12.4.36}$$

where k is the coefficient of friction and p(r) is given by equation (12.4.18). This is the subject of Example 10.11.4. The particular case of complete slip was later considered in detail by Hetenyi and McDonald (1958).

A review of recent work on contact compliances and an analysis of the effect of a layer pressed between the contacting bodies may be found in Goodman and Keer (1975).

Examples 12.4

1. Show that the compliance $C_y = \delta_y/P_y$ of a single elastic body when the contact region is the ellipse S is

$$C_y = \frac{1}{2\pi\mu a}\,[K(e)-\nu D(e)]\ .$$

2. Show that the tangential compliance given by equation (12.4.22) is $C_x = (2-\nu)/[8\mu a']$. [Remember that the compliance is $\partial u/\partial P_x$ and that a' is a function P_x.]

3. Show that the shear stress $\tau_{xz} = -kp$, where p is the Hertzian normal stress, acting over the ellipse S produces the following displacements on S

$$u = \frac{3kP_z}{2\pi a\mu}\left\{(K-\nu B)\left(1-\frac{x^2}{2a^2}-\frac{y^2}{2b^2}\right)-[(2D-K)-\nu(D-C-B/2)]\left(\frac{x^2}{a^2}-\frac{y^2}{b^2}\right)\right\}$$

$$v = 3kP_z\nu Cxy/(\pi\mu a^3)\ ,$$

and hence find the displacements satisfying (12.4.16) - (12.4.1 where S' is the ellipse $x^2/a'^2+y^2/b'^2 \le 1$ and $a'/a = b'/b$. Hence show that the tangential compliance C_x is $(K-\nu B)/(2\pi\mu a\ )$.

4. Show that the shear stress in the region $z = 0$, $r \le a'$ corresponding to the boundary conditions (12.4.35) - (12.4.36) is

$$\tau_{\theta z}(r,0) = -\frac{4\mu\gamma r}{\pi}(a'^2-r^2)^{-1/2} +$$

$$+\frac{3kP_z}{2\pi^2a^3}(a'^2-r^2)^{-1/2}\int_{a'}^{a}\frac{(s^2-a'^2)^{1/2}(a^2-s^2)^{1/2}}{s^2-r^2}\,ds$$

and that if there is to be no singularity as $r \to a'-$, then

$$\frac{\mu a^2}{kP_z}\gamma = \frac{3}{4\pi}\{K(e)-E(e)\} , \qquad e^2 = 1 - \frac{a'^2}{a^2} .$$

12.5 *The inhomogeneous half-space*

Contact problems for the inhomogeneous isotropic elastic half-space have great practical importance, and many attempts, starting with Borowicka (1943b), have been made to find simple solutions for such problems. A review of early works may be found in Olszak (1959) and a later review concerned with media with variable Poisson's ratio may be found in Golecki and Knops (1969). It will be shown below that, even in the comparatively straightforward case in which the elastic constants μ, ν vary according to

$$\mu = mz^{\lambda} , \qquad \nu = \nu_0 , \tag{12.5.1}$$

the equations governing an inhomogeneous medium are difficult; for this reason researchers have concentrated on a few special cases. (Note that λ in (12.5.1) is an arbitrary constant satisfying $0 \leq \lambda < 1$, and is *not* the Lamé constant λ of §1.6.)

It may easily be verified, by a direct calculation of the strains and stresses and subsequent substitution into the equilibrium equations, that

$$(u,v,w) = \left(\frac{z\partial\psi}{\partial x} , \frac{z\partial\psi}{\partial y} , -\psi + \frac{z\partial\psi}{\partial z}\right) \tag{12.5.2}$$

is a possible displacement field in the medium with μ, ν given by (12.5.1), provided that $\nu = 1/(\lambda+2)$, and ψ satisfies (Example 12.5.1)

$$\nabla^2\psi + \frac{\lambda}{z}\frac{\partial\psi}{\partial z} = 0 , \qquad 0 \leq \lambda < 1 . \tag{12.5.3}$$

The field (12.5.2) corresponds to shearing stresses τ_{xz}, τ_{yz} which are zero on the plane $z = 0$ (Example 12.5.3). This particular case was investigated by Klein (1956), Korenev (1957), Mossakovskii (1958), Popov (1959), Rakov and Rvachev (1961) and Belik and Protsenko (1967). Korenev and Mossakovskii did not state explicitly that ν had to have the particular value stated above. Rostovtsev (1964b) showed that this case was the only (non-singular) one in which a medium with $\mu = \mu(z)$, $\nu = \nu_0$ had a purely (spherically) radial state of stress. For such a stress state $\tau_{RR} = \sigma$, there is only one equilibrium equation

$$\frac{1}{R}\frac{\partial}{\partial R}(R^2\sigma) = 0 \ , \tag{12.5.4}$$

and by virtue of the axial symmetry, it follows that

$$\sigma = R^{-2}S(\theta) \ . \tag{12.5.5}$$

The strains are

$$e_{RR} = e = \sigma/E \ , \quad e_{\phi\phi} = e_{\theta\theta} = -\nu e \ , \quad e_{R\theta} = 0 = e_{R\phi} = e_{\theta\phi} \ , \tag{12.5.6}$$

and these must be related to displacements u_R, u_θ by the equations

$$\left.\begin{aligned} e_{RR} &= \frac{\partial u_R}{\partial R} \ , \quad e_{\theta\theta} = \frac{u_R}{R} + \frac{1}{R}\frac{\partial u_\theta}{\partial\theta} \ , \quad e_{\phi\phi} = \frac{u_R}{R} + \frac{u_\theta}{R}\cot\theta \ , \\ e_{R\theta} &= \frac{1}{2}\left[\frac{1}{R}\frac{\partial u_R}{\partial\theta} - \frac{u_\theta}{R} + \frac{\partial u_\theta}{\partial R}\right] \ . \end{aligned}\right\} \tag{12.5.7}$$

The equation $e_{\theta\theta} = e_{\phi\phi}$ yields $u_\theta = f(R)\sin\theta$, while $e_{R\theta} = 0$ gives $u_R = [Rf'(R)-f(R)]\cos\theta+g(R)$. The equation $e_{\theta\theta} = -\nu e_{RR}$ now gives

$$f'(R)\cos\theta+R^{-1}g(R) = -\nu Rf''(R)\cos\theta-\nu g'(R) \ , \tag{12.5.8}$$

and, since ν is constant, this implies

$$f'(R) = -\nu Rf''(R) , \qquad g(R) = -\nu Rg'(R) . \tag{12.5.9}$$

The strain e is

$$e = Rf''(R)\cos\theta + g'(R) . \tag{12.5.10}$$

If the rigid displacement, f(R) = constant, is omitted, then the solution of equations (12.5.9) may be written

$$f'(R) = AR^{-\lambda-2} , \quad g(R) = BR^{-\lambda-2} , \quad \lambda = \nu^{-1}-2 , \tag{12.5.11}$$

so that E, the ratio σ/e, is

$$E = -\nu R^{\lambda+1} S(\theta)/(AR\cos\theta + B) . \tag{12.5.12}$$

If E is to be a function of $z \equiv R\cos\theta$ only, then $S(\theta) = -\nu^{-1}C(\cos\ )^{\lambda+1}$, giving

$$E = Cz^{\lambda+1}/(Az+B) . \tag{12.5.13}$$

The corresponding displacements are

$$u_R = \frac{A}{1-\nu} R^{-\lambda-1}\cos\theta + BR^{-\lambda-2} , \quad u_\theta = -\frac{\nu AR^{-\lambda-1}}{1-\nu}\sin\theta . \tag{12.5.14}$$

If these displacements are to be meaningful in the case of a distributed load at the origin, the exponents of R must be greater than -2. Thus if $A \neq 0$ then $-\lambda-1 > -2$, or $\nu > 1/3$, which means that ν must lie in the range $1/3 < \nu < 1/2$. On the other hand, if $B \neq 0$, then $-\lambda-2 > -2$, or $\nu > 1/2$, which is impossible. Thus B = 0 and $\mu = mz^\lambda$, $\lambda = \nu^{-1}-2$, i.e., $\nu = 1/(2+\lambda)$, and $m = C/[2A(1+\nu)]$.

Rostovtsev excluded the two limiting cases in which the exponent in (12.5.14) equals -2. These are (i) $\nu = 1/3$, B = 0, for which $\lambda = 1$ and

$$u_R = \frac{3A}{2} R^{-2}\cos\theta , \quad u_\theta = \frac{A}{2} R^{-2}\sin\theta , \quad E = \frac{Cz}{A} , \tag{12.5.15}$$

(ii) $\nu = 1/2$, for which $\lambda = 0$ and

$$u_R = 2AR^{-1}\cos\theta + BR^{-2} , \quad u_\theta = -AR^{-1}\sin\theta , \quad E = Cz/(Az+B) . \tag{12.5.16}$$

These will be considered at the end of this section.

The only possible *non-singular* state is

$$u_R = \frac{AR^{-(\lambda+1)}}{1-\nu} \cos\theta , \quad u_\theta = \frac{-\nu AR^{-(\lambda+1)}}{1-\nu} \sin\theta , \tag{12.5.17}$$

where $\nu = 1/(2+\lambda)$. These correspond to a radial stress

$$\tau_{RR} = -2m(1+\nu)\nu^{-1}AR^{-2}(\cos\theta)^{\lambda+1} , \tag{12.5.18}$$

which in turn relates to a normal load at the origin, of magnitude

$$P = -2\pi \int_0^{\pi/2} R^2 \sin\theta\cos\theta\,\tau_{RR}\,d\theta = 4\pi m\,\frac{1+\nu}{\nu}\cdot\frac{A}{\lambda+3} = 4\pi mA . \tag{12.5.19}$$

When $\theta = \pi/2$, the displacement $-u_\theta$ is the normal displacement $w(x,y,0)$. Thus the surface displacement due to a distribution of normal load $p(x',y')$ over a region S of the x',y' -plane is

$$w(x,y,0) = \iint_S H_{33}(x-x',y-y')p(x',y')dx'dy' \tag{12.5.20}$$

where (Popov (1961b))

$$H_{33}(x,y) = \vartheta^0/[2\pi(x^2+y^2)^{(\lambda+1)/2}] , \quad \vartheta^0 = 1[2m(\lambda+1)] . \tag{12.5.21}$$

This reduces to $\vartheta/(2\pi r)$ when $\lambda = 0$, $\nu = 1/2$. When the region S is a circle of radius a then (12.5.20) may be used to give the relation between the Hankel transforms of w and p, as described by Mossakovski (1958) for the axisymmetric case and by Kassir (1972) in general

(Example 12.5.7). Kassir and Chuaprasert (1974) considered $\mu = \mu_0(z+c)^\lambda$, $\nu = 1/(2+\lambda)$. Ter-Mkrtich'ian (1961) considers $\mu = \mu_0 \exp(\lambda z)$, $\nu = \nu_0$; for this the governing equations are only trivially different from those for constant μ, ν; see also Puro (1973).

A general solution for torsional (Reissner-Sagoci) problems may be found when $\mu = \mu(z)$. In this case (Protsenko (1967), Kassir (1970)) the only non-zero displacement in cylindrical polar coordinates is $u_\theta = v(r,z)$ and

$$\tau_{z\theta} = \frac{\mu \partial v}{\partial z}, \qquad \tau_{r\theta} = \mu r \frac{\partial}{\partial r}\left(\frac{v}{r}\right). \tag{12.5.22}$$

Now the equilibrium equation (1.5.3), i.e.,

$$\frac{\partial \tau_{r\theta}}{\partial r} + \frac{\partial \tau_{z\theta}}{\partial z} + \frac{2\tau_{r\theta}}{r} = 0 , \tag{12.5.23}$$

leads to the equation

$$\nabla^2 v - \frac{v}{r^2} + q(z)\frac{\partial v}{\partial z} = 0 , \qquad q(z) = \frac{1}{\mu(z)}\frac{d\mu(z)}{dz} \tag{12.5.24}$$

which has the solution

$$v = \frac{\partial \psi}{\partial r}, \qquad \nabla^2 \psi + q(z)\frac{\partial \psi}{\partial z} = 0 . \tag{12.5.25}$$

For the power law variation (12.5.1), this yields (12.5.3), and the solution $\psi(r,z) = z^\sigma K_\sigma(sz) J_0(rs)$, where $\sigma = (1-\lambda)/2$, (see Example 12.5.2) leads to

$$v(r,z) = z^\sigma \int_0^\infty s^{-\sigma} A(s) K_\sigma(sz) J_1(rs)\, ds , \tag{12.5.26}$$

$$\tau_{r\theta}(r,z) = m z^{1-\sigma} \int_0^\infty s^{1-\sigma} A(s) K_{1-\sigma}(sz) J_1(rs)\, ds . \tag{12.5.27}$$

In this case the Reissner-Sagoci problem leads to a pair of integral equations of the form (10.2.24), (10.2.25); see Example 12.5.4. Kassir (1970) also considered $\mu = \mu_0 \exp(\lambda z)$, for which $q(z) = \lambda$; this is only trivially different from the homogeneous case. Kolybikhin (1971) and Singh (1973) analyzed the case $\mu = \mu_0 r^{\kappa} z^{\lambda}$; Singh (1972) considered $\mu = \mu_0 r^{\kappa}$; see also Dhaliwal and Singh (1978b), (1978c). The variation (12.5.1) is unrealistic in that $\mu(0) = 0$; Chuaprasert and Kassir (1973) therefore considered $\mu = \mu_0 (z+c)^{\lambda}$: this is equivalent to taking (12.5.1) in the half-space $z \geq c > 0$.

Gibson and Sills (1969) analyzed the axisymmetric Boussinesq problem for $\mu = \mu_0$, $\nu = \nu(z)$. When the equilibrium conditions are expressed in terms of (cylindrical polar) displacements we find that

$$\left(\nabla^2 - \frac{1}{r^2}\right) u + \frac{\partial}{\partial r}(g\varepsilon) = 0 = \nabla^2 w + \frac{\partial}{\partial z}(g\varepsilon) , \qquad (12.5.28)$$

where $g = 1/(1-2\nu)$ and ε is given by (1.6.13). These equations admit a fairly simple Hankel function solution which leads, as a special case, to the surface displacement due to a uniform normal pressure p over a circle $r \leq a$, $z = 0$, to be given by

$$w(r,0) = \frac{pb}{2\mu} \int_0^{\infty} \frac{J_0(r\xi) J_1(b\xi) d\xi}{\xi^2 \Lambda(\xi)} , \qquad \Lambda(\xi) = \int_0^{\infty} \frac{\exp(-2\xi z) dz}{1-\nu(z)} . \qquad (12.5.29)$$

It was shown in §6.10 that when $\mu = my^{\lambda}$, $\nu = \nu_0$, it is possible to obtain the Cartesian displacements u, v in the plane $y \geq 0$ due to stresses applied on the boundary; see (6.10.26), (6.10.27). This solution is not subject to the restriction $\nu = 1/(2+\lambda)$ encountered in the half-space problem above. Rostovtsev (1964b) used this to find the spatial displacements on $z = 0$ due to an axisymmetric normal loading when $\mu = mz^{\lambda}$, $\nu = \nu_0$ and ν is not necessarily equal to $1/(2+\lambda)$. He argued as follows. It was shown in (10.6.16) that an x-symmetric state in the x, z-plane is related to an axisymmetric state in the half-space $z \geq 0$ by the equations

$$\left.\begin{aligned} w(r,z) &= A_1[w^p(x,z);x \to r] , \\ A_1^{-1}[\tau_{zz}(r,z);r \to x] &= \tau^p_{zz}(x_1 z) , \end{aligned}\right\} \qquad (12.5.30)$$

due to A. Ia. Aleksandrov (1961a). These results were proved in §10.6 by using the Papkovich-Neuber representations in 2- and 3-dimensions, and these do not hold for inhomogeneous media. However, the operator A_1 does not depend on z, so that the results will still hold if μ, ν depend on z in the same manner in the two cases. (Gibson et al (1971) used this device also, making use of results of McNamee and Gibson (1960).) Suppose that in the plane problem

$$w^p(x,0) = \int_{-\infty}^{\infty} K^p(t-x)p^p(t)dt . \qquad (12.5.31)$$

(For the case considered in §6.10, $K^p(t-x) = g_{22}|t-x|^{-\lambda}$). Use equations (5.7.26) and (12.5.30) to write

$$P^p(x) = \int_0^x p^p(t)dt = \left(\frac{2}{\pi}\right)^{1/2} \int_0^x \frac{rp(r)dr}{(x^2-r^2)^{1/2}} , \qquad (12.5.32)$$

then formally

$$w^p(x,0) = -\int_{-\infty}^{\infty} \frac{\partial K^p}{\partial t} (t-x)P^p(t)dt . \qquad (12.5.33)$$

Introduce the Fourier transforms of K^p, P^p, i.e.,

$$\left.\begin{aligned} K_*^p(s) &= (2\pi)^{-1/2}\int_{-\infty}^{\infty} K^p(t)\exp(ist)dt , \\ P_*^p(s) &= (2\pi)^{-1/2}\int_{-\infty}^{\infty} P^p(t)\exp(ist)dt ; \end{aligned}\right\} \qquad (12.5.34)$$

then the convolution theorem (5.3.1), and equation (12.5.31), give

$$w^p(x,0) = -\int_{-\infty}^{\infty} isK_*^p(s)P_*^p(s)\exp(-ixs)ds \ , \qquad (12.5.33)$$

$$w(r,0) = -\left(\frac{2}{\pi}\right)^{1/2}\int_0^r \frac{dx}{(r^2-x^2)^{1/2}}\int_{-\infty}^{\infty} isK_*^p(s)P_*^p(s)\exp(-ixs)ds \ , \qquad (12.5.34)$$

and on interchanging the order of integration we find

$$w(r,0) = -\left(\frac{\pi}{2}\right)^{1/2}\int_{-\infty}^{\infty} isK_*^p(s)P_*^p(s)J_0(rs)ds \ . \qquad (12.5.35)$$

This may be used either to find w(r) for an axisymmetric load, or for a general load, the latter as follows. For a load P uniformly distributed over $r \leq a$

$$p(r) = \frac{P}{\pi a^2}H(a-r) \ ,$$

$$P^p(x) = \left(\frac{2}{\pi}\right)^{1/2} \cdot \frac{P}{\pi} \cdot \frac{1}{x+\mathrm{sign}x(x^2-a^2)^{1/2}} \ , \quad x \geq a \ , \qquad (12.5.36)$$

where it must be remembered that $P^p(x)$ is an odd function; as $a \to 0$,

$$P^p(x) = \left(\frac{2}{\pi}\right)^{1/2} \cdot \frac{P}{\pi} \cdot \frac{1}{2x} \ , \qquad P_*^p(s) = \frac{iP}{2\pi}\,\mathrm{sign}(s) \ , \qquad (12.5.37)$$

$$w(r,0) = K(r)P \ , \qquad K(r) = (2\pi)^{1/2}\int_0^{\infty} sK_*^p(s)J_0(rs)ds \ . \qquad (12.5.38)$$

The function K(r) is the influence function H_{33} so that

$$w(x,y,0) = \iint_S K(R)p(x',y')dx'dy' \ , \qquad (12.5.39)$$

where now $R^2 = (x-x')^2+(y-y')^2$. For the power law medium of §6.10 the integral (12.5.38) may be evaluated to give

$$2\pi K(r) = \vartheta^0(\nu,\lambda) r^{-\lambda-1} , \qquad (12.5.40)$$

$$\vartheta^0(\nu,\lambda) = \frac{1-\nu}{m} \cdot \left[\frac{\Gamma\left(\frac{\lambda+1}{2}\right)}{\pi^{1/2}}\right] \cdot \left[\frac{qB(\lambda+2)\sin\beta}{(\lambda+1)\Gamma(1+\lambda/2)}\right] , \qquad (12.5.41)$$

where the notation is that of §6.10. For the homogeneous medium, both the bracketed terms are unity giving $\vartheta = (1/\nu)/\mu$; when $\nu = 1(\lambda+2)$, (12.5.41) reduces to (12.5.21). Popov (1973b) found the influence coefficients k_{31}, k_{33} of equation (2.6.5) for the power-law axisymmetric medium, and found an appropriate integral relation, in the sense §6.11. Except in special circumstances (Gibson et al (1971)) the expressions for displacements off the plane $z = 0$ are extremely complicated; see Rostovtsev and Kranevskaia (1971).

For the power-law medium, equation (12.5.39) states

$$w(x,y,0) = \frac{\vartheta^0(\nu,\lambda)}{2\pi} \int\int_S \left[\frac{p(x',y')dx'dy'}{[(x-x')^2+(y-y')^2]^{(1-\sigma)}}\right]. \qquad (12.5.42)$$

where $\sigma = (1-\lambda)/2$. Rostovtsev (1961) showed that Galin's theorem (§12.3) may be extended to such a medium, in the sense that if *S* is the ellipse $x^2/a^2+y^2/b^2 \leq 1$ and if

$$P(x',y') = \left(1 - \frac{x'^2}{a^2} - \frac{y'^2}{b^2}\right)^{-\sigma} P_n(x',y') , \qquad (12.5.43)$$

where $P_n(x',y')$ is a polynomial of total degree n in x', y', then $w(x,y,0)$ will also be a polynomial $Q_n(x,y)$ within S. Rostovtsev (1964a) developed a complete theory using ellipsoidal coordinates for such a medium.

The elastic displacement field due to axisymmetric normal loading of a medium in which $\mu = \mu_0+mz$, $\nu = \nu_0$ was given by Gibson (1967) for an incompressible medium, for which $\nu = 1/2$ (see below). P.T. Brown and Gibson (1972) considered the general compressible medium $\nu \neq 1/2$. They noted the simplifications that occur when $\nu = 1/3$ or $1/2$. When $\mu_0 = 0$, $\nu = 1/3$ the displacement field is that given above, in (12.5.15). The complete solution for $\mu = \mu_0 + mz$, $\nu = \nu_0$ was given by Awojobi and Gibson (1973). Awojobi (1972), (1975) set up the equations of a nonhomogeneous medium in which $\mu = \mu(z)$, $\nu = \nu_0$ in terms of the dilatation E and the local rotation ω_θ in cylindrical polar coordinates. He noted that the equations simplify if (a) $\mu = \mu_0 + mz$, (b) $\mu = mz^\lambda$, where $\lambda = \nu^{-1} - 1$; this is the case A = 0, B ≠ 0 given by (12.5.13) and is unacceptable unless $\nu = 1/2$, (c) $\mu = m/(h-z)$. He obtained the solution for case (c). See also Kruijtzer (1976).

We now return to the singular case (ii) given in (12.5.16). Practically, there is little to be gained by taking A ≠ 0, since E(0) = 0 whether A = 0 or A ≠ 0. If A = 0, then

$$u_R = BR^{-2} , \quad u_\theta = 0 , \quad \mu = E/3 = Cz/(3B) = mz . \qquad (12.5.44)$$

This is the so-called *Gibson medium*, and has a purely radial state of stress,

$$\tau_{RR} = -6BmR^{-2}\cos\theta . \qquad (12.5.45)$$

But this is the elastic field due to a normal point force P at the origin, where

$$P = -2\pi \int_0^{\pi/2} R^2 \sin\theta\cos\theta\,\tau_{RR}\,d\theta = 4\pi mB , \qquad (12.5.46)$$

so that

$$\tau_{RR} = -3P\cos\theta/(2\pi R^2) . \qquad (12.5.47)$$

This result was obtained by Calladine and Greenwood (1978) and show that the stress field due to a normal point load P at the origin in the Gibson medium is the same as that in an incompressible homogeneous medium (Example 12.5.5); see also Alexander (1977). Again the Gibson medium behaves like a Winkler foundation. In fact the normal displacement $w(r,z)$ due to a load normally distributed over a circle $0 < r < b$ is (Example 12.5.6)

$$w(r,z) = (2m)^{-1}Pb \int_0^\infty J_1(b\xi)J_0(r\xi)\exp(-\xi z)d\xi \ , \qquad (12.5.48)$$

so that

$$w(r,0) = (2m)^{-1}P \ , \quad 0 < r < \beta \ , \quad w(r,0) = 0 \ , \quad r > b \ . \qquad (12.5.49)$$

Awojobi (1974) showed that this result holds also for a layer of Gibson soil of arbitrary depth h, provided that it rests on a frictionless rigid bed.

The full solution of the equations of inhomogeneous isotropic elasticity when μ, ν are function of z was given by Plevako (1971). The solution is reduced to the solution of two linear differential equations, one of second order and one of fourth order. In the special case in which μ, ν are constant, the Papkovich-Neuber solution is recovered. Plevako considered special cases of this solution (1972a), (1972b), (1973b), (1973c), (1974); see also Rappoport (1976) and Rogers and Clements (1978). Further references and theoretical and numerical results may be found in the journal *Geotechnique;* see for example Carrier and Christian (1973), Belloni and Jamiolowski (1972), Poulos and Davis (1974), P.T. Brown and Gibson (1973), Gibson (1974) and Selvadurai (1979).

Examples 12.5

1. Show that the strains corresponding to (12.5.2) are

$$e_{xx} = z\,\frac{\partial^2\psi}{\partial x^2} \ , \quad e_{xz} = z\,\frac{\partial^2\psi}{\partial x\partial z} \ , \quad e_{zz} = z\,\frac{\partial^2\psi}{\partial z^2} \ , \text{ etc.}$$

Show that these satisfy the equilibrium equations (1.4.11) - (1.4.13) provided that ψ satisfies (12.5.3).

2. Show that the solutions of (12.5.25) in cylindrical polar coordinates which lead to bounded stresses and displacements as z or r tends to infinity have the form $\psi(r,z,\theta) = z^p K_\sigma(sz)J_n(sr)\cos n\theta$, where $2\sigma = 1-\lambda$.

3. Find the displacements and stresses due to the ψ of Example 12.5.2 and show that $\tau_{rz} = 0 = \tau_{\theta z}$ on $z = 0$.

4. Show that the Reissner-Sagoci problem for the half-space $z \geq 0$ and the power-law variation $\mu = mz^\lambda$ leads to the dual integral equations

$$\int_0^\infty s^{-2\sigma}A(s)J_1(rs)ds = f(r)/[2^{\sigma-1}\Gamma(\sigma)] , \qquad r \leq a ;$$

$$\int_0^\infty A(s)J_1(rs)ds = 0 , \qquad r > a .$$

5. Show that (12.5.47) is the stress-distribution in a homogeneous compressible medium due to a normal concentrated load P applied at the origin. (Use (2.5.17).)

6. Derive (12.5.48), either by using the corresponding plane-strain displacement, or by using (12.5.44) and the identity

$$\int_0^{2\pi} (\rho^2+r^2+z^2-2\rho r\cos\phi)^{-1/2}d\phi = 2\pi[(\rho+r)^2+z^2]^{-1/2}{}_2F_1(1/2,1/2;1;u^2)$$

$$= 2\pi\int_0^\infty \exp(-z\xi)J_0(\rho\xi)J_0(r\xi)d\xi$$

where $u^2 = 4\rho r/[(\rho+r)^2+z^2]$.

7. Show that if $p(x',y')$ and S are axisymmetric, then equation (12.5.42) may be written

$$w(r) = \vartheta^* H_0[\xi^{\lambda-1}p^*(\xi);r] , \qquad p(r) = H_0[p^*(\xi);r]$$

where (Popov (1961b)) $\vartheta^* = \vartheta^0 2^{-\lambda}\Gamma[(1-\lambda)/2]/\Gamma[(1+\lambda)/2]$. Use Example 5.7.11.

12.6 *Miscellaneous contact problems*

There are a number of areas of research which fall within the general scope of 'classical' contact problems which have not been covered in the preceding pages. Some of these will now be listed with a brief, fragmentary, bibliography.

(1) Problems associated with loading over a region of the surface of the half-space which is neither circular nor elliptic.

Foremost here is the work of Kalker (1967), (1971), (1972), (1977b) on 'elastic line' contact. This relates to loading over a long slender region, and has practical importance for rolling contact, a topic not considered at all in this book. Panek and Kalker (1977), in particular, present a simple approximate solution for the deformation produced by a narrow rectangular punch with rounded ends; earlier solutions had been given by Borodachev and Galin (1974), Sivashinsky (1975) and Borodachev (1976b).

This problem is related to those for a strip-shaped punch discussed by Rvachev (1956a), Borodachev (1962) and Protsenko (1972), and to those for an infinite beam or strip resting on the half-space (with the beam considered not to be a simple one-dimensional body), for which see Wieghardt (1922), Galin (1943), Rvachev (1956b), (1958), (1959b), Lekkerkerker (1960), Conway and Farnham (1970), V.S. Protsenko and Rvachev (1976). V.S. Protsenko and V.G. Protsenko (1972) have considered the problem of two strip-shaped punches while V.G. Protsenko, V.S. Protsenko and Sinekop (1973) have considered two parallel beams on a half-space. Rakov (1962) uses the analysis of Rvachev (1956a) to find an approximate result for a system of parallel beams on a half-space. Medovnik and Rakov (1970) analyzed a periodic system of parallel beams.

(2) Approximate solutions relating to punches of arbitrary cross-section indenting half-spaces or layers. Here may be men-

tioned Gorbunov-Posadov (1939), Potelezhko (1971), Golikova and Mossakovskii (1976), and the work of Conway and his collaborators in Conway, Vogel, Farnham and So (1966), Conway and Farnham (1967a), (1968), Conway and Engel (1969), Engel and Conway (1971). These latter papers assume that the contact pressure is constant over each elemental region into which the contact region is divided. The papers may be considered as precursors of the finite element method; the application of this method to contact problems is described in Selvadurai (1979 p. 375f).

(3) The contact problem for a flat-faced punch of rectangular cross-section, with particular reference to the nature of the singularities on the edges and at the corners, has been studied by Borodachev (1967b). Extensive analytical and numerical results for both frictionless and adhesive cases may be found in Brothers (1976) and Brothers, Sinclair and Segedin (1977).

(4) Contact problems for an elastic rectangle, for which see Abramian (1957), Galfaian and Chobanian (1966), Prasad and Chatterjee (1973), Dundurs, Kiattikomol and Keer (1974), Prasad and Dasgupta (1975), Galfaian (1975), Dasgupta and Prasad (1976) and Chiu, Prasad and Dasgupta (1976).

(5) Problems associated with the compression or a rigid or elastic body between two half-planes, strips, half-spaces or slabs, for which see Okubo (1951), Khadem and O'Connor (1969a), (1969b), Alblas (1974), (1975), Aihara (1975), Gladwell (1977), (1979).

(6) Contact problems for a sphere or a spherical shell. This is the subject of many research papers, among which are the following. Sternberg and Rosenthal (1952), Sternberg, Eubanks and Sadowsky (1952), Mindlin and Deresiewicz (1953), Golecki (1957), Luré (1964), Abramian, Arutiunian and Babloian (1964), Arutiunian and Abramian (1964a), (1964b), Goodman and Keer (1965), Wang (1965), DeSilva and Tsai (1969), Noble and Hussain (1969a), Bondareva (1969) (1971), Hussain and Pu (1970), (1971b), Updike and Kalnins (1970),

(1972a), (1972b), Karpenko (1971), (1976), Guerrero and Turteltaub (1972), Grigorovich (1973), Gladwell and England (1975), Golecki (1976).

(7) Problems associated with a block or cylinder embedded in a semi-infinite elastic medium, for which see Poulos and Davis (1968), Muki and Sternberg (1969), (1970), Parlas and Michalopoulos (1972), Suriyamongkal, Karasudhi and Lee (1973), Haritos (1978), Dhaliwal, Singh and Sneddon (1979), Luk and Keer (1979), Haritos and Keer (1979).

Rolling contact, numerical methods such as the finite element method, and methods based on mathematical programming are outside the scope of this book.

JOURNALS CITED IN THE REFERENCES

1. Academie des sciences. Paris. Comptes rendues.

2. Academie Polonaise des sciences. Bulletin. Classe quatrième.

3. Accademia nazionale dei Lincei, Rome. Classe di scienze fisiche, matematiche e naturali. Rendiconti.

4. Acta mechanica.

5. Akademia górniczo-hutnicza w Krakowie. Zeszyty naukowe: Elektryfikacja i mechanizacja górnictwa i hutnictwa.

6. Akademie van wetenschappen, Amsterdam. Afdeeling natuurkunde. Proceedings. Series B. Physical sciences.

7. Akademiia nauk Armianskoi S.S.R. Izvestiia. Mekhanika.

8. Akademiia nauk Armianskoi S.S.R. Izvestiia. Seriia fizikc matematicheskikh nauk.

9. Akademiia nauk S.S.S.R. Doklady.

10. Akademiia nauk S.S.S.R. Doklady. International edition. Akademiia nauk S.S.S.R. Comptes rendues de l'académie des sciences de l'U.R.S.S..

11. Akademiia nauk S.S.S.R. Institut mekhaniki. Inzhenernyi sbornik.

12. Akademiia nauk S.S.S.R. Izvestiia. Mekhanika i mashinostroenie.

13. Akademiia nauk S.S.S.R. Izvestiia. Mekhanika tverdogo tela. [Mechanics of Solids].

14. Akademiia nauk S.S.S.R. Izvestiia. Otdelenie matematicheskikh i estestvennykh nauk.

15. Akademiia nauk S.S.S.R. Izvestiia. Otdelenie tekhnicheskikh nauk. Mekhanika i mashinostroenie.

16. Akademiia nauk U.R.S.R., Kiev. Dopovidi.

17. American journal of mathematics.

18. American society of civil engineers. Engineering mechanics division. Journal.

19. American society of civil engineers. Soil mechanics and foundations division. Journal.

20. Applied mechanics reviews.

21. Applied scientific research.

22. Archive for rational mechanics and analysis.

23. Archiwum mechaniki stosowanej. Archives of mechanics.

24. Arkiv för matematik astronomi och fysik.

25. Bauingenieur.

26. Cambridge philosophical society. Proceedings.

27. Canadian geotechnical journal.

28. Communications on pure and applied mathematics.

29. Computers and mathematics with applications.

30. Defence science journal.

31. Edinburgh mathematical society. Proceedings.

32. Forschung im Ingenieurwesen.

33. Franklin institute. Journal.

34. Geotechnique.

35. Glasgow mathematical association. Proceedings.

36. Ingenieur - archiv.

37. Institute of mathematics and its applications. Journal.

38. International computational centre. Bulletin.

39. International journal of engineering science.

40. International journal of fracture mechanics.
International journal of fracture.

41. International journal of mathematical education in science and technology.

42. International journal of mechanical sciences.

43. International journal of solids and structures.

44. Inzhenernyi zhurnal.

45. Inzhenernyi zhurnal-mekhanika tverdogo tela.

46. Iowa state college journal of science.

47. Journal d'analyse mathématique.

48. Journal de mécanique.

49. Journal für die reine und angewandte Mathematik.

50. Journal of applied mechanics.

51. Journal of applied physics.

52. Journal of elasticity.

53. Journal of engineering mathematics.

54. Journal of mathematical analysis and applications.

55. Journal of mathematical and physical sciences.

56. Journal of mathematics and mechanics.
Supersedes
Journal of rational mechanics and analysis.

57. Journal of mathematics and physics.

58. Journal of sound and vibration.

59. Journal of strain analysis.
Journal of strain analysis for engineering.

60. Journal of the mechanics and physics of solids.

61. Kansas state college. Bulletin.

62. Letters in applied and engineering sciences.

63. London mathematical society. Journal.

64. London mathematical society. Proceedings.

65. Matematicheskii sbornik.

66. Mathematika.

67. Mechanics research communications.

68. Moscow. Inzhenerno-stroitel'nyi institut. Sbornik trudov.

69. Moscow. Moskovskii elektromekhanicheskii institut inzhenerov zheleznodorozhnogo transporta. Trudy.

70. Moscow. Universitet. Uchenye zapiski.

71. National academy of sciences, Washington. Proceedings.

72. Nederlandse akademie van wetenschappen. Afdeeling natuurkunde. Proceedings of the section of sciences. Series A. Mathematical sciences.

73. Nederlandse akademie van wetenschappen. Afdeeling natuurkunde. Proceedings of the section of sciences. Series B. Physical Sciences.

74. Nihon Kikaigakkai. Japanese society of mechanical engineers. Bulletin.

75. Oesterreichische bauzeitschrift.

76. Osterreichisches ingenieur-archiv.

77. Philosophical magazine; a journal of theoretical, experimental and applied physics.

78. Physical society of London. Proceedings.

79. Physics; a journal of general and applied physics.

80. Prikladnaia matematika i mekhanika. [Journal of applied mathematics and mechanics].

81. Prikladnaia mekhanika. [Soviet applied mechanics].

82. Quarterly journal of mathematics.

83. Quarterly journal of mechanics and applied mathematics.

84. Quarterly of applied mathematics.

85. Revue Romaine des sciences techniques. Série de mécanique appliquée.

86. Royal institution of naval architects. Transactions.

87. Royal society of Edinburgh. Proceedings. Section A. Mathematical and physical sciences.

88. Royal society of London. Philosophical transactions. Section A. Mathematical and physical sciences.

89. Royal society of London. Proceedings. Series A.

90. Seismological society of America. Bulletin.

91. Sibirskii matematicheskii zhurnal. [Siberian mathematical journal].

92. Society for industrial and applied mathematics. Journal.

93. Society for industrial and applied mathematics. S.I.A.M. journal on applied mathematics.

94. Soils and foundations.

95. Soviet physics - Doklady.

96. Technical physics of the U.S.S.R.

97. Theoretical and applied mechanics. (Japan National Congress for Applied Mechanics).

98. Tohoku university. Technology reports.

99. Tokyo university. College of science. Journal.

100. Ukrainskii matematicheskii zhurnal. [Ukrainian mathematical journal].

101. Zeitschrift fuer angewandte Mathematik und Mechanik.

102. Zeitschrift fuer angewandte Mathematik und Physik.

103. Zhurnal technicheskoi fiziki. [Soviet physics - technical physics].

REFERENCES

Abramian, B.L., On the plane problem of the theory of elasticity for a rectangle.
80,Vol.21(1957)89-100. 630

Abramian, B.L., Arutiunian, N.Kh., and Babloian, A.A., On two contact problems for the elastic sphere.
80,Vol.28(1964)622-629,[769-777]. 630

Abramian, B.L., Arutiunian, N.Kh., and Babloian, A.A., On symmetric pressure of a circular stamp on an elastic half-space in the presence of adhesion.
80,Vol.30(1966)143-147,[173-179]. 458,486

Abramian,B.L. and Arutiunian, N.Kh., On some contact problems for composite half-spaces.
80,Vol.31(1967)1001-1008,[1007-1013]. 570

Abramian, B.L., Contact Problems of the Theory of Elasticity. Raleigh: North Carolina State University, Department of Mathematics (1971). 2

Abramian, B.L., Certain asymmetric contact problems for semi-space. In de Pater, A.D. and Kalker, J.J. (Eds.), The Mechanics of the Contact between Deformable Bodies. Delft University Press (1975) 84-98. 491

Abramov, V.M., The problem of contact of an elastic infinite half-plane with an absolutely rigid rough foundation.
10,Vol.17(1937)173-178. 159

Abramov, V.M., Étude d'un cas d'action non-symétrique d'un poinçon circulaire appliqué à un corps élastique d'epaisseur infinie et limité par un plan.
10,Vol.23(1939)759-764. 458,476

Abramowitz, M. and Stegun, I., Handbook of Mathematical Functions. 2nd ed. New York: Dover Publications (1972).
260,261,324,389,463,547,592

Adams, G.G. and Bogy, D.B., The plane solution for the elastic contact problem of a semi-infinite strip and half-space.
50,Vol.43(1976a)603-607. 455

Adams, G.G. and Bogy,D.B., The plane solution for bending of joined dissimilar elastic semi-infinite strips.
43,Vol.12(1976b)239-249. 455

Adams, G.G. and Bogy, D.B., The plane symmetric contact problem for dissimilar elastic semi-infinite strips of different widths.
50,Vol.44(1977)604-610. 455

Aihara, T., On the stress distribution in elastic contact between flat end of a circular cylinder and a semi-infinite body.
74,Vol.18(1975)570-578. 630

Alblas, J.B., A note on a special difference equation.
6,Vol.68(1965)39-45. 346

Alblas, J.B. and Kuypers, W.J.J., On the diffusion of load from a stiffener into an infinite wedge-shaped plate.
21,Vol.15(1965/66)429-439. 350

Alblas, J.B. and Kuipers, M., Contact problems of a rectangular block on an elastic layer of finite thickness. Part I: The thin layer.
4,Vol.8(1969)133-145. 271

Alblas, J.B. and Kuipers, M., Contact problems of a rectangular block on an elastic layer of finite thickness. Part II: The thick layer.
4,Vol.9(1970a)1-12. 271

Alblas, J.B. and Kuipers, M., On the two-dimensional problem of a cylindrical stamp pressed into a thin elastic layer.
4,Vol.9(1970b)292-311. 271

Alblas, J.B., On the two-dimensional contact problem of a rigid cylinder, pressed between two elastic half-planes.
67,Vol.1(1974)15-20. 630

Alblas, J.B., On the two-dimensional contact problem of a rigid cylinder, pressed between two elastic layers. In de Pater, A.D. and Kalker, J.J. (Eds.), The Mechanics of the Contact between Deformable Bodies. Delft University Press (1975)110-126. 630

Aleksandrov, A.Ia., Some relations between the solutions of the plane and axially symmetric elasticity problems for the infinite plate.
9,Vol.128(1959a)57-60 and 95,Vol.4(1959a)1121-1124. 511

Aleksandrov, A.Ia., Some relations between the solutions of plane and axially symmetric problems in the theory of elasticity and the solution of axially symmetric problems by means of analytic functions.
9,Vol.129(1959b)754-757 and 95,Vol.4(1959b)1378-1382. 513,514,516

Aleksandrov, A.Ia., Solution of axisymmetric problems of the theory of elasticity with the aid of relations between axisymmetric and plane states of stress.
80,Vol.25(1961a)912-920,[1361-1375]. 513,623

Aleksandrov, A.Ia., Solution of some axially symmetric problems in the theory of elasticity by means of analytic functions.
9,Vol.139(1961b)337-340 and 95,Vol.6(1962)630-633. 513,516

Aleksandrov, A.Ia. and Solov'ev, Iu.I., One form of solution of three-dimensional axisymmetric problems of elasticity theory by means of functions of a complex variable and the solution of these problems for the sphere.
80,Vol.26(1962)138-145,[188-198]. 516

Aleksandrov, A.Ia. and Solov'ev, Iu.I., The solution of the three-dimensional axisymmetric problems in the theory of elasticity by means of line integrals.
80,Vol.28(1964a)914-919,[1106-1112]. 486,516

Aleksandrov, A.Ia. and Solov'ev, Iu.I., On a generalization of the method of solution of axially symmetric problems in the theory of elasticity by means of analytic functions to spatial problems without axial symmetry.
9,Vol.154(1964b)295-297 and 95,Vol.9(1964b)99-102. 516

Aleksandrov, A.Ia., Solution of three-dimensional problems of the theory of elasticity for solids of revolution by means of analytic functions.
43,Vol.4(1968)701-721. 516

Aleksandrov, V.M. and Vorovich, I.I., The action of a die on an elastic layer of finite thickness.
80,Vol.24(1960)323-333,[462-476]. 556

Aleksandrov, V.M., On the approximate solution of a certain type of integral equation.
80,Vol.26(1962)934-943,[1410-1424]. 251,252,255,264,266,269

Aleksandrov, V.M., On the solution of certain contact problems of the theory of elasticity.
80,Vol.27(1963a)970-972,[1490-1494]. 268,270

Aleksandrov, V.M., Some contact problems for the elastic layer.
80,Vol.27(1963b)758-764,[1164-1174]. 556

Aleksandrov, V.M., On the approximate solution of a class of integral equations. (In Russian).
8,Vol.17(1964a)3-8. 265

Aleksandrov, V.M., The solution of a type of two-dimensional integral equation.
80,Vol.28(1964b)579-581,[714-717]. 556

Aleksandrov, V.M. and Vorovich, I.I., Contact problems for the elastic layer of small thickness.
80,Vol.28(1964)350-351,[425-427]. 270,271

Aleksandrov, V.M. and Babeshko, V.A., Contact problems for an elastic strip of slight thickness. (In Russian).
13,No.2(1965)95-107. 271

Aleksandrov, V.M., Babeshko, V.A., and Kucherov, V.A., Contact problems for an elastic layer of slight thickness.
80,Vol.30(1966)124-142,[148-172]. 556

Aleksandrov, V.M. and Solov'ev, A.S., Some mixed spatial problems in the theory of elasticity. (In Russian).
45,No.2(1966)134-139. 595

Aleksandrov, V.M., Contact problems for an elastic wedge. (In Russian).
45,No.2(1967a)120-131. 447

Aleksandrov, V.M., On a contact problem for an elastic wedge. (In Russian).
7,Vol.20,No.1(1967b)3-14. 447

Aleksandrov, V.M., On the approximate solution of some integral equations of the theory of elasticity and mathematical physics.
80,Vol.31(1967c)1117-1131,[1122-1136]. 557,558

Aleksandrov, V.M., Axisymmetric problem of the effect of an annular die on an elastic half-space. (In Russian).
45,No.4(1967d)108-116. 519

Aleksandrov, V.M. and Belokon, A.V., Asymptotic solution of a class of integral equations and its application to contact problems for cylindrical elastic bodies.
80,Vol.31(1967)704-710,[718-724]. 271

Aleksandrov, V.M., Asymptotic methods in contact problems of elasticity theory.
80,Vol.32(1968)672-683,[691-703]. 265-266,271

Aleksandrov, V.M. and Kopasenko, V.V., Contact problems for an elastic wedge with a rigidly clamped face.
81,Vol.4,No.7(1968)75-82,[47-50]. 447

Aleksandrov, V.M., Asymptotic solution of the contact problem for a thin elastic layer.
80,Vol.33(1969)61-73,[49-63]. 558

Aleksandrov, V.M., On plane contact problems of the theory of elasticity in the presence of adhesion or friction.
80,Vol.34(1970)246-257,[232-243]. 167,278

Aleksandrov, V.M. and Kucherov, V.A., On the method of orthogonal polynomials in plane mixed problems of elasticity theory.
80,Vol.34(1970)643-652,[612-621]. 266

Aleksandrov, V.M. and Smetanin, B.I., On an effective method of solving non-classical mixed problems of the theory of elasticity.
80,Vol.35(1971)80-87,[53-60]. 271

Aleksandrov, V.M. and Babeshko, V.A., On the pressure on an elastic half-space by a wedge-shaped stamp.
80,Vol.36(1972)88-93,[78-83]. 455

Alexander, L.G., Comment on Gibson, R.E., (1967).
34,Vol.27(1977)253-254. 626

Arscott, F.M., Periodic Differential Equations: an introduction to Mathieu, Lamé and allied functions. New York: MacMillan (1962). 597,604

Arutiunian, N.Kh. and Abramian, B.L., On the impression of a rigid die into an elastic sphere.
80,Vol.28(1964a)1101-1105,[1322-1327]. 630

Arutiunian, N.Kh. and Abramian, B.L., On pressing a rigid punch into an elastic sphere. 11th International Congress of Applied Mechanics. Proceedings. Munich (1964b). 630

Arutiunian, N.Kh. and Babloian, A.A., Contact problems for a half-space with an inclusion.
80,Vol.30(1966)1050-1056,[1249-1255]. 570

Arutiunian, N.Kh., Contact problem for a half-plane with elastic reinforcement.
80,Vol.32(1968)632-646,[652-665]. 332

Arutiunian, N.Kh. and Mkhitarian, S.M., Periodic contact problem for a half-plane with elastic laps (cover plates).
80,Vol.33(1969)813-843,[791-819]. 351

Arutiunian, N.Kh. and Mkhitarian, S.M., Some contact problems for a semi-plane with elastic stiffeners. In Czarnota-Bojarski, R.E., et al (Eds.), Trends in Elasticity and Thermoelasticity, Witold Nowacki Anniversary Volume. Groningen: Wolters-Noordhoff (1971). 332,334,339

Arutiunian, N.Kh. and Mkhitarian, S.M., Certain contact problems for a half-space reinforced by elastic gussets.
80,Vol.36(1972)770-787,[727-743]. 374

Arutiunian, N.Kh. and Mkhitarian, S.M., Contact problem of pressing a stamp into an elastic half-plane with a thin reinforcing covering.
80,Vol.39(1975)857-875,[823-841]. 339

Atkinson, K.E., A Survey of Numerical Methods for the Solution of Fredholm Integral Equations of the Second Kind. Philadelphia: Society for Industrial and Applied Mathematics (1976). 257

Awojobi, A.O., Vertical vibration of a rigid circular foundation on Gibson soil.
34,Vol.22(1972)333-343. 626

Awojobi, A.O. and Gibson, R.E., Plane strain and axially symmetric problems of a linearly non-homogeneous elastic half-space.
83,Vol.26(1973)285-302. 312,313,626

Awojobi, A.O., The invariance of Gibson's Law for a stratum on a frictionless base.
34,Vol.24(1974)359-366. 626

Awojobi, A.O., The settlement of a foundation on Gibson soil of the second kind.
34,Vol.25(1975)221-228. 626

Awojobi, A.O., On the hyperbolic variation of elastic modulus in a non-homogeneous stratum.
43,Vol.12(1976)739-748. 313

Babeshko, V.A., On an asymptotic method applicable to the solution of integral equations in the theory of elasticity and in mathematical physics.
80,Vol.30(1966)732-741,[870-880]. 271

Babeshko, V.A., Certain types of integral equations appearing in contact problems of the theory of elasticity.
80,Vol.33(1969)1034-1041,[1005-1012]. 271

Babeshko, V.A. and Berkovich, V.N., On the theory of mixed problems for the three-dimensional wedge.
80,Vol.36(1972)943-947, [892-896]. 447

Baker, C.T.H., The Numerical Treatment of Integral Equations. London: Oxford University Press (1977). 257

Bantsuri, R.D., Contact problem for a wedge with an elastic support.
9,Vol.211(1973)797-800 and 95,Vol.18(1973)561-562. 350

Bantsuri, R.D., Contact problem involving an elastically reinforced anisotropic wedge.
9,Vol.222(1975)568-571 and 95,Vol.20(1975)368-370. 350

Barber, J.R., Determining the contact area in elastic indentation problems.
59,Vol.9(1974)230-232. 520

Barber, J.R., Indentation of the semi-infinite elastic solid by a concave rigid punch.
52,Vol.6(1976)149-159. 520

Bardanov, Iu.M. and Popov, G.Ia., The bending of an infinite plate adhering to an elastic half-space. (In Russian).
45,No.1(1966)100-107. 546

Barnes, E.W., On functions generated by linear difference equations of the first order.
63,Ser.2,Vol.2(1904)280-292. 346

Bažant, Z.P., Three-dimensional harmonic functions near termination or intersection of gradient singularity lines. A general numerical method.
39,Vol.12(1974)221-243. 455

Bažant, Z.P. and Keer, L.M., Singularities of elastic stresses and of harmonic functions at conical notches or inclusions.
43,Vol.10(1974)957-964. 456

Begiashvili, A.I., (Beguiachvili, A.I.), Solution du problème de la pression d'un système de profiles rigides sur la frontière rectiligne d'un demiplan élastique.
10,Vol.27(1940)914-916. 167

Belen'kii, M.Ia., The mixed problem of the theory of elasticity for an infinitely long strip. (In Russian).
80,Vol.16(1952)283-292. 249

Belen'kii, M.Ia., Some axially symmetric problems of elasticity theory.
80,Vol.24(1960)282-283,[868-873]. 513

Belik, G.I. and Protsenko, V.S., The contact problem of a half-plane for which the modulus of elasticity of the material is expressed by a power function of the depth.
81,Vol.8,No.6(1967)137-140,[80-82]. 618

Belloni, L. and Jamiolkowski, M., Comment on Carrier, W.D. and Christian, J.T., (1973).
34,Vol.23(1973)596-600. 627

Belokon', A.V., Contact problem on the interaction between an elastic disk and two different rigid stamps.
80,Vol.33(1969)136-142,[132-139]. 403

Belyaev, N.M., Studies on Elasticity and Plasticity Theory. (In Russian). Moscow: GITTL (1957). 98

Benscoter, S.V., Analysis of a single stiffener on an infinite sheet.
50,Vol.16(1949)242-246. 320,322,330

Benthem, J.P. and Van der Vooren, J., On the diffusion of load from an edge-stiffener into a wedge-shaped plate. Amsterdam: Nat. Lucht. Lab. Technical Note S. 541(1959). 350

Benthem, J.P., On the stress distribution in anisotropic infinite wedges.
84,Vol.21(1963)189-198. 447

Benthem, J.P., Note on the Boussinesq-Papkovich stress-functions.
52,Vol.9(1979)201-206. 35

Berkovich, V.N., Some contact problems for a three-dimensional wedge with a finite number of contact regions.
80,Vol.38(1974)373-377,[344-350]. 447

Bickford, W.B., Generalization of the Boussinesq-Papkovich functions to an anisotropic medium.
102,Vol.28(1977)723-728. 583

Bielak, J. and Martinez, B., An asymmetric mixed boundary-value problem of the elastic layer: translation of a rigid circular disc parallel to the contact plane.
39,Vol.12(1974)821-835. 558

References

Biot, M.A., Bending of an infinite beam on an elastic foundation.
50,Vol.4(1937)A1-A7. 279

Blackketter, D.O. and Christensen, H.D., Contact stress between two-dimensional finite elastic bodies.
50,Vol.36(1969)397-402. 189

Bogy, D.B., Edge-bonded dissimilar orthogonal elastic wedges under normal and shear loading.
50,Vol.35(1968)460-466. 452

Bogy, D.B. and Sternberg, E., The effect of couple-stresses on the corner singularity due to an asymmetric shear loading.
43,Vol.4(1968)159-174. 452

Bogy, D.B., On the problem of edge-bonded elastic quarter-planes loaded at the boundary.
43,Vol.6(1970)1287-1313. 452

Bogy, D.B., Two edge-bonded elastic wedges of different materials and wedge angles under surface tractions.
50,Vol.38(1971)377-386. 452

Bogy, D.B. and Wang, K.C., Stress singularities at interface corners in bonded dissimilar isotropic elastic materials.
43,Vol.7(1971)993-1005. 452

Bogy, D.B., Solution of the plane end problem for a semi-infinite elastic strip.
102,Vol.26(1975a)749-769. 455

Bogy, D.B., The plane solution for joined dissimilar elastic semistrips under tension.
50,Vol.42(1975b)93-98. 455

Bondareva, V.F., On the effect of an axisymmetric normal loading on an elastic sphere.
80,Vol.33(1969)1029-1033,[1001-1005]. 630

Bondareva, V.F., Contact problems for an elastic sphere.
80,Vol.35(1971)61-70,[37-45]. 630

Borodachev, N.M., The plane contact problem for an elastic body of finite width. (In Russian).
15,No.6(1962). 629

Borodachev, N.M. and Borodacheva, F.N., Impression of an annular stamp into an elastic half-space. (In Russian).
45,No.1(1966)94-99. 519,520

Borodachev, N.M. and Galin, L.A., Contact problem for a stamp with a narrow rectangular base.
80,Vol.38(1974)125-130,[108-113]. 629

Borodachev, N.M., On the nature of the contact stress singularities under an annular stamp.
80,Vol.40(1976a)372-376,[347-352]. 520

Borodachev, N.M., Contact problem for a stamp with a rectangular base.
80,Vol.40(1976b)554-560,[505-512]. 520,629

Borodachev, N.M., On a particular class of solutions of triple integral equations.
80,Vol.40(1976c)655-661,[605-611]. 520

Borowicka, H., Influence of rigidity of a circular foundation slat on the distribution of pressures over the contact surface. 1st International Conference on Soil Mechanics and Foundation Engineering. Proceedings. Cambridge, Mass. (1936) Vol.1,145-149. 547

Borowicka, H., Druckverteilung unter elastisch Platten.
36,Vol.10(1939)113-125. 547

Borowicka, H., Über ausmittig belastete, starre Platten auf elastisch-isotropen Untergrund.
36,Vol.14(1943a)1-8. 547

Borowicka, H., Die Druckausbreitung im Halbraum bei Linear zunehmendem Elastizitätsmodul.
36,Vol.14(1943b)75-82. 617

Boussinesq, J., Application des potentials à l'étude de l'équilibre et du mouvement des solides élastiques. Paris: Gauthier-Villars (1885). 33

Brahtz, J.H.A., Stress distribution in wedges with arbitrary boundary forces.
79,Vol.4(1933)56-65. 438

Brothers, P.W., The rigid rectangular punch on the elastic half-space. M.E. Thesis, University of Auckland, New Zealand (1976). 630

Brothers, P.W., Sinclair, G.B., and Segedin, C.M., Uniform indentation of the elastic half-space by a rigid rectangular punch.
43,Vol.13(1977)1059-1072. 630

Brown, E.H., The diffusion of load from a stiffener into an infinite elastic sheet.
89,Vol.239(1957)296-310. 319

Brown, P.T. and Gibson, R.E., Surface settlement of a deep elastic stratum whose modulus of elasticity increases linearly with depth.
27,Vol.9(1972)467-476. 313,625

Brown, P.T. and Gibson, R.E., Rectangular loads on inhomogeneous elastic soil.
19,Vol.10(1973)917-920. 627

Brown, S.N. and Stewartson, K., Flow near the apex of a delta wing.
37,Vol.5(1969)206-216. 455

Budiansky, B. and Wu, T.T., Transfer of load to a sheet from a rivet-attached stiffener.
57,Vol.40(1961)142-162. 322

Bueckner, H.F., Field singularities and related integral representations. In Sih, G.C. (Ed.), Mechanics of Fracture I, Methods of Analysis and Solutions of Crack Problems. Leyden: Noordhoff (1973)239-314. 455

Buell, E.L., On the distribution of plane stress in a semi-infinite plate with partially stiffened edge.
57,Vol.26/27(1948)223-233. 319,342

Buell, J., Kagiwada, H., Kalaba, R., Ruspini, E., and Zagustin, E., Solution of a system of dual integral equations.
39,Vol.10(1972)503-510. 558

Büfler, H. Einige strenge Lösungen für den Spannungszustand in ebenen Verbundkörpern.
101,Vol.39(1959)218-236. 322,334,340

Büfler, H., Scheibe mit endlicher, elastischer Versteifung.
32,No.485(1961). 322,334

Büfler, H., Zur Krafteinleitung in Scheiben über geschweisste oder geklebte Verbindungen.
76,Vol.18(1964)284-292. 322,356

Büfler, H., Theory of elasticity of a multilayered medium.
52,Vol.1(1971)125-143. 559

Burlak, J., On the solution of certain dual integral equations.
35,Vol.6(1963a)39-44. 463

Burlak, J., On Gordon's method of solving dual integral equations.
35,Vol.6(1963b)117-122. 463

Busbridge, I.W., Dual integral equations.
64,Ser.2,Vol.44(1938)115-129. 462

Byrd, P.F. and Friedman, M.D., Handbook of Elliptic Integrals. Berlin: Springer Verlag (1971). 593,597

Calladine, C.R. and Greenwood, J.A., Line and point loads on a non-homogeneous incompressible elastic half-space.
83,Vol.31(1978)507-529. 313,314,626

Carleman, T., Sur la résolution de certaines équations intégrales.
24,Vol.16,No.26(1922). 165

Carothers, S.D., Plane strain in a wedge, with applications to masonry dams.
86,Vol.33(1913)292-306. 436

Carrier, W.D. and Christian, J.T., Rigid circular plate resting on a non-homogeneous elastic half-space.
34,Vol.23(1973)67-84. 627

Cattaneo, C., Sul contatto di due corpi elastici: distribuzione locale degli sforzi.
3,Vol.27(1938)342-348,434-436,474-478. 595,613

Chang, D.J. and Muki, R., Stress distribution in a lap joint under tension-shear.
43,Vol.10(1974)503-517. 307

Chantaramungkorn, K. and Keer, L.M., Stress analysis for a layer of finite length bonded to a half-space of identical material.
43,Vol.11(1975)1079-1096. 307

Chaplygin, S.A., Pressure of a rigid die on an elastic foundation. (In Russian). Collected Works. Vol. 3(1950)317-323. 140

Chemeris, V.S., On integral equations in the axisymmetrical theory of elasticity. (In Russian).
81,Vol.1,No.5(1965)36-46. 517

Chin, P.K., Prasad, S.N., and Dasgupta, S., Indentation of an elastic rectangle by a pair of rough punches with finite friction.
43,Vol.12(1976)619-634. 630

Chong, F., Indentation of a semi-infinite medium by an axially symmetric rigid punch.
46,Vol.26(1952)1-15. 458

Chuaprasert, M.F. and Kassir, M.K., Torsion of nonhomogeneous solid.
18,Vol.99(1973)703-714. 622

Civelek, M.B. and Erdogan, F., The axisymmetric double contact problem for a frictionless elastic layer.
43,Vol.10(1974)639-659. 558

Civelek, M.D. and Erdogan, F., The frictionless contact problem for an elastic layer under gravity.
50,Vol.42(1975)136-140. 302,306

Civelek, M.B. and Erdogan, F., Interface separation in a frictionless contact problem for an elastic layer.
50,Vol.43(1976)175-177. 306

Civelek, M.B., Erdogan, F., and Cakiroglu, A.O., Interface separation for an elastic layer loaded by a rigid stamp.
39,Vol.16(1978)669-679. 306

Cochran, J.A., Analysis of Linear Integral Equations. New York: McGraw-Hill (1972). 256

Collins, W.D., On the solution of some axisymmetric boundary value problems by means of integral equations.

I. Some electrostatic and hydrodynamic problems for a spherical cap.
83,Vol.12(1959a)232-241. 499,509

II. Further problems for a circular disc and a spherical cap.
66,Vol.6(1959b)120-133. 509

III. Some scalar diffraction problems for circular disks.
83,Vol.14(1961)101-117. 509

IV. Potential problems for a circular annulus.
31,Ser.2,Vol.13(1963)235-246. 509,519,520

Comninou, Maria, Stress singularity at a sharp edge in contact problems with friction.
102,Vol.27(1976)493-499. 455

Conway, H.D., Vogel, S.M., Farnham, K.A., and So, S., Normal and shearing contact stresses in indented strips and slabs.
39,Vol.4(1966)343-359. 630

Conway, H.D. and Farnham, K.A., The contact stress problem for indented strips and slabs under conditions of partial slipping.
39,Vol.5(1967a)145-154. 630

Conway, H.D. and Farnham, K.A., Shearing stresses induced on an axis of symmetry by circular sliding contact on a transversely isotropic body.
50,Vol.34(1967b)756-757. 582

Conway, H.D., Farnham, K.A., and Ku, T.C., The indentation of a transversely isotropic half-space by a rigid sphere.
50,Vol.34(1967)491-492. 582

Conway, H.D. and Farnham, K.A., The relationship between load and penetration for a rigid, flat-end punch of arbitrary cross-section.
39,Vol.6(1968)489-496. 630

Conway, H.D. and Engel, P.A., Contact stresses in slabs due to round rough indenters.
42,Vol.11(1969)709-722. 630

Conway, H.D. and Farnham, K.A., Bending of a finite beam in bonded and unbonded contact with an elastic foundation.
42,Vol.12(1970)997-1005. 629

Cooke, J.C., A solution of Tranter's dual integral equations problem.
83,Vol.9(1956)103-110. 463

Cooke, J.C., Triple integral equations.
83,Vol.16(1963a)193-203. 519,524

Cooke, J.C., Some further triple integral equation solutions.
26,Ser.2,Vol.13(1963b)303-316. 519,520

Cooke, J.C., The solution of triple integral equations in operational form.
83,Vol.18(1965)57-71. 519

Cooper, D.H., Hertzian contact-stress deformation coefficients.
50,Vol.36(1969)296-303. 97

Copson, E.T., On the problem of the electrified disk.
31,Ser.2,Vol.8(1947)14-19. 462,509

Copson, E.T., On certain dual integral equations.
35,Vol.5(1961)21-24. 463

Dahan, M. and Zarka, J., Elastic contact between a sphere and a semi-infinite transversely isotropic body.
43,Vol.13(1977)229-238. 582

Dasgupta, S. and Prasad, S.N., On the phenomenon of separation during compression and frictional sliding of an elastic rectangle.
50,Vol.43(1976)268-273. 630

Deich, E.G., On an axially symmetrical contact problem for a non-plane die circular in plan.
80,Vol.26(1962)931-934,[1404-1409]. 467

Delves, L.M. and Walsh, J.(Eds.), Numerical Solution of Integral Equations. Oxford: Clarendon Press (1974). 257

Dempsey, J.P. and Sinclair, G.B., On the stress singularities in the plane elasticity of the composite wedge.
52,Vol.9(1979)373-391. 448,452,455

Deresiewicz, H., The half-space under pressure distributed over an elliptical portion of its plane boundary.
50,Vol.27(1960)111-119. 588

DeSilva, C.N. and Tsai, P.J., On the contact problem of thin elastic shells.
101,Vol.49(1969)267-73. 630

Deutsch, E., Indentation of a transversely isotropic half-space by a cylindrical, conical and paraboloidal punch.
2,Vol.11(1963)311-320. 583

Dhaliwal, R.S., Punch problem for an elastic layer overlying an elastic foundation.
39,Vol.8(1970)273-288. 558

Dhaliwal, R.S. and Rau, I.S., The axisymmetric Boussinesq problem for a thick elastic layer under a punch of arbitrary profile.
39,Vol.8(1970)843-856. 558

Dhaliwal, R.S. and Singh, B.M., Annular punch on an elastic layer overlying an elastic foundation.
39,Vol.15(1977a)263-270. 558

Dhaliwal, R.S. and Singh, B.M., Axisymmetric contact problem for an elastic layer on a rigid foundation with a cylindrical hole.
39,Vol.15(1977b)421-428. 558

Dhaliwal, R.S. and Singh, B.M., Torsion by an annular die of an elastic layer bonded to a semi-infinite elastic medium.
55,Vol.12(1978a)127-140. 567

Dhaliwal, R.S. and Singh, B.M., On the theory of elasticity of a non-homogeneous medium.
52,Vol.8(1978b)211-219. 622

Dhaliwal, R.S. and Singh, B.M., Torsion by a circular die of a nonhomogeneous elastic layer bonded to a nonhomogeneous half-space.
39,Vol.16(1978c)649-658. 567,622

Dhaliwal, R.S., Singh, B.M., and Sneddon, I.N., A problem of Reissner-Sagoci type for an elastic cylinder embedded in an elastic half-space.
39,Vol.17(1979)139-144. 570,631

Dinnik, A.N., Selected Works. (In Russian). Kiev: Izd.-vo Akademiia nauk U.R.S.R., Vol. 1(1952). 2,98

Doetsch, G., Handbuch der Laplace Transformation. Basel: Birkhäuser (1950). 431

Dovnorovich, V.I., On the effect of a die of circular planform on a finite thickness elastic layer on a rigid base. (In Russian).
12,No.2(1964)119-122. 555

Du Ching Hua, The two-dimensional problem of the nonhomogeneous isotropic medium. In Radok, J.R.M. (Ed.), Problems of Continuum Mechanics. Philadelphia: Society of Industrial and Applied Mathematics (1961)104-108. 308

Dundurs, J., Dependence of stress on Poisson's ratio in plane elasticity.
43,Vol.3(1967)1013-1021. 182

Dundurs, J., Discussion of paper, Bogy, D.B., Edge-bonded dissimilar orthogonal elastic wedges under normal and shear loading.
50,Vol.36(1969)650-652. 182,188,291,452

Dundurs, J., Some properties of elastic stresses in a composite. In Eringen, A.C. (Ed.), Recent Advances in Engineering Science. New York: Gordon and Breach (1970)203-216. 188

Dundurs, J. and Stippes, M., Role of elastic constants in certain contact problems.
50,Vol.37(1970)965-970. 182

Dundurs, J. and Lee, M-S., Stress concentrations at a sharp edge in contact problems.
52,Vol.2(1972)109-112. 454

Dundurs, J., Tsai, K.C., and Keer, L.M., Contact between elastic bodies with wavy surfaces.
52,Vol.3(1973)109-115. 181

Dundurs, J., Kiattikomol, K., and Keer, L.M., Contact between plates with sagged supports.
18,Vol.100(1974)445-456. 630

Dundurs, J., Properties of elastic bodies in contact. In de Pater, A.D. and Kalker, J.J. (Eds.), The Mechanics of the Contact between Deformable Bodies. Delft University Press (1975)54-66. 182

Dundurs, J., Tsai, K.C., and Keer, L.M., Some observations on the smooth contact between a layer and a half-space.
50,Vol.42(1975)221-222. 295

Elliott, D.A., An algorithm for the numerical solution of linear integral equations.
38,Vol.6(1967)207-224. 258

Elliott, H.A., Three-dimensional stress distributions in hexagonal aeolotropic crystals.
26,Vol.44(1948)522-533. 583

Elliott, H.A., Axial symmetric stress distributions in aeolotropic hexagonal crystals. The problem of the plane and related problems.
26,Vol.45(1949)621-630. 583

Engel, P.A. and Conway, H.D., Contact stress analysis for an elastic cylinder indenting a slab in the presence of friction.
42,Vol.13(1971)391-402. 630

England, A.H., A punch problem for a transversely isotropic layer.
26,Vol.58(1962)539-547. 558,567

England, A.H. and Green, A.E., Some two-dimensional punch and crack problems in classical elasticity.
26,Vol.59(1963)489-500. 181

England, A.H., Complex Variable Methods in Elasticity. London, New York: Wiley-Interscience (1971a). 42,141

England, A.H., On stress singularities in linear elasticity.
39,Vol.9(1971b)571-585. 439,448,451,456

England, A.H. and Shail, R., Orthogonal polynomial solutions to some mixed boundary value problems in elasticity II.
83,Vol.30(1977)397-414. 85,499,518,525

England, A.H., Love's integral and other relations between solutions to mixed boundary problems in potential theory. Private communication (1979). 518

Erdelyi, A. (Ed.), Tables of Integral Transforms. (2 Vols.). New York: McGraw-Hill (1954). 213

Erdelyi, A. (Ed.), Higher Transcendental Functions. (3 Vols.). New York: McGraw-Hill (1955). 134,597

Erdogan, F., Approximate solutions of singular integral equations.
93,Vol.17(1969)1041-1059. 262

Erdogan, F., Analysis of elastic cover plates. In Developments in Mechanics, Vol. 6. 12th Midwestern Mechanics Conference. Proceedings. Notre Dame, Indiana: University of Notre Dame Press (1971)817-829. 335,351

Erdogan, F. and Gupta, G.D., The problem of an elastic stiffener bonded to a half-plane.
50,Vol.38(1971)937-941. 333,334

Erdogan, F. and Gupta, G.D., On the numerical solution of singular integral equations.
50,Vol.29(1972)525-534. 262

Erdogan, F., Gupta, G.D., and Cook, T.S., Numerical solution of singular integral equations. In Sih, G.C. (Ed.), Mechanics of Fracture I, Methods of Analysis and Solutions of Crack Problems. Leyden: Noordhoff (1973)368-425. 262

Erdogan, F. and Civelek, M.B., Contact problem for an elastic reinforcement bonded to an elastic plate.
50,Vol.41(1974)1014-1018. 364

Erdogan, F. and Ratwani, M., The contact problem for an elastic layer supported by two elastic quarter planes.
50,Vol.41(1974)673-678. 447

Erdogan, F. and Arin, K., Fracture and contact problems for an elastic wedge.
52,Vol.6(1976a)57-71. 447

Erdogan, F. and Arin, K., Effect of friction in wedging of elastic solids.
52,Vol.6(1976b)261-275. 447

Erdogan, F. and Gupta, G.D., Contact and crack problems for an elastic wedge.
39,Vol.14(1976)155-164. 447

Essenburg, F. and Gulati, S.T., On the contact of two axisymmetric plates.
50,Vol.33(1966)341-346. 551

Eubanks, R.A. and Sternberg, E., On the completeness of the Boussinesq-Papkovich stress-functions.
56,Vol.5(1956)735-746. 34

Fabrikant, V.I., Effect of shearing force and tilting moment on a cylindrical punch attached to a transversely isotropic half-space.
80,Vol.35(1971)178-182,[147-151]. 582

Filon, L.N.G., On an approximate solution for the bending of a beam of rectangular cross-section under any system of load, with special reference to points of concentrated or discontinuous loading.
88,Vol.201(1903)63-155. 41

Filonenko-Borodich, M.M., Certain approximate theories of elastic foundation. (In Russian).
70,Vol.46(1940). 537

Filonenko-Borodich, M.M., The simplest model of an elastic foundation which is capable of distributing a load. (In Russian).
69,Vol.53(1945). 102,104,529

Flavin, J.N. and Gallagher, J.P., A rigid elliptic inclusion in an anisotropic elastic whole space.
43,Vol.12(1976)671-682.

Florence, A.L., Two contact problems for an elastic layer.
83,Vol.14(1961)453-459. 559,566

Fox, L. and Parker, I.B., Chebyshev Polynomials in Numerical Analysis. London: Oxford University Press (1968). 61

Frasier, J.T. and Rongved, L., Force in the plane of two joined semi-infinite plates.
50,Vol.24(1957)582-584. 235

Frederick, D., On some problems of bending of thick circular plates on an elastic foundation.
50,Vol.23(1956)195-200. 546

Freeman, N.J. and Keer, L.M., Torsion of elastic cylinders in contact.
43,Vol.3(1967a)799-808. 570

Freeman, N.J. and Keer, L.M., Torsion of a cylindrical rod welded to an elastic half-space.
50,Vol.34(1967b)687-692. 570

Galfaian, P.O. and Chobanian, K.S., Solution of a contact problem for an elastic rectangle.
80,Vol.30(1966)569-575,[676-684]. 630

Galfaian, P.O., Contact problem of the theory of elasticity for a rectangle.
81,Vol.11,No.4(1975)54-64,[387-395]. 630

Galin, L.A., On the Zimmerman-Winkler hypothesis for beams. (In Russian).
80,Vol.7(1943)293-300. 629

Galin, L.A., Pressure of a punch with friction and adhesion domains. (In Russian).
80,Vol.9(1945)413-424. 272

Galin, L.A., Three-dimensional contact problems in the theory of elasticity for punches of circular plan form. (In Russian).
80,Vol.10(1946)425-448. 467

Galin, L.A., Pressure of a stamp with a flat base in the form of an infinite wedge on an elastic half-space.
9,Vol.58,No.2(1947a). 455

Galin, L.A., On the pressure of a punch of elliptic shape upon the elastic half-space. (In Russian).
80,Vol.11(1947b)281-284. 588,595,596

Galin, L.A., Estimate of the displacements in spatial contact problems of the theory of elasticity. (In Russian).
80,Vol.12(1948)241-250. 89

Galin, L.A., Contact Problems in the Theory of Elasticity. (In Russian). Moscow: Gostekhizdat, 1953. English translation, Sneddon, I.N. (Ed.). Raleigh: North Carolina State College, Department of Mathematics (1961).
2,89,93,140,165,182,272,455,505,588,589,595,596

Galin, L.A., Contact problems of the theory of elasticity in the presence of wear.
80,Vol.40(1977)981-986,[931-936].

Gdoutos, E.E. and Theocaris, P.S., Stress concentrations at the apex of a plane indenter acting on an elastic half-plane.
50,Vol.42(1975)688-692. 455

Gecit, M.R. and Erdogan, F., Frictionless contact problem for an elastic layer under axisymmetric loading.
39,Vol.14(1978)771-785. 558,574

Gibson, R.E., Some results concerning displacements and stresses in a non-homogeneous elastic half-space.
34,Vol.17(1967)58-67 [errata in Vol.18(1968),275-276; Vol.19(1969)160-161]. 313,625

Gibson, R.E. and Sills, G.C., On the loaded elastic half-space with depth-varying Poisson's ratio.
102,Vol.20(1969)691-695. 622

Gibson, R.E., Brown, P.T., and Andrews, K.R.F., Some results concerning displacements in a non-homogeneous elastic layer.
102,Vol.22(1971)855-864. 308,313,623,625

Gibson, R.E., The analytical method in soil mechanics.
34,Vol.24(1974)115-140. 99,627

Gibson, R.E. and Kalsi, G.S., The surface settlement of a linearly inhomogeneous cross-anisotropic elastic half-space.
102,Vol.25(1974)843-847. 313-314

Gibson, R.E. and Sills, G.C., Settlement of a strip load on a non-homogeneous orthotropic incompressible elastic half-space.
83,Vol.28(1975)233-243. 314

Gladwell, G.M.L., The calculation of mechanical impedances relating to an indenter vibrating on the surface of a semi-infinite elastic body.
58,Vol.8(1968)215-228. 457,496

Gladwell, G.M.L., A contact problem for a circular cylindrical punch in adhesive contact with an elastic half-space: the case of rocking, and translation parallel to the plane.
39,Vol.7(1969)295-307. 458,491,566

Gladwell, G.M.L. and Iyer, K.R.P., Unbonded contact between a circular plate and an elastic half-space.
52,Vol.4(1974)115-130. 562

Gladwell, G.M.L., Unbonded contact between a circular plate and an elastic foundation. In de Pater, A.D. and Kalker, J.J. (Eds.), The Mechanics of the Contact between Deformable Bodies. Delft University Press (1975)99-109. 562,563,564

Gladwell, G.M.L. and England, A.H., Contact problems for the spherical shell. 12th Annual Meeting of the Society of Engineering Science. Proceedings. University of Texas at Austin (1975)869-877. 631

Gladwell, G.M.L., On some unbonded contact problems in plane elasticity theory.
50,Vol.43(1976)263-267. 238,268,283

Gladwell, G.M.L., The contact problem for a rigid cylinder pressed between two elastic layers.
50,Vol.44(1977)36-40. 630

Gladwell, G.M.L. and England, A.H., Orthogonal polynomial solutions to some mixed boundary value problems in elasticity theory.
83,Vol.30(1977)175-185. 518

Gladwell, G.M.L., Polynomial solutions for an ellipse on an anisotropic elastic half-space.
83,Vol.31(1978)251-260. 589,595

Gladwell, G.M.L., A note on a three-part contact problem, Chebyshev polynomials and elliptic integrals.
39,Vol.17(1979)(to appear). 630

Gladwell, G.M.L. and Gupta, O.P., On the approximate solution of elastic contact problems for a circular annulus.
52,Vol.9(1979)335-348. 526

Glagolev, N.I., Elastic stresses along the bottom of a dam.
10,Vol.34(1942)187-191. 160,165

Glagolev, N.I., Determination of the stresses for the pressure of systems of rigid profiles. (In Russian).
80,Vol.7(1943)383-388. 160

Glauert, H., Elements of Aerofoil and Airscrew Theory. London: Cambridge University Press (1948). 66

Glover, R.E., Discussion of Benscoter (1949).
50,Vol.17(1950)218. 320

Goland, M. and Reissner, E., The stresses in cemented joints.
50,Vol.11(1944)A17-A27. 307

Golecki, J.J., Analogy between boundary value problems for regions bounded by concentric circles and axially symmetrical boundary value problems for regions bounded by concentric spherical surfaces.
2,Vol.5(1957)327-333. 630

Golecki, J.J., On the foundations of the theory of elasticity of plane incompressible non-homogeneous bodies. In Olszak, W. (Ed.), Non-homogeneity in Elasticity and Plasticity. London: Pergamon Press (1960)39-51. 308

Golecki, J.J. and Knops, R.J., Introduction to a linear elastostatics with variable Poisson's ratio.
5,Vol.30(1969)81-92. 617

Golecki, J.J., Divergent integral appearing in Fourier-transform applications in elasticity.
41,Vol.6(1975)321-326. 207

Golecki, J.J., Solutions in terms of homogeneous harmonic functions in three-dimensional elastostatics and the viscous flow theory.
36,Vol.45(1976)69-78. 35,631

Golibykhin. See Kolybikhin.

Golikova, S.S. and Mossakovskii, V.I., Pressure of a plane stamp of nearly circular cross-section on an elastic half-space.
80,Vol.40(1976)1143-1145,[1085-1088]. 630

Golub, V.K. and Mossakovskii, V.I., The bending of a circular plate on an elastic half-space in the presence of adhesion. (In Russian).
12,No.2(1960)88-94. 551

Golub, V.K., The bending of a circular plate of variable rigidity on an elastic foundation in the presence of adhesion. (In Russian).
12,No.4(1964)92-95. 551

Goodier, J.N. and Hsu, C.S., Transmission of tension from a bar to a plate.
50,Vol.21(1954)147-150. 321,370

Goodman, L.E., Contact stress analysis of normally loaded rough spheres.
50,Vol.29(1962)515-522. 482

Goodman, L.E. and Keer, L.M., The contact stress problem for an elastic sphere indenting an elastic cavity.
43,Vol.1(1965)407-415. 630

Goodman, L.E. and Keer, L.M., Influence of an elastic layer on the tangential compliance of bodies in contact. In de Pater, A.D. and Kalker, J.J. (Eds.), The Mechanics of the Contact between Deformable Bodies. Delft University Press (1975)127-151. 558,616

Gorbunov-Posadov, M.I., On beams and rectangular plates supported by an elastic half-space.
10,Vol.24(1939)421-425. 528,630

Gorbunov-Posadov, M.I., Analysis of beams and plates on an elastic half-space. (In Russian).
80,Vol.4,No.3(1940)61-80. 528

Gorbunov-Posadov, M.I., Slabs on Elastic Foundations. (In Russian). Moscow: Gosstroiizdat (1941). 528

Gorbunov-Posadov, M.I., Beams and Slabs on Elastic Foundations. (In Russian). Moscow: Mashstroiizdat (1949). 528

Gorbunov-Posadov, M.I. and Malikova, T.A., Calculation of Structures on Elastic Foundation. (In Russian). Moscow: Stroiizdat (1973). 2,528

Gordon, A.N., Dual integral equations.
63,Vol.29(1954)360-363. 461,462

Goree, J.G., Approximate determination of contact stresses in an infinite plate containing a smooth rigid elliptic insert.
50,Vol.32(1965)437-439. 395,408

Gradshteyn, I.S. and Ryzhik, I.M., Tables of Integrals, Series and Products. 4th. ed. Moscow 1963. English translation, Jeffrey, A. (Ed.). New York: Academic Press Inc. (1965).
163,229,310,411,414,415,473,497,502,593,595

Green, A.E., On Boussinesq's problem and penny-shaped cracks.
26,Vol.45(1949)251-257. 509

Green, A.E. and Sneddon, I.N., The distribution of stress in the neighbourhood of a flat elliptical crack in an elastic solid.
26,Vol.46(1950)159-163. 596

Green, A.E. and Zerna, W., Theoretical Elasticity. Oxford: Clarendon Press (1954). 1,141,499,509,583

Greenwood, J.A., The elastic stresses in the mid-plane of a slab by pressures applied symmetrically at its surface.
26,Vol.60(1964)159-169. 559

Griffith, A.A., The phenomena of rupture and flow in solids.
88,Vol.221(1920)163-198. 498

Grigorovich, V.M., The contact of a spherical shell with a rigid punch in the presence of adhesion.
81,Vol.9,No.5(1973)24-30,[483-487]. 631

Grilitskii, D.V. and Kizyma, Ia.M., Concomitant torsion of a rod and a half-space.
81,Vol.3,No.2(1967)89-98,[54-59]. 570

Grinberg, G.A. and Kuritsyn, V.N., Diffraction of a plane electromagnetic wave by an ideally conducting plane ring and the elastostatic problem for such a ring.
103,Vol.31(1961)1017-1027,[Vol.6(1962)743-749]. 520

Gubenko, V.S., Pressure of an axisymmetric annular die on an elastic layer and half-space.
15,No.3(1960)60-64. 519

Gubenko, V.S. and Mossakovskii, V.I., Pressure of an axially symmetric circular die on an elastic half-space.
80,Vol.24(1960)334-340,[477-486]. 519,523

Gubenko, V.S., Grabko, G.K., and Nakashidze, G.M., The contact problem for a circular punch and a half-space in the presence of friction forces.
81,Vol.7,No.3(1971)119-121,[332-334]. 492

Gubenko, V.S., Kiselev, M.Ia., and Lamzyuk, V.D., Action of an annular punch with a plane base on a layer lying on an elastic half-space.
81,Vol.11,No.7(1975)11-16,[694-698]. 558

Guerrero, I. and Turtleltaub, M.J., The elastic sphere under arbitrary concentrated loads.
52,Vol.2(1972)21-33. 631

Guidera, J.T., The general non-symmetrical punch and crack problems. Ph.D. Thesis, Simon Fraser University (1975). 583

Guidera, J.T., Lardner, R.W., and Tupholme, G.E., The indentation of a half-space of hexagonal elastic material by a circular punch of arbitrary end-profile.
53,Vol.12(1978)77-82. 583

Gurtin, M.E., Linear Theory of Elasticity. In Flügge, S. (Ed.), Handbuch der Physik ,Vol. 6a/2. Berlin, Heidelberg, New York: Springer (1972). 1,32,34

Hannah, M., Contact stress and deformation in a thin elastic layer.
83,Vol.4(1951)94-105. 264

Hara, T., Shibuya, T., Koizumi, T., and Nakahara, I., An axisymmetric mixed boundary value problem of the thick elastic plate: moment applied to the annular rigid stamp.
97,Vol.25(1975)505-512. 520

Hara, T., Shibuya, T., Koizumi, T., and Nakahara, I., The bending of an elastic layer with an annular crack.
97,Vol.27(1977)47-53. 520

Hara, T., Shibuya, T., Koizumi, T., and Nakahara, I., An axisymmetric mixed boundary value problem of the elastic half-space subjected to moment by an annular rigid punch.
74,Vol.21(1978a)566-571. 520

Hara, T., Shibuya, T., Koizumi, T., and Nakahara, I., The effect of bending on the infinite elastic solid with an annular crack.
74,Vol.21(1978b)1517-1575. 520

Harding, J.W. and Sneddon, I.N., The elastic stresses produced by the indentation of the plane surface of a semi-infinite elastic solid by a rigid punch.
26,Vol.41(1945)16-26. 460,466

Haritos, G.K., Stress Analysis of a Rigid Block Embedded in an Elastic Half Space. Ph.D. Thesis, Northwestern University (1978). 631

Haritos, G.K. and Keer, L.M., Stress analysis of a rigid block embedded in an elastic half-space.
43,(1979)(to appear). 631

Harrington, W.J. and Ting, T.W., Stress boundary-value problems for infinite wedges.
52,Vol.1(1971)65-81. 441

Hassan, H.A.Z., Reissner-Sagoci problem for a non-homogeneous large thick plate.
48,Vol.18(1979)197-210. 537

Hearmon, R.F.S., An Introduction to Applied Anisotropic Elasticity. London: Oxford University Press (1961). 27

Hein, V.L. and Erdogan, F., Stress singularities in a two-material wedge.
40,Vol.7(1971)317-330. 452

Heins, A.E. and MacCamy, R.C., On the scattering of waves by a disk.
102,Vol.11(1960a)249-264. 513

Heins, A.E. and MacCamy, R.C., On mixed boundary value problems for axially symmetric potentials.
54,Vol.1(1960b)331-333. 513

Hertz, H., Über die Berührung fester elastischer Körper.
49,Vol.92(1882a)156-171. In Hertz, H., Gesammelte Werke, Band 1. Leipzig: Johann Ambrosius Barth (1895)155-173. 93

Hertz, H., Über die Berührung fester elastischer Körper und über die Harte. Berlin (1882b). In Hertz, H., Gesammelte Werke, Band 1. Leipzig: Johann Ambrosius Barth (1895)174-196. 93

Hertz, H. Über die Verteilung der Druckfrafte in einem elasteschen Kreiscylinder. (1883). In Hertz, H., Gesammelte Werke, Band 1. Leipzig: Johann Ambrosius Barth (1895)283-287. 395

Hertz, H., Über das Gleichgewicht shwimmender elastischer Platten. (1884). In Hertz, H., Gesammelte Werke, Band 1. Leipzig: Johann Ambrosius Barth (1895)288-294. 542,546

Hetenyi, M., Beams on Elastic Foundation. Ann Arbor: University of Michigan Press (1946). 2,98,99,100,102,104,279

Hetenyi, M., A general solution for the bending of beams on an elastic foundation of arbitrary continuity.
51,Vol.21(1950)55-58. 102,529

Hetenyi, M. and McDonald, P.H., Jr., Contact stresses under combined pressure and twist.
50,Vol.25(1958)396-401. 616

Hetenyi, M., A method of solution for the elastic quarter-plane.
50,Vol.27(1960)289-296. 447

Hetenyi, M., Beams and plates on elastic foundations and related problems.
20,Vol.19(1964)95-102. 2,99,281,528

Hobson, E.W., The Theory of Spherical and Ellipsoidal Harmonics. London: Cambridge University Press (1931). 498,500,503,596

Hogg, A.H.A., Equilibrium of a thin plate, symmetrically loaded, resting on an elastic foundation of infinite depth.
77,Vol.25(1938)576-582. 542

Hogg, A.H.A., Equilibrium of a thin slab on an elastic foundation of finite depth.
77,Vol.35(1944)265-276. 542

Holl, D.L., Equilibrium of a thin plate, symmetrically loaded, on a flexible subgrade.
46,Vol.12(1938)455-459. 542

Holl, D.L., Thin plates on elastic foundations. 5th International Congress of Applied Mechanics, Cambridge, Mass., 1938 Proceedings. New York: John Wiley (1939)71-74. 542

Hooper, J.A., Analysis of a circular raft in adhesive contact with a thick elastic layer.
34,Vol.24(1974)561-580. 559

Hooper, J.A., Elastic settlement of a circular raft in adhesive contact with a transversely isotropic medium.
34,Vol.25(1975)691-711. 559

Hooper, J.A., Parabolic adhesive loading of a flexible raft foundation.
34,Vol.26(1976)511-525. 559

Hussain, M.A., Pu, S.L., and Sadowsky, M.A., Cavitation at the ends of an elliptic inclusion inside a plate under tension.
50,Vol.35(1968)505-509. 417

Hussain, M.A. and Anderson, G., Lifting of a plate from a foundation due to axisymmetric pressure. In Weiss, J.H., et al (Eds.), Developments in Mechanics, Vol.5. 11th Midwestern Mechanics Conference. Proceedings. (1969)577-590. 574

Hussain, M.A. and Pu, S.L., A note on an elastic sphere indenting an elastic spherical cavity. 15th Conference of Army Mathematicians. Transactions. (1970)1-24. 630

Hussain, M.A. and Pu, S.L., Slip phenomenon for a circular inclusion.
50,Vol.38(1971a)627-633. 405

Hussain, M.A. and Pu, S.L., Non-Hertzian contact of an elastic sphere indenting an elastic cavity. In Developments in Mechanics, Vol.6. 12th Midwestern Mechanics Conconference. Proceedings. Notre Dame, Indiana: University of Notre Dame Press (1971b)647-658. 630

Huth, J.H., The complex-variable approach to stress singularities.
50,Vol.20(1953)561-562. 448

Inglis, C.E., Stresses in a plate due to the presence of cracks and sharp corners.
86,Vol.55(1913)219-230. 498

Ishkova, A.G., An exact solution of the problem of bending of a round plate supported by an elastic half-space and subjected to an axially symmetrical uniformly distributed load.
11,Vol.56,No.2(1947)129-132. 548

Ishkova, A.G., Some generalizations concerning solutions of problems of bending of a circular plate and an infinite strip lying on an elastic half-space. (In Russian).
80,Vol.21(1957)287-290. 549

Ishkova, A.G., The bending of a plate lying on an elastic half-space, under the action of a certain non-axisymmetric load. (In Russian).
12,Vol.165(1958)78-84. 549

Ishkova, A.G., Bending of a strip and a circular plate lying on the elastic half-space. (In Russian).
11,Vol.28(1960)171-181. 549

Ishkova, A.G., The bending of a strip and a circular plate on a combined foundation. (In Russian).
44,Vol.2(1962)270-279. 549

Iyer, K.R.P., Unbonded Contact of a Circular Plate and an Elastic Half-Space. Ph.D. Thesis, University of Waterloo (1973). 2,550,562

Jabali, H.H., Torsional stiffness of a layer bonded to an elastic half-space.
43,Vol.11(1975)1219-1221. 566

Jain, D.L. and Kanwal, R.P., An integral equation method for solving mixed boundary value problems.
93,Vol.20(1971)642-658. 519

Jain, D.L. and Kanwal, R.P., Three-part boundary value problems in potential and generalised axially symmetric potential theories.
47,Vol.25(1972)107-158. 519

Jung, H., Druckverteilung unter elastisch gelagerten Kreisplatten.
36,Vol.20(1952)8-12. 561

Kalaba, R., Yakush, A., and Zagustin, E.A., A new method for the Lebedev-Ufliand integral equation for contact problems of elasticity.
52,Vol.3(1972)181-189. 558

Kalaba, R. and Zagustin, E.A., An initial value method for dual integral equations.
39,Vol.10(1972a)603-608. 558

Kalaba, R.E. and Zagustin, E.A., Initial value treatment for torsion of an elastic layer.
33,Vol.293(1972b)345-353. 558

Kalaba, R. and Zagustin, E.A., An improved initial value method for a contact problem with a nonplane punch.
39,Vol.11(1973)459-476. 558

Kalaba, R. and Zagustin, E.A., On penetration of a rigid axisymmetric punch into an elastic layer.
29,Vol.1(1975)13-25. 558

Kalaba, R.E., Yakush, A., and Zagustin, E.A., On a conical punch pressing into an elastic layer.
52,Vol.6(1976)441-449. 558

Kalandia, A.I., On contact problems in elasticity theory. (In Russian).
80,Vol.21(1957)389-398. 421,427

Kalandia, A.I., On the approximate solution of a class of singular integral equations.
9,Vol.125,No.4(1959). 263

Kalandia, A.I., Remarks on the singularity of elastic solutions near corners.
80,Vol.33(1969a)132-135,[127-131]. 448

Kalandia, A.I., Stress conditions in plates reinforced by stiffening ribs.
80,Vol.33(1969b)538-543,[523-529]. 320,350

Kalker, J.J., On the rolling contact of two elastic bodies in the presence of dry friction. Ph.D. Thesis, Technische Hogeschool, Delft (1967). 629

Kalker, J.J., On the uniqueness of contact problems in the theory of elasticity. Laboratorium voor Technische Mechanica, Delft, Netherlands: Technological University. Report No. 397(1968). 2

Kalker, J.J., On Elastic Line Contact. Complete report. Institute of Mathematics, Delft University of Technology (1971). 49,629

Kalker, J.J., On elastic line contact.
50,Vol.39(1972)1125-1132. 49,629

Kalker, J.J., Aspects of contact mechanics. In de Pater, A.D. and Kalker, J.J. (Eds.), The Mechanics of the Contact between Deformable Bodies. Delft University Press (1975) 1-25. 2

Kalker, J.J., A survey of the mechanics of contact between solid bodies.
101,Vol.57(1977a)T3-T17. 2

Kalker, J.J., The surface displacement of an elastic half-space loaded in a slender, bounded, curved surface region with application to the calculation of the contact pressure under a roller.
37,Vol.19(1977b)127-144. 629

Kantorovich, L.V. and Krylov, V.I., Approximate Methods of Higher Analysis. English translation, Benster, A.D. (Ed.). New York: Interscience (1958). 259,333

Kapshivyi, A.A. and Maslyuk, G.F., The solution of the mixed axisymmetric problem from the theory of elasticity for a half-space by the method of p-analytic functions.
81,Vol.3,No.7(1967)21-27,[13-16]. 517

Karp, S.N. and Karal, F.C., The elastic-field behaviour in the neighbourhood of a crack of arbitrary angle.
28,Vol.15(1962)413-421. 452

Karpenko, L.N., Approximate solution of a singular integral equation by means of Jacobi polynomials.
80,Vol.30(1966)564-569,[668-675]. 263

Karpenko, V.A., Axisymmetric contact problem for an elastic sphere.
7,Vol.24,No.4(1971)16-19. 631

Karpenko, V.A., Axisymmetric impression of two stamps into an elastic sphere.
80,Vol.40(1976)763-766,[711-714]. 631

Kassir, M.K., The Reissner-Sagoci problem for a non-homogeneous solid.
39,Vol.8(1970)875-885. 621,622

Kassir, M.K., Boussinesq problems for non-homogeneous solid.
18,Vol.98(1972)457-469. 620

Kassir, M.K. and Chuaprasert, M.F., A rigid punch in contact with a non-homogeneous elastic solid.
50,Vol.41(1974)1019-1024. 621

Keer, L.M., The contact stress problem for an elastic sphere indenting an elastic layer.
50,Vol.31(1964a)143-145. 499,553

Keer, L.M., The torsion of a rigid punch in contact with an elastic layer where the friction law is arbitrary.
50,Vol.31(1964b)430-434. 567

Keer, L.M., A class of non-symmetrical punch and crack problems.
83,Vol.17(1964c)423-436. 476,509

Keer, L.M., Nonaxisymmetric punch and crack problems for initially stressed bodies.
84,Vol.23(1965)97-108. 517

Keer, L.M., Mixed boundary-value problems for an elastic half-space.
26,Vol.63(1967)1379-1386. 509

Keer, L.M. and Chantaramungkorn, K., Loss of contact between an elastic layer and half-space.
52,Vol.2(1972)191-197. 294

Keer, L.M., Dundurs, J., and Tsai, K.C., Problems involving receding contact between a layer and a half-space.
50,Vol.39(1972)1115-1120. 287,571,573

Keer, L.M. and Silva, M.A.G., Two mixed problems for a semi-infinite layer.
50,Vol.39(1972)1121-1124. 297

Keer, L.M., Dundurs, J., and Kiattikomol, K., Separation of a smooth circular inclusion from a matrix.
39,Vol.11(1973)1221-1233. 405,420

Keer, L.M. and Freedman, J.M., Static response of a rigid strip bonded to an elastic layer.
4,Vol.17(1973)1-15. 334

Keer, L.M., Stress analysis of bonded layers.
50,Vol.41(1974)679-683. 307

Keer, L.M. and Chantaramungkorn, K., Stress analysis of a double lap joint.
50,Vol.42(1975)353-357. 307

Keer, L.M. and Goodman, L.E., Tangential loading of two bodies in contact.
50,Vol.42(1976)513-514. 612

Keer, L.M. and Luk, V.K., Stress analysis of an elastic layer attached to an elastic half-space of the same material.
39,Vol.14(1976)735-747. 558,574

Keer, L.M. and Parihar, K.S., A note on the singularity at the corner of a wedge-shaped punch or crack.
93,Vol.34(1978)297-302. 456

Kerr, A.D., Elastic plates on liquid foundations.
18,Vol.89,EM3(1963)59-71. 547

Kerr, A.D., Elastic and viscoelastic foundation models.
50,Vol.31(1964)491-498. 102

Kerr, A.D., A study of a new foundation model.
4,Vol.1(1965)135-147. 102

Kerr, A.D., On the derivation of well-posed boundary value problems in structural mechanics.
43,Vol.12(1976)1-11. 102

Khadem, R., Some Contact Problems in Elasticity. D. Phil. Thesis, Oxford University (1967). 458

Khadem, R. and O'Connor, J.J., Adhesive or frictionless compression of an elastic rectangle between two identical elastic half-spaces.
39,Vol.7(1969a)153-168. 630

Khadem, R. and O'Connor, J.J., Axial compression of an elastic circular cylinder in contact with two identical elastic half-spaces.
39,Vol.7(1969b)785-800. 630

Kil'chevskii, N.A. and Kostyuk, E.N., 20th century developments in the theory of contact interactions between solids.
81,Vol.2,No.8(1966)32-39,[19-23]. 98

Kim, I.V. and Suncheleev, R.Ia., A contact problem for an orthotropic half-space.
81,Vol.6,No.7(1970)40-47,[719-724]. 579

King, L.V., On the acoustic radiation pressure on circular discs, inertia and diffraction corrections.
89,Vol.153(1935)1-16. 462

Kizima, Ia.M., Contact stresses in the case of adhesion of a circular die to an elastic layer.
45,Vol.4,No.2(1964)376-382. 558

Klein, G.K., Taking into account inhomogeneity, discontinuity of strains and other mechanical properties of soil, when calculating structures on solid foundations. (In Russian).
68,Vol.14(1956). 618

Klubin, P.I., Calculation of plate-beams and circular plates on an elastic foundation. (In Russian).
11,Vol.12(1952)95-135. 66,83,281,549,587

Knein, M., Zur Theorie des Druckversuchs.
101,Vol.6(1926)414-416. 448

Koiter, W.T., Approximate solution of Weiner-Hopf type integral equations with applications.
6,Vol.57(1954)558-579. 447

Koiter, W.T., On the diffusion of load from a stiffener into a sheet.
83,Vol.8(1955)164-178. 319,342,346,348

Koiter, W.T. and Alblas, J.B., On the bending of cantilever, rectangular plates.
6,Vol.57(1954)250-269,548-557;Vol.60(1957)173-181. 452

Koiter, W.T., Solution of some elasticity problems by asymptotic methods. IUTAM Symposium on Applications of the Theory of Functions in Continuum Mechanics. Proceedings. Tbilisi, 1963. Moscow: Nauka Publishing House (1965)15-31. 271

Kolybikhin, Iu.D. and Popov, G.Ia., On the contact problem for annular punch. (In Russian).
7,Vol.22,No.6(1969)10-21. 526

Kolybikhin, Iu.D., The contact problem of the torsion of a half-space due to an elastic disk.
81,No.12(1971)25-31,[1311-1316]. 622

Korenev, V.G., The Calculation of Beams and Plates on Elastic Foundations. (In Russian). Moscow: Gosstroiizdat (1954). 541

Korenev, V.G., A stamp lying on an elastic half-space whose modulus of elasticity is a power function of depth. (In Russian).
9,Vol.112(1957)823-826. 618

Korenev, V.G., Certain Problems of the Theory of Elasticity and Heat Conductance Solvable in Terms of Bessel Functions. (In Russian). Moscow: Fizmatgiz (1960a).

Korenev, V.G., Structures resting on an elastic foundation. In Rabinovich, I.M. (Ed.), Structural Mechanics in the U.S.S.R. 1917-1957. English translation, Herrmann, G. (Ed.). New York: Pergamon Press (1960b)160-190. 99,528

Krahn, E., Transmission of tension from a bar to a plate. Discussion of Goodier and Hsu (1954).
50,Vol.22(1955)139-140. 321,370

Krein, M.G., A new method of solving linear integral equations of the first and second kinds. (In Russian).
9,Vol.100(1955)413-416. 270

Krenk, S., On the use of the interpolation polynomial for solutions of singular integral equations.
84,Vol.32(1975a)479-484. 263

Krenk, S., On quadrature formulas for singular integral equations of the first and second kinds.
84,Vol.33(1975b)225-232. 263

Krenk, S., Periodic contact and crack problems in plane elasticity.
62,Vol.4(1976)343-353. 181

Krenk, S., Quadrature formulae of closed type for solution of singular integral equations.
37,Vol.22(1978)99-107. 263

Krolevets, M.S., A critical survey of methods of solving the contact problem of elasticity theory.
81,Vol.2,No.11(1966)68-82,[36-46]. 2,98

Kruijtzer, G.F.J., Dynamics of a linearly inhomogeneous incompressible isotropic elastic half-space.
21,Vol.32(1976)1-29. 626

Kucherov, V.A., Method of orthogonal polynomials in plane antisymmetric mixed problems of elasticity theory with two contact sections.
80,Vol.38(1974)331-338,[302-309]. 174

Kuczma, M., Functional Equations in a Single Variable. Warsaw: PWN - Polish Scientific Publishers (1968). 346

Kurshin, L.M., Mixed plane boundary value problems of the theory of elasticity for a quadrant.
80,Vol.23(1959)981-984,[1403-1408] 447,452

Kuz'Min, Iu.N. and Ufliand, Ia.S., The contact problem of an elastic layer compressed by two punches.
80,Vol.31(1967)711-715,[725-730]. 559

Kuznetsov, E.A., Periodic fundamental mixed problem of elastic theory for a half-plane.
81,Vol.12,No.9(1976)89-97,[942-948]. 181

Lamziuk, V.D. and Privarnikov, A.K., On the maximum value of the radius of the contact area between a punch and a layer.
80,Vol.35(1971)1047-1052,[994-999]. 574

Lebedev, N.N., The functions associated with a ring of oval cross-section.
96,Vol.4(1937)1-24. 511

Lebedev, N.N. and Ufliand, Ia.S., Axisymmetric contact problem for an elastic layer.
80,Vol.22(1958)320-326,[442-450]. 463,553

Lefevre, E.J., Mudge, D.R.J., and Dickie, J.F., An analysis of the distribution of tension in a bar attached to a plate.
59,Vol.1(1966)389-393. 322,370,373

Lekhnitskii, S.G., Radial distribution of stresses in the wedge and in a half-plane with variable modulus of elasticity.
80,Vol.26(1962a)146-151,[199-206]. 308

Lekhnitskii, S.G., The elastic equilibrium of a transversely isotropic layer and a thick plate.
80,Vol.26(1962b)687-696,[1026-1039]. 559,582

Lekhnitskii, S.G., Theory of Elasticity of an Anisotropic Elastic Body. English translation, Brandstatter, J.J. (Ed.). San Francisco: Holden-Day (1963a). 254,582

Lekhnitskii, S.G., On the problem of the distribution of stresses in an elastic half-plane with a variable modulus of elasticity.
Studies in Elasticity and Plasticity. Leningrad: State University Press (1963b). 308

Lekkerkerker, J.G., Bending of an infinite beam resting on an elastic half-space.
6,Vol.63(1960)484-497. 629

Leonov, M.Ia., On the theory of the analysis of elastic foundations. (In Russian).
80,Vol.3,No.2(1939a)52-78. 541,549

Leonov, M.Ia., Calculation of the horizontal forces for an unlimited plate resting on an elastic half-space. (In Russian).
80,Vol.3,No.3(1939b)67-74. 541,550

Leonov, M.Ia., On the analysis of foundation plates. (In Russian).
80,Vol.4,No.3(1940)80-98. 550

Linz, P. and Noble, B., A numerical method for treating indentation problems.
53,Vol.5(1971)227-231. 491

Loo, Tsu-tao, Effect of curvature on the Hertz theory for two circular cylinders in contact.
50,Vol.25(1958)122-124. 395

Love, A.E.H., A Treatise on the Mathematical Theory of Elasticity. 4th ed. London: Cambridge University Press (1927).
1,5,7,14,17,26,27,31,43,75,140,232

Love, A.E.H., The stress produced in a semi-infinite solid by pressure on part of the boundary.
88,Vol.228(1929)377-420. 78,466

Love, A.E.H., Boussinesq's problem for a rigid cone.
82,Vol.10(1939)161-175. 466,498,509

Love, E.R., The electrostatic field of two equal circular co-axial conducting disks.
83,Vol.2(1949)428-451. 509,518

Low, R.D., On a doubly mixed boundary value problem for an elastic layer.
84,Vol.22(1964a)153-157. 558

Low, R.D., On a mixed boundary value problem for an infinite elastic layer.
83,Vol.17(1964b)395-400. 558

Lowengrub, M. and Sneddon, I.N., The solution of a pair of dual integral equations.
35,Vol.6(1963)14-18. 463

Lowengrub, M. and Sneddon, I.N., The distribution of stress in the vicinity of an external crack in an infinite elastic solid.
39,Vol.3(1965)451-460. 503

Lubkin, J.L., The torsion of elastic spheres in contact.
50,Vol.18(1951)183-187. 567,615

Luk, V.K. and Keer, L.M., Stress analysis for an elastic half-space containing an axially-loaded rigid cylindrical rod.
43(1979)(to appear). 631

Luke, Y.L., Integrals of Bessel Functions. New York: McGraw-Hill (1962). 213

Lur'e, A.I., (Lourié, A.), Investigation of asymmetrical pressure of a rigid plane stamp of elliptic cross-section on an elastic semi-space.
10,Vol.28(1940)106-109. 596

Lur'e, A.I., Contact problems of the theory of elasticity. (In Russian).
80,Vol.5(1941)383-408. 466,498,589,596

Lur'e, A.I., Three-Dimensional Problems of the Theory of Elasticity. Moscow 1955. English translation, Radok, J.R.M. (Ed.). New York: Interscience (1964).
35,94,466,554,587,595,596,630

Lutchenko, S.A., On pressing a die into the surface of an elastic foundation in the form of a wedge.
81,Vol.2,No.12(1966)61-66,[40-43]. 445

Lutchenko, S.A. and Popov, G.Ia., Some planar-contact problems of elastic theory for a wedge.
81,Vol.6,No.3(1970)64-71,[276-281]. 445

Lutchenko, S.A., The exact solution of some mixed problems for the elastic wedge.
81,Vol.8,No.3(1972)117-120,[323-325]. 447

MacDonald, H.M., The electrical distribution induced on a circular disc placed in any field of force.
77,Ser.1,Vol.26(1895)257-260. 462

Mansfield, E.H., The diffusion of load into a semi-infinite sheet. Parts I and II. Reports and Memoranda 2670, Aeronautical Research Council. London: H.M.S.O.(1948). 321

Margetson, J. and Morland, L.W., Separation of smooth circular inclusions from elastic and viscoelastic plates subjected to uniaxial tension.
60,Vol.18(1970a)295-309. 419

Margetson, J. and Morland, L.W., A simplified procedure for singular integral equations with logarithmic kernels.
37,Vol.6(1970b)302-309. 263,419

Marguerre, K., Druckverteilung durch eine elastische Schicht auf starrer rauher Unterlage.
36,Vol.2(1931)108-117. 264

Marguerre, K., Spannungsverteilung und Wellenausbreitung in der kontinuierlich gestützten Platte.
36,Vol.4(1933)332-353. 542,552

Marguerre, K., Über den Träger auf elastischer Unterlage.
101,Vol.17(1937)224-231. 279,511,542

Martynovich, T.L. and Yurinets, V.E., A non-homogeneous isotropic half-plane with reinforced edge.
81,Vol.11,No.10(1975)63-69,[1075-1079]. 327

Martynovich, T.L. and Yurinets, V.E., Elastic anisotropic half-plane with a non-symmetric beam reinforcement.
81,Vol.12,No.5(1976)79-84. 327

Martynovich, T.L. and Yurinets, V.E., Heterogeneous isotropic half-plane with a non-symmetrically reinforced edge.
81,Vol.13,No.3(1977)48-56,[246-253]. 327

Maslyuk, G.F., The solution of axisymmetric contact problems for a half-space by the method of p-analytic functions.
81,Vol.3,No.3(1967)100-106,[62-65]. 517

Matczyński, M., Elastic wedge with discontinuous boundary conditions.
23,Vol.15(1963)833-855. 447

McNamee, J. and Gibson, R.E., Displacement functions and linear transforms applied to diffusion through porous elastic media.
83,Vol.13(1960)98-111. 623

Medovnik, M.I. and Rakov, A.Kh., Calculation of a periodic system of parallel beams on elastic foundation.
16,Ser.A,No.4(1970). 629

Meijers, P., The contact problem of a rigid cylinder on an elastic layer.
21,Vol.18(1968)353-383. 265,266,267,269

Melan, E., Der Spannungszustand der durch eine Einzelkraft im Innern beanspruchten Halbscheibe.
101,Vol.12(1932a)343-346. 235,318

Melan, E., Ein Beitrag zur Theorie geschweisster Verbindungen.
36,Vol.3(1932b)123-129. 318

Michell, J.H., On the direct determination of stress in an elastic solid, with application to the theory of plates.
64,Vol.31(1899)100-124. 182,435

Mikhlin, S.G., Integral Equations. English translation, Armstrong, A.H. (Ed.). London: Pergamon Press (1957). 256

Milne-Thomson, L.M., Plane Elastic Systems. Berlin: Springer Verlag (1960). 27,141

Milovic, D.M. and Touzot, G., Bidimensional stress distribution in an anisotropic layer of finite thickness.
34,Vol.20(1970)198-203. 254

Mindlin, R.D., Force at a point in the interior of a semi-infinite solid.
79,Vol.7(1936)195-202. 35

Mindlin, R.D., Compliance of elastic bodies in contact.
50,Vol.16(1949)259-268. 458,477,587,595,608

Mindlin, R.D. and Deresiewicz, H., Elastic spheres in contact under varying oblique forces.
50,Vol.20(1953)327-344. 630

Morar' G.A. and Popov, G.Ia., On the contact problem for a half-plane with finite elastic reinforcement.
80,Vol.34(1970)412-421,[389-399]. 331,332

Morar' G.A. and Popov, G.Ia., On a periodic contact problem for a half-plane with elastic coverings.
80,Vol.35(1971)172-178,[140-147]. 179,351

Morar', G.A. and Popov, G.Ia., On the theory of contact problems for cylindrical bodies with allowance for friction forces.
13,Vol.11,No.2(1976)87-96,[75-83]. 419

Morland, L.W., Singular integral equations with logarithmic kernels.
66,Vol.17(1970)47-56. 263,419

Morrison, J.A. and Lewis, J.A., Charge singularity at the corner of a flat plate.
93,Vol.31(1976)233-250. 456

Morse, P.M. and Feshbach, H., Methods of Theoretical Physics. New York: McGraw-Hill (1953). 498

Morton, W.B. and Close, L.J., Notes on Hertz' theory of contact problems.
77,Ser.6,Vol.43(1922)320-329. 2,605

Moss, L.W. and Kobayashi, A.S., Approximate analysis of axisymmetric problems in fracture mechanics with application to a flat toroidal crack.
40,Vol.7(1971)89-99. 519

Mossakovskii, V.I., On the question of the estimation of the displacements in spatial contact problems. (In Russian).
80,Vol.15(1951)635-636. 89

Mossakovskii, V.I., Application of the reciprocity theorem to the determination of the resultant forces and moments in three-dimensional contact problems. (In Russian).
80,Vol.17(1953)477-482. 89

Mossakovskii, V.I., The fundamental mixed problem of the theory of elasticity for a half-space with a circular line separating the boundary conditions. (In Russian).
80,Vol.18(1954a)187-196. 458,467,481,511

Mossakovskii, V.I., Pressure of a die, approximately circular in plan form, on an elastic half-space. (In Russian).
80,Vol.18(1954b)675-680. 511

Mossakovskii, V.I., Pressure of a circular die (punch) on an elastic half-space whose modulus of elasticity is an exponential function of depth.
80,Vol.22(1958)123-125,[168-171]. 618,620

Mossakovskii, V.I., Compression of elastic bodies under conditions of adhesion. (Axisymmetric case).
80,Vol.27(1963)418-427,[630-643]. 458,467,482,492

Mossakovskii, V.I. and Rybka, M.T., Generalization of the Griffith-Sneddon criterion for the case of a nonhomogeneous body.
80,Vol.28(1964)1061-1069,[1277-1286]. 513

Muki, R., Asymmetric problems of the theory of elasticity for a semi-infinite solid and a thick plate. In Sneddon, I.N. and Hill, R. (Eds.), Progress in Solid Mechanics. Amsterdam: North Holland (1960)401-439. 476

Muki, R. and Sternberg, E., Transfer of load from an edge-stiffener to a sheet - A reconsideration of Melan's problem.
50,Vol.34(1967)679-686. 322,356,361

Muki, R. and Sternberg, E., On the diffusion of load from a transverse tension bar into a semi-infinite elastic sheet.
50,Vol.35(1968a)737-746. 322,364,366,369,370,372

Muki, R. and Sternberg, E., On the stress analysis of overlapping bonded elastic sheets.
43,Vol.4(1968b)75-94. 321,322

Muki, R. and Sternberg, E., On the diffusion of an axial load from an infinite cylindrical bar embedded in an elastic medium.
43,Vol.5(1969)587-605. 322,631

Muki, R. and Sternberg, E., Elastostatic load transfer to a half-space from a partially imbedded axially loaded rod.
43,Vol.6(1970)69-87. 322,631

Muskhelishvili, N.I., Some Basic Problems of the Mathematical Theory of Elasticity. 3rd ed. Moscow 1949. English translation, Radok, J.R.M. (Ed.). Groningen: Noordhoff (1953a).
1,14,30,39,43,107,140,146,151,159,160,164,165,167,182,376,380, 395,418,486

Muskhelishvili, N.I., Singular Integral Equations. 2nd ed. Moscow 1946. English translation, Radok, J.R.M. (Ed.). Groningen: Noordhoff (1953b).
107,116,117,140,155,165,167,182,259,331,350,380,413,414

Naghdi, P.M. and Rowley, J.C., On the bending of axially symmetric plates on elastic foundations. 1st Mid Western Conference on Solid Mechanics. Proceedings. Urbana, Illinois (1953)119-123. 545

Nakashidze, G.M., Taking account of frictional forces in the contact region between a circular punch and an elastic half-space.
81,Vol.11,No.4(1975)110-113,[432-434]. 492

Naumov, Iu.A. and Shevlyakov, Iu.A., Flexure of circular plates on a multilayered foundation. (In Russian).
45,No.1(1967)154-163. 559

Nayak, M. and Srinivasan, R.J., Stresses and displacements under a flexible circular foundation in the presence of friction or adhesion.
101,Vol.54(1974)225-232. 492

Neuber, H., Ein neuer Ansatz zur Lösung räumlicher Probleme der Elastizitätstheorie. Der Hohlkegel unter Einzellast als Beispiel.
101,Vol.14(1934)203-212. 33,34

Noble, B., On some dual integral equations.
82,Ser.2, Vol.6(1955)81-87. 463

Noble, B., Certain dual integral equations.
57,Vol.37(1958)128-136. 463

Noble, B., The solution of Bessel-function dual integral equations by a multiplying-factor method.
26,Vol.59(1963)351-362. 463

Noble, B. and Hussain, M.A., Angle of contact for smooth elastic inclusions. 10th Midwestern Mechanics Conference. Proceedings. (1967)459-476. 463

Noble, B. and Hussain, M.A., A variational method for inclusion and indentation problems.
37,Vol.5(1969a)194-205. 405,630

Noble, B. and Hussain, M.A., Exact solution of certain dual series for indentation and inclusion problems.
39,Vol.7(1969b)1149-1161. 409

Noble, B. and Spence, D.A., Formulation of two-dimensional and axisymmetric problems for an elastic half-space. Mathematics Research Centre, United States Army, The University of Wisconsin. Technical Summary Report No.1089(1971). 83,513

Nuller, B.M., Some contact problems for an elastic infinite wedge.
80,Vol.36(1972)157-163,[146-152]. 447

Nuller, B.M., Deformation of an elastic wedge reinforced by a beam.
80,Vol. 38(1974)876-882,[823-829]. 447

O'Connor, J.J., Compliance under a small torsional couple of an elastic plate pressed between two identical elastic spheres.
50,Vol.33(1966)377-383. 566

Ōkubo, H., The stress distribution in a semi-infinite domain having a plane boundary and compressed by a rigid body.
101,Vol.20(1940)271-276; correction Vol.21(1941)383-384. 164

Ōkubo, H., On the two-dimensional problem of a semi-infinite elastic body compressed by an elastic plane.
83,Vol.4(1951)260-270. 630

Olesiak, Z., Annular punch on elastic semi-space.
23,Vol.17(1965)633-648. 519

Olszak, W. (Ed.), Non-homogeneity in Elasticity and Plasticity. New York: Pergamon Press (1959). 617

Palmov, V.A., The contact problem of a plate on an elastic foundation.
80,Vol.24(1960)416-422,[609-618]. 551

Panek, C. and Kalker, J.J., A solution for a narrow rectangular punch.
52,Vol.7(1977)213-218. 629

Pao, Y.C., Wu, T.-S., and Chiu, Y.P., Bounds on the maximum contact stress of an indented elastic layer.
50,Vol.38(1971)608-614. 271,559

Papkovich, P.F., Solution générale des équations différentielles fondamentales d'élasticité exprimée par trois fonctions harmoniques.
1,Vol.195(1932a)513-515. 33,34

Papkovich, P.F., Expression for the general integral of the principal equations of the theory of elasticity by harmonic functions . (In Russian).
14,No.10(1932b)1425-1435. 33,34

Parihar, K.S. and Keer, L.M., Stress singularity at the corner of a wedge-shaped crack or inclusion.
50,Vol.45(1978)791-796. 456

Parihar, K.S. and Keer, L.M., The singularity at the apex of a bonded wedge-shaped stamp.
50,Vol.46(1979)(to appear). 456

Parlas, S.C. and Michalopoulos, C.D., Axisymmetric contact problem for an elastic half-space with a cylindrical hole.
39,Vol.10(1972)699-707. 519,631

Paskalenko, A.A. and Popov, G.Ia., Plane problem of bending of a semi-infinite beam resting on a linearly deformable foundation.
80,Vol.36(1972)94-105,[84-94]. 546

Pasternak, P.L., The Basic Features of a New Method of Calculating Foundations on an Elastic Base by Means of Two Bearing Coefficients. (In Russian). Moscow-Leningrad: Gosstroiizdat. (1954). 102,104,529,537

Payne, L.E., On axially symmetric punch, crack and torsion problems.
92,Vol.1(1953)53-71. 498

Pflüger, A., Halbscheibe mit Randglied. Ein Spannungsgleichnis Zum Problem der tragenden Linie.
101,Vol.25/27(1947)177-185. 320

Pickett, G., Raville, M.E., Janes, W.C., and McCormick, F.J., Deflections, moments and reactive pressures for concrete pavements.
61,No.65(1951). 550

Pickett, G. and McCormick, F.J., Circular and rectangular plates under lateral load and supported by an elastic solid foundation. 1st U.S. National Congress of Applied Mechanics. Ann Arbor, Michigan: J.W. Edwards (1952)331-338. 550

Plevako, V.P., On the theory of elasticity of inhomogeneous media.
80,Vol.35(1971)853-860,[806-813]. 627

Plevako, V.P., On the possibility of using harmonic functions for solving problems of the theory of elasticity of nonhomogeneous media.
80,Vol.36(1972a)886-894,[834-842]. 627

Plevako, V.P., The state of stress of a nonhomogeneous layer resting on an elastic half-space.
81,Vol.8,No.4(1972b)64-76,[397-403]. 627

Plevako, V.P., Equilibrium of a nonhomogeneous half-plane under the action of forces applied to the boundary.
80,Vol.37(1973a)905-913,[858-366]. 308

Plevako, V.P., The deformation of a nonhomogeneous half-space under the action of a surface load.
81,Vol.9,No.6(1973b)16-23,[593-598]. 627

Plevako, V.P., A problem concerned with the action of shear forces applied to the surface of an inhomogeneous half-space.
81,Vol.9,No.11(1973c)49-55,[1191-1195]. 627

Plevako, V.P., Inhomogeneous layer bonded to a half-space under the action of internal and external forces.
80,Vol.38(1974)865-875,[813-823]. 627

Polozhii, G.N., The Generalized Theory of Analytic Functions of a Complex Variable. P-analytic and (p,q) - analytic functions and certain applications of them. (In Russian). Kiev University (1965). 527

Polozhii, G.N., On a problem of the theory of axisymmetric potential and on a system of circular and annular dies.
81,Vol.3,No.12(1967)16-27,[13-19]. 527

Polya, G. and Szego, G., Inequalities for the capacity of a condenser.
17,Vol.67(1945)1-32. 93

Popov, G.Ia., Bending of an unbounded plate supported by an elastic half-space with a modulus of elasticity varying with depth.
80,Vol.23(1959)1095-1100,[1566-1573]. 618

Popov, G.Ia., On an approximate method of solution of certain plane contact problems of the theory of elasticity. (In Russian).
8,Vol.14,No.3(1961a)81-96. 66

Popov, G.Ia., On a method of solution of the axisymmetric contact problem of the theory of elasticity.
80,Vol.25(1961b)76-85,[105-118]. 314

Popov, G.Ia., Bending of a semi-infinite plate resting on a linearly deformable foundation.
80,Vol.25(1961c)342-355,[502-520]. 546

Popov, G.Ia., The contact problem of the theory of elasticity for the case of a circular area of contact.
80,Vol.26(1962)152-164,[207-225] 83,587

Popov, G.Ia., Some properties of classical polynomials and their application to contact problems.
80,Vol.27(1963a)821-832,[1255-1271]. 66,314,315

Popov, G.Ia., On the solution of a planar contact problem of elasticity theory in the presence of couple or frictional forces.
8,Vol.16(1963b)15-32. 167,278,316

Popov, G.Ia., Certain properties of classical polynomials and their application in contact problems.
80,Vol.28(1964)442-451,[544-555]. 317

Popov, G.Ia., Plane contact problem of the theory of elasticity with bonding or frictional forces.
80,Vol.30(1966)551-563,[653-667]. 167,278,316

Popov, G.Ia., Impression of a punch on a linearly-deforming foundation taking into account friction forces.
80,Vol.31(1967a)337-343,[360-367]. 317

Popov, G.Ia., Some new relations for Jacobi polynomials.
91,Vol.8(1967b)1399-1404,[1059-1061]. 317

Popov, G.Ia., On an approximate method of solution of the contact problem for an annular stamp. (In Russian).
7,Vol.20,No.2(1967c)19-36. 317,526

Popov, G.Ia., A remarkable property of Jacobi polynomials.
100,Vol.20(1968)540-547,[467-472]. 316,332

Popov, G.Ia., On the method of orthogonal polynomials in contact problems of the theory of elasticity.
80,Vol.33(1969)518-531,[503-517]. 66,69,241,314

Popov, G.Ia., On integral equations of the theory of elasticity with difference and summation kernels.
80,Vol.34(1970)603-619,[575-590]. 317

Popov, G.Ia., On the reduction of integral equations of the theory of elasticity to infinite systems.
80,Vol.36(1972a)672-681,[633-642]. 317

Popov, G.Ia., Plates on a linearly elastic foundation (A survey).
81,Vol.8,No.3(1972b)3-17,[231-242]. 2,528,533

Popov, G.Ia., Plane contact problem for a linearly-deformable foundation in the presence of adhesion.
80,Vol.37(1973a)254-261,[237-245]. 317

Popov, G.Ia., Axisymmetric contact problem for an adhesive inhomogeneous half-space in the presence of cohesion.
80,Vol.37(1973b)1109-1116,[1052-1059]. 625

Popov, G.Ia. and Tikhonenko, L.Ia., Two-dimensional problem of the contact between a semi-infinite beam and an elastic wedge.
80,Vol.38(1974)312-320,[284-292]. 447

Popov, G.Ia. and Tikhonenko, L.Ia., Exact solution of plane problems on the contact between semi-infinite beams and an elastic wedge.
80,Vol.39(1975)1100-1109,[1054-1063]. 447

Poritsky, H., Stresses and deflections of cylindrical bodies in contact with application to contact of gears and of locomotive wheels.
50,Vol.17(1950)191-201. 60

Potelezhko, V.P. and Filippov, A.P., The contact problem for a slab lying on an elastic base.
81,Vol.3,No.1(1967)87-91,[49-51]. 558

Potelezhko, V.P., The pressure of a rigid unbounded cylinder on an elastic layer.
81,Vol.7,No.3(1971)122-125,[335-338]. 271,630

Poulos, H.G. and Davis, E.H., The settlement behaviour of single axially-loaded piles and piers.
34,Vol.18(1968)351-371. 631

Poulos, H.G. and Davis, E.H., Elastic Solutions for Soil and Rock Mechanics. New York: John Wiley (1974). 99,627

Pouyet, J. and Lataillade, J.L., Elastic strain of a transversely isotropic half-space caused by a rigid punch.
62,Vol.17(1979)765-772. 582

Prager, W., Zur Theorie elastisch gelagerter Konstruktionen.
101,Vol.7(1927)354-360. 279

Prager, W., Introduction to Mechanics of Continua. New York: Dover Publications (1961). 16

Prasad, S.N. and Chatterjee, S.N., Some mixed boundary value problems of elasticity in a rectangular domain.
43,Vol.9(1973)1193-1210. 630

Prasad, S.N. and Dasgupta, S., Effect of sliding friction on contact stresses in a rectangle compressed by rigid planes.
50,Vol.42(1975)656-661. 630

Privarnikov, A.K., Symmetric contact problem for a layer of finite thickness. (In Russian).
16,No.3(1963)358-361. 271

Privarnikov, A.K., On the contact between a plate and an elastic half-space.
13,Vol.7(1972)163-167. 560

Protsenko, V.G., Protsenko, V.S., and Sinekop, N.S., On the calculation of two infinite beams on elastic foundation.
81,Vol.9,No.7(1973)104-108,[781-785]. 629

Protsenko, V.S., Torsion of an elastic half-space with the modulus of elasticity varying according to a power law.
81,Vol.3(1967)127-130,[82-83]. 621

Protsenko, V.S. and Protsenko, V.G., The solution of the integral equation of the contact problem in the case of two punches.
81,Vol.8,No.7(1972)60-65,[743-747]. 629

Protsenko, V.S., The pressure on an elastic half-space due to a strip-shaped punch.
81,Vol.8,No.6(1974)45-50,[618-622]. 629

Protsenko, V.S. and Rvachev, V.L., Plate in the form of an infinite strip on an elastic half-space.
80,Vol.40(1976)298-305,[273-280]. 629

Pu, S.L. and Hussain, M.A., Note on the unbonded contact between plates and an elastic half-space.
50,Vol.37(1970)859-861. 584

Pupyrev, V.A. and Ufliand, Ia.S., Certain contact problems for an elastic layer.
80,Vol.24(1960)683-690,[1015-1026]. 553

Puro, A.E., Application of Hankel transforms to the solution of axisymmetric problems when the modulus of elasticity is a power function of depth.
80,Vol.37(1973)945-948,[896-900]. 621

Pykhteev, G.N., On the evaluation of certain singular integrals with a kernel of the Cauchy type.
80,Vol.23(1959)1074-1082,[1536-1548]. 66

Rakov, A.F., Analysis of parallel beams on elastic foundation.
80,Vol.26(1962)588-592,[883-891]. 629

Rakov, A.Kh. and Rvachev, V.L., Contact problems of the theory of elasticity for a half-space whose modulus of elasticity is a power function of the depth.
16,Vol.3(1961)286-290. 618

Rao, A.K., Stress concentrations and singularities at interface corners.
101,Vol.51(1971)395-406. 452

Rao, H.S. and Shadley, J.R., An initial value method for an axisymmetric contact problem.
39,Vol.8(1970)219-229. 558

Rapoport, I.M., On one class of singular integral equations.
9,Vol.59(1948)1403-1406. 486

Rappoport, R.M., On the construction of general solutions of the elasticity theory equations for transversely isotropic inhomogeneous bodies.
80,Vol.40(1976)956-958,[909-912]. 627

Ratwani, M. and Erdogan, F., On the plane contact problem for a frictionless elastic layer.
43,Vol.9(1973)921-936. 296,297

Rau, I.S. and Dhaliwal, R.S., Further considerations on the axisymmetric Boussinesq problem.
39,Vol.10(1972)659-663. 558

Reismann, H., Bending of circular and ring-shaped plates on an elastic circular foundation.
50,Vol.21(1954)45-51. 544,550

Reissner, E., On the theory of beams on a yielding foundation.
71,Vol.23(1937)328-333. 279

Reissner, E., Note on the problem of the distribution of stress in a thin stiffened elastic sheet.
71,Vol.26(1940)300-305. 318,364

Reissner, E., On the theory of bending of elastic plates.
57,Vol.23(1944)184-191. 545

Reissner, E. and Sagoci, H.F., Forced torsional oscillations of an elastic half-space. I.
51,Vol.15(1944)652-654. 458,468,498

Reissner, E., The effect of transverse shear deformation on the bending of elastic plates.
50,Vol.67(1945)A69-A77. 545

Reissner, E., On a variational theorem in elasticity.
57,Vol.29(1950)90-95. 545

Reissner, E., Stresses in elastic plates over flexible subgrades.
18,Vol.81(1955)690-1 - 690-29. 542

Reissner, E., A note on deflections of plates on a viscoelastic foundation.
50,Vol.25(1958)144-145. 102

Riedel, W., Beiträge zur Lösung des ebenen Probleme eines elastischen Körpers mittels der Airyschen Spannungsfunktion.
101,Vol.7(1927)169-188. 235

Robinson, N.I., Polynominal solutions of elastic layer problems.
93,Vol.13(1973)29-43. 559

Robinson, N.I., Green's function for bending of a strip on a Pasternak foundation.
101,Vol.55(1975)182-183. 104

Rogers, C. and Clements, D.L., Bergman's integral operator method in inhomogeneous elasticity.
84,Vol.36(1978)315-321. 627

Rostovtsev, N.A., Complex stress function in the axisymmetric contact problem of the theory of elasticity. (In Russian).
80,Vol.17(1953)611-614. 509

Rostovtsev, N.A., On the problem of torsion of an elastic half-space. (In Russian).
80,Vol.19(1955)55-60. 509

Rostovtsev, N.A., Complex potentials in the problem of a punch of circular section. (In Russian).
80,Vol.21(1957)77-82. 509

Rostovtsev, N.A., An integral equation encountered in the problem of a rigid foundation bearing on nonhomogeneous soil.
80,Vol.25(1961)164-168,[238-246]. 625

Rostovtsev, N.A., On certain solutions of an integral equation of the theory of linearly deformable foundation.
80,Vol.28(1964a)111-127,[127-145]. 625

Rostovtsev, N.A., On the theory of elasticity of a nonhomogeneous medium.
80,Vol.28(1964b)601-611,[745-757]. 308,311,618,622

Rostovtsev, N.A. and Khranevskaia, I.E., The solution of the processing problem for a half-space whose modulus of elasticity is a power function of the depth.
80,Vol.35(1971)1053-1061,[1000-1009]. 625

Rvachev, V.L., The pressure on an elastic half-space of a punch, strip-shaped in the plane. (In Russian).
80,Vol.20(1956a)248-254. 629

Rvachev, V.L., On the analysis of an infinite beam lying upon an elastic half-space. (In Russian).
80,Vol.20(1956b)667-669. 629

Rvachev, V.L., On the bending of an infinite beam on an elastic half-space.
80,Vol.22(1958)698-700,[984-988]. 629

Rvachev, V.L., The pressure on an elastic half-space of a stamp with a wedge-shaped planform.
80,Vol.23(1959a)169-171,[229-233]. 455

Rvachev, V.L., On the pressure under an infinite beam resting on an elastic half-space.
80,Vol.23(1959b)399-400,[565-567]. 629

Rvachev, V.L., Studies on the contact problems of elasticity theory in the Ukraine.
81,Vol.3,No.10(1967)109-116,[76-81]. 2

Rybakov, L.S. and Cherepanov, G.P., Discrete interaction of a plate with a semi-infinite stiffener.
80,Vol.41(1977)322-328,[313-319]. 350

Sack, R.A., Extension of Griffith's theory of rupture to three dimensions.
78,Vol.58(1946)729-736. 498

Sadowsky, M.A., Zweidimensionale Probleme der Elastizitätstheorie.
101,Vol.8(1928)107-121. 140,155,181

Sadowsky, M.A., Stress concentration caused by multiple punches and cracks.
50,Vol.23(1956)80-84. 174

Sagoci, H.F., Forced torsional oscillations of an elastic half-space II.
51,Vol.15(1944)655-662. 468

Sarkisian, V.S. and Avanesian, R.G., On load transfer from an infinite stringer to two elastic strips with fixed edges.
7,Vol.64,No.3(1977)188-192. 327

Schindler, K., Der Träger auf elastischer Bettung unter Ausschluss von Zugkraften zwischen Träger und Unterlage.
75, Vol.11,No.3(1956)47-53. 101

Schleicher, F., Zur Theorie des Baugrundes.
25,Vol.48(1926a)931-935. 545,547

Schleicher, F., Zur Theorie des Baugrundes.
25,Vol.49(1926b)949-952. 545,547

Schleicher, F., Kreisplatten auf elastischer Unterlage. Berlin: Springer Verlag (1926c). 545,547

Schwartz, J. and Harper, E.Y., On the relative approach of two-dimensional elastic bodies in contact.
43,Vol.7(1971)1613-1626. 403

Segedin, C.M., The relation between load and penetration for a spherical punch.
66,Vol.4(1957)156-161. 467

Segedin, C.M., A note on geometric discontinuities in elastostatics.
39,Vol.6(1968)309-312. 595

Selvadurai, A.P.S., The interaction between a rigid circular punch on an elastic half-space and a Mindlin force.
67,Vol.5(1978)57-64. 468

Selvadurai, A.P.S., Elastic Analysis of Soil-Foundation Interaction. Amsterdam-Oxford-New York: Elsevier (1979).
99,104,528,537,550,627,630

Shail, R., Lamé polynomial solutions to some elliptic crack and punch problems.
39,Vol.16(1978)551-563. 589,597

Shepherd, W.M., Stress systems in an infinite sector.
88,Vol.148(1935)284-303. 438

Sheremet'ev, M.P., The elastic equilibrium of an infinite plate with an inserted absolutely rigid or elastic disc. (In Russian).
80,Vol.16(1952)437-448. 405

Sheremet'ev, M.P., The solution of the equations of certain contact problems of the theory of elasticity. (Equations of Prandtl type.) In Radok, J.R.M. (Ed.), Problems of Continuum Mechanics. Philadelphia, Pennsylvania: Society for Industrial and Applied Mathematics (1961)419-439. 414

Shibuya, T., Koizumi, T., and Nakahara, I., An elastic contact problem for a half-space indented by a flat annular rigid stamp.
39,Vol.2(1974)759-771. 519,526

Shibuya, T., Koizumi, T., Nakahara, I., and Tanaka, S., An axisymmetric contact problem for a thick elastic plate indented by a pair of flat annular rigid punches.
62,Vol.3(1975)177-187. 558

Shibuya, T., A mixed boundary value problem of an elastic half-space under torsion by a flat annular stamp.
74,Vol.19(1976)233-238. 520,526

Shibuya, T., Some elastic contact problems of an elliptical punch.
74,Vol.20(1977)903-914. 589

Shibuya, T., Koizumi, T., Nakahara, I., and Tanaka, S., The stress distribution in a thick elastic plate under torsional displacements by a pair of annular rigid punches.
74,Vol.20(1977)675-679. 558

Shibuya, T., Koizumi, T., Nakahara, I., and Hara, T., Elastic contact problem of a half-space pressed onto a rigid base with a cylindrical protrusion or pit.
39,Vol.17(1979)349-356. 520

Shield, R.T., Notes on problems in hexagonal aeolotropic materials.
26,Vol.47(1951)401-409. 583,588

Shield, R.T., Load-displacement relations for elastic bodies.
102,Vol.18(1967)682-693. 89

Shtaerman, A.Ia., (E. Steuermann), To Hertz's theory of local deformations in compressed elastic bodies.
10,Vol.25(1939)359-361. 97

Shtaerman, A.Ia., (E. Steuermann), A generalization of Hertz's theory of local deformations in elastic bodies pressed against each other.
10,Vol.29(1940)179-181. 97

Shtaerman, A.Ia., (E. Steuermann), On the question of local deformations in elastic bodies pressed against one another.
10,Vol.31(1941)738-741. 97

Shtaerman, A.Ia., Some special cases of the contact problem.
10,Vol.38(1943)197-200. 97

Shtaerman, A.Ia., Contact Problems of the Theory of Elasticity. (In Russian). Moscow, Leningrad: Gostekhteoretizdat (1949).
2,140,155,167,182,252

Singh, B.M., Reissner-Sagoci problem for a nonhomogeneous solid.
30,Vol.22(1972)81-86. 622

Singh, B.M., A note on the Reissner-Sagoci problem for a nonhomogeneous medium.
101,Vol.53(1973)419-420. 622

Sivashinsky, G.I., The problem of a slender die.
52,Vol.5(1975)161-166. 629

Smirnov, V.I., Course of Higher Mathematics. Sneddon, I.N. (Ed.). Vol. 4. Oxford: Pergamon Press (1964). 256

Smith, J.O. and Liu, C.K., Stresses due to tangential and normal loads on an elastic solid with applications to some contact problems.
50,Vol.20(1953)157-166. 605

Smith, S.F., On a flat punch indenting an elastic layer in plane strain.
82,Ser.2,Vol.15(1964)223-237. 271

Smithies, F., Integral Equations. London: Cambridge University Press (1958). 256

Sneddon, I.N., The stress distribution due to a force in the interior of a semi-infinite elastic medium.
26,Vol.40(1944)229-238. 235

Sneddon, I.N., Boussinesq's problem for a flat-ended cylinder.
26,Vol.42(1946)29-39. 467

Sneddon, I.N., Note on a boundary value problem of Reissner and Sagoci.
51,Vol.18(1947)130-132. 458,468

Sneddon, I.N., Boussinesq's problem for a rigid cone.
26,Vol.44(1948)492-507. 467

Sneddon, I.N., Fourier Transforms. New York: McGraw-Hill (1951). 197,199,206,213,235,460,466,468

Sneddon, I.N. and Berry, D.S., The Classical Theory of Elasticity. In Flügge, S. (Ed.), Handbuch der Physik, Vol.6, Elasticity and Plasticity. Berlin: Springer Verlag(1958)1-126. 1,7

Sneddon, I.N., The elementary solution of dual integral equations.
35,Vol.4(1960)108-110. 463,464

Sneddon, I.N., Dual integral equations with trigonometric kernels.
35,Vol.5(1961)147-152. 463

Sneddon, I.N., The relation between load and penetration in the axisymmetric Boussinesq problem for a punch of arbitrary profile.
39,Vol.3(1965)47-57. 464,467

Sneddon, I.N., Mixed Boundary Value Problems in Potential Theory. Amsterdam: North Holland (1966a). 463,508,519,570

Sneddon, I.N., The Reissner-Sagoci problem.
35,Vol.7(1966b)136-144. 468

Sneddon, I.N. and Lowengrub, M., Crack Problems in the Classical Theory of Elasticity. New York: John Wiley (1969). 501

Sneddon, I.N., Fourier Transform Solutions of Quarter-Plane Problems in Elasticity. Raleigh: North Carolina State University, Department of Mathematics (1971). 447

Sneddon, I.N., The Use of Integral Transforms. New York: McGraw-Hill (1972). 197,199,430,431,432

Sneddon, I.N., Gladwell, G.M.L., and Coen, S., Bonded contact of an infinite plate and an elastic foundation.
62,Vol.3(1975)1-13. 102,281,542

Sneddon, I.N., The stress intensity factor for a flat elliptical crack in an elastic solid under uniform tension.
39,Vol.17(1979a)185-191. 229

Sneddon, I.N., A note on the Boussinesq problem for a flat-ended cylinder with elliptical cross-section.
52,Vol.9(1979b)215-219. 229

Sokolnikoff, I.S., Mathematical Theory of Elasticity. New York: McGraw-Hill (1956). 1,7,141

Solov'ev, A.S., An integral equation and its application to contact problems in the theory of elasticity with friction and cohesion forces.
80,Vol.33(1969)1042-1050,[1012-1020]. 167,278

Solov'ev, Iu.I., The fundamental mixed problem of the axisymmetric theory of elasticity.
80,Vol.30(1966)956-962,[1137-1145]. 517

Spence, D.A., Self-similar solutions to adhesive contact problems with incremental loading.
89,Vol.305(1968a)55-80. 161,458,482,484,486,490,491,492

Spence, D.A., A Wiener-Hopf equation arising in elastic contact problems.
89,Vol.305(1968b)81-92. 458,486,492

Spence, D.A., An eigenvalue problem for elastic contact with finite friction.
26,Vol.73(1973)249-268. 272,278

Spence, D.A., Similarity considerations for contact between dissimilar elastic bodies. In de Pater, A.D. and Kalker, J.J. (Eds.), The Mechanics of the Contact between Deformable Bodies. Delft University Press (1975a)67-76. 83

Spence, D.A., The Hertz contact problem with finite friction.
52,Vol.5(1975b)297-319. 492

Srivastav, R.P., An axisymmetric mixed boundary value problem for a half-space with a cylindrical cavity.
56,Vol.13(1964)385-393. 570

Srivastav,R.P. and Prem Narain, Certain two-dimensional problems of stress distribution in wedge-shaped elastic solids under discontinuous load.
26,Vol.61(1965)945-954. 447

Srivastava, K.N. and Lowengrub, M., Finite Hilbert transform technique for triple integral equations with trigonometric kernels.
87,Vol.68(1970)309-321. 174

Sternberg, E., Eubanks, R.A., and Sadowsky, M.A., On the axisymmetric problem of two concentric spheres. 1st U.S. National Congress of Applied Mechanics. Proceedings. New York: American Society of Mechanical Engineers (1952)209-215. 630

Sternberg, E. and Rosenthal, F., The elastic sphere under concentrated loads.
50,Vol.19(1952)413-421. 630

Sternberg, E. and Koiter, W.T., The wedge under a concentrated couple: a paradox in the two-dimensional theory of elasticity.
50,Vol.25(1958)575-581. 437,441

Sternberg, E., Load-transfer and load-diffusion in elastostatics. 6th U.S. National Congress of Applied Mechanics. Proceedings. New York: American Society of Mechanical Engineers (1970)34-61. 322

Sternberg, E. and Turteltaub, M.J., Compression of an elastic roller between two rigid plates. In Sedov, L.I., et al (Eds.), Continuum Mechanics and Related Problems of Analysis. Muskhelishvili Anniversary Volume. Moscow: Nauka Publishing House (1972)495-515. 395,404

Stippes, M., Wilson, H.B., Jr., and Krull, F.N., A contact stress problem for a smooth disk in an infinite plate. 4th U.S. National Congress of Applied Mechanics. Proceedings, Vol.2.New York: American Society of Mechanical Engineers (1962)799-806. 405,412,427

Stippes, M., A unified derivation of displacement potentials in classical elasticity.
52,Vol.1(1971)93-94. 35

Suriyamongkal, S., Karasudhi, P., and Lee, S.L., Axially loaded rigid cylindrical body embedded in an elastic half-space. In Developments in Mechanics. Vol. 7. 13th Midwestern Mechanics Conference. Proceedings. Pittsburgh: School of Engineering (1973)333-347. 631

Svelko, V.A., Boussinesq type problems for the anisotropic half-space.
80,Vol.28(1964)908-913,[1099-1105]. 579

Svelko, V.A., Concentrated force in a transversally-isotropic half-space and in a composite space.
80,Vol.33(1969)532-537,[517-523]. 583

Sveklo, V.A., The action of a stamp on an elastic anisotropic half-space.
80,Vol.34(1970)172-178,[165-171]. 579

Sveklo, V.A., Hertz problem on compression of anisotropic bodies.
80,Vol.38(1974)1079-1083,[1023-1027]. 579

Szabo, I., Die achsensymmetrisch belastete dicke Kreisplatte auf elastischer Unterlage.
36,Vol.19(1951a)128-142. 550

Szabo, I., Beiträge zur Theorie der achsensymmetrisch belasteten dicken Kreisplatte insbesondere bei elastischer Lagerung.
36,Vol.19(1951b)342-354. 550

Szabo, I., Die in Achsenrichtung rotationssymmetrisch belastete dicke Kreisplatte auf nachgiebiger und auf starrer Unterlage.
101,Vol.32(1952a)145-153. 550

Szabo, I., Beiträge zur Theorie der achsensymmetrisch belasteten schweren dicken Kreisplatte.
101,Vol.32(1952b)359-371. 550

Szelagowski, F., The influence of a bolt inserted in a hole of an infinite plate subjected to a one-directional tension or compression on the stresses in the plate.
2,Vol.4,No.1(1956)9-20. 405

Takakuda, K., Shibuya, T., Koizumi, T., and Hara, T., Elastic contact problems of a half-plane pressed on a rigid base with a protusion or excavation.
97,Vol.27(1977)91-97. 520

Tamate, O. and Kamada, T., On the simple tension of a semi-infinite plate with partially stiffened edge.
98,Vol.28,No.1(1963)39-47. 322

Temple, G., Whittaker's work on the integral representation of harmonic functions.
31,Ser.2.,Vol.2(1958)11-24. 513

Teodorescu, P.P., One hundred years of investigations in the plane problem of the theory of elasticity.
20,Vol.17(1964)175-186. 141

Terazawa, K., On the elastic equilibrium of a semi-infinite solid.
99,Vol.37(1916),Article 7. 86

Ter-Mkrtich'ian, L.N., Some problems in the theory of elasticity of non-homogeneous elastic media.
80,Vol.25(1961)1120-1125,[1667-1675]. 621

Theocaris, P.S., The order of singularity at a multi-edge corner of a composite plate.
39,Vol.12(1974)107-120. 455

Theocaris, P.S. and Gdoutos, E.E., Stress singularities at vertices of composite plates with smooth or rough interfaces.
23,Vol.28(1976)693-704. 455

Theocaris, P.S. and Ioakimidis, N.I., The inclusion problem in plane elasticity.
83,Vol.30(1977a)437-448. 419

Theocaris, P.S. and Ioakimidis, N.I., Numerical integration methods for the solution of singular integral equations.
84,Vol.35(1977b)173-183. 263

Thomas, H.R. and Hoersch, V.A., Stresses due to the pressure of one elastic solid upon another. Engineering Experiment Station, University of Illinois, Urbana, Illinois. Bulletin 212(1930).
595,605,608

Timoshenko, S. and Woinowsky-Krieger, S., Theory of Plates and Shells. 2nd ed. New York: McGraw-Hill (1940)259f. 547

Timoshenko, S. and Goodier, J.N., Theory of Elasticity. New York: McGraw-Hill (1951). 1,2,5,14,98,460

Ting, E.C., Unified formulation of two-parameter foundation model.
101,Vol.53(1973a)636-637. 104

Ting, E.C., Unbonded beams on a two-parameter foundation.
33,Vol.296(1973b)77-89. 105

Titchmarsh, E.C., Introduction to the Theory of Fourier Integrals. Oxford: Clarendon Press (1937)337. 199,462

Todhunter, I. and Pearson, K., History of the Theory of Elasticity. Vol.2,Part 2. Cambridge University Press (1893).
33,75,78,94

Tonoian, V.S., On the indentation of the elastic half-plane by two identical rigid stamps. (In Russian).
8,Vol.17,No.2(1964)81-90. 174

Torvik, P.J., Note on the boundary conditions for contact problems.
50,Vol.34(1967)761-763. 95

Tranter, C.J., The use of the Mellin transform in finding the stress distribution in an infinite wedge.
83,Vol.1(1948)125-130. 439

Tranter, C.J., On some dual integral equations occurring in potential problems with axial symmetry.
83,Vol.3(1950)411-419. 552

Tranter, C.J., On some dual integral equations.
82,Ser.2,Vol.2(1951)60-66. 462

Tranter, C.J., A further note on dual integral equations and on application to the diffraction of electromagnetic waves.
83,Vol.7(1954)317-325. 462

Tranter, C.J., Integral Transforms in Mathematical Physics. New York: John Wiley (1956). 430

Tranter, C.J., Some triple integral equations.
35,Vol.4(1960)200-203. 519

Trefftz, E., Mathematische Elastizitätstheorie. In Grammel, R. (Ed.), Handbuch der Physik, Vol. 6, Mechanik der Elastischen Körper. Berlin: Julius Springer (1928). 481

Tsai, K.C., Dundurs, J., and Keer, L.M., Contact between an elastic layer with a slightly curved bottom and a substrate.
50,Vol.39(1972)821-823. 293

Tsai, K.C., Dundurs, J., and Keer, L.M., Elastic layer pressed against a half-space.
50,Vol.41(1974)703-707. 295

Tsai, N.C. and Westmann, R.A., Beam on tensionless foundation.
18,Vol.93,EM5(1967)1-12. 101

Tsamasphyros, G.J. and Theocaris, P.S., Le premier problème aux limites pour un coin infini.
48,Vol.15(1976)615-630. 441

Tu, Yih-O and Gazis, D.C., The contact problem of a plate pressed between two spheres.
50,Vol.31(1964)659-666. 574

Tu, Yih-O, A numerical solution for an axially symmetric contact problem.
50,Vol.34(1967)283-286. 574

Turteltaub, M.J. and Sternberg, E., Elastostatic uniqueness in the half-space.
22,Vol.24(1967)233-242. 2

Ufliand, Ia.S., The contact problem of the theory of elasticity for a die, circular in its plane, in the presence of adhesion. (In Russian).
80,Vol.20(1956)578-587. 458,481,496

Ufliand, Ia.S., Torsion of an elastic layer.
9,Vol.129(1959)997-999 and 95,Vol.4(1959)1383-1386. 566

Ufliand, Ia.S., Survey of Articles on the Applications of Integral Transforms in the Theory of Elasticity. English translation, Sneddon, I.N. (Ed.). Raleigh: North Carolina State University, Department of Mathematics (1965). 2,441,467,481,499

Ufliand, Ia.S., Some three-dimensional problems of the theory of elasticity for a wedge. In Sedov, L.I., et al (Eds.), Continuum Mechanics and Related Problems of Analysis. Muskhelishvili Anniversary Volume. Moscow: Nauka Publishing House (1972)549-553. 447

Updike, D.P. and Kalnins, A., Axisymmetric behaviour of an elastic spherical shell compressed between rigid plates.
50,Vol.37(1970)635-630. 630

Updike, D.P. and Kalnins, A., Axisymmetric postbuckling and nonsymmetric buckling of a spherical shell compressed between rigid plates.
50,Vol.39(1972a)172-178. 631

Updike, D.P. and Kalnins, A., Contact pressure between an elastic spherical shell and a rigid plate.
50,Vol.39(1972b)1110-1114. 631

Valov, G.M., Infinite elastic layer and half-space under the action of a ring-shaped die.
80,Vol.32(1968)894-907,[917-930]. 519,558

Vesic, A.B., Bending of beams resting on isotropic elastic solid.
18,Vol.87(1961)35-53. 281

Vlasov, V.Z. and Leont'ev, N.N., Beams, Plates and Shells on Elastic Foundation. Jerusalem:Israel Program for Scientific Translations (1966). 2,102,105,279,528,529,537

Vorob'ev, V.L. and Popov, G.Ia., Contact problem for an elastic half-plane and a semi-infinite elastic rod adhering to it.
80,Vol.34(1970)354-359,[335-341]. 320,350

Vorob'ev, V.L., Popov, G.Ia., and Serdyuk, G.N., Bending of two semi-infinite plates lying on linearly deformable foundation of general type.
13,Vol.12,No.3(1977)58-64,[50-56]. 546

Vorovich, I.I. and Ustinov, Iu.A., Pressure of a die on an elastic layer of finite thickness.
80,Vol.23(1959)445-455,[637-650]. 556

Wang, I.C., A contact stress problem for a rigid smooth sphere in an extended elastic solid.
50,Vol.32(1965)651-655. 630

Watson, G.N., A Treatise on the Theory of Bessel Functions. 2nd ed. London: Cambridge University Press (1958). 213

Way, S., Some observations on the theory of contact pressures.
50,Vol.7(1940)A147-A157. 2,595,608

Weber, C., Achsensymmetrische Deformation von Umdrehungskörpern.
101,Vol.5(1925)466-468. 510

Weber, C., Zur Umwandlung von rotationssymmetrischen Problemen in zweidimensionale und umgekehrt.
101,Vol.20(1940)117-118. 511

Weber, H. , Ueber die Besselschen Functionen und ihre Anwendung auf die Theorie der elektrischen Strome.
49,Vol.75(1873)75-105. 462

Weitsman, Y., On the unbonded contact between plates and an elastic half-space.
50,Vol.36(1969)198-202. 562,563,573

Weitsman, Y., On foundations that react in compression only.
50,Vol.37(1970)1019-1030. 102,104,543

Weitsman, Y., A tensionless contact between a beam and an elastic half-space.
39,Vol.10(1972)73-81. 101,283,286

Westman, R.A., Axisymmetric mixed boundary-value problems of the elastic half-space.
50,Vol.32(1965)411-417. 477

Whittaker, E.T. and Watson, G.N., A Course of Modern Analysis. 4th ed. London: Cambridge University Press (1952). 347,508,607

Wieghardt, K., Über den Balken auf nachgiebiger Unterlage.
101,Vol.2(1922)165-184. 629

Williams, M.L., Surface stress singularities resulting from various boundary conditions in angular corners of plates under bending. 1st U.S. National Congress of Applied Mechanics. Proceedings. (1951)325-330. 448

Williams, M.L., Stress singularities resulting from various boundary conditions in angular corners of plates in extension.
50,Vol.19(1952)526-528. 448

Williams, M.L., Complex-variable approach to stress singularities - II.
50,Vol.23(1956)477-478. 448

Williams, M.L., The stresses around a fault or crack in dissimilar media.
90,Vol.49(1959)199-204. 452

Williams, W.E., The solution of certain dual integral equations.
31,Ser.2,Vol.12(1962)213-216. 463

Williams, W.E., Integral equation formulation of some three part boundary value problems.
31,Ser.2,Vol.13(1963)317-323. 519

Willis, J.R., Hertzian contact of anisotropic bodies.
60,Vol.14(1966)163-176. 575,576,579,588

Willis, J.R., Boussinesq problems for an anisotropic half-space.
60,Vol.15(1967)331-339. 589

Willis, J.R., The stress field around an elliptical crack in an anisotropic elastic medium.
39,Vol.6(1968)253-263. 589

Willis, J.R., The distribution of stress in an anisotropic elastic body containing an exterior crack.
39,Vol.8(1970)559-574. 589

Wilson, H.B., Jr., Approximate determination of contact stresses in an infinite plate with a smooth circular insert. 2nd South-Eastern Conference on Theoretical and Applièd Mechanics. Proceedings. (1964a)147-163. 405,414,417

Wilson, H.B.,Jr., Contact stresses in an infinite plate containing a smooth rigid elliptic insert.
50,Vol.31(1964b)703-704. 405,417

Wilson, H.B.,Jr. and Hill, J.L., Non-Hertzian contact stresses in a smoothly cradled heavy cylinder.
52,Vol.2(1972)143-151. 421,429

Winkler, E., Die Lehre van der Elastizität und Festigkeit. Prag: Dominicus (1867). 99

Woinowsky-Krieger, S., Über die Biegung dünner rechteckiger Platten durch Kreislasten.
36,Vol.3(1932)236-250. 542

Woinowsky-Krieger, S., Berechnung einer auf elastischem Halbraum aufliegenden, unendlich erstreckten Platte.
36,Vol.17(1949)142-148. 542

Wu, T.-S. and Chiu, Y.P., On the contact problem of layered elastic bodies.
84,Vol.25(1967)233-242. 271,559

Yi-Yuan Yu, Axisymmetrical bending of circular plates under simultaneous action of lateral load, force in the middle plane, and elastic foundation.
50,Vol.24(1957)141-143. 550

Zak, A.R. and Williams, M.L., Crack point stress singularities at a bi-material interface.
50,Vol.30(1963)142-143. 452

Zakorko, V.N., Axisymmetric strain of an elastic layer with a circular line of separation of the boundary conditions on both faces.
80,Vol.38(1974)131-138. 559

Zhemochkin, B.N. and Sinitsyn, A.P., Practical Methods of Calculation of Foundation Beams and Slabs on Elastic Foundation without the Winkler Hypothesis. (In Russian). Moscow: Stroiizdat (1948). 536

Zimmermann, H., Die Berechnung des Eisenbahnoberbaues. Berlin: (1888). 99

Zlatin, A.N. and Ufliand, Ia.S., Axisymmetric contact problem on the impression of an elastic cylinder into an elastic layer.
80,Vol.40(1976)79-84,[67-72]. 559

NOTATION

INDEX